GLOBALIZATION AND TIME

The process of globalization has brought about countless changes in societies, communities, regions and economies across the globe. It has been analyzed from many perspectives as a result and much has been written to muddy the waters of our understanding of this important concept. In going back to the real origins of the global economy, this book demonstrates that understanding this phenomenon as a 'battle against time' will bring a new clarity to the subject.

The process of globalization was accompanied by the mastering of 'social time', thereby producing a progressive increase in the speed of business transactions, both in manufacturing and in services. The context is the development of international trade in western societies and the creation of business institutions to drive forward growth. The account takes a 'long view', beginning with early European exploration in the BC period, and ending with the establishment of multinational enterprises in the twentieth century.

Using an impressive range of sources this unique book will be valuable reading for students and academics involved with the study of international business, economic history, business history and politics, among other disciplines.

Luchien Karsten is Professor of International Business and Management at the University of Groningen, the Netherlands, as well as visiting professor at many other institutions, including most recently, Newcastle Business School in the United Kingdom. His research interests focus on time and management, the transfer of management concepts and the history of changes in multinationals. He is a member of the editorial board of *Personnel Review* and the *South East Asian Journal of Management*.

GLOBALIZATION AND TIME

Luchien Karsten

Routledge
Taylor & Francis Group
LONDON AND NEW YORK

First published 2013
by Routledge
2 Park Square, Milton Park, Abingdon, Oxon OX14 4RN

Simultaneously published in the USA and Canada
by Routledge
711 Third Avenue, New York, NY 10017

Routledge is an imprint of the Taylor & Francis Group, an informa business

British Library Cataloguing in Publication Data
A catalogue record for this book is available from the British Library

Library of Congress Cataloging in Publication Data
Karsten, Luchien.
Globalization and time / Luchien Karsten.
p. cm.
Includes bibliographical references and index.
1. Economic history. 2. Time measurements. 3. Globalization—Economic aspects. 4. Industrialization. 5. International trade. I. Title.
HC21.K37 2012
337—dc23
2011048604

ISBN: 978-0-415-68312-8 (hbk)
ISBN: 978-0-203-11676-0 (ebk)

Typeset in Bembo
by Book Now Ltd, London

MIX
Paper from responsible sources
FSC® C004839
FSC www.fsc.org

Printed and bound in Great Britain by the MPG Books Group

CONTENTS

ILLUSTRATIONS

Figures

Tables

Boxes

ACKNOWLEDGMENTS

In preparing this book, I have been very conscious of the unstinting help generously given by many persons. The interest for the topic under scrutiny began during my stay at the Ecole des Hautes Etudes en Sciences Sociales in Paris. My PhD supervisor, Ger Harmsen, must be singled out for special thanks. He introduced me to the topic of the Eight Hour Day as a Western achievement to regulate working hours and maintain a humane civil society. The support of the librarians from the International Institute of Social History (IISG) in Amsterdam has been indispensable for the research on the Eight Hour Day.

I am deeply grateful to John Leopold for his inspiring support and the arrangements he made for visiting professorships at Nottingham Business School and Newcastle University Business School where I was able to present parts of the topic covered by this book. He and I also jointly lectured during early springtime throughout the late 1990s at the Institut d'Administration des Enterprises (IAE) of the University of Rennes 1 (France) following an invitation by Dean David Alis. Together with Bernard Ricciardi (Ceram, Nice), we established a consortium which offered a master program in International Business which, for some years, was presided over by Pervez Ghauri.

I owe thanks to Robert Locke, emeritus professor of Hawaii, who inspired me to continuously focus on the historical developments of management thought and the role of business schools. With Arndt Sorge, who was head of the Department of International Business at the University of Groningen for many years, I shared his insistence on the relevance of an historical focus on international management. He frequently amazed me with his extensive knowledge about European armies and the aviation industry.

With historian Peter van Baalen (EUR, Rotterdam), I shared a fascination for Dutch management thought which culminated in publications about Dutch business schools. Colleague Kees van Veen proved a reliable source of expertise and friendship while jointly exploring the role of management concepts as devices for managers to coordinate time within organizations. With colleague Ben Gales, who was willing to read the whole text and suggested many improvements, we instructed economics and business students for more than fifteen years about the historical developments of multinational enterprises.

I enjoyed the collaboration with Colin Bryson (Nottingham Trent / Newcastle) while working together on a text concerning the latest development in managing human resources. I am also indebted to Suzanne Tietze from Nottingham Trent (now in Sheffield) who pointed out different issues the book should cover. I am indebted to Ida Sabelis (Vrije University Amsterdam) who instructed me about the relevance of the anthropological perspective to understand temporalities in organizations.

Many other colleagues also deserve my thanks. Colleague Miriam Wilhelm critically proofread the chapter about Japan. My long-term colleague and friend, Pim Kooy, was also willing to read the whole manuscript and suggest improvements. I thank them all for their serious comments on the draft version.

Understanding the intricacies of international management will be accomplished not only by writing about it, but also even more by traveling to so many different places with other cultures. Over the last twenty years, I shared with my colleagues and friends, Bart-Jan Pennink and Clemens Lutz, many of the outer-European stays to assist colleagues in Burkina Faso, Tanzania, Russia, and Indonesia in improving teaching and training programs.

Some colleagues must be singled out for special thanks: Mady Koanda, Florent Hien, Honorine Illa, Marcel Krabbendam, and Serge and Geeta Bayala from Burkina Faso, and Sari Wayuni and Poppy Ismalina from Indonesia. These contacts assisted us in learning how to appreciate the immense variety of temporalities. The thrill of supervising PhD students from Eritrea, Indonesia, Burkina Faso, and Tanzania is difficult to express yet made multicultural collaboration a thrilling experience.

Research for this book would have seriously suffered without the help of others. Dutch graduate student Willem-Jan van der Stok assisted in tracing innumerable sources and German graduate student Maria Baumann sorted out some of the literature references. Without the support of the librarians of the Faculty of Economics and Business of the University of Groningen that would have been an insurmountable task. Dutch air pilot Niels Woltsky attended the course on the dynamics of the aviation industry and shared his knowledge with us. I learned much from Dutch graduate student Stephanie Petitjean who interviewed expats about their experience working in high-context cultures. Chinese graduate student Phoenix Lee did a challenging project on Asian time concepts and is now working on a PhD about the same topic. Consultants Marc Muntinga allowed me to tap into his considerable experience of quality thinking, and Martijn de Wildt enlightened me about temporalities in the context of work–life balance.

The topic under scrutiny was itself more time-consuming than I had expected. German graduate assistant Verena Girschik, who assisted me for several years in the first-year course "History of Management," critically reviewed all the chapters, suggesting improvements in structure and content. Yet without the massive support of American secretary Jenny Hill who worked long days at the office and at home to meet the tight editing deadline and did tireless work on picture research, this book would not have been finished in due time. Terry Clague and Alexander Krause of Routledge deserve a special thank-you for their interest in publishing this book.

Production Editor, Bonita Glanville-Morris and Book Now Publishing Services proofread the whole book and suggested improvements in content and style. It is more than conventionally necessary to absolve them of any responsibility for the errors, of fact or interpretation, which the book may contain.

The book is dedicated to my wife, Geer, from whom I have taken away so many hours which otherwise could have been used for other joyful activities.

ABBREVIATIONS

AD	Anno Domini (After Birth)
AFL	American Federation of Labor
ANDB	Algemene Nederlandse Diamantbewerkersbond (Dutch Diamond Workers' Union)
AOL	America Online
AT&T	American Telephone and Telegraph
B&O	Baltimore & Ohio Railway
BC	Before Christ
BOAC	British Overseas Airways Corporation
BPR	Business Process Re-engineering
BST	British Summer Time
CBS	Centraal Bureau voor de Statistiek (Central Bureau of Statistics)
CEO	Chief Executive Officer
CGT	French Confédération Générale du Travail (General Confederation of Labor)
CMA-CGM	Compagnie Maritime d'Affrètement-Compagnie Générale Transatlantique
CME	Coordinated Market Economics
CMP	Controlled Materials Plan
CPT	"Colored People's Time"

CTRC	Computing-Tabulating-Recording Company
DARPA	Defense Advanced Research Projects Agency
DC1	Douglas Commercial
DEC	Digital Equipment Corporation
DHL	Division of the *Deutsche Post named after the founders Larry Hillblom, Adrian, Dalsey and Robert Lynn*
DST	Daylight Saving Time
ECT	European Container Terminus
EES	European Employment Strategy
EIC	English East India Company
ENIAC	Electronic Numerical Integrator and Computer
EPA	Economic Planning Agency
EU	European Union
FedEx	Federal Express
FTC	Fixed Term Contract
G-20	Group of Twenty Finance Ministers and Central Bank Governors
G-7	Group of Seven (Canada, France, Germany, US, Italy, Japan, United Kingdom
GATT	General Agreement on Tariffs and Trade
GDP	Gross Domestic Product
GM	General Motors
GMAC	General Motors Acceptance Corporation
GMT	Greenwich Mean Time
GWR	Great Western Railway
HP	Hewlett Packard
HRM	Human Resource Management
IATA	International Air Transport Association
IBM	International Business Machines Corporation
ICT	Information and Communications Technology
ILO	International Labor Office

IT	Information technologies
ITRC	International Time Recording Company
ITT	International Telephone and Telegraph
ITU	International Telecommunication Union
JIT	Just-In-Time
JUSE	Japanese Union of Scientists and Engineers
JVC	Victor Company of Japan
KLM	Koninklijke Luchtvaart Maatschappij (Royal Dutch Airlines)
LME	Liberal Market Economics
LNWR	London and North Western Railway
LOT	Polskie Linie Lotnicze (Polish Airlines)
MBO	Management by Objectives
MIT	Massachusetts Institute of Technology
MITI	Ministry of International Trade and Industry
MNE	Multination Enterprise
MSC	Mediterranean Shipping Company
NBC	National Broadcasting Company
NEC	Nippon Electric Company
NRA	National Recovery Administration
NRC	National Research Council
NUMMI	New United Motor Manufacturing Incorporation
OECD	Organization for Economic Cooperation and Development
OMC	Open Method of Coordination
PC	Personal Computer
POF	Parti Ouvrier Français (French Workers' Party)
QDC	Quick Die Change
QWL	Quality of Work Life
RCA	Radio Corporation of America
SABRE	Semi-Automated Business Reservations Environment

SCP	Sociaal en Cultureel Planbureau (Netherlands Institute for Social Research)
SIA	Singapore Airlines
SMH	Société Suisse de Microélectronique et d'Horlogerie (Swiss Corporation for Microelectronics and Watchmaking Industries)
SPD	Sozialdemokratische Partei Deutschlands (Social Democratic Party of Germany)
SPSM	Society for the Promotion of the Science of Management
TAT	Transcontinental Air Transport
TBC	Time Based Competition
TMC	Toyota Motor Company
TPS	Toyota Production System
TQM	Total Quality Management
TTM	Time To Market
TUC	British Trades Union Congress
TWA	Transcontinental and Western Air
TWF	Transport Workers Federation
TWI	Training-Within-Industry
UAT	Union Aéromaritime de Transport (Air France)
UK	United Kingdom
UNDP	United Nations Development Program
UPS	United Parcel Service
UPS	Universal Product Scanning
UTC	Coordinated Universal Time
VDI	Verein Deutscher Ingenieure (Association of German Engineers)
VOC	Dutch East India Company
VW	Volkswagen (Volkswagon)
WIC	Dutch West India Company
WTD	Working Time Directive
WTO	World Trade Organization

INTRODUCTION

Handbooks written about international business generally pay scant attention to the historical developments which have led to the rise of international trade, commerce, and business and the impact this rise has had on Western society. In particular, the role of Western multinationals in creating the global economy of today is often taken as simple fact and perceived as a linear process and the outcome of a "natural" order. Some argue that this globalization was set in motion with the thirteenth-century Pax Mongolica when international trade of goods and cultural exchange of ideas moved relatively freely (O'Rourke and Williamson 2002). Others share a broader perspective and stress that globalization is a combined process of political, economic, and cultural changes resulting from the massive transnational circulation of goods, capital, ideas, and information through long-distance connections (Waters 1995). Others believe that a single global world economy based on the capitalist mode of production dates back to 1500, and inaugurated a global epoch of world history (Vries 2010).

Without an historical understanding of the origins of international trading companies, the rise of multinational enterprises, and their embeddedness in Western societies, it is difficult to recognize that even today shifts in globalization are taking place and will continue to do so (Levitt 1983; Buckley 2009).

As globalization is such a vast topic in today's world, it is important to indicate at which point this book begins. Among economic historians who see globalization as a recurring phenomenon, there is some debate about how far back in history globalization processes can be identified. Generally speaking, globalization began with international trade between different physical spaces with time required to move goods and human beings. The principle of *time is money* as the encapsulation of the spirit of capitalism has placed globalization in a specific context and today is the key driving force connecting even the most recent worldwide technological changes with global shifts of international trade, commerce, and production.

Whatever its origins, globalization became stimulated by the increasing interaction and interconnectivity of trading and manufacturing activities by internationally operating economic organizations. Within that context, the coordination of space and time has become an essential feature. This book's central focus, therefore, is the role of coordinating time as it has been, and still is, playing a hegemonic role in structuring economic endeavors whether in a specific locality or between localities within the capitalist market economy. The mentality of Western social life has ingrained a pattern of orientation in hours, days, and months as much as an orientation in speed using landmarks, routes, and distance. Yet there is one thing more which has made Europe so remarkable: the invention of the mechanical clock.

While the process leading to today's globalization has a long history, the book tries to put in perspective the applications of clock time in so many places such as towns, factories, and regions, as well as the development of the chronometer which has eased the traveling across the Atlantic and other oceans over large distances.

There is an intriguing parallel development to be noticed between phases of capitalism and the application and diffusion of clock time. International commercial, military, and industrial progress in different parts of the world have all contributed in different forms to the crucial role of clock time even today. Through the manufacturing and diffusion of increasingly accurate timepieces in Europe, the capacity to measure and divide time was constantly being revolutionized. In the process of globalization, clock time began to impose ever widening uniformity in a continuously expanding geographical area. The faster any information about these connections is transmitted from one locality to another, the more advantages economic actors can obtain.

The initial steps towards a continuous ruling of clock time can be evidenced at the end of the medieval period when Western Europe became familiar with the application of clock time which subsequently became, for example, applied by postal services for royal powers. Networks of city-states as well as empires put infrastructures in place which facilitated the reliable transportation of printed material and persons based on strict time schedules. Over the centuries, mobility slowly, but undeniably, became a modern credo in the Western world. Historically, however, its development took place in different and geographically determined regions, as well as countries, before it began to encompass almost the entire world in the twentieth century. Therefore, certain historical epochs of globalization need to be distinguished. The influence of particular city-states, nation-states, and empires which were regarded as leading or dominant facilitated the appearance of specific periods during which firms created "the First Industrial Revolution." Moreover, the appearance of multinationals facilitated the coming about of periods such as "the Age of Taylorism" or "the Era of Flexibility."

Waves of capital accumulation

The Belgian historian H. Pirenne (1967: 106) noticed that the development of capitalism has passed through several different stages characterized by an alteration of classes of capitalists who created different epochs in history. "The growth of capitalism is not a movement proceeding along a straight line, but has been marked, rather, by a series of separate impulses not forming continuations one after another, but interrupted by crises." Phases of "economic regulation" have been replaced by periods of "economic freedom." Each transition to a new stage of capitalist development involved a change in leadership in the world-scale processes of capital accumulation. The French historian F. Braudel (1984) has further elaborated Pirenne's thesis and pointed out that the recurrent financial expansion of capitalism was the result of specific capitalist actors and institutions that were able to "breed" their money systematically and persistently in a specific period of time. Like Pirenne, the historian Braudel asserted that the pre-industrial world trade was urban centred. A complex network with multifunctional "world-cities" and less-specialized regions serviced and drew on their respective hinterlands. The initial steps of capitalism originated in the Italian city-states and further advanced in the system of European territorial nation-states of the sixteenth and seventeenth centuries.

Braudel began his description of the development of the capitalist world economy with the fundamental role of Italian city-states such as Venice and Genoa at the end of the medieval period. Whereas Venice developed a well-organized and regulated form of capitalism by investing mainly in international commerce, Genoa created an informally organized and less regulated form of capitalism when the Genoese capitalist oligarchy withdrew from international commerce and invested its capital in the territorial expansion of the rulers of Portugal and Spain. Genoese bankers soon became the most important financiers in the Iberian region and turned the city into a place from where modern finance capitalism developed. Genoese bankers settled in Cordoba, Cadiz, Granada, and Seville and established close links with Iberian ruling groups. Port cities created the nexus of the maritime power of the Iberian state-run enterprises and northern-European merchants and traders. The influx of Italian capital remained essential to the election of Charles V, King of Spain. He gained the title of Emperor in 1519 due to the critical role of Genoese capital (Arrighi 2010).

However, tensions between the economic regulation of Venice and the economic freedom of Genoa led to a form of financial expansion which eventually ended the first phase of capitalist development when Spain and Portugal lost their dominant political, military, and economic positions. With their naval explorations and promotion of international trade, these Italian city-states, but more in particular Portugal and Spain, initiated the first phase of terrestrial globalization. Braudel observed that during the period of financial expansion, the Genoese oligarchy claimed "economic freedom" which he viewed as a sign of maturity in the first phase of capitalist development. The financial expansion of Genoa with its capital surpluses went to the empire of Spain but also announced the autumn of Spain's hegemonic military and political system.

Braudel has described the expansions and contractions of financial capital as waves of specific economic and financial activities which took place primarily in selective geographical areas in particular periods of time, eventually leading to the dominance of a new region (the Low Countries) over an old region (the Italian city-states as well as Spain). The second phase of terrestrial globalization, therefore, began when the Dutch economy took control and launched a huge commercial expansion in the seventeenth century. Yet again, a similar process of regulation and freedom – this time within one nation – occurred when Dutch capitalists began to transform into the "bankers of Europe" and invested their capital in the material expansion of the productive forces in Great Britain. This facilitated the appearance of the epoch of industrial globalization when industrial products were shipped to ever increasing international markets. The vast amount of accumulated capital created the springtime of Britain's industrial globalization within a new world encompassing empire.

During the eighteenth century, improved technology facilitated the rise of the Industrial Revolution and, with an increased ease of communications and transportation, opened the way for Britain to shape a global empire. The interstate competition for mobile capital in the following centuries constituted the world-historical specificity of the modern era when more European nations began to industrialize and compete economically with one another.

The comparative-historical sociologist G. Arrighi (2010) has taken up the Pirenne/Braudel approach of alternation of material and financial expansions by stressing a fourth fundamental transition in the spatial configuration of capital accumulation: after the Italian city-states, the Dutch Republic, and the empire of Great Britain, the continent-sized United States took over. As the previous waves of accumulation were accompanied by disruptive and financial crises, this also happened in the British case. Accumulated capital in Great Britain was eventually invested in the former American colony, and the international credit system created by Britain favored the formation of the United States.

During the fourth episode, the United States took over the lead position from Great Britain and eventually launched the epoch of electronic globalization based on a new series of interconnected developments in the domain of information and communication technologies (ICT).

> The development of historical capitalism as a world system has been based on the formation of ever more powerful cosmopolitan-imperial (or corporate-national) blocs of governmental and business organizations endowed with the capability of widening (or deepening) the functional and spatial scope of the capitalist world-economy.
>
> (Arrighi 2010: 225)

This book will follow the four spatial configurations of capitalism but reconfigure them into a process of globalization which passed through three different epochs of terrestrial, industrial, and electronic globalization. The beginning of terrestrial globalization can be pinpointed with the rise of Italian city-states and the eagerness of Portuguese explorers to broaden their horizons. At the end of the medieval period, both the Italian and Portuguese regions played a central role in the extension of international trade and commerce. Spain, with primarily financial support from Genoese bankers, became the empire which advanced terrestrial globalization further by occupying newly gathered territories. Political, merchant, and military leaders played a dominant role in determining the expansion of the Spanish empire. Notwithstanding its overwhelming military power to maintain its empire and occupy other territories, Spain, however, lost out on the advancement of commercial capitalism.

In the seventeenth century, this role of advancement was gradually taken over by the Dutch Republic which aligned its commercial and military interests primarily in economic terms. The Dutch introduced the first public stock international trading company in 1602 and established inland transportation improvements by constructing waterways which operated according to strict travel schedules. They also became famous for synchronizing mobility of soldiers in the army by rigid exercise and imposing strict discipline. These new arrangements in patterns of temporality reinforced the power interests of military and political as well as economic elites and modernized the Dutch nation-state (Scheuerman 2003).

By the end of the eighteenth century, Great Britain not only assumed a new leading commercial role but also succeeded in combining international trade with industrial production, introducing the epoch of industrial globalization.

> In 1750, more than 50 percent of the world's industrial output was produced in China and India, compared to some 18 percent in Western Europe. The following eighty years saw the Industrial Revolution with Western Europe's industrial output more than doubling and that of the United Kingdom increasing by a factor of seven.
>
> (Crafts and Venables 2003: 335)

During the initial period of industrial globalization, Great Britain became the first society broadly ingrained by clock time patterns. Manufacturing for anonymous (inter)national markets facilitated the establishment of a factory system where operating according to strict working-time schedules increased the scale of production. Along with the coordination of time in local places such as these factories, the ability to control time across geographical spaces with the building of railroads and steamships advanced in a spectacular way. After the first trains and telegraph systems introduced an increasing practice of abiding to clock time, this practice advanced in the sphere of distribution and communication as well as in the sphere of production. The Industrial Revolution increased production output, and steam engines driving trains reduced inland transport costs while improvements of ships reduced ocean transport costs.

At the end of the nineteenth century, however, Britain was beginning to be displaced as the leading industrial power by the emergence of the United States as the main driver in industrial globalization. The advent of large-scale, infrastructural provisions in transportation and communication facilities covering almost the entire American continent, as well as the arrival of fully clock-time-driven production facilities, intensified and expanded on an unprecedented scale. The coevolution of a spatiotemporal infrastructure based on railroads connecting East and West America with the temporal infrastructure of assembly lines in manufacturing companies increasingly reinforced each another. Faster transport and transmission reduced the price of goods and information and further facilitated the growth of international trade and business. Railroads, telegraphs, steamships, and automobiles facilitated a process of convergence in the period between 1870–1914 among Western European economies and the temperate economies of European settlements such as Canada, Australia, and New Zealand in terms of technologies, institutions, and productivity levels (Dowrick and DeLong 2003).

In 1955, North America remained the leading exporter and "accounted for over a quarter of manufactured exports" (Crafts and Venables 2003: 327). Its international industrial presence further expanded with revolutionary developments in electronic transmission of information and improved means of transportation by air and sea. The introduction of the radio, containerization, and jet cargo transport increased the speed of transmission of information and transportation of goods. Due to the impact of television and telecommunication in the 1990s, this expansion reached an unimaginable level of acceleration. A new era of electronic globalization became apparent which, since then, has begun to encompass almost every part of the world. Multinationals have played a dominant role in the advancement of globalization.

Familiar features of modern capitalism such as the reduction of time necessary to produce a specific product by speeding up production in order to optimize the use of labor power, decreasing the throughput time in transmission and distribution, accelerating the pace of technological innovations, and adjusting production to rapidly changing fluctuations by introducing processes of flexibilization can all be contributed to the influence of multinationals. Transnational corporations are currently proving to be adept at making use of new opportunities for speed and acceleration and are exploiting economic advantages. Transnational forms of production and cross-border economic flows and exchange, simultaneously manufacturing in distant corners of the world due to policies of subcontracting, outsourcing, 'Quick Die Change', just-in-time inventory flows and delivery systems have allowed the acceleration of the pace of production and allows multinationals to adjust rapidly to changes in consumption patterns.

This book traces the origins of the first multinationals, such as the Dutch East India Company, and their influence in the subsequent stages of globalization. Their increasing influence throughout the stages of terrestrial, industrial, and electronic globalization will be illustrated by utilizing descriptions of some renowned entrepreneurs and companies. Driven by the adagium of *time is money*, currently extending to "around-the-world and around-the-clock," the book attempts to illustrate how multinationals have continually been improving their ways of coordinating time by defending a "cosmocratic" perspective which favors the expansion of the clock time regime in a borderless world.

What will be discovered in the following chapters? The book is organized into three closely related, but distinct, parts based on three different epochs of globalization.

Each part will cover a particular time period during which elements of globalization developed and expanded. It will begin with terrestrial globalization followed by industrial and electronic globalization, each subsequent epoch reflecting an increasingly intensifying and deepening process of interconnected links and integration. All three epochs have been qualitatively different and will provide different insights into the nature and degree of a growing interconnectedness in global activities. All three epochs also emphasize the role large international trading companies, factories, and multinationals have played in the process of increasing *tempo, speed, acceleration, and timing.* This being the focal point of this study, attention will be paid to the way forms of clock time and concrete temporalities developed, became more and more interconnected, were sometimes temporally opposed to each other, and sometimes caused inter-temporal conflicts.

The book attempts to provide a comprehensive account of the driving force of *coordination* and *controlling of time* in the different epochs of terrestrial, industrial, and electronic globalization. The mode of capitalist production with its own clock time has, over the years, become exemplified by multinationals. Their operations have generated a wide impact on many areas of social existence which require a broad analysis for their understanding. *Globalization and Time*, therefore, is a cross-disciplinary and multilevel text book and deliberately draws from a range of academic disciplines such as history, sociology, economic geography, and economics, as well as business and management.

Chapter 1 provides insights into the geographical shifts which have been taking place during the process of globalization and the way spatiotemporal developments have been studied in the social sciences.

Part I Terrestrial globalization

Part I focuses on the contours of terrestrial globalization; that is, the exploration of regions unknown within Europe before the subsequent military occupation and economic exploitation of new areas by city-states, empires, and nation states. Some nation-states in particular developed well-equipped military forces which gave them the ability to swiftly move their armies and surprise their enemies. These nation-states became fully acquainted with the applicability and integration of the mechanical clock in their society and imposed this clock-driven temporality in their colonies.

Chapter 2 is an attempt to explain the rise of international trade as it was initiated by Italian city-states and subsequently expanded by Portugal and Spain as they began to occupy new territories in order to exploit trading facilities. With the decay of the medieval system of rule, Italian city-states had gained more freedom and independence. The new system of city-states was mainly established in places such as Venice, Genoa, Florence, and Milan. Whereas Florence and Milan primarily became engaged in manufacturing and overland trade, Venice and Genoa specialized in maritime trade with the East. Long distance trade increased and was primarily based on a quasi-monopolistic control of information over a large economic space.

Venice became the "prototype of a capitalist state with a merchant capitalist oligarchy firmly keeping state power in its grip with the objective to promote state and war making activities" (Arrighi 2010: 38). This inter-capitalist competition frequently led to wars to take over markets or territories of competitors.

While Venice constituted the prototype of state (monopoly) capitalism and a high level of cooperation, Genoa was operating on the principle of competition and established "the prototype of all future forms of cosmopolitan finance capitalism" (Arrighi 2010: 153). Surplus capital was no longer used for profitable investment in trade but instead to finance the growing public debt of other socio-political communities. This, in itself, created the "supply" for further financial expansion with the Genoese elite switching from trade in commodities to trade in money, in particular to Spain. The drive for wealth and new land as well as the desire to Christianize the rest of the world mobilized the empire of Spain into a power where "the sun never sets."

Chapter 3 analyzes the introduction and the diffusion of the mechanical clock in Western Europe as it was initiated within Italian city-states. It assesses the impact of the mechanical clock on other temporalities in Western society. Within the wealthy Italian city-states, mechanical clocks became the technical thrill of the time. Public clocks became a symbol of political autonomy, economic wealth, military power, and cultural richness. Although clock time was not created by capitalism, it did become a central phenomenon of the mode of capitalism (Whitrow 1975).

Italian astronomer and physician Giovanni de Dondi, in Padua, developed an astrarium, or astronomical clockwork, with his father, Jacopo, and exposed it in 1364 to the public at the Palazzo del Capitano of Padua. The astrarium was more than a clock as it used a year wheel and a geared assembly to track the movements of the planets. Due to this invention, mechanical timekeeping became "the single most important practical innovation of the entire Middle Ages, both scientifically and socially" and as such, was "one of the great turning points of history" leading to the European economic and technological pre-eminence (North 2005: xv; 143).

Soon, the list of public clocks not only included Florence, Genoa, Venice, and Milan but coupled with cities such as Monza, Vicenza, and Trieste (Dohrn-van Rossum 1992). Within these cities, the interest for calculation and measurement grew (Le Goff 1977). The pressure for a precise application of time signals was especially strong in Florence which was engaged in textile manufacture, the first and greatest of medieval industries. Ordinary people began to distinguish between the hour of the day in traditional ways and regular, mechanical clock time. Public clocks, especially those with dials, made the verification of time a continuous option for the inhabitants of the city, especially in the case of the cloth-producing industry. In Florence, the powerful guilds in the textile industry instituted statutes registering work hours. "Because workers were paid by the day, conflict became focused on the length and the definition of the workday" (Postone 1993: 210).

The diffusion of the mechanical clock across monasteries and other Continental European cities created a feverish horological activity. This led to the introduction of the standardization of time as an abstract and homogeneous, as well as divisible and measurable, entity in the cities. "The introduction of equal hours and the habituation of urban populations to public time announcements had profound consequences for the European mentality" (Landes 1983: 77). "The historical transition from a mode of time reckoning based on variable hours to one based on constant hours implicitly marks the emergence of abstract time, of time as an independent variable" (Postone 1993: 203). The abstract, de-contextualized measure of time became a material force that began to exert "a disciplinary pressure" within capitalism (Castree 2009: 36–37). Soon the mechanical clock, such as the computer today, became the technical sensation of the time (Landes 1983: 57; Rispens 2005).

The use of the mechanical clock, however, was not only contained within the walls of monasteries and cities. As different transportation systems developed, the use of clock time also began to pertain to the distance between cities. The geographer Vance (1986) explained how specific new "technologies of movement" from carriages and coaches to new ship designs, together with the improvement of "facilities" (i.e., infrastructure), improved transportation in Europe and reduced travel time. Prior to the invention of electric transmission, information could only travel at the same speed as the transportation facilities allowed.[1] The transportation revolution began with the introduction of a system of mail coaches in Europe which applied specific timetables based on the use of clocks. For the establishment of reliable, scheduled departures and arrivals, time measurement became an important topic. Postal services improved possibilities of communication by delivering mail in due time. Clock time became integrated in the infrastructural provision of mail services. In their effort to speed up the transfer of information and to enhance the appropriate arrival and delivery of mail, European postal agencies, such as Taxis, implemented time schedules, but the speed never exceeded the gallop of a horse. Land routes and postal services created a network of communication within a new spatio-temporal configuration in Europe. Yet overall movement over land was slow and difficult, and the frequency of the mail coach was regularly interrupted.

A similar development unfolded at sea. The great discoveries by the Portuguese and Spanish captains and sailors stimulated the expansion of trade and facilitated the penetration of the East Indies and colonization of the Americas. These explorers delivered new maps about parts of the world no one had known about before. "The mapping of the world opened up a way to look upon space as open to appropriation for private uses" (Harvey 1989: 228). Maps which provided access to the riches of the Indies became valuable objects, and national rulers were willing to pay high prices for copies which described new land in detail.

Yet "bad measurements made bad maps" (Landes 1983: 110). Therefore, a tremendous interest grew for more precise instruments to measure time at sea and to improve the *reliability* of the transport of precious spices to European markets. The initial discussions on how the precise location of longitude – how far east or west a ship was positioned – should prevent ships on the Atlantic getting lost increased the interest in the applicability of reliable mechanical clocks at sea. Since the early sixteenth century, the problem of geographical longitude had been known to be potentially soluble by means of accurate timekeeping.

In 1530, Dutch astronomer Gemma Frisius launched the idea of using a portable timekeeper to identify a ship's longitude. He knew that the earth rotates on its axis once in 24 hours and that there are 360 degrees of longitude. By dividing 360 degrees by 24 hours, in one hour the earth rotates 15 degrees of longitude. Due to this relationship between longitude and time, it became essential to develop a method to keep the time at a place of known longitude on a clock aboard the ship. This chronometer had to keep time accurately throughout the ship's voyage so that the time at the reference location could be read at any moment during the voyage. The local time was then compared with the time at the reference location, and the difference between the two of them could then be converted to the ship's precise location at sea. By accurately measuring the longitude, it was possible to make truly accurate maps and gain an upper hand in the race of European powers to claim lands.

Despite the collective efforts of artisan clockmakers, clocks remained for long a quite inaccurate and unreliable instrument to determine a ship's position at sea. Yet the mechanical clock created awareness among scientists that the passage of time is subject to mathematical treatment. Bright scientists such as Copernicus, Kepler, and Galileo began to study time and, at a symbolic level, the clock introduced an abstract measurement of time based on its quantification. Their work promoted a shift from time being one with the eternal cosmos, as the Greek philosophers perceived in the age of cosmic globalization, to a perception of time which refers to the results of scientific and technological systems of time measurement (Hongladarom 2002).[2] This became especially essential in the context of seafaring. The emergence of an early modern science "became the embodiment of the quest for precise and reliable knowledge" (Cohen 1994: 353). The rulers of European seafaring nations, such as the Spanish emperor, offered large awards to those who would be able to develop a reliable clock which would make it easier to determine one's position in open sea. But until the eighteenth century, such an instrument was not available nor any to indicate meridians that specify east–west positions. After the first explorations, the Treaty of Tordessilas (1494), which the papacy sponsored, divided the world beyond Europe into separate zones of influence. The prime meridian ratified a monopoly of African trade to Portugal and trade with the newly discovered Indies to Spain (Blackburn 1997).

Chapter 4 addresses the way in which international trading companies expanded their influence. When Charles V made his son Philip II regent of Spain, he then also became ruler of the Low Countries. His empire grew to such a global size that a chronicler labeled it as so far extended and wide that: "the sun always shines upon me," "*sol mihi semper lucet*" (Parker 1998; Kamen 2002). Indeed, Philip II ruled the first global empire in history, but governing such a territorialist global monarchy was complicated. Although he lived by the clock – working day and night with two illuminated clocks – he lacked the ability to maintain control over his empire. With the deterioration of this empire, the Genoese bankers lost their position as influential bankers and financiers, which led to the end of the first wave of capital accumulation (Braudel 1984).

The Dutch Republic took over that position. The Low Countries had become wealthy through the commercial trade within the Baltic Sea, and Amsterdam had turned into an influential trading place. From the seventeenth century forward, the Dutch combined the commercial capabilities of Venice excelling in international trade and commerce with the financial capabilities of Genoa excelling in financial expansion. When Spanish troops were sent to occupy the Low Countries in 1566, Dutch rebels resisted the occupation, took to the sea, and organized themselves to fight against the Spanish occupation. For 80 years (1568-1648), imperial Spain was continually weakened in its territorialist domination. Contrarily, the ensuing competitive power struggles increased the fortunes of Dutch capitalism (Arrighi 2010).

For a significant period of time, commerce had largely been the domain of individual merchants who were shipping their goods between local communities. The seafaring Dutch Republic, however, began to cultivate an increasing inclination toward international trade and commerce. This will, in particular, be illustrated by the Dutch United East India Company (VOC) which was established in 1602. It was the first public joint-stock international trading company. The VOC is viewed by Arrighi as the best example for international business students to learn and appreciate the functioning of joint-stock, internationally operating organizations. The newly chartered merchant enterprises were part governmental, part business organization specializing territorially in the exclusion of other similar organizations. They traded between Europe and the rest of the world based on a monopoly on government contracts. The Golden Age of the Dutch Republic relied to a large extent on its international commercial activities. "From 1610 to circa 1730, the upper strata of the Dutch merchant class remained the leaders and governors of the European capitalist engine" (Arrighi 2010: 144). Dutch Prince Maurits was highly involved in the establishment of the VOC which facilitated a Dutch version of a formally organized and regulated capitalism. Although the Dutch were not alone, nor indeed the first, in such a venture, the VOC became by far their greatest success.

Within the Dutch nation-state, a particular form of a rational application of clock time firmly established itself in two areas: water transportation and military warfare. In the Dutch Republic, an extensive system of barges crossing canals was pioneered which serviced the transfer of passengers and information between different cities on precise time schedules. The Low Countries had a unique ubiquity of waterways, but the interconnected canals had been only marginally exploited for transportation facilities. Around the 1580s, Amsterdam settled agreements with Utrecht, Dordrecht, Hoorn, Leiden, Haarlem, and other cities to maintain an intercity, common water carrier transportation system and create an efficient network of routes for regulated sailboats (*beurtveren*) transporting their cargoes.

The length of the journey of sailing vessels through these canals was mainly determined by the availability of wind. Drawn boats or pulling barges (*trekschuiten*) being pulled by horses along a towpath next to the canal did not have this problem (Dehé 2005). A network of canals with these drawn barges introduced temporal coordination and increased the interdependence of cities on an integrated transportation system. The system was the answer to an increasing desire for a speedy (i.e., reliable), year-round means of travel. Ordinances were issued which fixed the exact time of departure and the maximum duration of the trips with these barges with bells ringing to announce departure time. To facilitate the maintenance of the schedules and routes, many barges were equipped with hourglasses to assure observation of these requirements. "The skippers faced heavy penalties for failure to depart or arrive on schedule" (Vries 1982: 24).

The network of the drawn barges was an economic innovation and created a particular transportation practice which was based as much on commercial as religious interests in terms of not wasting one's time. The increasing economic wealth provided the Dutch society with opportunities to introduce an infrastructural transportation system by building a network of canals which functioned on the basis of strict time schedules. Next to reliability, punctuality, and speed, military interests played a role as well: some cities which functioned as fortresses guarding against Spanish threats of invasion could rely on a quick transportation of soldiers.

In 1588, the States-General decided to decrease the size of the army, ensuring a decent and regular pay for the soldiers in exchange for obedience. Apart from being the patron for the international Dutch trade, Maurits was most famous as a military genius, introducing some remarkable novelties within the Dutch army. The Dutch Republic had been the progenitor of new ways of fighting based on gunpowder weaponry. Maurits, youngest son of William of Orange, was striving for more and looked for a transition from obedience to discipline. To increase combat skills, he began training soldiers by introducing ways to quickly and flexibly maneuver smaller army units to assert Dutch independence against the military might of Catholic Spain. He also designed the contra-march which enabled soldiers to fire continuously. For the correlative fine tuning within the ranks, it was necessary for the soldiers to not only watch their own footsteps but those of their colleagues as well. Their bodily rhythms became visible as symmetry in time, and symmetry became rhythm in space. Sociologist Simmel described this "symmetrical-rhythmic formation" as the first emergence of rationalism which created a certain style of life among soldiers and commanders alike (Simmel 1990: 489).

Maurits built an army which allowed a more-or-less intelligent response to unforeseen circumstances by relying on a properly drilled habitus amongst the soldiers. "Every movement attained a new level of exactitude and speed. The individual movements of soldiers when firing and marching as well as the movements of battalions across the battlefield could be controlled and predicted as never before" (McNeill 1984: 130). Maurits' military revolution improved discipline and was copied throughout Europe. "He achieved for the Dutch Army and Navy what scientific management would achieve for the American industry two centuries later" (ibid.). During the 80-year war (1568–1648) against the Spanish ruler, the Dutch military commander Maurits introduced this military revolution, which was based on the application of specific choreographic instructions and special military techniques to improve the effectiveness of army units in the battlefields. This new approach predated scientific management by two centuries and was soon imitated by other European military leaders.

The Dutch Republic became famous for its international expansion and its internal water transportation system, as well as its military advancements. Throughout the seventeenth century, the Dutch Republic fused the advantages of commercial capitalism and territorialism "far more effectively than any of the northern Italian city-states had ever managed to do" (Arrighi 2010: 139). Yet the narrow territorial and demographic base of the Dutch nation-state limited the expansion of their fortunes. During the Anglo-Dutch wars, the Dutch capitalist class turned away from trade and focused on a more exclusive specialization in high finance. The accession of William of Orange to the English throne in 1689 made it possible for Dutch capital to shift from Dutch to British ground. Surplus capital was now not only invested in British stock and government securities, but the Dutch capitalist strata also invested in British enterprises which boosted British textile and metal industries.

During the times of the joint-stock trading companies, Europe saw the creation of modern nation-states which triumphed over an array of competing models of political organization (empires and city-states, as well as loose networks of city-states). These European nation-states were heavily involved in supporting the exploitation of human beings with no remorse, forcing them to work ceaselessly on plantations arising with the successful exploitation of new territories. With the increasing reliance on slaves, European joint-stock trading companies, as much as private trading companies, were involved in slave trade.

For several centuries, commercial and proto-industrial capitalism relied on slavery. Although the effect of the slave trade on European wealth remains controversial, many have argued that all the overseas trade was central to the transformation of European commercial capitalism. Yet even when slavery was abolished, capitalism was able to survive and develop further on the basis of exploiting the rest of the world in the form of colonialism.

Eventually, it was in Great Britain that the transition from terrestrial to industrial globalization took place. Its leading position in international trade generated challenges to introduce new industrial practices in their factories based on a free labor force.

The British commercial expansion during the eighteenth century urged the country to refine the mechanical clock into a precise instrument called the chronometer to be used at sea and measure the longitude in order to let the ships return home safely. As ships did not yet have the most reliable means of establishing their specific locations, untold numbers of sailors died when rock formations suddenly loomed out of the sea and took them by surprise.

With the increasing hegemony at sea, the British navy pressed for reliable instruments to determine the exact position of its vessels to assure safe journeys. In 1675, King Charles II ordered the building of an observatory on Greenwich Hill. On October 22, 1707, four British warships ran aground at the Scilly Isles off the coast of Cornwall and nearly 2,000 men lost their lives. The warship accident catapulted the longitude question into the forefront of British national affairs (Sobel and Andrewes 1998b). The Longitude Act of 1714 asked for an accurate clock regulated to the time at Greenwich to indicate 1:00 p.m. when the sun is directly to the north (local noon time), pinpointing the ship's location on the longitude 15 degrees west of Greenwich. In 1753, the British amateur clockmaker John Harrison (1693–1776) completed a marine clock which looked like a pocket watch, an achievement recognized in 1773 by the Board of Longitude. The miniaturized version turned the marine chronometer into an object of industrial manufacture and commercial use. The British horological revolution led to a common use of the chronometer.

Part II Industrial globalization

Part II explores the closely connected development of industrial globalization with the application and intensification of mechanical clock time in factories, transportation, and communication systems. New production technologies, novel forms of organizing production processes, and extension of transportation networks created favorable conditions to intensify the application of clock time and increase speed within transportation and production facilities.

Chapter 5 deals with the way the accumulation of capital in Britain was mainly spurred by capitalist enterprises focusing on the improvement of production processes (Arrighi 2010). J. Wedgwood became one of the first entrepreneurs to introduce a division of labor in his pottery works that was arranged in sequence of order to provide an efficient production line and to avoid unnecessary movement of workers and goods. He redesigned the production process and introduced a rudimentary form of bureaucracy involving task specialization, formalization of authority, standardization, and hierarchy.

While artisanal work was still very episodic, Wedgwood tried to impose a strict discipline. The work of craftsmen became transformed into acts of labor indicating that it was no longer performed for its own sake or for the sake of creating a lasting product but merely as a means to maintain the dynamics of the factory system. This approach became a common feature in the textile industry. Spinning, weaving, and finishing became separate and specialized domains of different factories.

The consequences were harsh for the textile workers. When the first villagers and errant agricultural families were brought into the British textile factories to work, they were not customized to the factory bell and linear scheduled time patterns. The new factory workers were subordinated to the rules of the factory owners. Moralizers and moralists influenced the understanding of the proper meaning of the regulations as they were imposed onto the workforce. A strict working-time discipline which contained features similar to those Maurits had developed was imposed in manufactures and factories. The factories became the familial *landscape* of disciplined industrial capitalism.

Once time had been reduced to a quantifiable unit of money, it could be used, saved, and gained. Techniques and technologies such as the steam engine instilled new temporal norms for factory workers which were defined by clock time. The development of the factory system in the textile sector will be illustrated by the case of Robert Owen, who was the first to put limits on working time to prevent exploitation of the adult workforce.

Still, Britain became the birthplace of the Industrial Revolution and created the largest trading and manufacturing economy of the nineteenth century. This allowed Britain to redesign the world map. The size of the international trade and commerce that had been made possible by the penetration of the East Indies and Chinese markets, as well as the colonization of the Americas, created the conditions for the emergence of modern industry. The third British systemic cycle of accumulation developed in the form of industrial globalization. Steam and machinery revolutionized industrial production and eventually integrated trading and manufacturing activities, maturing into a large economic integration of the worldwide markets. Subsequent renewal in transportation and communication facilities further diminished the time it took to get people and materials from one place to another.

Great Britain became the first nation to build a mechanical infrastructure of train transportation, applying strict time schedules and making the general public familiar with the advantages of a specific time regime. The introduction of railroads was soon paralleled by the technological advance from sailboats to steamships. In 1825, Britain obtained the first 40-kilometer railroad service between Stockton with its Durham coal fields and the market town of Darlington in the north-east of England. In 1833, a group of Bristol financiers founded the Great Western Railway (GWR), which provided a straight connection with London. Punctuality was introduced as a new word in the language of travelers, and passengers began to appreciate railroad speed. This growing use of railroad transportation changed the mentality about standard clock time. Railroad timetables were introduced and seriously integrated in the infrastructural provisions of the railroads. By 1855, almost all public clocks in Britain were set to Greenwich Mean Time. The railroads reinvented the landscape and revolutionized spatial relations. They cut through hills and were carried across valleys and estuaries.

Throughout the nineteenth century, Britain developed into a nation where industrialism merged with imperialism. British hegemony was as much a world power as it was a world economy. Although its economic and technological leadership was unquestionable, the ruling class turned its interest away from industry and used their surplus capital for managing the empire. "The result of the acceleration in the material expansion of capital was the globalization of the capitalist world economy" (Arrighi 2010: 165). During and after the Great Depression of 1873–1896, the tremendous expansion of the Industrial Revolution in Britain ended, and its overabundance of money capital was primarily invested in the United States. The third systemic cycle of capital accumulation with British high finance in the lead came to a close, yet it had created a modern world which was no longer space bound but time bound instead.

However, the British elites preferred focusing on jobs in the public sector and aspired to obtain lucrative civil servants' jobs (Brendon 2009). The cosmocratic network of high finance which had initially invested in industrial Britain moved into the banking and financing of imperial Britain.

Chapter 6 addresses the issue of standardization of the measurement of distance. Whereas in Britain the application of standard time reckoning became broadly introduced, the same did not occur for the measurement of distance. It was during the French Revolution that such an initiative was taken. The revolutionary French government wanted to present to the world an example of the clearing up of ancient muddles and disenchant the world by destroying the long-lasting influence of the old orders of Church, royal families, and trade guilds. It instructed the Académie des Sciences (Academy of Science) to devise a new, rational unit of measurement which should regulate the society in a modern manner. Thereupon, the Academy introduced the meter as the new unit of length, with divisions and multiples of ten. Its introduction coincided with the adoption of the decimal system. The metric system was seen as a scientific incarnation of the values of standardization, universality, and equality with science being the rational mediator between humanity and nature.

Chapter 7 emphasizes how, during the end of the nineteenth century, the center of industrial gravity moved from Great Britain to the USA and how this nation intensified the speed of transportation, communication, and manufacturing. At the world exhibition in London in 1851, some American manufacturers demonstrated their skills and practices to produce standardized products.

One such company, The Colt Patent Firearms Manufacturing Company in Hartford (Connecticut), played a significant role in enhancing the process of synchronization within its factory by refining and improving the production process. The gun factory was fitted with 400 machines producing around 24,000 pistols per year. This approach was inspired by the work of the watchmaker Eli Terry who was one of the first to improve the production process for making timepieces in his firm in Plymouth (Connecticut) where precision watch-making machines created the American system of watch manufacturing. The American Springfield Armory was also among the first to gain experience with this new production method and changed the production of weapons from skilled, manual labor to factory production. With the new approach, the company was able to produce 30,000 muskets per year. The gun manufacturer became the example of a single-unit factory developing low-cost, high-productivity manufacturing with time discipline at one location, which was later imitated by the Ford Motor Company.

In the half century following the American Civil War (1861–1865), the American business community underwent a remarkable organizational revolution. Whereas the British industrial regime primarily remained a system of small and medium-sized enterprises, American companies developed into a large number of vertically integrated, bureaucratically managed corporations (Chandler 1977). The emergence and swift expansion of modern corporations was based on the pioneering efforts of American railroad companies.

American railroads emerged as the nation's first big business. The meeting of the Union Pacific and the Central Pacific in May 1869 turned America into one continental nation. After the end of the Civil War, however, each American railroad company still applied its own railroad time. The difference between East and West amounted to more than three and a half hours. An effective use of a complex railroad network and prevention of accidents forced the companies to consider standardization of time reckoning at a level higher than that of the individual railroad company. The railroad companies covered such distances that several time zones were necessary to secure the efficiency of the timetables. In April 1883, 56 railroads inaugurated a uniform railroad time with four time zones. On Sunday, November 18, 1883, (the Sunday of the two noons), the railroads introduced a single uniform railroad time: the Standard Railroad Time. All local solar times were changed into one common railroad time. Railroad time became firmly established as a regulator of public life. The railroads became, at one and the same time, an economic activity and an essential public good with a natural monopoly.

The managerial practice of the American railroads led to the standardization of railroad times and of civic time in the United States as a whole. People began to rewrite locally embedded social times into this new version of abstract time. Sanford Fleming, director of the Canadian Pacific Railway, insisted on the necessary introduction of a uniform time for the whole world: terrestrial time. Whereas Dowd's national system only encompassed North American railroads, Fleming's terrestrial time extended around the world and was called cosmopolitan time (Bartky 2007). Most Americans set their clocks and watches to a new version of abstract time, and the United States introduced one synchronized time pattern. It was the most momentous development in the history of uniform public time since the invention of the mechanical clock. An empire of chrono-technology dawned.

By the beginning of the twentieth century, civil times based on exact hours from the Greenwich meridian also became the norm throughout much of the European continent (Bartky 2007, 120). The entire industrialized world became organized according to standard railroad time, even more so than ever attained by the Gregorian calendar. For a smooth operation of the American national telegraph system, a single standard, i.e. the Standard Time Zone, was required to which all local times could be referred. Together with the introduction of a standard meridian system and a standard time system, the general diffusion of cheap pocket watches began to instill a sense of punctuality, calculability, and speed in business transactions as well as human relations (Simmel 1990).

Chapter 8 focuses on the way management and control of time became the key feature of American industrial capitalism. F. W. Taylor (1856–1915) used the stopwatch and introduced chrono-technology in the factories. The stopwatch symbolized a heightened "time thrift," a greater intensity of production. Time compression and intensification of manufacturing processes were soon viewed as inevitable. Taylorism used techniques to study human movement more so than the humans themselves and had physical labor as its subject. In theory, it intended to achieve low average costs per unit of output. The easiest way to do so was by producing more units. The need to produce more in order to cut costs led to larger and faster machines operating on an ever greater scale. Its ideology was efficiency and its result managerialism.

Time-and-motion studies ultimately aimed at determining standards for maximizing labor efficiency for the benefit of the company and were meant to be a managerial tool for the policing of workers. Whenever the term "scientific" was used, it was meant to disguise an engineer's ideology implying a range of techniques for interfering in the attitudes and mentalities of the workers. It aimed to eliminate the conflict between capital and labor by paying strictly on the basis of work done; that is, by the piece.

Scientific management became "the first of many twentieth century management fads to spill over into the general culture" (Hoopes 2003: 59) and equated efficiency with speed. "Speed means more money" became an increasing popular adage and an absolute imperative for business. Taylor's discipline-based managerial approach constructed an economically efficient work discipline which resulted in twentieth century managerialism. Timekeepers became the new regulators and controllers of the tasks of the workmen. They looked for the maximization of productivity time either by lengthening of the working day or, when this became impossible, by intensifying the work done in the maximum time available. "As an engineer, Taylor considered the body as a machine, which either operated efficiently or it did not" (Schor 1992: 58).

This study looks in particular at the impact of Taylorism (scientific management) and Fordism (assembly line production) in the sphere of production. Mechanization as well as automation within factories sped up production. Trains, cars, buses, and subways facilitated mobility, ensuing an intensification of traffic circulation. The twentieth century itself turned into the age of mobility.

The ensuing industrial corporations came to enjoy competitive advantages based on their drive to attain economies of speed founded through scientific management and a tremendous growth of managerial hierarchies and bureaucratic structures. Initially, these companies mainly grew by domestic territorialism due to the expansion of the railroads throughout the North American continent. As soon as they had completed their continent-wide integration within the United States, they eventually began expanding internationally.

Chapter 9 takes into consideration the consequence of the single-minded focus of management on a strict application of mechanical clock time in production facilities. American companies searched for ways to improve industrial output. The efficiency drive, which had been enforced during World War I to improve industrial output, however, created resentment about too much control and lowered morale amid the workforce. There was a simultaneous growing interest in a more humanistic strand of management thinking. This became particularly clear in the high-intensity electric-lighting and information-processing sectors. The dramatic growth in demand for telephones in the last decade of the nineteenth century and the early twentieth century placed tremendous pressures on the Bell Company.

Since 1882, Bell used Western Electric as its main manufacturer. As Western Electric had no competitor, Bell and Western Electric jointly coped with the crisis of capacity and built a new plant, the Hawthorne Works. The decision of the board to construct Hawthorne indicated a change in the strategy of the corporation: it departed from the electrical power business and put its energy in the telephone business. In February 1905, the Hawthorne Cable Plant opened its doors and soon joined the ranks of giant works in American industry. In 1913, the plant employed 14,000 workers and was the single biggest employment site in the Chicago area. Western Electric had initially responded to the capacity crisis by renting additional space and operating around the clock – when workers were available. The company, however, was concerned about negative publicity concerning night working and overtime in their plants. A strike of the machinists in 1900 over the recognition of their union and the reduction of working time to nine hours had worried the board of management.

Western Electric and General Electric eventually became well-suited for policies which could improve their overall control and involvement of employees which would reach beyond a policy of employee welfare. After some grievances expressed by the workers, the board of Western Electric became interested in the art of human engineering. In 1924, Hawthorne conducted the first experiments and focused on the usefulness of electrical lighting for workers on assembly lines. Yet an increase in production could not be credited to the increase of illumination. In April 1927, a team of Hawthorne engineers initiated new research to determine whether the variables of rest periods and various hours of work might increase the efficiency of the operators. Again the results were not positive. Eventually, Elton Mayo (1880–1949) was asked to look at the results of the research projects. Mayo was convinced that physiological states due to fatigue could transform into psychological states rendering the worker irrational and apt to reveries.

By 1929, Western Electric was so pleased with Mayo's contributions that the company offered him a job to interview all supervisors at the plant and design a technique of social control that could deal with problems of industrial unrest. As many workers did not know what their real problems were, it would take educated personnel managers to assess what the workers really needed. Mayo initiated a supervisory training program intended for 2,500 supervisors to teach them how to handle workers in a friendlier way, respecting their emotional side. This approach expanded into Human Relations, an approach which looked at the workers less as machines and more as sensible and sensitive human beings.

Chapter 10 addresses how, with the introduction of the railroad, the steamship, and the automobile, people have become able to move to many more places much faster and cheaper than ever before. In the early twentieth century, the railroads had created an era of machine-driven transportation; the steamship was outperforming fast sailing ships such as the clipper, and cars became a popular means of transportation once an appropriate infrastructure was constructed. The economy was dominated by huge industries – steel mills, assembly plants, and even railroads which provided critical masses of stable jobs. New transportation technologies such as automobiles, buses, tramways, subways, and improved infrastructures strengthened the expansion of industrial globalization. The fascination for mobility was translated into words such as speed and velocity, which became the emblems of modern times. The advent of faster transport by air shipping and the introduction of containers on ocean vessels created a spirit of time saving in the aviation industry and of increasing transfer of cargo attained overseas throughout the twentieth century.

During the twentieth century, aviation became – as will be shown in this chapter – the emblem of technological acceleration. The Age of Aviation began in a bicycle repair shop of the brothers Wilbur and Orville Wright in Dayton (Ohio) who opened their shop in 1893. They became increasingly interested in flying and studied *Animal Mechanisms* by Frenchman Etienne-Jules Marey, the same man who had inspired Gilbreth for his motion studies. From that time forward, the brothers began designing gliders with the conviction that only by becoming lighter than air was flying possible. On December 17, 1903, they succeeded in keeping their *Flyer* in the air for twelve seconds. Their construction, being "heavier-than-air" but using the density of air as a lift mechanism, finally made it happen. Their *Flyer* became the first powered airplane that could take off and land without crashing. The *physical* experience of challenging time and space in a new form of flying was realized. Ever since that moment, there has been a craze for ever faster planes and flying across longer distances. This chapter will reveal the role of some of the main players in turning the aviation business into a popular and successful way of crossing long distances at high speed.

Part III Electronic globalization

Part III sketches the contours of intensifying social acceleration in the western World. After the Second World War, Western Europe quickly became a very fertile ground for the transnational expansion of American corporations. Together with the international position of the dollar and military supremacy, multinational corporations became the cornerstones of American hegemony. General Motors became the symbol of The American Century, the era of American economic and military power. The post-war boom in Western Europe was the result of playing catch-up with the United States.

Chapter 11 addresses the introduction and impact of the revolution in information and communication technologies. IBM's role in the speedup of transmission of information is highlighted. Their and other corporations' technological improvements facilitated the creation of an empire of speed. By introducing new technologies in the transmission of information, standardizing procedures and routines, as well as introducing specific management concepts to keep employees committed, conglomerates attempted to prevent a crisis of control and accelerate their activities by designing an administration as an efficient timesaving instrument.

In 1969, Arpanet was introduced, the predecessor of today's internet. Once data could be packaged and sent efficiently and after Netscape went public in 1995, the Internet was born. Computers became capable of processing in *real time* by shrinking existing process cycle time as small as possible. Worldwide infrastructures facilitated the transport of information on a global scale. Yet real time, or time-less time as Castells has called it, describes the full potential of what information technologies can attain to speed up information flows while primarily serving the interest of financial institutions, ICT companies, and other multinational corporations. With ICT, multinational corporations have been able to institute worldwide transportation and communications networks which have evolved into a global telecommunications infrastructure. They create an electronic globalization which hints at a process of integration across frontiers of market economies at a time of rapidly falling costs of transport and communication, suggesting the world is becoming smaller, faster, and more connected (Wolf 2005). Globalized production systems began to present a confounding pattern: multiple time zones were to be taken into account even in local production schedules. Eventually, Japanese corporations contributed to this development by introducing *timing* as a key new concept.

Chapter 12 focuses on the role Japan began to play in the global economic arena. Although the USA was able to maintain its world economic leadership, in the 1990s Japan became the second largest economy in the world by introducing a new "just-in-time" production system. This chapter addresses the Japanese miracle and the manner in which the car manufacturer, Toyota, introduced the concept of timing as a novel version of the application of clock time, which paved the way for flexible production systems and a flexible workforce. The Japanese introduced a new challenge with their concept of timing. Speed alone was no longer a guarantee for success. Instead, timing has become the key to further advance internationalization of manufacturing, commerce, and trade.

Chapter 13 addresses the way flexibility became such a central topic during the last quarter of the twentieth century. In spite of a substantial increase in technology inputs and acceleration in the pace of technological change, American businesses were struggling with the slowing down of productivity growth rates. At the same time, Japan was introducing the flexible firm. Due to institutional differences, there was, however, a remarkable divergence to be noticed between the Japanese setting, the American national circumstances, and the European Union in facilitating the coming about of a flexible labor market and a flexible firm. The issue of flexibility highlighted an urgently felt need for transformation of work organizations in firms, subsidiaries, and conglomerates. The industrialized world began moving away from a once generally embraced combined notion of regularity, standardization, and central coordination with uniformly arranged working-time patterns. A shift appeared from invariant rhythms within the organization of social life to variable ones. A frantic drive for greater organizational efficiency and flexibility through quality improvement and customer satisfaction, the long-established standard workweek and standard workday came under tremendous pressure. The outcome was a greater diversity in flexible working-time arrangements. Temporary organizational forms appeared in sectors such as construction, software development, biotechnology, consulting, fashion, and advertising.

Notions of regularity, standardization, and central coordination became liable to shift and the permanent, full-time, weekday, daytime hours arrangements are in decline – albeit more slowly than politicians and management gurus would have us believe.

Chapter 14 contains a broader look at the consequences of the hegemonic clock time fixation on other temporalities which one encounters in the world. This chapter will address the need for international managers such as expatriates to develop a sense of cultural intelligence to actually be able to put the overemphasis of clock time practices in international business in a broader cultural context especially within those cultures which are not yet fully dominated by the discipline of the clock time regime. Its impact is increasingly noticeable in non-Western parts of the world which are continually subduing the constant pressure of the capitalist market economy. The question will be addressed to what extent the social acceleration of the Western world as pursued by multinationals will force the rest of the world to catch up.

Chapter 15, the epilog, reviews the historical perspective on international business and the impact of the drive to coordinate time which thus far has gone through four distinct but interrelated periods of globalization. The main outcomes in terms of speed-up and of acceleration processes will be linked to the challenges that cosmopolitan managers are facing when dealing with today's globalization. Timescape management will be proposed to mitigate the effects of the clock time regime.

1

THE COORDINATION OF TIME

In 1990, two consultants from the Boston Consulting Group, Stalk and Hout, presented a compelling case for time as the key topic for international business. In their study *Competing Against Time*, they stressed the importance of the link between time and international business:

> Time is a fundamental business performance variable. Listen to the ways in which managers talk about what is important to the success of their companies: response *time*, lead *time*, up *time*, on *time*. Time may sometimes be a more important performance parameter than money. In fact, as a strategic weapon, time is the equivalent of money, productivity, quality, and even innovation.
>
> (Stalk and Hout 1990: 39)

Their point was that, in international business, a timely response to the desires of customers leads to a competitive advantage. They were convinced that the best way for Western businesses to improve their practice of managing time should be to imitate Japanese multinationals, which pioneered the use of *timing* as a competitive weapon in the globalizing economy.

Timing has become a recent translation of clock time replacing speed as the dominant marker of the second Industrial Revolution (1870–1960). Since multinational corporations began to apply timing to survive in international competitive markets, this latest version of clock time has obtained a global character. This book has the intention to demonstrate how clock time is historically embedded and, even though we perceive it as abstract time, it is still socially constituted. It has, therefore, a history that implies that its development and application changes continuously. Ever since the invention of the mechanical clock, abstract time may have become uniform, continuous, homogeneous, and empty, but it is also historically unique. Its application became socially accepted somewhere during the period between the fourteenth to the seventeenth centuries before industrial capitalism began to dominate Western Europe.

We will explore the processes which have led to the perception that clock time as a universal phenomenon has become hegemonic, has become integrated in infrastructural provisions like railroads and telecommunications, and has been a driving force behind several different forms and stages of globalization. Many of the products we use in our daily lives are derived from a complex geography of production, distribution, and consumption on an ever increasing scale.

Why does this book focus on time in relation to international business and globalization? During the eighteenth century, America's oldest Founding Father, Benjamin Franklin, coined the phrase *Time is Money*. This dictum implements itself today on a global level. Presently, talk about globalization seems to implicitly imply the validity everywhere of the statement that *Time is Money*. Indeed, the current stage of internationalization, primarily of economic activities, is commonly equated with the process of electronic globalization where the movement of capital and the sharing of information within and between businesses move faster than ever before. "Advances in information and communications technology now allow vast currents of capital to circuit the globe at the blink of an eye and firms to produce different components of a single commodity in distant corners of the globe" (Scheuerman 2004: xiii).

By the end of the 1980s, the traditional distinction between domestic and multinational companies became blurred. International cooperation and competition no longer seemed the prerogative of industrial giants alone. At the turn of the 1990s, globalization surfaced as the new buzzword. The steady dismantling of trade barriers in Western Europe and in the Americas, the increasing availability of global capital, the unexpected collapse of communism in Eastern Europe, and the adoption of social market policies in some parts of China stimulated this process. The disappearance of economic barriers such as national borders, the drive for deregulation and privatization of state-owned companies, as well as the progressive convergence of consumer tastes has opened new opportunities for multinational enterprises operating internationally on a large scale as well as opportunities for small and medium-sized businesses, otherwise known as *born globals*. Indeed, globalization presently describes the actual reality of an economic world which is neither fully contained nor constrained by national boundaries (Spicer 2006).

Within that context, globalization is currently seen as the worldwide movement towards commercial, economic, financial, and communications integration. It implies opening up beyond local perspectives to a broader outlook of an interconnected and interdependent world with free transfer of capital, goods, and services, as well as people, across national frontiers.[1] Globalization today reflects increasing competition in larger markets, creating a pressure which is not only felt by multinational players who invest overseas to gain competitive advantage but also by those who operate in domestic markets who may feel threatened by the new players from abroad. The faster and more efficient this movement rolls along, the more gains the international players are supposed to realize. However, this vision of globalization does not contain any historical connotation. It simply implies that this movement of globalization is dawning. Yet where did this movement come from, what direction has it taken, and what are the driving forces behind it? To answer these questions, the book will put globalization in its historical spatio-temporal context.

The geographer Dicken (2007: 7) has stressed that the "newness" of globalization has been "grossly exaggerated." The world economy was already extensively integrated in the period before the First World War. However, it was mainly a shallow integration which manifested itself through an arm's length trade in goods between independent firms and international movement of capital for simple, direct investments with low interconnection.

In *The Human Web: a Bird's-Eye View of World History*, father and son McNeill have indicated, however, that initial globalization is much older and that the "Old World Web" may have even had its origins 2,000 years ago when a handful of successful imperial societies increased their power and imposed their cultures on new lands (McNeill and McNeill 2003). Some civilizations within Eurasia and Africa homogenized the "Old World Web" and advanced quickly, leaving many others behind. The advancement towards a modern form of globalization began after 1450. "In the three and a half centuries after 1450, the peoples of the earth increasingly formed a single community" (ibid.: 155). After 1450, oceanic navigation united several different metropolitan webs together "into a single cosmopolitan web [...] based on a unitary maelstrom of cooperation and competition" (ibid.: 5). This single cosmopolitan web became the first actual stage of globalization beginning with the factual discoveries of new territories and the economic rebound of the late fifteenth and early sixteenth centuries.

The global economy of today is, therefore, the outcome of a long historical process whereby technological infrastructural development and military, as well as economic relationships, helped shape the interconnected changes and relationships of subsequent periods. Cartographers were able to design the globe by depicting the new world on artificial globes which reflected the process of shaping the Old World Web into a new form of terrestrial globalization. This book takes its starting point here while terrestrial globalization coincides with the rise of international business in Europe. As will be shown, terrestrial globalization opened up specific applications of abstract time, attempting to diminish the influence of various local, cultural, and national societal systems with their own temporal complexity based on all sorts of concrete event times. Yet the continuous tension between manifestations of abstract time and concrete time cannot be understood simply "in terms of the supersession by abstract time of all forms of concrete time" (Postone 1993: 216). The tension is a rather significant cause of the different forms of globalization that developed over the course of Western history.

Over the last few decades, however, globalization has also become a heavily contested issue. Some, like the Japanese management author K. Ohmae (1990), have developed a very favorable view on globalization. It is perceived as the outcome of a fast growth of world trade and commerce with the liberalization of national markets through global competition with an outcome of daunting technological advances and increasing interdependence. He presupposes that there is increasing geographical spreading and increasing concentration as well as a growing functional integration between different economic activities. Defenders of this view believe that we will soon all live in a borderless world where "national" limits are disappearing.

Others, however, see the *hyper*-globalists as spreading a myth. They emphasize that although globalization is currently being presented as a universally profitable stage of modernity, it is, in fact, mainly a project defended by elitist Western neoliberal interests in an attempt to subjugate even the remotest corners of the world and primarily safeguard Western economic benefits and prosperity. The mass protests of *anti*-globalists during the World Trade Organization (WTO) meeting in Seattle in 1999 which derailed the launch of another round of trade liberalization and the G-7 meeting in Genoa in 2001 have been a clear indication of a growing resistance against a one-sided globalization process. The protesters during the G-20 summit in London in 2009 demonstrated peacefully against global capitalism which was struggling with the shadow of its latest financial crisis (James 2009). They did not ask for a restoration of national hegemonies but stressed the establishment of global citizenship as a way to further embed advances of a better form of globalization based on a new politics of solidarity. The *pro*-globalizers look at the increasing political-economic expansion as a succession of stages which calls for a legal and institutional framework based on universal values and norms to embed and regulate globalization which more individuals and social groups might enjoy within their own range of temporalities. Globalization in that sense offers opportunities in terms of improving world economics and the general quality of life without undermining human security, social equity, freedom for the individual, and democracy. The *pro*-globalists look for approaches which actually reinforce positive developments of this kind of globalization. One of them is a more humane approach of social time.

Spatio-temporal approaches

The sociologist A. Giddens (1979) has placed the concept of time and space at the heart of social theorizing about societies and economic organizations to understand the processes that led to modern society. He stresses that Marx primarily focused on capitalist production relations and the commoditization of time and that Weber has mainly analyzed the processes of rationalization and bureaucratization to understand modern society and the coordination of time. However, both authors studied this modernization process primarily from the time perspective. Giddens, however, stresses that this process has been the outcome of the dissolution of social systems, i.e., social communities like city-states, from local contexts of interaction and have become re-embedded in social systems across large distances in time *and* space. He points out that the extension of social life in time and space is an overall characteristic of the capitalist development which advances into what he calls an *extensive system integration*.

Capitalism is constituted temporally and spatially, and both concepts are core components of modernization in which administrative power became exerted as much in capitalistic production processes as in public administrative procedures of nation-states. Giddens called this power the practice of *surveillance,* i.e., "the control of information and the superintendence of the activities of some groups by others" (Giddens 1987: 175). The effectiveness of administrative power became nested within a particular temporal and spatial context. He, therefore, criticizes authors who still define time and space in general terms as having no history themselves but simply see them as setting the stage for actions, i.e., the "environment of actions." They implicitly apply them without recognizing that abstract time itself has a history which in a particular way became characteristic for modern Western culture and society.

The development of economic organizations like international businesses, which played and still play such an important role as drivers to speed-up economic and social changes, can only be clearly understood if they are studied in the context of modernization and are analyzed from an historically determined and integrated spatio-temporal perspective.

The geographer D. Harvey (1989) has similarly contributed to a better understanding of capitalism's trajectory by stressing its geographical context. Like time, space is materially produced by different societies in a variety of ways. Space comprises qualitatively different physical forms which are the outcome and creator of particular political, economic, as well as cultural processes. Capitalism has had a tremendous influence on the way these geographical forms and outcomes have been shaped. Harvey introduced the notion of time–space compression to indicate that the increased pace of change due to new technologies and the advancement of globalization has enhanced the awareness of far distant places having an impact on our own lives instantaneously.

Both authors have stressed the importance of studying the contextualities of institutionalized patterns of interactions across time and space that are viewed as inherent in the investigation of social reproduction of capitalism. During the era of industrial capitalism, general economic processes began to characteristically affect the disembedding of individuals and institutions from their traditional contexts which are primarily determined by concrete time, trying to re-embed them in a setting which is dominated by abstract time.[2] Concrete time is always registered differentially in different settings depending on local customs, practices, and traditions (Postone 1993). Within capitalism, abstract clock time was eventually used to subsume all manner of space–time differences like differences in work cultures, production processes, as well as the styling of products and services into its own regime and turning them into profitable outcomes. Yet, as long as concrete time is embedded in local socio-cultural patterns, it can still operate alongside abstract time. In that sense, the space–time domain of globalizing capitalism shapes "territorially expressed, sometimes juxtaposed, sometimes distantiated relations that are dynamic and differentiated" (Castree 2009: 33–34).

The sociologist M. Castells (1996) has integrated this space–time approach in his analysis about the emerging informational capitalist network society. He stresses that, due to the revolutionary development of Information and Communication Technology (ICT), the network society has dissociated itself from the place-centered formation of industrial society and is primarily focusing on the space of flows of information which is driven by speed, flexibility, and timing. In the context of the global economy and due to the impact of ICT, time and space have obtained a new meaning. The latest industrial space that is organized around flows of information and capital has created a global economy which is different from a world economy:

> It is an economy with the capacity to work as a unit in real time on a planetary scale. A space of flows in the form of networks transmitting information has created a setting in which distance becomes meaningless. The space of flows ultimately is a space-less space. This not only has serious consequences for the way the global economy is spatially operating but also impacts on the concept of time within societies in the world. At the level of time, we have entered the era of time-less time, a modular time where the traditional rhythms of night and day have disappeared. Time zones, holidays, bank holidays, etc. have disappeared as markers. [...] Financial markets operate 24/7 throughout the year. The biological rhythms of human beings have become obsolete [...]. Time-less time belongs to the space of flows while time discipline, biological time, and socially determined sequencing characterize places around the world, materially structuring and destructuring our segmented societies.
>
> (Ibid.: 465)

Real time describes the full potential of what these technologies might achieve. Castells defines real time as space-less time; "What I call time-less time is only the emerging, *dominant* form of social time in the network society" (ibid.: 434). The space of flows induces time-less time as the latest form of abstract time which even becomes detached from the traditional clock time concept and only exists in the form of sequence.

Hope (2006), however, has questioned the validity of this concept. "Castells appears to be saying that time-less time contains no temporal contradictions. But the history of globalization illustrates the opposite tendency (...). Under global capitalism, the ruling myth of progress is synonymous with the ideology of real time" (ibid.: 289–291). Time-less time seems to imply that "this temporal realm is beyond human influence" and that time-less time is now "born to rule forever." Yet this time-less time concept owes its existence to information technologies and networks, which are related to particular historical processes driven by globalizing capitalism. The latest form of globalization may transcend existing economic relations and facilitate the appearance of economic, cultural, and political processes operating at the level of a global order. But this still involves "a transition from a geographically ordered world, in which the basis for economic and political organization was sovereign territorially, to an a-territorial, networked mode of organization whose present and evolving form is not yet clear" (Kobrin 2005: 219). It becomes necessary to analyze globalizing capitalism with its own social acceleration caused by technological improvements within a broader and more historical context and describe what has happened within the capitalist space–time context to better appreciate the tension between the global and the local in terms of abstract and concrete time.

The German sociologist Hartmut Rosa has developed such an encompassing approach to understand the intensifying speed-up of processes which have led to the impression of a general acceleration of Western society. Social acceleration is a constitutive trait of modernization which is not only an outcome of speed-up processes taking place in transportation, communication, and production as Giddens and Harvey have stressed but also manifests itself in quickly changing patterns of stability in society as well as in the pace of life. To clarify this point, Rosa (2003) has drawn a distinction between technological acceleration, acceleration of social change, and acceleration of the pace of life. Technological acceleration is primarily driven by the introduction of new technologies in communication, transportation, and production at a faster pace than in earlier historical periods, bringing about a whole range of changes in social practices, communication structures, and corresponding forms of life.[3] Giddens and Harvey primarily addressed the outcome of this development in terms of "time–space distanciation" and "time–space compression."

Rosa, however, points at a paradox: although we might expect that technological acceleration would lead to a decrease of the time needed to carry out everyday processes and actions in the sphere of transportation, communication, production, and reproduction, the opposite seems to be the case. People in Western societies constantly have the feeling of being under time pressure, and they complain about scarcity of time. Rosa (2005) finds an explanation by looking at the two other forms of acceleration in social change and pace of life. He notices that Western people are inclined to constantly intensify their actions: distances are not only being crossed faster, but also more frequently; production has not only been sped up to increase output but more production runs are introduced which intensify the variety of production outcomes to satisfy the needs of potential consumers. Accelerated social changes can be noticed in altered workplace structures, changes in family relations, as well as in new modes of communication or cultural life.

Yet the measuring of rates of social change remains an unresolved challenge. The heightened tempo of everyday life becomes noticeable in the complaints of people with busy lives who not only intend to execute their activities faster like multitasking but also fill their lives with more activities and experiences. They feel their lives becoming too hectic because they cram an ever greater number of activities into smaller units of time and accept the blurring of temporal borders between the weekend and normal workweeks. This threatens to overwhelm their human capacities to absorb information and coordinate one's own life in a meaningful way.

The acceleration of activities in transportation, production, distribution, and consumption could potentially have led to significant time savings and more leisure and free time, yet in practice it seems to have led to increasing acceleration of social changes and the pace of life. American political scientist Scheuerman (2004: 17), however, warns that the relentless revolutionizing of modern technology should not be conflated with accelerated social changes. "Social acceleration is a multi-sided phenomenon with significant non-economic as well as economic origins." He points in particular to the slower processes of social change within the political and law-making sphere which follow its own temporal rhythm compared to the accelerated processes in the economic sphere.

Rosa (2003) has also signaled that there is a flipside to social acceleration which is noticeable in corresponding phenomena of social deceleration and slowdown like traffic jams and movements for environmental protection and "slow food" which might fulfill a functional necessity in terms of maintaining a pattern of de-synchronization within an accelerated society instead of being just a residual to it. Although the three forms of social acceleration can be analytically distinguished, they can be plausibly interpreted as constituting elements of a single social trend. Scheuerman and Rosa both believe that processes of globalization have generated a wave of social acceleration, which seems to intensify the paradox of the simultaneity of technological acceleration and an increasing scarcity of time.

One of the latest solutions to cope with this paradox seems to be flexibilization, which has recently become a popular slogan in modern Western economies. Flexibilization is required for the labor market, with respect to labor processes as well as organizational structure of the modern factory and office and attitudes towards global challenges. Yet flexibilization seems to be seen primarily as a way to address social acceleration and the way people in the Western world try to cope with the paradox (Rosa 2005; Scheuerman 2004). There are, however, still "territorial as well as social and cultural 'niches'" (Rosa 2003: 15) that have until now been minimally influenced by the dynamics of social acceleration in modern society. This issue of the impact of multinational corporations figuring out how to accelerate production and optimize flexibility even in non-Western societies will be elaborated in Chapter 14. It will raise the question whether globalization with its almost hegemonic fixation on linear time and acceleration of everyday life will colonize any other form of temporality of human activity.

However, globalization and global corporations are not identical. Globalization can also indicate a spreading and interpenetration of multiple local activities each with their local times which may obtain a world wide range and significance. Globalization may, therefore, lead to the intensification of the sharing of information within worldwide social relations which link distant localities in such a way that local happenings are shared by events occurring many kilometers away. That process of globalization embodies a transformation in the spatio-temporal organization of social relations and transactions which will be characterized by an increasing heterogenization. This view on globalizing processes stresses that in different historical circumstances, spatio-temporal changes have taken place leading to transformations in the worldwide organization of social relations and transactions which have been embedded, disrupted, and re-embedded into different spatio-temporal horizons. Many people still live in multiple places located within their own regions and act within their own organizational approaches based on their own time preferences and time arrangements. In that sense, there are still forms of time and space which remain outside the capitalist spatio-temporal configuration which in itself may help broaden the landscape of time. Will these temporalities be able to resist the abstract clock time as it is used in capitalism and maintain a world full of contingency and difference? Whatever the conclusion may be, "the announcement of new eras should not fool us into assuming that the new order crowd out established behavior" (Sorge 2005: 233).

PART I

Terrestrial globalization

During Christmas 2008, there was a remarkable newsreel about the Chinese marines joining Western forces in military action in the Gulf of Aden to capture Somalian pirates who raided international vessels taking the route through Suez. The news reported that this was the first time after 600 years that the Chinese naval power sailed with three ships from the harbor of the island of Hainan to become involved in military action beyond its own waters. The news commemorated how, in 1421, the Ming emperor Zhu Di had sent his eunuch admiral Zheng He with an armada to explore unknown waters and improve trade. It indicated that China had established links with African coastal civilizations long before the Europeans approached the same borders.

The next chapters will deal with the way Chinese and Europeans began to explore the globe on a scale never seen before. Although the Chinese had preceded the Europeans in their explorations of unknown seas in the fifteenth century, European trading companies, in close collaboration with their nation-states, eventually took more advantage of their discoveries than the Chinese. The open seas offered opportunities which could not have been attained via land routes. Of course, there was the Silk Road, an ancient trade route that linked Europe, the Middle East, and Asia, and the civilizations of the Romans and the Chinese. The road – which could better be called a system of trails – was composed of different caravan tracks that skirted Asian deserts and ran through different cities and oases. Once the Tang dynasty took the Silk Road under greater control and protection, the traveling of foreigners and their goods on this system of trails increased and endured through the establishment of Islam as well as the rise and fall of the Mongol Empire.

There were, additionally, other routes over land which made civilizations scattered throughout the globe become familiar with each other. Northern Africa, and especially the great ancient Egyptian civilization, had established large-scale trade with Europe and Asia. A number of African kingdoms were well developed and the routes to the Sudan and the Horn of Africa frequently taken. For centuries, there existed primitive trans-Saharan trade routes in the eastern part of Africa, as well as along the West African Niger and Congo rivers that led the trade in iron, copper, cattle, and other goods. With the spread of Islam, Trans-Saharan trade in salt, slaves, horses, and other goods increased.

Beyond the Trans-Saharan trade routes were sea routes along the Indian Ocean well suited for long-distance sea trade. India traded with Africans, Arabs went to China, and Chinese went to Africa. The regularity of the seasonal monsoon winds aided Arab and Persian sailors to trade in gold, ivory, and slaves from East Africa as well as horses, salt, and textiles from the areas surrounding the Red Sea and the Persian Gulf. In Asia, the spice islands of Sumatra, Borneo, Java, and the surrounding islands provided rare spices of all sorts. Silks and ceramics made in China became available elsewhere. In Europe, the Mediterranean Sea offered far-reaching opportunities for the transport of raw materials, goods, and luxury products which came from afar. So, one could say that by the thirteenth century, "a clearly recognizable system of horizontal communications between the principal markets of Eurasia and Africa was already in place" (Arrighi 2010: 11). However, worldwide trade and commerce across the big oceans to establish free markets had not yet been explored.

2

EXPLORING THE GLOBE

Amidst several big territorial empires that supervised international trading activities from a dominant central power, the Greeks forged history's first free market economy (Moore and Lewis 1999). During the period 800–500 BC. when the urban pattern of the classical Greek civilization slowly crystallized, long-distance trade with Syria and the East increased. The introduction of coins for payments aided in the spread of a money economy. Like the Phoenicians in the ninth and eighth centuries sailing from Sidon and Tyrus along the coasts of North Africa to establish new trading posts, Greek seafarers intensified the trading around the Mediterranean Sea on their own without the support or restriction of a dominant state or empire.

Around the mid-sixth century, there were approximately 1,500 Greek cities in the Hellenic homelands and abroad. One of these cities, Corinth, developed into a community with a gift of inventiveness and a dwelling of good government. Another like Corinth, Athens, grew into a community with a particular civic cohesiveness.[1] Yet the majority of the big cities were congeries of landowners who drew their material wealth mainly from the countryside and not from trade. Agriculture furnished the main fortunes of the cities, and slaves were deemed essential for the maintenance of that agriculture (Anderson 1974).

> Most often, slaves were obtained after aggressive rivalry with their foes and military conquests of city-states of classical Greece. These military successes and subsequent collection of slaves became easier when the citizens of Cyprus moved the iron-working economy to Greece, which then helped them to improve the making of military weaponry.
>
> The change from bronze to iron swords was the beginning of a change in both the tactics and strategy of warfare which would also have profound political and economic implications for Greece. Large standing armies

> replete with generals, chariots, cavalry, bowmen, and battering rams were an Oriental, not a Hellenic, phenomenon. The hoplite revolution in warfare, together with the division of Greece into scores of tiny city-states, would help guarantee the existence of independent farming and commercial enterprise.
>
> (Moore and Lewis 1999: 145)

During the East Hellenistic era, the mercantile space between Rhodes, Antioch, and Alexandria developed into a strong trading axis. In contrast to the Greek tradition, however, monarchs in the Hellenistic world became divinized, and a considerable royal bureaucracy developed. As it did not establish any legal system to stabilize its functions, its political and military position was eventually weakened.

From 200 BC onward, Roman imperial power advanced eastward and overthrew Hellenistic powers. The Roman economy became increasingly oriented toward agriculture as its army was made up of farmers who were rich enough to enrol in the organization. These farmers/soldiers were paid for their services by obtaining landed property. Yet the people who owned the land did not necessarily cultivate it. The rich citizens were usually absent from their estates and, as landowners, employed managers to conduct business on their behalf. These managers coordinated the input of skilled and unskilled labor, both slaves and freedmen. The introduction of the large-scale slave mode of agrarian production (*latifundium*) was a decisive innovation (Anderson 1974).

Although slavery was, by definition, an employment arrangement of an involuntary kind, the slave-worked agriculture grew at a remarkable rate. The bailiff or manager (*vilicus*) was usually a slave or freedman connected to the land, and though his position could still be revoked at will, he was usually treated with respect.[2] He performed a wide range of activities such as keeping track of the yields, recording the tasks to be executed, registration of transactions, buying and selling of tools, animals, fodder, and slaves (Aubert 1994). Dependent peasant tenants, who could themselves employ slave labor, were tied to the landlord's estate and were sometimes forced to involuntary servitude, though they worked the land in their own rhythm while the slaves were exploited full-time.

In Italy, commercial progress advanced regardless of the fact that the Roman elite affected a prejudice toward trade and commerce, which included the business people (institores) dealing with it. Etruria, in particular, had large deposits of iron, copper, and silver and became the metallurgical hub of the Mediterranean. Its metal-smiths produced bronze and iron utensils, tools, weapons, and luxury goods using advanced techniques. The young Roman Republic took over the system of independent enterprise. "Most firms remained in independent hands led by bosses who were largely entrepreneurs and often foreigners, freedmen, or even slaves" (Moore and Lewis 1999: 219). The early Roman state in general refrained from interference in the market but stimulated international trade along the routes (*viae publicae*) they had constructed. The Roman legal system regulated trading relationships and exchange between private citizens on the basis of contracts. The imperial state rested on a system of private and public law which remained in Western Europe as a strongly codified body of civil jurisprudence.

This strong coherence, however, could not prevent the intrusion of Germanic invasions that erupted within the Empire. Germanic migrations changed the Roman agrarian system, and the new practice of European feudalism emerged which forced peasants into serfdom. The Medieval period was characterized by the dissolution of the Roman Empire, the period of Germanic invasions, and the establishment of the medieval Carolingian Empire.

During this period, the Muslim conquests in the Middle East resulted in the creation of the Umayyad caliphate which enforced itself onto southern Europe. Many local European kingdoms (*regna*) also appeared after the disintegration of the Carolingian empire and were usually based on dominant warrior groups. During the period in which trade and commercial activities continued and expanded elsewhere, Europe lagged behind, joined in late, and had to catch up with Asian, Chinese, and African trading communities and empires. Eventually, however, it was in Europe where terrestrial globalization found its seedbed.

A European miracle

In 1981, the historian E. L. Jones wondered for the first time about *The European Miracle* (reprinted 2003), and tried to explain how Europeans were able to set sail from home, discover, and establish bases around the world, settle in some parts, and conquer others. This publication belongs to a wide range of different historical narratives that explain how Europe eventually moved into unknown seas to find unexplored territories. Along with specific geographical and ecological circumstances, which provided certain windfalls, it was the relative autonomy of the economic system and the establishment of independent nation-states in Europe that eventually created a decentralized and flexible continent where, as Jones (2003: 110) emphasized, "unity in diversity offered some of the best of both worlds."

The most important factor in understanding European success is, according to the French historian Braudel, the unique merger between city-states and capital interests which began at the end of the medieval period. Capitalism only triumphed when it became identified with the state.

> In its first phase, power of the Italian city-states of Venice, Genoa, Milan, and Florence lay in the hands of the moneyed elite. In seventeenth century Holland, the aristocracy of the Regents governed for the benefit of, and even according to, the directives of the businessmen, merchants, and moneylenders. Likewise in England, the Glorious Revolution of 1688 marked the accession of business similar to that in Holland.
>
> (Braudel 1977: 64–65)

The historical development of capitalism can, therefore, only be understood by realizing that it moved through different waves of capital accumulation within Europe and eventually arrived in the United States. But how did Europe become a recognizable unity after some independent Italian city-states began to dominate international trade?

According to the historian L. Lewis (2008), there were two phenomena that occurred which made Europeans eventually think of themselves as a people and could begin to consider themselves as Europeans: one was the creation of a vast Holy Roman Empire by Charlemagne in the ninth century; the other was the interaction with the Muslim culture in Spain, which the Arabs called al-Andalus. Although European civilization is the cultural legacy of the Catholic Franks, the interactions with al-Andalus have seriously influenced this legacy.

Following the Barbarian invasions and the Islamic invasions in southern Europe, the Roman transport infrastructure that had linked different parts of Europe to each other had fallen into disuse. The economic space had fragmented and offered little opportunity for general development prior to the emergence of some cities as centers of industry and commerce. During the seventh century, Islam burst out of Arabia, spreading with astonishing speed in every direction and seized Arabia, Persia, and Egypt as well as North Africa. The Muslims built a formidable military power. They possessed the ambition to enforce themselves as the rulers of Constantinople with its boat-building and trading capacities, its tremendous wealth and expertise, as well as being the pinch point where Europe met Asia. Between 674 and 718, they besieged Constantinople but did not capture it.[3]

On the other side of the Mediterranean in North Africa, a large Berber army was guided by a freed slave, known as Tariq,[4] who had become a general. Tariq crossed the Straits of Gibraltar into Spain, where the army overthrew the Visigoths who had ruled much of the Roman province during a significant part of two centuries. By 720, most of the Iberian Peninsula had fallen under Muslim rule. Abd al-Rahman reshaped Cordoba into a center of culture and science with a large library rivaling Abbasidian Bagdad. The vast, disciplined, and experienced Muslim army, however, was hungry for more conquest. The army crossed the Pyrenees to attack France but when they marched on to Poitiers, they were defeated in 732 by Charles Martel, Charlemagne's grandfather. A Christian scribe in a Latin chronicle written in 754 called the victors at Poitiers "*Europenses*." It is the first recorded use of a Latin word for the people of Europe (Lewis 2008).[5] A further military expansion of Muslim influence was blocked by internal divisions among the Muslim rulers themselves when they established several different caliphates, but this did not prevent exchanges of ideas and techniques between Muslims, Christians, and Jews in the Hispanic peninsula. Astrology, astronomy, and hydraulics for agricultural use advanced and were broadly shared. Some of the Islamic intellectual achievements were clear evidence of a kind of civilization which was brought to Spain and reinforced similar developments in Europe.[6]

Interactions with the Arab world

While Muslim rulers were dominating Spain, Charlemagne (748–814) created a vast Carolingian monarchy. As a great scholar and soldier, a patron of learning, and an ambitious administrator, he annexed Lombardy in Italy and conquered Saxony. He created new bureaucratic structures, sending his royal officials to each of the 350 counties of his realm to deliver his commands, hear cases, and summon his people to war. He became ruler of the Christian continent, took on an imperial mantle, and was crowned by Pope Leo III as the Holy Roman Emperor – a title he had invented himself – to succeed the previous Roman Empire on Christmas Day of the year 800. Monastic centers of learning were established, drawing scholars from across his Empire and beyond. His cathedrals, castles, and locally administrated manors for kings and nobles established organized communities in Western Europe.

Charlemagne extended existing social relations and converted them into vertical ties of authority. As there was no national system of taxation, he lived off plunder and the product of his own estates. What his lords owed him was military service. Sometimes in late spring, they were obliged to show up for a military campaign. Although there was not yet a network of big cities, extensive commercial connections continually expanded his kingdom. Out of fear that Muslim forces might further enter Europe, Charlemagne assembled a vast army to enter Spain in 778, but this assault was without success.

During his reign, civil servants were actively turning his regnum of political, cultural, and ethnic diversity into a whole. Charlemagne relied on writers, copyists, lawyers, bishops, and diplomats to expand his influence (McKitterick 2008). Civil servants traveled along Roman routes as far as the routes still existed. Charlemagne handed fiefs (or the medieval word *feudum*), which were delegated grants of land, to his loyal elite, vested with juridical and political powers. He thus acted as "the real incubator for European feudalism" (Anderson 1974: 168). Feudal lords became the masters and managers of agricultural production and the entire process of social life.

One of the issues of growing importance was the establishment of a proper calendar system in his empire. Charlemagne saw uniformity in the calendar as instrumental to the functioning of his European kingdom. He wanted an "imperial" calendar based on a standard cycle of Christian festivals and required reliable calculations which were necessary to establish the date of Easter. The story is told that Charlemagne acquired an hourglass from an Arab king large enough to run a full 12 hours before it required turning. Such timepieces represented learning and progress "much like a model T or an early Remington typewriter once signaled modernity in small isolated towns across America" (Duncan 1998: 129).

Charlemagne proposed to rename the Roman calendars in the Frankish language and call them after the seasons of the year, festivals, and holy celebrations:

> At the heart of this calendar was a cult of the Saints, but it was decreed that no new saints were to be venerated, and only those already chosen based on their passions or lives would be used. The Christian orientation of time was extended across the realm as the Frankish territory expanded in an attempt to determine more general perceptions of the monthly, seasonal, and agricultural cycle of the year.
>
> (McKitterick 2008: 324)

The major Saints' days – some 81 Saints – became well known and were widely observed. During the medieval period, Saints' days remained an informal method of dating years and centuries.

The Carolingian empire, however, mostly remained a landlocked area with minimal foreign trade. When he died in 814, Charlemagne left his empire divided among the children of his only surviving son, Louis. During Charlemagne's reign, the medieval monarch was considered "*primus inter pares*" and continually vied with powerful princes and barons for political ascendancy. After Charlemagne's death, however, the socio-political order decentralized and became unstable. Continuous warfare among local lords increased and new sovereignties appeared. The number of states in continental Western Europe increased from fewer than ten in the year 800 to more than 200 by the year 1300. This increase was primarily due to the establishment of many city-states in Italy and Flanders.

After the dismemberment of Charlemagne's empire into several different political centers, control over land remained an effective instrument to control all the essential resources for survival. Europe declined to an agrarian subsistence level, the Roman route infrastructure slowly fell into despair, and people lacked the integrated system of routes the Roman engineers had built (Anderson 1974; Vance 1986). Feudalism created a retreat to the local entity in most activities. The feudal economy became essentially manorial and was built around non-monetary exchanges that were local in nature. Peasants were bound to a subservient rank in life by divine ordinance. Kings became endowed with quasi-divine powers, but in some cases, the concept of the "King in Parliament" reinstituted some power to the clan region at the expense of the fief. Collaboration gave way to competition. However, the limits of the landlocked Carolingian kingdoms with their "castellar" landscape designed to protect the rural population did not limit further social and economic prosperity. On the contrary, there appeared a paradoxical situation of political disintegration with economic expansion. Two different explanations have been developed as to why this remarkable progress occurred.

The first explanation is a top-down approach which states that, in the political vacuum created after the Carolingian empire disintegrated, the Catholic Church took over and extended its dominance by creating an extensive network of monasteries. This created a new institutional framework for Europe by developing a systematized set of laws governing the Catholic Church and many associated secular areas such as urban law, feudal law, and manorial law. "In what was the core area of feudal Europe – France, Belgium, Germany, Northern Italy, and the adjacent parts of Spain and England – the number of monasteries increased spectacularly" (Zanden 2009: 43). The Catholic Church imposed an administrative hierarchy to assure the implementation and extension of the new legal framework.

The other explanation is a bottom-up approach stressing how rising independent city-states in Northern Italy such as Venice, Pisa, Genoa, Bologna, and Flanders, plus Northern France with places such as Le Mans, Cambrai, Brugge Gent, and Leon became governed by independent communes which soon established corporate bodies such as craft guilds. These guilds fostered market exchange by instigating a practice of quality controls and becoming highly involved in the government of the city-states.

Although the fear of the Islamic drive for expansion continued and cut off the northern European economy from an important part of international trade, some medieval towns in Europe began to practice trade and manufacture and evolved into self-governing communities. They enjoyed political and military autonomy from the local nobility which created a "dynamic opposition" between towns and country (Anderson 1974: 150). Agricultural expansion, as well as urbanization and expansion of rural settlements, progressed during the end of the medieval times. From 950 to 1300, a period of remarkable prosperity facilitated the establishment of a highly urbanized European region. The combination of these responses to political fragmentation explains the specific institutional framework that emerged in Europe from 1000 to 1300 (Zanden 2009).

Europeans on the move

As was mentioned earlier, Jones (1981/2003) was puzzled by the European Miracle. Why was Europe able to conquer the world rather than the long-lasting civilization of China, the Ottoman Turk Empire, or the empire of the Mughals in India? (Cf. Meier 2005.) One argument is that these were land-based empires with territorial armies and political authority relations between a central power and many diverse and differentiated entities. This hub-and-spoke network structure preferred expansion on land with armies of foot soldiers and horsemen and felt less at ease on big seas. The exception was the great voyages of Admiral Zheng He in the early fifteenth century. Within the Ottoman Turk Empire, the seaborne trade was largely the province of individual Christian and Arab subjects who were well known as great seafarers. They transported merchants to the Mediterranean and Black Sea waters as well as the Indian Ocean.

The evolutionary biologist J. Diamond (1997) has stressed that there were ecological advantages which certainly eased international trade and economic exchange for Europe: the middle latitudes which stretch all the way from Europe to China allowed for a supply of similar foodstuffs and grains discovered in the Middle East to be grown in Spain and China. These ecological advantages reinforced the exchange of goods through east–west trade routes between Europe and China. Travelers following the Silk Route with its many caravanserais also took with them ideas and techniques that were important and quickly applied them in Europe. Inventions such as gunpowder, which was discovered in China, became deployed in European arms.

Although the Chinese art of papermaking was first known to the Muslim world in the eighth–ninth centuries (Masood 2009) and block printing was first recorded in Chinese history in the twelfth–thirteenth centuries, the invention of the printing practice by German Johann Gutenberg in mid-fifteenth century tremendously facilitated the transfer of knowledge in Europe (Mokyr 2002). According to this view, it was the superior mobility of thoughts, more particularly in the form of scientific discoveries, which eventually aided Europe in establishing itself and in conquering other parts of the world (cf. Leamer 2007).[7]

There are, however, more nautically technical considerations which have to be included such as the incremental improvements in many products and processes such as ships and navigational techniques to understand the European miracle. Entrepreneurs and guilds began taking advantage of the new navigational techniques and flourished in medieval cities. Whereas in northern Europe the cog and hull were designed to transport more goods over the Baltic and North Sea, the Portuguese had been involved in further promoting the design of the caravel as a highly maneuverable ship to be used for long voyages of exploration. This most prominent improvement of shipbuilding took place during the fifteenth century with the introduction of the three-masted, full-rigged ship that the Portuguese called a carrack and was useful for both war and trade. This type of ship "was not replaced as the prime sailing vessel of Europe until the arrival of the clipper ships of the nineteenth century" (Hunt and Murray 1999: 176). By then, the remarkably faster clippers with more sail to catch the winds were able to clip off the time of a journey while reaching a speed of 14 knots.

Max Weber (1956) has pointed out that interactions with outer Europe have been unique in creating Europe. Urbanization developed particularly in places that were linked to the outer-European economy. Medieval European cities with their particular autonomy gained a special character where the transfer of knowledge as well as military rivalry and commercial competition found a fertile seedbed. Thanks to the tradition of Roman law, inhabitants enjoyed security in legal matters and a right to participate in local government and appropriate institutions. With the progress of a corporate autonomy of medieval cities, a particular conception of citizenship developed in the West. These cities became centers of trade and places for the exchange of knowledge which survived amidst overall political fragmentation. They facilitated the appearance of new patrician strata that began to control the urban economy. The concentration of power in the hands of particular blocks of governmental and business elites reinforced competition as much as cooperation.

In the thirteenth century, Italian city-states such as the maritime cities of Venice and Genoa, and the commercial-industrial Florence, Siena, and Milan rose into communities where a civilized trade between urban and rural areas was established. Major Italian towns such as Venice and Genoa maintained a municipal urban life by acting as trading centers for commercial traffic across the Mediterranean. Regional specialization was sustained by long-distance trade, and mercantile traffic quickly accumulated large capital sums in Italian hands. These city-states "were able to organize collective efforts" on a very advanced level (McNeill 1974: 12) and as maritime republics developed into powers of worldwide importance.

The city-states were granted juridical statehood but lacked the capability to carry out all kinds of governmental activities usually associated with statehood. A key feature of most successful city-states was that political power no longer remained in the hands of only the landowning class. The de facto autonomy of Italian cities generally led to de jure autonomy of merchants as powerful elites (Gipouloux 2011).

Urban merchant classes became the town's capitalists and were highly involved in city politics. Whereas in Genoa, competition was the norm amongst the merchants and traders, in Venice there was much more cooperation and control. "Cosmopolitan merchants, diplomats and diligent informers mingled in courtesans' salons" (Gipouloux 2011: 31).

Venetian state authorities were, for example, highly involved in shipbuilding facilities. Venice became governed by an oligarchy of businessmen who maintained land and commercial interests (Hunt and Murray 1999). The city effectively turned into the entrepot, the staple market for the northern Adriatic. Artisanal craft guilds were allowed but did not have an influential say in city politics.

Magistrates in city-states invested their spare funds regularly in trading ventures. Ad hoc corporations were established for specific trade arrangements and sometimes involved a large number of wealthy citizens. The capital costs for ships to be used in overseas trade were usually divided into a number of shares and organized in a partnership (commend). Responsibility was divided between silent and active partners. Some cities stationed agents, comparable to the expatriates of today, at other trading places, who stockpiled suitable goods in anticipation of the ship's arrival. Merchants did not establish lasting partnerships to safeguard their activities but usually dissolved them with the demise of many of its members, and set up other business associations.

On the other hand, merchants were generally oriented toward market activities which were meant to make local authorities commit to the security of foreign merchants and helped trade to thrive. Guilds guaranteed the protection of merchants abroad, ensured trade rights and privileges, and designed strategies to promote capital accumulation by establishing their own guilds. In Florence, the situation was completely different. Here, 40 landowning families had been banned from political life, enabling political and economic interests to collude in a way that created widespread wealth. Guilds obtained powerful political positions, and leading businessmen held high positions in the dominant guilds. In other cities, cohabitation between merchants and aristocracy or kings was maintained as a convenience between unequal actors. At times, the trade expansion led to conflicts with the landed aristocracy who were more interested in building rural domains into exploitable "fiefs" than they were in promoting urban commercial expansion.

However, with the increase of international trade, merchants and traders who spearheaded the development of techniques that transformed medieval commerce became increasingly involved in the government of major Italian city-states (Hunt and Murray 1999). A case in point is Venice. For a long period of time, the Venetians made voyages to Egypt and Constantinople to gain profits from long-distance trade. Venice had a strategic location at the head of the Adriatic Sea and early on developed a strong seafaring and trading tradition. Venetians sold spices and other luxurious goods which came via Egypt from India to the wealthy classes of north Italy and transalpine Europe. The Venice maritime Republic built an overseas empire based on its remarkable Arsenal, an Arabic word for "place of manufacture" (Friedel 2007).

> Modeled on Byzantine precedents, the Arsenal was a shipyard under public management that specialized in the construction of galleys for use in war. In later times, mass production of a standard design became the hallmark of the Venetian Arsenal. Workmen developed unusual skill and efficiency as a result of specialization; standardized replicable parts made repairs easy; stockpiling such parts allowed the state to maintain a cadres of skilled men always available, who, in case of need, could direct the efforts of a suddenly enlarged work such as might be required for building a new fleet in a hurry [...]. In its heyday, the Arsenal was capable of assembling a completely equipped galley in less than an hour.
>
> (McNeill 1974: 5–6)

The total number of craftsmen on occasion reached into the thousands and a newly established Arsenal became an "industrial wonder" (Friedel 2007: 100). With a professional elite of permanent employees, Venetian shipbuilders created a reputation for designing and building war ships (Zan 2004).

Yet the Arsenal was not a production site meant for profit maximization but rather manufactured ships for the common good and public interest (ibid.). One can still imagine how the size and formidability of the Venetian navy impressed any enemy.[8] It exercised an extraordinary influence on European life and trade which was soon imitated by other Italian cities. Under the spell of the crusades and international trade opportunities, Genoa and Pisa also developed naval powers which grew rapidly in size and force.

The city became well known as a center of shipbuilding and operated an early assembly line devoted to turning out galleys at an intensive rate from standardized, prefabricated components such as keels and masts. Each galley was towed down a canal between rows of windowed buildings. Workers stood at the windows handing out equipment and supplies as the vessel moved past. Guilds in the city were able to maintain self-regulation over their work practices which created a hybrid organization where even the timekeepers had to comply with guild rules (Zan 2004). The city also introduced early banking laws in Europe to regulate a nascent banking sector. Though Venice was celebrated for its merchants, it was also feared for its ruthless naval warriors.

Via complex trade routes from Africa, the Middle East, and Western Europe, Venetian merchant families poured minerals, salt, wax, drugs, camphor, Arabic gum, myrrh, sandalwood, cinnamon, nutmeg, grapes, figs, pomegranates, weapons, ivory, wool, ostrich and parrot feathers, pearls, iron, copper, gold dust, gold bars, silver bars, and Asian slaves into Venice. They formed partnerships with distant governments and merchants that cut across racial and religious divisions. Arabs, Jews, Turks, and Greeks became trading partners with Venice. "Venice was highly structured, fiercely independent and commercial, and based on a unique combination of feudal obligation and global outlook" (Bergreen 2007: 13).

In 1204, during the fourth crusade, Venice had combined with European forces and captured Constantinople, still the capital of the Byzantine Empire. The occupation lasted until 1261, a period during which the hostile rival Genoa supported the

Byzantine Empire financially and militarily. Similar developments of arrangements with other powerful rulers took place in other mercantile – maritime centers that participated in the outer – European trade and commerce. This reflected a tendency for major economic changes to occur not only in the Mediterranean but also in other regions.

The closer involvement of the noble class in the management of the Venetian state caused tensions. This Venetian oligarchy composed of a small clique of renters was more interested in maintaining their noble status within the Venetian Republic as active management of industry and commerce mainly passed into the hands of domiciled foreigners. The interests and opinions of the merchant class were hardly respected, although this class maintained intensive contacts with other parts of the world (McNeill 1974; Arrighi 2010). Another advantage was that profits of trade accumulating in the hands of the merchants and bankers was passed as credits to weavers who were willing to produce a larger supply of goods which the traders could sell at distant markets. In addition to raw materials, Venetian merchants also imported a different principal bulk import from Syria and Asia Minor which also turned out to be profitable: slaves.

> Genoese and Venetian merchants carried many thousands of at least nominally Christian slaves from Black Sea ports to Egypt and elsewhere in the Mediterranean. Most of these people had been captured by Tatar raiding parties that harassed the Russian lands incessantly.
>
> (McNeill 1974: 54)

Along with the Silk Route, there was another route by which information, ideas, and techniques from other civilizations came to Europe. In southern Europe, internal struggles among Muslim rulers offered Alfonso VI, the Christian ruler of the United Kingdoms of León and Castile, a chance to capture Toledo in 1085. This was an important center of Islamic civilization where Greek and Arabic texts on medicine, astronomy, mathematics, and philosophy were translated into Latin and diffused through Europe. Toledo became the center where Christian, Muslim, and Jewish intellectuals met and discussed a rational understanding of the world.

In the twelfth century, one such intellectual, Al-Zarqali, manufactured astronomical instruments for the astronomers in Toledo, and his simple astrolabe became a common instrument for astronomy and land measuring. His most famous device was a water clock that projected the phases of the moon and the hours of the day. His precise measurements led to the production of the Toledo tables and an almanac (*al-manakh*) which made a comparison between Islamic and other calendars easier (Masood 2009).

Throughout the time of these advancements in other parts of Europe, the political and military situation in Spain remained complex. Although the Christians and Moors experienced long periods of peace with harmonious coexistence, there also existed a porous frontier zone between the north and the south of Spain that expanded and contracted with the tides of conflict and settlement. Christian and

Islamic kingdoms established and broke alliances as necessity dictated. Eventually, an alliance among Catholic princes in the thirteenth century stimulated the gradual movement of Christians southward, wrestling territory back from the Moors. This movement, which has been called the *reconquest*, extended over a long period of time.

On the other side of the Pyrenees, thirteenth century Europe was still fragmented by sectarian rivalry and consumed with the crusades, the last one taking place around 1272. In the east, there was a permanent threat from the Mongols ravaging Eastern Europe. Rumors about the destruction of Kiev in 1240, the religious heart of the Slavic world, terrified European inhabitants (Chua 2007). Nevertheless, the core European population fostered trade and production for commerce but also propagated military conquest and shaped their territories into different regna with state administration and European law. Some kingdoms stimulated arts and sciences as well as technological change, which favored the civilization of subjected populations.

European expansion

European cities became entangled in a competitive struggle which certainly contributed to the rise of Western capitalism. Concentration of capitalist power went along with intercity competition. Although the West European economy experienced a period of downturn between 1320 and 1520, two areas of European navigation continued to be active in international trade: the Mediterranean Basin and North West Europe. The Mediterranean Basin had been the site of layers of civilization – Greek, Roman, and Ottoman – "each of which had contributed to the richness, texture, and local color of the canvas of the modern world" (Barkey 2008: 8). Ships powered by sails instead of oars appeared in the thirteenth century. The compass was taken into widespread use between 1270 and 1300 and eroded the old seasonality of sailing, since it made safer voyages possible in the Mediterranean Sea in winter.

Whereas the Mediterranean sailors used the magnetic compass and a portolan,[9] the North West European navigators in the North and Baltic Sea mainly relied on the magnetic compass together with a rutter, as well as a lead-and-line to periodically sound the nature of the seabed. In both cases, navigation was still depending upon observable geographical features, and sailing close to the seashore was a common practice.

Beginning in 1291, the Straits of Gibraltar gave free access to the Atlantic and North Sea after Benedetto Zaccaria, a Genoese buccaneer and business man, crushed a Moroccan fleet which tried to maintain Muslim control over the straits (McNeill 1974: 48). Italian ships no longer faced serious political and military obstacles in passing freely from the Mediterranean Sea, and voyages from the Strait of Gibraltar to the English Channel during all seasons were finally possible.

As passing through Gibraltar became more common and access to Spanish harbors on the Atlantic side were secured, trade between Italian producers and merchants within the Mediterranean area and the North Sea intensified. "Genoa nurtured active commercial entrepreneurs in the western Mediterranean–North Sea context" (McNeill 1974: 57). Genoa became a community of adventurers and traders. In the fourteenth century, Genoa inaugurated a new maritime route to Portugal, Flanders, and England (Gipouloux 2011). Important city building efforts resulted in wealthy zones such as the Parisian Basin, the London Basin, and Flanders. Medieval Europe offered considerable diversity in terms of terrain as well as historical traditions in places such as Gaul, England, and Scandinavia (Meier 2005).

After the devastating Black Death in the 1340s,[10] which had reduced the population of Europe and the Middle East by about a third, new towns appeared, sometimes promoted or protected by feudal lords who favored the development of local markets. In places such as the Italian city-states, as well as Flemish and English cities, increasing commercial and manufacturing activities in the textile industry created a sizeable wage-earning class of urban laborers beneath the elite of artisans (Anderson 1974).

Yet international trade and commerce between the Mediterranean and Nordic seas were not always seen as favorable. British producers felt at a disadvantage to Italian manufacturers. Whereas the British delivered the raw materials, the final products were manufactured in Italian cities and sold to the Levant. In 1485, British businessmen used their political connections to King Henry VII in order to obtain subsidies and an export duty on raw wool that increased the raw material prices for their competitors on the continent. These measures had a negative impact on the continental wool industry, especially in places such as Florence, but stimulated local wool manufacturing in England. Beyond these rigid measures, however, trade connections between the Mediterranean and Nordic as well as Baltic seas (the Hansa) intensified.[11] Northern Europe saw the development of several different seafaring countries with wealthy trade cities such as Antwerp, Lübeck, and Amsterdam.

The historian J. Blaut estimated that, during the Italian Renaissance, the Italian trading city-states had barely risen above the level of many non-European centers in economic terms. Yet, "within a few decades after 1492, the rate of growth and exchange speeded up dramatically, and Europe entered a period of rapid development" (Blaut 1992: 136).

The towns with freedom to offer space within their own walls had assured the disintegration of serfdom in the West. In free city-states, people gained freedom to strive for personal advancement. A type of active entrepreneurship developed and specialization and competition increased with far-reaching consequences. Apparently, Europe provided a particular institutional setting which embedded their role in a favorable form. Rivalry between city-states and regions contributed to the city's independence and the possibilities for economic initiatives. The increased civic activity promoted the development of new techniques, research, and crafts. The outer-European orientation became clearly visible in harbors and ports in Italian city-states. The drive of merchants, manufacturers, and profit-oriented aristocrats as a proto-capitalist class certainly played a key role.

Fernandez-Armato (2006) sees the oligarchies of merchants, manufacturers, and traders as the beginning of a bourgeois class. They favored one feature of the European culture which made Europeans particularly conducive to breeding explorers: the idealization of adventure. Many explorers strove to embody the great aristocratic ethos of their day: the code of chivalry. The explorers wanted to become heroes with their risky seaborne adventures. Wandering at sea would ennoble the soul.

But the idealization of adventure in commercial city centers certainly would not have led to much success without the availability of specific techniques and instruments, some of which were merely imported, while others were internally developed. The historian David Landes (1983) stresses that, although the West invented the waterwheel, spectacles, the mechanical clock, and printing with movable type, the Chinese developed the magnetic compass, the astronomical clock, and a wonderful system of artificial canals. So, medieval Europe was no more ahead than maybe "by a nose." The prior existence of intellectually advanced civilizations created important external opportunities for Europe. Some of the borrowed techniques were directly useful in the European voyages of discovery: navigational aids, the compass and astrolabe, and gunpowder for cannons to subdue the enemy. Europe assimilated technologies that had originated as far east as China, but new instruments and techniques cannot have been the sole explanation.

The interactions between western European and Asian populations and kingdoms had created combined resources that the Europeans turned into dynamic capabilities that provided them tremendous political power and economic wealth. The European expansion and the overseas imperial systems it created were as much the outcome of European technological improvements, for instance in shipbuilding, leading to more efficient ways to transport and produce goods than any other empire had done before, as the outcome of interactions between so many different societies and cultures (Kamen 2002). When the commercial spirit with a rationalistic approach to technical innovation and scientific inquiry seized control of powerful city-states, a growing market economy transformed the life of many citizens. But, above all, an important and particularly European driving force pushed toward the European Miracle: the existence of multiple and diverse political entities in competition with one another.

After the discoveries of new territories by Columbus and other courageous sea captains, Catholic Portugal and Spain combined expansion with domination. Spain turned into the first global empire, yet it did not do so merely by its own forces alone; its empire was the joint result created by interactions between Europeans, Asians, Africans, and Native Americans alike. With their expansion and domination, the powers of the Iberian peninsula moved out across the globe, found new territories, and made trade and commerce synonymous with the initial stage of terrestrial globalization.

Parallel to these explorations and discoveries another development took place. European cities had become seedbeds for the birth of a new order which contested the existing orders dominated by the Catholic Church and royal families. This led to modernism or, as Weber labeled it, to "the disenchantment of the world." This disenchantment became an existential condition for Europe and generated the Reformation, the Enlightenment, and the French Revolution. Eventually, modern European nation-states became more important in facilitating this process as they triumphed over other competing modes of political organizations such as city-states and their networks because of superior manipulation of speed. Their growing influential role was due to their superiority in preparing for war. "The modern nation-state military and administrative success was primarily based on ever more mobile and rapid-fire forms of military organization" (Scheuerman 2004: 20).

Military rivalry and commercial competition frequently went hand in hand. The political fragmentation in late medieval Europe led to decentralized authority and spurred continuing competition among rulers. In this respect, the European civilization was unique and clearly a miracle.

> There was a continuous interplay between the physical needs of the state and its credibility in its relationship with merchants and the citizenry in general. In particular, the evolution of capital markets was critically influenced by the policies of the state because, to the extent that the state was bound by commitments that it would not confiscate assets or in any way use its coercive power to increase uncertainty in exchange, it made possible the evolution of financial institutions and the creation of more efficient capital markets.
>
> (North, quoted in Wolf 2005: 27)

Some of Europe's rulers paired up with merchants who paid taxes in money which allowed them the ability to afford armies of their own instead of being dependent on unreliable feudal vassals. The direction of business and trade on the Iberian Peninsula remained mainly in the hands of political rulers. Yet the overall institutional context with the presence of independent city-states and some powerful states run by the aristocracy created within Europe a long-lasting state-system. Next to international trade and urbanization, an internal rivalry fed the process to European superiority over other civilizations. Eventually, the process of nation-state formation in Spain, England, France, and the Netherlands appealed to concepts of nationhood and liberty of the subject.

Within the upcoming western world with its many separate political entities, particular regions established close cultural and intellectual ties. Scholars at universities, artists, and literati were in communication with their counterparts in cities and regions using the shared language of Latin (Mokyr 2002). This intellectual development also contributed to the European Miracle by stimulating an interest in civilizations beyond European borders. The information they obtained about other civilizations, however, was initially the result of courageous travelers and traders.

Marco Polo

The West has become familiar with far-reaching civilizations such as the Chinese empire through Marco Polo's *Description of the World*, which was published around 1300 and motivated future generations of explorers.[12] "In the aftermath of the Mongol conquest of the Abbasid Calif's territory in the first half of the 13th century, the Venetian merchants, such as the Polos, were able to travel from Europe to China through the Middle East" (Gipouloux 2011: 46). Marco Polo (1254–1324) was born into one of those families of Venetian merchants who roamed the four corners of the world in search of valuable spices, gems, and fabrics. In 1271, at the age of 17, he began his journey to China with his father and uncle. During their trip, the three Polo men stayed at Buddhist monasteries, where they learned about the use of the lunar calendar that had proper days dedicated to certain saints or idols. After three-and-a-half years of traveling along the Silk Road, they finally arrived in China, where Marco Polo stayed for almost 17 years in Beijing at the court of the Mongol emperor, Khubilai Khan, the grandson of Genghis Khan.

The Mongol emperor, who kept the throne from 1260 to 1294, favored commerce with the world and imposed a Pax Mongolica, which permitted Venetian and Genoese, as well as other European merchants, to safely travel to Persia and China. The Pax Mongolica facilitated an easier transfer of knowledge and techniques across large distances but also the transmission of horrifying diseases such as the deadly plague germ, which originated in China and ravaged Europe between 1346 and 1354.

The Silk Road facilitated trade caravans packed with spices and silk to crisscross the Eurasian land mass. These caravans connected Constantinople to the Chinese capital. The Mongol emperor, however, did not hesitate to subdue the southern Chinese Sung dynasty (960–1279), reunifying North and South China into the Middle Kingdom and forging the largest and most populous empire of that time.[13] He moved the capital of his empire from Khara Khorum in Mongolia to Shangdu, where he received Marco Polo, and then to the former Jin Dynasty capital of Zhongdu, currently Beijing.[14]

To improve orderliness, Khan developed a postal system to transmit vital information across great distances to administer his empire. Thousands of posts stabling fresh horses dotted the landscape. The horsemen carried a girdle of sounding bells so that, while on the move, they could be heard coming from quite a distance. Along the main trade routes, garrisons were installed to provide a safe journey. Watchtowers used smog and flag signals to warn travelers of danger along the routes. Marco Polo was impressed by the network of posts, messengers, and fast horses reaching everywhere throughout the Mongol empire.

Khan ordered the Chinese to give up their coins made of copper, gold, and silver in favor of paper currency. As the Mongols lagged behind the Chinese in technology, literature, and architecture, he welcomed artisans, traders, and merchants to his court. He recruited experienced non-Mongols to build and man his fleet. He invited a Persian astronomer who brought with him new devices such as the sundial, the astrolabe, and celestial as well as terrestrial globes.[15]

The Su Sung clock

When Polo visited the former Su Sung (also known as Song) capital city of Quinsai (currently Hangzhou) he was impressed by its refinement. It had flourished as a large commercial city and seemed to have operated beyond the rigid bureaucratic mold of Emperor Khan, allowing foreign merchants and financiers some free commercial space. Marco also marveled at the striking of the town clock of the city Cambulac three times every evening, after which no one was allowed to be outside in the street.[16]

Polo failed to mention, however, that during the Sung dynasty, astronomers had constructed a marvellous astrarium in Quinsai dating from 1088.[17] As a mechanism for the study and display of the movements of heavenly bodies, this astronomical clockwork was intended to mechanically reproduce heavenly motions and act as a "natural clock" support for decision-making by the emperor (Jami 1995).[18] Astronomers were the only persons who were permitted to use timekeeping and astronomical instruments or to engage in astronomical studies, and they used the astrarium to predict the movements of the heavenly bodies and to compile statements of these movements. These statements would, perhaps, serve as a guide to the Emperor to state when the time had come to sow and reap, or "a time to fight and a time to stop fighting, a time to have sex with the empress and a time to couple with a concubine" (Landes 1983: 29). As astronomy was believed to be something mysterious, the astrarium became a symbol of the mysterious power of the emperor who became perceived as master of the heavens and the master of time.

Attached to the outside of the Su Sung waterwheel tower was a clock which showed time by means of tablet-carrying jacks. There were bells or gongs sounding at intervals, but all of these functions were accessory. This Chinese waterwheel clock was primarily an astronomical device with no diffusion or exchange of knowledge to the public. Knowledge of the right time and season was political power, and mastering this knowledge governed the acts of everyday life and decisions of the state. The Chinese bureaucracy set up an Imperial Astronomical Bureau devoted to the measurement of time and heavenly space. Time was announced by sound: a small bell struck automatically at the beginning of each hour. For those living in the vicinity of the building, the time of the clock might have been an instrument similar to the sundial for measuring time but otherwise had no impact in the daily life of the Chinese population.[19] The astrarium, along with the calendar, were pre-requisites of sovereignty equal to the right to mint coins or issue paper money. Each emperor inaugurated his reign with the promulgation of a calendar which was often different from the one that had preceded it.

Although the Su Sung dynasty disappeared, the Mongol Khubilai demonstrated ethnic tolerance and surrounded his court with Chinese advisors and intellectuals. He allowed Chinese literature and scholarship to flourish. His astronomers and cartographers produced sophisticated maps, nautical charts, and terrestrial globes. He attempted to introduce a universal calendar which would unify the Arab lunar calendar, the European solar calendar, and the Chinese twelve-year animal cycle.

He favored international trade, overland as much as maritime, between China and the Mongol-dominated Persia, Central Asia, and Europe. "Throughout the Middle Kingdom, merchants of every religion and ethnicity made fortunes under the Pax Mongolica" (Hua 2007: 119). After Khan's death, the empire crumbled away and, as the Chinese Ming rulers rose in 1368, the Mongol power ebbed into insignificance.[20] Knowledge about time measurement and astronomy disappeared. "Its disappearance led to a 'magnificent dead end'" (Landes 1983: 17). A new dynasty took over without maintaining the new timekeeping technique.

After 17 years in service at the Mongol's court, Marco Polo returned with his father and uncle by naval route back to Venice. After his arrival, he became involved in a sea battle with Genoa and was captured. During his imprisonment, he dictated his memories about his travels in the *Description of the World*. It was widely copied, and his experiences were extensively discussed at the courts of many European rulers. Though Polo's record of the widespread use of coal and paper money fell strangely on European ears, there was no mention of the Su Sung clock. It remains a mystery why Polo did not mention the astrarium.

Admiral Zheng He

"China developed into a great naval power during the final years of the Song dynasty (960–1279) under the Yuan (1279–1368) and at the beginning of the Ming dynasty" (Gipouloux 2011: 74). Naval battles between the Sung and the Mongol frequently took place. Kublai Khan had commissioned the construction of thousands of ships for raids in the South China Sea. The Ming dynasty continued in this endeavor, but also turned the South China Sea into an important shipping lane.

The Ming dynasty (1368–1644) not only became famous for the building of the Great Wall but also undertook a series of trading and diplomatic expeditions by sea. Between 1405 and 1433, Emperor Zhu Di sponsored seven formidable expeditions under the leadership of a Turkish eunuch, Admiral Zheng He, who ranged the Indian Ocean and further beyond. The admiral had been appointed commander-in-chief of one of the largest fleets ever built.

His first expedition with 62 enormous, nine-masted, treasure ships carrying a huge quantity of goods in their vast holds and a crew of 28,000 men set sail to Malacca, Cambodia, Java, and finally the west coast of India. Lasting on average two years each, the first three voyages between 1405 and 1411 sailed as far as the Malabar Coast, the principal source of the world's pepper supply, with excursions along the coasts of Siam, Malaya, Java, Sumatra, and Ceylon (Sri Lanka). On the fourth voyage from 1413 to 1415, ships visited the Maldives, Ormuz, and Jeddah. In 1421, the sixth voyage departed with the reconnaissance of the east coast of Africa as its main objective and visiting Mogadishu, Mombasa, Zanzibar, and Dar Es Salaam. During this visit to east Africa, the sultan of Malindi offered the Chinese travelers a giraffe for the Chinese emperor. This *chang jing lu* (or *quilin* in Japanese) became an important animal in Chinese mythology. Zheng He's armada also visited the African west coast and supposedly reached the Cape Verde Islands.

Royal Navy officer and commander of a submarine, Gavin Menzies became fascinated by the Chinese explorations and pointed out the tremendous success of their maritime voyages. He is convinced that part of the armada under the command of Zhou Wen sailed up the coast of North America, reached Cuba, and further circumnavigated the Eurasian landmass. His belief is that the Chinese discovered America 75 years before Columbus did. Although it might have been perfectly possible for Chinese ships to visit Europe or the Americas, it is not certain whether this actually happened. "Although the evidence that Chinese vessels ever crossed the Pacific to America is, at best, equivocal, it is perfectly possible that they may have done so" (Fernandez-Armato 2006: 116). Yet Menzies (2003) is convinced that the first settlers in North America were not actually European pioneers but Chinese in the junks of Admiral Zhou Wen who had split from Zhen He's fleet, actually crossed the Atlantic, and landed some time at the end of 1421.

However, his amalgamation of historical analysis, "travelogue," and detective story became contested, and historians found Menzies' claim unproven. Menzies has been unable to adequately answer some of the criticism and has been accused of writing pseudo-history (Fritze 2009). He defended himself by writing that if he were right, "history would need to be radically revised, but it seemed extremely presumptuous for a retired Royal Navy submarine captain to be the one initiating this process" (Menzies 2003: 269).[21]

Regardless of what this debate attempts to prove, Chinese merchants have been familiar with trade routes of the Indian Ocean for centuries. The Chinese had established trade with a region that was highly lucrative in products such as spices, hardwoods, valuable medicinal drugs, and exotic animal products (Gipouloux 2011). The motives for the expeditions of Zheng He must have transcended commerce. They were flag-waving missions, impressing the ports that were visited with Chinese power and stimulating the awe of the emperor's home constituency with exotica which the Chinese classified as a tribute of remote people (Fernandez-Armato 2006).

The voyages demonstrated that the Chinese mastered techniques to safely sail unknown waters. Ports in the northern hemisphere were marked with their latitude according to the elevation of the polestar above the horizon, which the sailors verified by means of "guiding star boards," which were ebony strips of various breadths held at a fixed distance from the observer's face to fill the space exactly between the star and the horizon. In the early eleventh century, the Chinese had invented the portable compass, a device that soon found its place in Chinese ocean-going fleets. Marine charts in the form of a portolan chart (*mappa mundi*) which presented information in graphic form were developed around the same time.[22] Timetables for determining latitude according to the hours of sunlight were easy to use but demanded more accurate timekeeping than most mariners could manage.

For a long period, navigators relied on the sheer accumulation of practical craftsmanship to guide them in unknown waters. Navigational manuals which had distilled various experiences into sailing directions genuinely assisted the navigator. During the Su Sung dynasty, instead of traveling along the coastal lines, Chinese ships had already been able to sail across the western part of the Indian Ocean directly from India to Qatar on the southeast coast of the Arabian Peninsula.

While Zheng He was organizing his voyages, China, however, became increasingly governed by scholars with indifference toward expansion and civil servants with contempt for trade. Traders and merchants were not respected as an independent category in society. Business persons who originated from families who had been involved in trade for centuries simply had to offer their products at fair prices, which were frequently checked by governmental bodies. After the death of emperor Zhu Di, his son Zhu Gaozhi ascended the throne in 1424 and restored Confucian officeholders whom his predecessor had dismissed. Orderly and well-mannered continuity was at the heart of Confucianism and the Mandarin Government.

In subsequent years, the balance of power at the Ming court shifted in favor of Confucian office holders and away from the Buddhists, eunuchs, Muslims, and merchants who had supported Zheng He's expeditions. The mandarins, who were steeped in Confucian ethics and maintained a strict code of moral values, looked for stability and had contempt for foreigners. In their point of view, merchants and foreigners continually upset the Tao, or in other words, the stable interaction of natural forces. Since the new emperor was less interested in the voyages of the treasure ships, he canceled Zheng He's next trip and ordered all the other ships back. As China's frontiers on land became insecure due to the revival of Mongol power, the new Ming emperor turned away from the sea and began to secure himself against the new threat by building big walls. After 1433, the magnificent fleet of Zheng He with its entire heavy merchant fleet of 3,500 ships was de-masted and left to decay (Menzies 2003).[23]

China: The Middle Kingdom

The hostility of the Chinese court toward maritime expansion ensured that the country did not build up the sort of wide-ranging global empire that Atlantic seaboard nations later acquired. In China, the new way of thinking in terms of Confucian order became very common and, in the end, imposed limits on individual enterprise. The leading group of merchants and manufacturers who had taken initiatives to promote maritime trade, commerce, and manufacturing submitted to the Chinese bureaucratic system. Chinese disengagement from the wider world was not the result of any deficiency of technology or curiosity.

The eccentric Cambridge biochemist Joseph Needham (1969) who, under the spell of his mistress Lu Gwei-djen, turned into a famous sinologist (Winchester 2008), extensively studied the Chinese society and the development of science and technology; he came to the conclusion that the Chinese were much more interested in practice than in theory with a primary focus on the natural and the organic reality of nature.[24] Needham stressed that the Confucian bureaucrats succeeded in preventing the merchants from becoming a social group with an autonomous power base in society. This hampered a close cooperation between scientists and artisans.

During the Sung dynasty, technological and commercial development had flowered, but due to this change in the political regime, scientific advance decreased sharply. During the Qing dynasty (1644–1912), China was poorly informed about the rising European powers (Hua 2007), stayed self-sufficient, and needed little information or products from outside. The "Country of the Middle" rarely initiated explorations. The Chinese emperor ordered sailors to sail close to the Chinese coastline and not move beyond the Chinese shores, a prohibition which prevented merchants from advancing in commercial activities. Confucian China remained dominated by the ever recurring figures of the warrior-ruler and the priest-scribe-bureaucrat. The power they held was supposed to be the most effective route to wealth.

Despite its technological and economic power, the Chinese state maintained a weakness in naval advancements. It cared little about maritime trade, and the Chinese maritime merchants did not receive any backing from the Chinese state. Yet, the Chinese state was powerful enough to stop maritime explorations (Jones 2003). China's position in technological sophistication was finally lost to forces outside Asia. These forces developed under the European mercantilist tradition and "were fuelled by zealotry for conquering the whole world with sea-based empires for material and/or spiritual reasons. The Chinese never had such a drive" (Deng 1999: 170). The free ocean was left to European explorers and traders. China's inward orientation facilitated Europe's leap into prominence.

Around 1430, the Ming dynasty was forced to defend itself against the intrusion of Mongols. Needham indicates that the sudden ending of China's great scientific burst of energy remained an unsolved issue. Yet while the Chinese society shielded itself from external influences, in Europe the outer world became more and more a challenge resulting from the fact that:

> Europeans suffered from a schizophrenia of the soul, oscillating forever unhappily between the heavenly host on one side and "the atoms and the void" on the other; while the Chinese, wise before their time, worked out an organic theory of the universe which included Nature and man, church and state and all things past, present and to come. It may well be that here, at this point of tension, lies some of the secret of the specific European creativeness when the time was ripe.
>
> (Needham 1969: 121–122)[25]

With the first audacious Italian explorers leaving the Mediterranean Sea and with the voyages of Marco Polo, Europe was no longer confined between the barriers of an impassable Atlantic Ocean to the west and of a remote Asian landmass to the east. Windows of opportunity stood wide open.

Portugal

From an institutional perspective, one might stress how Western sea-power nations such as Portugal and Spain became the pulling factor in economic development.

In that sense, the Chinese and Western European sea powers had different origins. Whereas traditional empires such as the Chinese, the Moghuls, or the Ottoman empires had been formed by straightforward military expansion with obvious rewards for those engaged in it and lasted for centuries, the young Western system of upcoming city-states and regna had been fragile and precarious. This system which developed at the end of the Middle Ages did not have any central commanding authority as an equivalent of the decision of the Ming Court in the 1430s to stop Chinese maritime explorations without any contestation. Conflicts between western rulers, however, were likely to draw others in and threatened the stability and survival of the whole. The checks and balances of decentralized power and a code of international law in Europe remained feeble safeguards against chaos.

Concurrently, an eagerness to explore the world developed. European sea powers, however, which relied on a convergence of socio-political and socio-economic forces, were mainly built on organized violence. While in Western Europe this pattern could be noticed, a similar pattern had been lacking in China (Deng 1999). Yet the Chinese voyages emphasize that a one-sided Eurocentric view on globalization must be belied.

Despite the effort of the crusades (1095–1270) to drive the Muslims out of the eastern and southern shores of the Mediterranean, these initiatives did not succeed but taught the Europeans much about riches such as tapestries, porcelain, and spices. Italian and Portuguese appetites to attain these treasures had been wetted by the tales about Polo's travels and inspired explorers to reach the East via a water route. Around the end of the thirteenth century, curiosity among European mariners about unknown waters intensified and drove them to take risks and initiate exploration. In 1291, the Italian brothers Ugolino and Vadino Vivaldi left the harbor of Genoa with two galleys – war vessels – with about 200 rowing men on board. They had the courage to pass through the Pillars of Hercules (Gibraltar), probably attempting to row down the west coast of Africa. However, "mere improvisation could not accomplish a feat" (Chaunu 1979: 82), and they vanished from the earth. The story about their failure raised questions among the Mediterranean sailors. Some believed that the world might end at the Atlantic Ocean.

But other Italian, as well as Portuguese, sailors were as courageous as the Vivaldi brothers. Of all the European peoples, the Portuguese were in the best geographical position to carry out the risky ambitions and find a new water route. With the help of English Crusaders, Lisbon had been captured in 1147, and the last Muslim stronghold fell in 1249. The Portuguese reconquest from Muslim occupation was carried out more swiftly than that of Spain. The kingdom of Portugal had its origin in an earldom established around Oporto, and "in the years leading to the capture of the Algarve, Italian ecclesiastics and French knights wielded influence" (Blackburn 1997: 97). While Muslim power was weakened due to internal struggles, the reconquest stimulated projects to invade North Africa and intensify the Atlantic trade.

In 1415, a fleet of Portuguese ships sailed off from Lisbon to cross the strait at the entrance of the Mediterranean Sea. On board were the King of Portugal with his three sons and military men of noble birth from Portugal, England, and France. The fleet dropped anchor upon reaching Ceuta, the North African port and trading center located directly across the strait from the Rock of Gibraltar. Ceuta was the northern terminus of trade routes bringing gold, ivory, and slaves across the Sahara. It was a city attracting a cosmopolitan population and home to some Islamic universities, which were called Madrasah (Abernethy 2000). The youngest son of the King of Portugal on the expedition to Ceuta was Prince Henry (Dom Henrique), romantically known as Henry the Navigator (1394–1460).[26] He shared chivalric sentiments with his entourage, and their courage brought the first Portuguese ships to Ceuta, not meaning to foster trade interests but only to loot and take tribute. After a pitched battle, the town and citadel were captured. It gave the Portuguese a secure harbor on African soil. The capture of Ceuta was the first European victory over the Muslims in their own territories.

The Portuguese victory marked a new phase in world history. It was the beginning of an era of European-centered empires which, step by step, extended over the globe. Portugal became one of the first countries to establish tight connections between nautical, commercial, military, and state power.

In Ceuta, the design of European ships became modernized and navigational instruments and techniques developed and improved, which stimulated the organization of voyages for exploration and colonization. The Portuguese built the caravel (its origin is unknown), a fast, small ship that the Portuguese adapted from Arab and Mediterranean vessels carrying a main mast and a mizzen mast which were usually lateen-rigged, which made it possible to convert winds from many different directions into forward motion (Law 1989). The caravels were a technological innovation and could sail close to the wind and, in combination with a hull design of a shallow draft, the caravels' rigging could make headway in regions with variable winds and adverse currents. However, in order to build these ships and equip them, Henry needed all the gold he could find to pay for the ships and reward the crews.

Upon Henry's return to Portugal, he succeeded his father to the throne. Encouraged by the success of the expedition, he built a castle and a center for explorers at Sagres on Cape Saint Vincent in the Algarve (southwest Portugal) in 1419. This city became a center which brought together Catalan sea captains, Jewish cartographers, Arab astronomers, and Portuguese knights as well. Portuguese princes stimulated the application of new techniques to celestial navigation and map-making (Blackburn 1997).[27] Among the learned men and mariners, Henry recruited and financed subsequent expeditions. Soon Portuguese sailors expanded their horizons, moved beyond the pillars of Hercules at the strait of Gibraltar, and rushed into what was known as the Great Outer Sea driven by a rush for silver and gold. This eventually amounted to a massive enslavement of African peoples, and an intensive deployment of military and maritime technology around the world.

Portugal began an oceanic system of commerce and navigation which preceded the epoch of Columbus and turned the nation into the first European country to acquire an overseas empire (Dias 2007). Henry's sea captains set sail to Madeira which was settled in 1419. There, he introduced sugar cane which he had brought in from Crete. It flourished well in the warm, damp climate. Sugar cane production showed spectacular profits, and soon entrepreneurs and contractors obtained commercial rights from Portuguese authorities to exploit the newly found grounds (Blackburn 1997; Sandler 2008).

In 1431, he ordered his sea captains to find the islands of Antilia which were indicated on a chart from 1424 and signed by the Venetian cartographer Zuane Pizzigano (Menzies 2003: 31).[28] The map suggested that China and the Spice Islands could be reached by sailing westward. According to Menzies (2003), the map "revolutionized" European exploration for all time. With a thrilling desire to find gold, the fever of exploration intensively progressed. Sea-republics such as Venice and Genoa did not wait for the Portuguese to conquer the world and began to build larger ships that were supported by marine armies.[29]

In 1434, Henry's sailors dared to pass the landmark Cape Bojador on the West-African coast that had long been called the Cape of Fear. In the meantime, they discovered the circle of winds and currents, *volta*, which no longer forced them to stay close to the coast but instead allowed them to sail their return journey via the Canary Islands to the Iberian Peninsula. The trade wind system dominating the Atlantic provided a regular pattern of prevailing winds continuously blowing in the same direction regardless of the seasons (Fernandez-Armato 2006).

The Portuguese approached the Azores around the 1430s, which offered opportunities to outflank the traditional trans-Saharan gold road with offshore staging posts near the fabled "river of gold," the Niger, in West Africa. Senegal was reached in 1444 and broke the Arabic monopoly of the Saharan land route. It was here where Portuguese captains quickly learned about cheaper ways to buy slaves. "West Africa was already a well-established source of slaves for the trans-Sahara slave trade at the time of the first Portuguese expeditions" (Blackburn 1997: 102). The Cape Verde Islands were reached in the period 1450–1460. Portuguese way stations were established, sown with wheat, and stocked with wild sheep.

The race to explore the sources of the West African gold trade gathered pace. Portuguese domination of the African coast became more commercially oriented when, during the 1450s, wealthy Genoese merchants followed the Portuguese example and sponsored several voyages by private groups of captains and traders along the West African coast in search of direct access to African gold and slaves. In 1452, Portugal requested the Pope to officially approve the right to acquire newly explored areas. Portugal justified its actions as an extension of the holy war against Islam. A Papal Bull gave the Portuguese permission to invade, capture, and subjugate all the unbelievers, reduce these persons into perpetual slavery, and convert them. Uninhabited archipelagos such as Madeira and the Azores were colonized. Portugal led an astonishing burst of global exploration to the west while the Eastern Mediterranean was still blocked by Islamic power.

In 1459, the Portuguese obtained another map which had been drawn by Fra Mauro, a Venetian cartographer who was working for the Portuguese court. His map contained a clear reference to the East Indies, and this stimulated the drive for further explorations.[30] The "Gold Coast" was reached in 1470–1471. Portuguese ships appeared off the coast of Ghana where gold was mined near the shore. Portuguese commercial interests acquired the rights of exploration: the Portuguese crown promoted the navigation of West Africa and commissioned merchants to be involved as entrepreneurs to establish trading posts and establish friendly relations with powerful coastal chieftains such as the Wolof chiefs of Senegambia, the Obas of Benin, and ultimately the Manicongos of Congo. This development inaugurated a new policy of permanent footholds, disciplined trading, and protection by royal initiatives.

A reconnaissance expedition under the command of Bartholomew Diaz rounded the Cape of Good Hope in 1487, which opened a maritime route that gave the Portuguese access to the spices of Asia. During the expedition, captains discovered that in the northern hemisphere, the earth's rotation caused winds to swirl counter-clockwise, while in the southern hemisphere winds moved clockwise. Increasing knowledge about the intersecting ellipses of seaborne communications in the North and South Atlantic, each with their prevailing winds and oceanic currents, facilitated the establishment of an Atlantic commercial system. The Portuguese learned that, with their own explorations, wealth and richness by trade in gold, spices, and slaves lay at their feet.

Ten years later, Vasco Da Gama left with four sailing vessels charged with cast bronze cannons from the Tagus in Lisbon and sailed beyond the Cape in 1497 to India. On May 20, 1498, he anchored near the Malabar Coast of Southwest India. The success of Da Gama's voyage upset the spice trade: the routes by which supplies of spices had come to the Mediterranean from the East Indies were diverted, and the initial steps for terrestrial globalization were set. Soon the Spanish kingdoms of Castilians and Aragonese followed the Portuguese pioneers in their footsteps with their drive for world domination.

Nautical knowledge

Next to new ship designs, another technological innovation based on the variations in the latitude of the sun could be used by captains to determine the ship's position. There was a complication, however, because the apparent annual path of the sun around the earth (the ecliptic) is not parallel to the plane of the equator. The sun moves higher and lower with the seasons. This angular deviation is known as the sun's declination and had to be taken into account when reckoning latitude from the sun's noontime altitude.[31] In the 1480s, Portuguese sailors developed a practical method for the astronomical determination of latitude on board their ships. The general idea was that if the height of the sun above the horizon (altura) can be determined and compared with the known altura of the port of destination, then the ship can sail north or south until it reaches that latitude and then sail East and West in the certainty of finding its point of destination.

The measurement of the altura was possible thanks to the work of astronomers who were able to apply a range of instruments such as the astrolabe and quadrant.

These instruments were used to depict the positions of heavenly bodies at a given point in time. The astronomers were already familiar with Ptolemy's view of the earth that every point on its surface can be described by a unique latitude and longitude, and by 1473, the astronomers in Lisbon had made up a list of latitudes, a table of solar declination which reached as far the equator. However, the techniques to apply these instruments were highly complicated and difficult for common sailors to understand.

A third technological innovation, therefore, focused on the design of a practical application of the measurements of the altura in the form of instruction tables. In 1484, Portuguese King John II convened a scientific committee to find an easy method for navigating outside European waters which would facilitate the measurement of the altura. The committee proposed three solutions. First, simplified versions of the astrolabe and the quadrant were designed which made it easier for sailors to apply them. Second, the semi-educated mariners were provided instructions about the practical calculation of the latitude by applying tables containing the northern latitudes of important coastal features and, in particular, the major ports and capes. In other words, a metric was produced containing observations which might give absolute north–south meaning and from which the position of the ship could be located.[32] A set of rules told the mariners precisely how they should use the instruments and data which were provided to determine the latitude. The printed instructions showed the sailors how to use these data for their sailing practice.

The third improvement was introduced some time later when it was discovered that sailors still struggled with the application of the new instruments and the tables. "The majority of good pilots still relied on their own experiences without using these tables of declination" (Landes 1983: 105). In the early sixteenth century, the Portuguese installed the Casa de Guinea e India at Lisbon to provide general instructions to the pilots. They obtained class-instructions about new navigation techniques and learned about the applicability of the tables. The Spanish soon followed this example (Law 1986). A trinity composed of simple devices, printed instructions, and disciplined navigators provided a reliable knowledge basis and increased the fidelity among the pilots and the court power to make the endeavor of exploration a success (Law 1989).

Nevertheless, bravery and a will to survive at sea were as important as new knowledge about the coordination of the ship at sea by the use of astronomy, timekeeping, and scrupulous measuring of water speeds. With their reliance on implicit, experiential knowledge as much as on explicit knowledge as learned in class, the Portuguese sailors set sail to continue their explorations to advance trading opportunities. The problem of the return voyage of the ships loaded with gold was solved by heading seaward in a north westerly direction toward the nearby Canary Islands and then taking the westerly winds from those islands back to Portugal. "The Portuguese were able to convert the currents, winds, and the rest from opponents into allies and were able to associate these elements with their ships and navigational techniques in an acceptable and usable manner" (Law 1989: 120).

The routes pioneered by the explorers of the 1490s cracked the code of the Atlantic wind system. These European sailors had realized how to use the south-east trade winds across the ocean in the southern hemisphere and to reach the belt of strong, fast westerly winds that circle the world south of the equator. The immediate effect was that this knowledge enabled them "to break into the world's richest and busiest arena of commerce: the Indian Ocean" (Fernandez-Armato 2006: 157).

The breakthrough of the 1490 voyages transformed the Atlantic into a potential arena of long-range cultural transmission. The navigators, traders, and merchants were able to transform the currents based on their geographical knowledge and navigational competence into forces that speeded rather than hindered their voyages (Law 1986). The findings of the sailors, however, would not have been appropriated in the correct manner if the institutional setting had not favored their discoveries. Europe became the seedbed facilitating a conjunction of windfall and entrepreneurship which "happened only once in history" (Jones 2003: 84).[33] The Iberian Peninsula harbored sailors, explorers, bankers, and financiers to undertake courageous new initiatives.

Lisbon and Seville attracted large colonies of Italian traders. Genoese merchants, in particular, who settled in Seville contributed their commercial skills with bills of exchange and credit and provided knowledge about the ways to raise large amounts of capital needed to deal with risky oceanic commerce. Genoese merchant-banking families such as the Spinola, Grimaldi, Pinelli, and Centurioni acted as contractors supplying naval stores, dealing with agricultural products, and getting involved in the new Atlantic Ocean trades. Italian merchants in Lisbon such as Lomellini, Affaitati, Giraldi, and Marchione became bankers and moneylenders to the court and its envoys, and invested in sugar production and slave trade (Braudel 1972).

The Portuguese empire, however, remained a dispersed and territorially small venture closely monopolized by the Portuguese themselves. Although it lacked the resources for imperial dominance and power, it established connections with China, Japan, and India. They pioneered maritime routes which ran around the coast of Africa and across the Indian Ocean to the South China Sea which competed directly with the overland spice and silk routes which for long had linked Europe with markets of the East. They were responsible in large part for a new way of viewing the world. Their cartography provided indispensable new graphic conceptions of global space which converted navigational knowledge into a strategic and commercial advantage.

While the Portuguese were involved in reaping the fruits of their discovery of the eastern route to India, Spanish rulers became determined to follow the path blazed by Columbus.

Spain

During the eleventh to the thirteenth centuries, the Crusades marked the interaction between Christians and Muslims, shifting the focus of the conflict to the East. This arena of religious and political conflicts remained unsettled until the

Ottomans breached the walls of Constantinople in 1453 by using a new invention: the cannon. After their victory, the city became known as Istanbul and five years later turned into the capital of the Ottoman Turk Empire. The conquest became a key event in the construction of the Ottoman empire, which became a robust and adaptive political entity with a strong bureaucratic apparatus that kept a patrimonial center in balance and a strong army (Barkey 2008: 75).

The land trade routes between the Christian West and the rich and exotic countries in the East which had large Christian populations within their borders became severely frustrated by the presence of the Ottomans. Venice concluded its war with the Ottomans (1463–1479) on terms that made the Turkish victory unmistakable, and it eventually sank to a secondary rank (McNeill 1974: 88–89). The Ottoman Empire reinforced its navy and built fortresses on both sides of the Bosporus, cutting off the Black Sea from the Mediterranean and the land routes. Its expansion was based on principles of conquest, but the empire respected toleration as one among many policies of incorporation such as persecution, assimilation, conversion, or expulsion. Non-Muslims were tolerated as long as they did not disturb the Islamic order.

In the cities, local production was based on a system of guilds. Trade between European countries and the Ottoman Empire concentrated mainly in the eastern Mediterranean where the main Ottoman ports had access to the hinterland that was blocked for the European merchants. The fall of Constantinople and the establishment of the Ottoman Empire brought humiliation to the Western world.[34]

To keep the Turks at bay, the European Habsburg ruler expanded his defence measures. Resurgent conflicts increased the interest in weaponry and, during the period of 1470–1520, changes in warfare were the result of a widespread adoption of handheld firearms. The Spanish as well as the Ottoman Empire became gunpowder empires. Soldiers were trained to use muskets while the horsemen used pistols. Infantry trained to fight in loose formations used a lighter musket called arquebus. New forms of warfare practice "sprang from changes in metallurgy and gunpowder milling that made firearms truly effective for the first time in Europe" (Wilson 2009: 85).

Once Ferdinand, King of Aragon, and Isabella, Queen of Castile, united Spain with their marriage, they prepared for a final assault on the last Islamic enclave in Spain south of the Sierra Nevada Mountains near the city of Granada. The end of the Muslim empire in Spain came with the submission in 1492 of the last emir when his kingdom of Granada surrendered to the Spanish armies. Although the city had capitulated on terms that respected its religion and its property, Spain became a religiously intolerant society with a zealous drive to impose Christianity (Kamen 2002). The victory over the Muslims and the expulsion from main Spanish cities of the Jews who were not willing to convert illustrated the Spanish crown's aspiration for expansion based on religious motives.

The Spanish rulers also began to expand their horizons for other lucrative spice routes as their Mediterranean ships no longer had access to the hinterland of Constantinople. The Catholic monarchs decided to launch a rivalry with Portugal about a contest for space. Explorers who were inclined to head off in unknown directions found a listening ear at the Spanish court. Columbus, a Genoese ex-slave trader (Stannard 1992: 200) who lusted after the gold of the Indies, tried to convince the Portuguese court about his intentions, but they showed no interest. He saw himself forced to look for other opportunities to travel the unexplored routes and went to the Spanish court.

Columbus

Columbus (1451–1506), seen by some as the "Don Quixote" of the world seas (Bucher 2006), planned a voyage westward, intending to break the monopoly of the spice trade, which was mainly in the hands of Egyptian rulers as well as merchants from Italian city-states such as Venice and Genoa, and approached the Spanish monarchy. Upon being informed about the Portuguese advancements in opening up the road to India, the Spanish monarchs became interested in downplaying the Portuguese monarch. They were in agreement with Columbus' project as they believed it might bring them the riches of the East when he sailed westward across the Atlantic. Eventually, he was able to persuade a royal official and some Genoese merchants living in Seville to finance his first exploration (Hunt and Murray 1999).

The second trip became a royal venture when the Spanish rulers presented enough money to finance Columbus' voyage. "Queen Isabella's coffers were swollen with confiscated Jewish property when she dipped into them to finance Columbus" (Jones 2003: 123). Columbus knew that his only chance lay with the Catholic monarchs who did not have the 1428 World Map and, thus, did not know about the shorter route that the Portuguese were exploring.[35]

Columbus, however, was a poor cartographer and sometimes used a crude speedometer (log) to locate his position. "On his first voyage, calculations of latitude were 20 degrees out" (Menzies 2003: 301). Nevertheless, he sighted land in the Bahamas (today the Dominican Republic and Haiti) in 1492. When Columbus went ashore, he offered the local people small bells used by the Spaniards when hunting with hawks (Koning 1991; Buecher 2006; Sandler 2008). He found the natives peaceable and obedient. They were without arms and naked and could eventually be enslaved (Kamen 2002). Columbus had sailed to the New World while carrying with him Polo's *Description of the World*, hoping to find China as well.

According to Menzies, the new knowledge about the world, which was "hard-won by the Chinese during the great voyages of exploration, was to become the driving force behind the European voyages of discovery" (Menzies 2003: 395). Although it is usually presupposed that Columbus with two caravels and his flagship, a barrack which had a larger capacity for cargo, was the explorer who discovered a new geographical area in 1492, the New World obtained its name from the Florentine Amerigo Vespucci who had followed in Columbus' footsteps a few years later and explored America as a continent.

Many have depicted the arrival on the new land from the perspective of the Europeans' side of the ocean. The Dutch painter Jan Mostaert (*c.* 1475–1555/56),

FIGURE 2.1 *West Indian Landscape*. Jan Mostaert, 1542. (Courtesy of Marei von Saher, the Heir of Jacques Goudstikker.)

however, made the earliest known painting of the New World. His *West Indian Landscape* (Figure 2.1) has been interpreted as a visual parable of some primitive society in the New World where man still lived innocently and peacefully in an unspoiled, pastoral paradise. What we see are Indians rushing to the sea, unsure of what to think of the large sailing ships arriving at their shore and wondering if the intruders would attempt to seize what was not rightfully theirs (Honour 1975).

Were the first encounters as a result of terrestrial globalization merely based on human curiosity, without fear for or denial of the presence of other cultures? One interpretation suggests that the painter might have been worried about the native Americans becoming governed, evangelized, and exploited by Spanish conquerors who would draw them into the orbit of a European hegemony. Would the white "visitors" nevertheless respect local cultures or aggressively demolish local practices and occupy the "free" territory for their own greedy interests?

The Treaty of Tordesillas

In the late fifteenth century, Mediterranean explorers had begun to venture into the Caribbean and the Americas and rapidly established settlements centered around mining. The flow of precious metals began immediately after the discovery of America. Gold and silver were the first of those flows that circulated the globe (Stannard 1992), and between 1492 and 1640, around 180 tons of gold and 17,000 tons of silver were transported back to Europe where Seville and Genoa became important gold- and silver-trading places. The silver market became a truly global market once the four continents were firmly connected by maritime routes. China would become the main buyer of silver, while Spanish America evolved into the main supplier (Gipouloux 2011).

After Columbus discovered the New World in 1492, it was clear that conflict would soon arise over land claims by Spain and Portugal. Soon after Columbus' return, Ferdinand and Isabella of Spain requested the same authority of subjugation from the Pope as the Portuguese had obtained. They were convinced that both crowns should be allowed to expand into new territories, place them under their jurisdiction, and place unbelievers and their lands into perpetual subjugation. This act would need to be sanctified by the Holy Roman Church. On May 4, 1493, Pope Alexander VI took action to clarify any confusion that might arise over territorial claims. He issued a decree which established an imaginary line running north and south through the mid-Atlantic, 100 leagues (480 km) from the Cape Verde islands. Spain would have possession of any unclaimed territories to the west of the line, and Portugal would have possession of any unclaimed territory to the east of the line.

The Portuguese King, John II, was not pleased with this arrangement as it prevented him from possessing India, his short-term goal. He opened negotiations with the Spanish monarchs to move the line to the west which would allow him to claim newly discovered lands east of the line. In June 1494, the line was renegotiated, and the agreement was officially ratified during a meeting in the Spanish town of Tordesillas.

The Spanish rulers chose the Florentine explorer Amerigo Vespucci to follow in Columbus' wake and explore other territories.[36] In May 1499, an expedition set sail from Spain but was forced to return due to serious damage to the ship from an infestation of sea worms. The Spanish King, however, was less interested in financing a second expedition, thus, Vespucci undertook a second trip under the Portuguese flag, discovering an entirely new continent and the Atlantic as a discrete body of water. In his account of the voyages, *Mundus Novus* (1503), he described the discovery of the new continent. The German cartographer Martin Waldseemuller produced a world globe and a large map inscribing the name *America*.

In 1500, Portuguese noble and military man Pedro Álvares Cabral, while en route to India, accidentally landed in Brazil. The new territory was within the Portuguese sphere according to the Treaty of Tordesillas which had reconfirmed the previous papal bulls but with a slight adaptation. It had introduced a precise line of demarcation with each country receiving title to half of the unconverted globe.

> Along the longitude running 100 leagues to the west of the Canary Islands, the lands were assigned to Spain. Everything to the east of the line was to belong to the Portuguese. In 1506, Pope Julius II reaffirmed these terms of the treaty.
>
> (Drescher 2009: 63)

The treaty divided the Atlantic zones of navigation between them by drawing a meridian through the Azores, dividing the ocean. Portuguese seafarers would be confined to Africa and Spain and would have a monopoly over access to most of maritime Asia. This line of demarcation was about halfway between the Cape Verde Islands (already Portuguese) and the islands discovered by Christopher Columbus on his first voyage (claimed for Spain), named in the treaty as Cipangu and Antilia (Cuba and Hispaniola). The lands to the east would belong to Portugal and the lands to the west to Spain. The treaty ratified the Portuguese monopoly of the African slave trade. As a result, Asian markets would be approached only via the West. The sovereignty over a great part of the new land was safeguarded.

Initially, the line of demarcation did not encircle the earth. Instead, Spain and Portugal could conquer any new lands they were the first to discover, Spain to the west and Portugal to the east, even if they passed each other on the other side of the globe. But Portugal's discovery in 1512 of the Moluccas with its rich spices such as cloves caused Spain to argue that the Treaty of Tordesillas divided the earth into two equal hemispheres. After the Portuguese navigator Ferdinand Magellan, operating in the service of the Castilian crown, circumnavigated the globe for the first time in 1519, Spain and Portugal discussed the exact position of the demarcation line again. They finally agreed with the Treaty of Saragossa, signed on April 22, 1529, that Spain would relinquish its claims to the Moluccas upon the payment of 350,000 ducats of gold by Portugal to Spain.

Henceforward, the other side of the world was divided via an anti-meridian to the line of demarcation specified in the Treaty of Tordesillas. Spain gained lands, including most of the Americas. The easternmost part of current Brazil was granted to Portugal. "It was religious proselytism, as much, if not more than, the quest for spices, that was responsible for the division" (Gipouloux 2011: 113). The line was not strictly enforced, and the Spanish did not resist the Portuguese expansion of Brazil across the meridian. Yet despite all the successes at sea, and its ability to control the trade routes between Asia and Europe, Portugal eventually lost its dominant position to Spain.

Imperial Spain

The discovery of the New World laid the foundations for Spain's global empire. The advantage of the north-east trade winds allowed the maritime communities around the mouth of the Tagus and Guadalquivir to obtain privileged access to much of the rest of the world. The prodigious reach of the Spanish and Portuguese empires was, in part, the result of this good fortune.

> The Gulf Stream was discovered in 1513 by a Spanish explorer. Europeans now had accurate and extensive, though still incomplete, knowledge of how the winds and currents of the north Atlantic worked. More significantly, they had become acquainted with the wind patterns of the South Atlantic as well.
>
> (Fernandez-Armato 2006: 157)

Soon, Spanish conquistadores organized many transatlantic missions to colonize the new territories and look for wealth in terms of gold and silver to enrich the Spanish Habsburg monarchy and its expanding empire (Winchester 2010; Sandler 2008).

The Habsburg dynasty began in Switzerland and emerged as a Holy Roman Empire when Rudolf I was crowned as emperor by the Pope in 1273. He subsequently elected his son as ruler of Austria. The new ruler of the Habsburg dynasty was asked to imitate the achievements of Charlemagne and unite all Christendom, maintaining links with strong Catholic institutions and values which fused with the political order and enhanced their view of being the guardians of Christendom. Like Islam, Christendom professed a world mission striving to possess God's final dispensation.

Through marriage and inheritance, Maximilian I extended the Habsburg Empire into the Burgundian Low Countries in 1477, to Spain in 1516, as well as to other territories in Europe. His son, Philip the Handsome, integrated the Spanish Kingdom with the House of Habsburg. Under the Habsburg ruler, Charles V, conquests continued but not without problems. Charles of Ghent, grandson of Maximilian of Habsburg, Duke of Burgundy,[37] became successor to the kingdom of Castile and Aragon in 1516 and was also elected Holy Roman emperor. As Charles V, he combined his Spanish and Habsburg inheritances into a single sprawling structure. Three years later, his reign became a dominant European power (Braudel 1972: ch. 4).

With the Spanish extension into other lands, institutions such as the University of Salamanca launched a so-called "controversy of the Indies" about the rights of the Spaniards to conquer and occupy other people's lands and subjugate the indigenous people as slaves. The question about the right of dominion over others agonized the empire for years and, until the eighteenth century, Spain's territories on the American mainland were referred to as plantations rather than "colonies" in the sense of land with a dependent status. Yet the drive to occupy and maintain political and military dominance in all parts of its empire drove the Spanish rulers to impose hegemonial presence.[38] In 1543, Charles V made his son, Philip II, regent of Spain and, subsequently, ruler of the Low Countries in 1555. As Philip was worried about the actual size of the empire, he issued two ordinances on July 14, 1573, concerning the measurement of longitude through the Spanish empire. The first of these stipulated that all longitudes should be computed from the meridian of the city of Toledo. Spanish longitudes should be measured westward because it conformed to the discovery of the Indies. Yet, the precise determination of the longitude remained a great mystery, a riddle to seamen, a challenge to scientists, and an obstacle to kings and political leaders.

The second ordinance ordered all colonial governors to take every opportunity to observe, wherever possible, events such as eclipses and other phenomena capable of yielding the difference of longitude and report the results to the cosmographers at Seville. The mapmakers were required to provide accurate information on the geography of his realm (Kamen 2002: 160). However, the determination of the prime meridian remained a highly politicized issue and did not bring about a solution.

When the ruler of Portugal, Cardinal Dom Henrique, died childless in 1580, Philip II, through his marriage with princess Isabella of Portugal in 1526, asserted a dynastic claim and attached Portugal and its overseas possessions to the kingdom of Castile. Between 1580 and 1640, the Treaty of Tordessilas was rendered meaningless as the Spanish King and the King of Portugal were one and the same.[39] The alliance of a dual kingdom, an arrangement in which Portugal retained its autonomy which lasted until 1640, offered Philip II an opportunity to extend the seaborne trade and exploit the great natural harbor in Lisbon. He became ruler of an immense combination of territories, and his empire had grown to such a global size with the support of many Europeans, native Asians, and natives in the Americas that a chronicler labeled it as so far extended and wide that "the sun always shines upon me" (*sol mihi semper lucet*) (Parker 1998; Kamen 2002).[40]

Portugal and Spain had been brought together with the spice, sugar, and silver trades under the united Iberian crown, but it was not a one-sided arrangement. According to historian Kamen, "the natives of the New World constructed the economy and society of the empire controlled there by Spain" (Kamen 2002: 124). The Iberian voyages and explorations shifted the previous commercial dominance of Venice and other Italian city-states toward the Atlantic Ocean and via the Cape route to the Indian Ocean.

By 1600, the Portuguese and Spanish had open sea routes that spanned the globe and approximately 500 trading corporations were active in commercial practices. The emperor continued to insist on improvement of the cartographic precision in order to accurately determine the precise spheres of his influence. In 1582, he installed a new Academy of Mathematics to precisely determine the longitude as had been fixed by the Treaty of Tordesillas. Until his death in 1598, Philip II ruled the first global empire in history. The complications of governing a global monarchy with its complex of territories, each with its own recognized laws, institutions, and privileges all but sharing allegiance to a common sovereign were tremendous, and the effectiveness of the government was dependent on the quality of the royal officials who staffed the imperial bureaucracy (Parker 1998).

Philip's primary occupation in his monarchy was to extensively deal with written reports that were required to be hand carried from Vienna to Italy and Brussels. In Brussels, his government had employed a postmaster from the Italian Tassis family (spelled Taxis in the German language) since the 1490s. The family earned wealth and a noble rank by organizing and financing postal communications in a reliable way. Mail links were organized for the emperor which created a postal service of unprecedented quality. Tassis also set up and maintained a direct courier link between Spain and the Low Countries through France with 106 relay stations (Kamen 2002). The majority of the many messages from the Low Countries took more than two weeks to reach the court of Spain. To be sure that mail would arrive safely, a second Spanish road connecting Brussels to Milan provided an alternative link with Spain whenever passage through France was rendered unsafe (Parker 1998: 48). Vital letters sometimes passed from Brussels via Besancon, Chambéry, and Turin to Genoa where they took a sea-crossing to Barcelona and from there went directly to Madrid. However, this mail usually only arrived there after two months. "The exchange of letters was, nevertheless, a massive investment as express courier could cost up to 400 ducats" (ibid.: 57). But the expenses were thought necessary to provide relevant information via a reliable and fast way.[41]

Although the provision of information seemed to have been abundant, Philip's managerial style was complicated. He rarely made any firm decisions, and his motto was "*Yo con el tiempo*" (Time and I are a match for anyone). Yet, though he expected everything to operate like clockwork (ibid.: 72), he himself lacked the ability to delegate and failed to design a clear strategic vision. "He believed that he always knew his business better than anyone else" (ibid.: 36). However, he did not understand his finances and though he lived by the clock, working day and night with two illuminated clocks (ibid.: 30), the information overload prevented him from managing his political business in a proper administrative way. He was mainly driven by other, more religious goals.

The Spanish Empire remained an empire of conquest and a vast entity in which non-Spaniards continuously played a crucial role (Kamen 2002). Spanish conquest and settlement of the silver-producing regions of Central and South America facilitated economic integration on a global scale and, as such, inaugurated terrestrial globalization.

Second to the desire for wealth, Philip's other preoccupation was the zealous defence of the Catholic faith which caused serious resentment and protest. The Spanish crown played a leading role in the evangelisation of the New World. The Protestants of the Low Countries soon discovered that the empire's chief mission was religious. Out of fear of the increasing Protestant rebellion in the Low Countries and England, Pope Gregory XIII tried to convince Philip to invade the area. The religious conflicts and the increasing economic power of these rising northern regna changed the pattern of international trade and commerce fundamentally.

The weight of the world's power was moving toward the countries of the Iberian-dominated Atlantic. The Dutch, the English, and the French initially arrived as interlopers in the Spanish and Portuguese dominated Atlantic but soon the protestant nations of the oceangoing north infiltrated with increasing success in the trade of silver, sugar, and slaves. "A cosmopolitan network became established which included the two largest channels of Europe's intercontinental trade" (Kamen 2002: 297), one to the Americas, the other to East Asia. For a time, imperial Spain continued its domination of this vast commercial enterprise, but its essential arteries became more and more controlled by non-Spaniards such as the Dutch. At the same time, the great artery of the Silk Road across Asia further fragmented and left the Ottoman Empire to stagnate at the dead end of the Mediterranean.

Eventually, one of the main reasons why Europe outstripped China, India, and the Islamic world was competition among rulers of nation-states. The state was a monopolist, and their command systems were the principal way of mobilizing resources in their societies where commerce and trade were significant but not dominant. Within Western Europe, however, trade and market-regulated behavior became dominant features of a new commercial and political society. The European Miracle was primarily an outcome of this international system of broad trading interests among competitors which stimulated technological and intellectual progress.

Spain, nevertheless, continued to assert its claims that had been determined by the prime meridian of the Treaty of Tordesillas. In 1559, the Treaty of Cateau-Cambrésis settled a peace arrangement between France, Spain, and England for Europe but left open the precise status of territories "beyond the line" (Blackburn 1997: 8). When the Spanish King Phillip III ascended the throne in 1598, he offered a princely reward to the "discoverer of the longitude": a perpetual pension of 6,000 ducats. When one perceives the earth as a clock, longitude can be inferred from time differences. To do so, a highly precise instrument was needed to measure time in an undisputable way and keep constant track of the time at a place of known longitude and compare it with local time. Spain, however, did not have the pool of knowledge or talents to solve the problem.

Although at the end of the sixteenth century "the globalising dynamic was unquestionably in the hands of the Spanish and Portuguese" (Gipouloux 2011: 300), French and English privateers, many of them being Protestants, disputed Spain's commercial monopoly. By the 1580s–90s, Flemish and Dutch Protestants ("sea-beggars") joined them in refusing to respect the prime meridian. Eventually, this attitude would drive northern Protestant nation-states to find a technical instead of a religious way to solve the positioning of meridians by applying a reliable mechanical clock on board of ships to determine the precise position of the longitude. The invention of the chronometer and concurrent mechanical thinking essentially contributed to the European Miracle. Although it took Europe some more time to reach that level of precision technologically, it eventually succeeded. In the meantime, Europe stumbled into the New World that had been discovered in 1492 and began its future hegemony over large parts of the globe.

3

THE EMERGENCE OF TEMPORALITIES

The purpose of this chapter is to clarify the links between the European miracle and the technical development of the mechanical clock, the influence of monasteries, and the way citizens in Italian city-states began to live according to clock time. From the fifteenth century onward, the application of the mechanical clock quickly diffused in Western Europe and, slowly but surely, changed the life of citizens. One might say that the West has developed an approach in which time has become "a basic element of human organization" (Hassard 1989: 80). Yet eventually, as historian Macey (1980: 17) said: "the advances of western man have been made at the price of an increasing servitude to time." You may wonder about the meaning of this statement, but this chapter will demonstrate that somewhere between the fourteenth and the seventeenth centuries, clock time began to impose a regime unknown to any other civilization. This chapter will, therefore, be confined to the question how this servitude to time has been developed by looking at the way clock time has become applied in its ordinary form, or man-in-the-street sense.

It is a social construction, which means that the views we hold about various times are the product of human interaction and converge into different cultural patterns dealing with time. This chapter will primarily focus on the techniques and methods that have been developed to reckon, measure, and master time, some of which, however, do not have an exclusively Western origin. The study of how time is conceived and used in Western organizations is, therefore, one of the most important categories for economic, political, as well as cultural analysis (Schein 1992). Chapter 14 will more explicitly deal with the way social time, as a multilayered concept in different historical periods and cultures, has introduced a variety of approaches and been applied in different cultural settings.[1]

To properly understand the remarkable European contribution to the measuring of time by the invention of the mechanical clock, we must start with the period when, within the ancient agriculture communities, the recording and coordination of human activities required device systems for relating social events to a sequence of regular and predictable natural recurrences. For millennia, human rhythmicity was patterned in close relationship to the rhythms of nature. In the ancient agricultural communities, priests studied astronomical phenomena at a remarkable level of sophistication. These visible phenomena were more easily studied than the things they observed at the earth's surface such as the flow of water. They noticed certain periodicities in the waxing and waning of the moon, and they believed that the sun revolved once a day around the celestial pole, tracing a path through the same constellations of stars every day, called the zodiac.

Primitive devices were developed to analyze these periodic changes with a fair degree of mathematical precision. The presupposed motion of the sun became translated into a calendar, and a crude clock such as the gnomon[2] or the sundial. Within ancient communities, timekeeping provided a ritual rhythm to human life. Priests and witchdoctors studied the cycle pattern of harvest and sowing based on systems of artificial contrivance.

Although a large variety of systems have been designed, the range of variation was limited by the facts of nature. In recorded history, a calendar, however crude, had been essential to virtually every religion (Adam 2004). Most calendars were either solar or lunar, but stars were also used as a calendar. The Egyptians anticipated the flooding of the Nile by observing the rising of the star Sirius. The presupposed annual movement of the sun around the earth provided a crude division of a man's life into years.[3] The orbital movement of the moon provided a crude division into months. However, lunar calendars had to be corrected for the discrepancy between lunar and solar year by adding extra days every few years.[4] Modern science began with studies of the earth's rotation on its axis, the moon's revolution around the earth, and the earth's revolution around the sun. These rotations and revolutions underlie the most widespread units for measuring time: the year, the month, and the day.

Calendars

The calendar expresses the rhythm of collective activities in a society. It creates a temporal framework for social coordination, synchronization, and temporal regulation.

> The calendar's unrepeatable succession of numbered years symbolically represents the unrepeatable of social and natural events. It thus serves as a means of orientation in the great continuum of change which is at once the natural and the social world.
>
> (Elias 1992: 6)

Archeological and historical records show that cultural structuring of social time with reference to cosmic and natural phenomena dates back to prehistoric times. Nearly every ancient culture worshipped the moon as the planetary source of cultural forms of time reckoning and associated rhythmic practices that integrated all the significant levels of existence. Calendars were designed to shape the pace of life in terms of rhythms, sequences, and synchronies, attuning social activities to one another. In ancient Egypt and Babylonia, India, and China, time reckoning was based on a culturally established unity between cosmos and nature, between the divine and social organization.

The origin of astronomy and the calendar can be traced in the construction of magnificent buildings which incorporated significant cosmic events. The pyramids and temples of ancient Egypt, as well as the temple structures of Inca, Maya, and Aztec cultures have been related to the analysis of extreme positions of the moon and other planets to bring heavenly and earthly activities into unity. The Egyptians used pyramids to measure shadows in order to determine the coming of the equinox.[5] Despite the fact that all these early societies had at their disposal the same physical data of the moon and other planets and the sun and stars from which to construct the systems of time *reckoning,* there were large differences between the knowledge systems which led to the calendars (Adam 2004).

The Sumerians used a calendar year of 360 days with the lunar month of 30 days. It is still not clear why the Sumerians and the Babylonians based their number system on six and 60.[6] "Babylonians refined the Sumerian system by dividing the day up into 24 hours, which is dividable by six, and also divides evenly into 360" (Duncan 1998: 15). Around 6,000 years ago, Egypt was the first ancient civilization to correct the error of the moon and embrace the sun as focal point for their calendar. The Egyptians were familiar with three seasons of life: flooding from late June to late October when the Nile brought rich silt; crops were grown from October to February; harvest took place from February until the end of June.[7] Based on the annual pattern of floods and harvesting, Egyptian priests designed a simple device called a Nilometer as the most accurate calendar to indicate the seasons. Egyptian astronomers supplemented the Nilometer with another discovery that made their solar year even more accurate: Sirius. This so-called Dog Star, i.e., the brightest star in the sky, ascends in the dawn sky once a year in a direct line with the rising Sun. Sirius' appearance coincided with the Nile's annual flood. This became the Egyptian New Year's Day, commemorated annually with elaborate ceremonies (Duncan 1998).[8] By timing Sirius' appearance exactly from year to year, the Egyptian astronomers eventually realized that the solar year was one fourth of the day longer than 360 days.

Egyptian priests were influential administrators who standardized weights and measures and acted as notaries. The temples became centers and protectors of trade (Moore and Lewis 1999). Once Julius Caesar became familiar with Egyptian practices, he reordered the Roman calendar in 46 BC, which became used as the empire wide system of time reckoning. "Caesar supposedly received the method of repairing the Roman calendar from an Egyptian astronomer, named Sasigenes of Alexandria" (Komzsik 2011: 38). The first day of his calendar, known as the Julian calendar, was January 1, 45 BC. Caesar's reform was based on a year of 365 days and 6 hours. But he did not reform the names of the 12 months which had been originally designed by Romulus, the first King of Rome in 735 BC. The first month was Martis, for the God of War, followed by Aprilis, for the raising of hogs, Maius, for a local Italian goddess, and Junius, for the Queen of the Latin gods. The subsequent months were simply counted, naming the fifth Quintilis, the sixth Sextilis, the seventh September, the eighth October, the ninth November, and the tenth December. In 700 BC, King Numa added Januarius and Februarius, which then brought the year to the standard of 354 days with twelve months (he later added another day to make 355). Each month was divided not into weeks but into day markers that fell at the beginning of the month, on the fifth day, and in the middle. These days were called *kalends* (the origin of our word calendar), *nones*, and *ides*. To round out his calendar reforms, Caesar reorganized the length of the months to add another 10 days required to bring the year from 355 to 365 days, arranging them to create a calendar of twelve alternating 30- and 31-day months with the exception of February. While priests and politicians continued to tinker with the calendar, Emperor Augustus restructured it in 8 BC, whereupon the Senate decided to honor this decision by renaming the month of Sextilis to Augustus.

The Julian calendar, which was inspired by the Egyptian calendar, injected a new spirit into how people thought about time. Originally, the calendar had been thought of as a cycle of recurring natural events or as an instrument of power.

> Now the calendar was available to everyone as a practical, objective tool to organize shipping schedules, grow crops, worship gods, plan marriages, and send letters to friends. Combined with the rising popularity of sophisticated sundials and water clocks, the new Julian calendar introduced the concept of human beings ordering their own individual lives along a linear progression operating independent of the moon, the seasons and the gods.
> (Duncan 1998: 50)

Although a seven-day week was respected in other civilizations and was inspired by an ordering of the seven planets of the astrological week (Saturn–Jupiter–Mars–Sun–Venus–Mercury and Moon), the ancient Romans – to whom we owe most of our temporal habits – respected an eight-day system.[9] The last day of the eight-day cycle was a market day. The eight-day system was experienced as an official and regular commercial calendar, and its use indicated some degree of official planning and attempted coordination of social life. Priests were involved in the fixing of some ritual moments on certain days. For example, to slaughter animals was an emblem of the hierarchy of the cosmos, with humans in the middle between the beasts on the one hand and the gods on the other.[10]

Usually the priests were not people with a special religious calling, i.e., not full-time religious officials, and they did not take any pastoral responsibility for the moral and religious needs of a congregation. They were usually the same men as those who conducted the political business of the city. As Cicero, who was himself both a political leader and a priest, put it: "among the many things [...] that our ancestors created and established under divine inspiration, nothing is more renowned than their decision to entrust the worships of the gods and the highest interests of state to the same men" (Cicero, quoted in Beard 2008: 280).

When Christianity gradually won out to the Roman Empire and its practices, Constantine the Great declared Sunday as a holy day in a new seven-day week. Christmas was celebrated on a fixed date,[11] but Easter was not as it was tied to the Jewish lunar calendar. The new calendar blatantly rejected the long-held observance of Saturday as the Sabbath by the Jews and by Roman pagans. Instead, a seven-day week with a Sunday Sabbath became respected. The dissociation of the seven-day week from a natural cycle such as the waxing and waning of the moon has been one of the most significant contributions of Judaism to civilization. It established a mechanical periodicity. "The invention of the continuous week was one of the most significant breakthroughs in human being's attempts to break away from being prisoners of nature and create a social world of their own" (Zerubavel 1985: 11).[12]

Beginning in the seventh century, Islam imported the seven-day cycle in Central Asia, to Sudan, to the east coast of Africa, and to a large part of northern and West Africa. The choice of Friday as the weekly day of Muslim public worship bears a striking resemblance to the Christian choice of Sunday as the Lord's Day. Whereas Christianity copied Judaism by taking Sunday, which was temporally proximate to Saturday, Islam similarly picked Friday as a form of imitation. Both tapped into an already familiar institution: a weekly holy day (Bluedorn 2002).

With the growth of international trade, merchants and traders needed to draw up contracts with dates of delivery, inventories, and accounting records.

> A serious point of confusion emerged among the practical minded traders on the docks and in the markets. Whose calendar and whose number symbols and counting systems should be applied? Arab merchants used Islam's lunar calendar and various versions of civil solar calendars, while Europeans conveniently used Caesar's Calendar. No consensus existed on basic matters such as when the year started, which could vary from town to town and from country to country. Some used the first of January, for others the new year started on Good Friday or on Easter day, still others started the New Year in March.
>
> (Duncan 1998: 200)

As international trade grew livelier, people attempted to sort out the babble of names and dates with little success because no central authority existed to standardize the calendar. Astronomers also became anxious about streamlining their techniques to calculate planetary positions and improve the applicability of the calendar. This continuously fed the desire to design suitable computing devices to improve astronomical observations.

Calendar reform

In 1514, Pope Leo X invited an expert on the calendar, the Dutch astronomer and Bishop Paul of Middelburg, to chair a commission on reforming the calendar.

A few years earlier, the Bishop had sent the Pope a tract demanding reform to the calendar which, since the thirteenth century, had shown a discrepancy exceeding a week. There was controversy over the question when the New Year and Easter should be celebrated . The church had a continuing concern to solve and systematize the dating of Easter. The principles of calculation, the science known as *computus*, were complex and gave rise to multiple solutions with different Christian rites. The practitioners of astronomy and *computus* were interested not only in the moon and the seasons but in the day and its divisions as well. City-states and royal rulers made their own decisions. In 1522, Venice adopted January 1 as the first day of the year.[13] "In 1563, Charles IX of France made a similar decision and determined that the New Year would start from the first of January and not with Easter. The French parliament accepted this decision to be implemented from January 1, 1567" (Montaigne 1972: 107).

The difficulties of Easter reckoning generated a host of debates and conflicts. No one precisely knew the correct meridian on which to base the calculation of Easter: Should it be Rome or Jerusalem? Since the third century, it had been agreed that Easter should be the Sunday after the fourteenth day of the first lunar month (Luna XIV). Bishop Paul had consulted a well-known astronomer Nicolaus Copernicus who lived on the Baltic coast of Poland and who had attempted to properly measure and calculate the length of the year.[14] His analysis initially attracted little controversy as few people understood what he had calculated. An exception was the vehement reaction of the Protestant Martin Luther (1483–1546) who, as a biblical purist and as one of the instigators of the Reformation, viewed any deviation from the scriptures and its geocentric world view as subversive. He refuted Copernicus' heliocentric planetary system (Duncan 1998: 255). For many years, the Catholic Church remained silent about the issue.[15]

Finally in 1578, Pope Gregory XIII instituted a process to improve the reckoning of the calendar. A member of Pope Gregory's commission on the calendar reform had pointed out the back slide between the Julian calendar and the year of the seasons. After several years of bickering, the commission took a decisive action and dropped 10 days out the calendar, thus putting artificial time and nature's time back in step. The equinox was moved from March 11, the date on which it had occurred in 1582, to March 21. Pope Gregory XIII wrote a decree declaring that the day after October 4, 1582, should be called the fifteenth. Although not the most accurate correction, it was the most convenient one. The decree promulgated a calendar that still bears his name.

The Italian city-states, Portugal, and Spain were prompt to comply. France, however, resented the papal interference. The new calendar was much disliked as it changed an age-old custom and upset the traditional farmer's year and weather proverbs. France waited with the implementation until December, when King Henry III ordered its implementation. The Catholic provinces of the Low Countries delayed until the end of the same year and implemented the calendar the day after December 21, which was followed by January 1 and took out

Christmas day. Protestants in the Low Countries and Germany, however, rejected the reform. With the exception of two Dutch provinces (Holland and Zeeland), the whole non-Catholic world retained the Julian calendar.

Hours

The subdivision of the daily cycle into 24 hours was a distinctively Babylonian, as well as Egyptian, mathematical practice. For astronomers, this division became an essential element of their measurements. To circumvent non-uniformities in their measurements, the Babylonian astronomers had invented artificial hours which they defined as one twenty-fourth part of the day. These hours were determined by the position and rate of movement of celestial bodies and as an artificial system of time units were equal and rational.[16] In the daily practice of the common people, it did not yet play an important role.[17] They were using temporal hours in the sense that they conveyed the sense of changing. Except at the equinox, the hours of the day differed from those of the night, just as the hours of one month differed from those of another month.

Caesar's calendar reform did not indicate any information about the daily time rhythm of Roman citizens. The Romans used sundials and water clocks to reckon the hours. The times of sunrise and sunset usually designated as beginning and ending of the day were variable, as was the length of daylight and night-time hours. During Roman times, day and night were joined as a unit that began and ended at midnight, a convenient time to make the transition, at least in the business world.

Still, the hours reckoned by the sun's casting shadow were unequal.[18] The Roman military introduced callers watching the position of the sun to announce the changing of the guard at the third hour (*tertia hora*), in the morning, at the sixth (*sexta hora*), at midday, and at the ninth (*nona hora*), in the afternoon. The length of the daylight period, therefore, varied with the rhythm of the seasons.[19] In Pompeii, the sundial in the many public baths allowed busy bathers "to keep an eye on the time" (Beard 2008: 30). Along with sundials as simulacra of time, the Romans were familiar with water clocks, which the ancient Greeks had designed as an aesthetic satisfaction for imitating the heavens. They were also used during discussions in the Senate. The term *aquam perdere* – to lose water – meant to waste time. When a speaker went on too long, "his colleagues would shout for his water to be taken away" (Levine 1997: 55–56). Seneca, however, did not care much about those simulacra as application of the hour: "I can't tell you the hour precisely. It is easier to get agreement among philosophers than among clocks" (Seneca, quoted in North 2005: 137).

Medieval times

By legend, church bells had been invented in the fifth century in the town of Nola in Campania. Bells probably spread first to monasteries where monks used handbells to signal canonical hours for prayer. However, in late medieval Europe, in particular during 1271–1283, the mechanical clock was *in the air*, becoming the new lord and arbiter of time. The heart of the mechanical clock is its escapement, the device which regulates the running down of the motive power. The application of this verge-and-foliot escapement was an invention of radical originality (North 2005). Once invented and displayed, there was a clear sense of excitement and pride about mechanical clocks. It became, like the computer today, a technical sensation of the period (Landes 1983). It has been suggested that this escapement was an import from the East, but "most historians of horology believe it to be an indigenous European invention" (Mayr 1986: 5).

Slowly but surely, mechanical clocks introduced a transition from an economy based on a labor time ruled by religious festivals, governed by agrarian rhythms, free of haste, careless of exactitude, and unconcerned by productivity, to a new commercial economy based on chronological clock time (Le Goff 1977). "Time was recognized as a commodity of great value and as a source of material gain" (Gurevich 1985: 150). In the medieval towns, institutions of government, church, trade, and proto-industry came together and increasingly constructed new artificial time frames. The mechanical clock became recognized as the most perfect apparatus ever conceived by the human mind. It seized the imagination with surprising ease and speed. Clerics, kings, and scholars marveled at it. Soon mechanical clocks were installed in City Halls, monasteries, and church towers, which fueled the interest for time measurement.

In 1317, the Italian city of Parma, for example, had a bell that struck the hours for laborers. In 1325, the French city of Valenciennes signaled the opening and closing of markets with a bell. Milan obtained a public clock in 1336 that struck the hours and is one of the first mechanical clocks of which we have reliable knowledge. Hour striking became of enormous social significance and could be heard everywhere by the inhabitants of large towns. The list of public clocks quickly multiplied with cities such as Padua, Monza, Vicenza, Trieste, Genoa, and Florence. Within these commercial and wealthy cities, the striking clocks spread with remarkable speed over the European mainland, the British Islands, and Scandinavia (Dohrn-van Rossum 1992: 152 and 161). An atmosphere of calculation and measurement grew (Le Goff 1977). The Salisbury Cathedral clock is thought to date from 1386; the Rouen clock in France dates from 1389; and the Wells clock in England from 1392. Both the English clocks had gong and striking mechanisms, and the Wells clock also chimed the quarters. Bells sounded for the start of work, meal breaks, and the end of work. They sounded for closing the gates, the start and closing of the market, for assemblies, emergencies, council meetings, end of drinks service, and time for street cleaning. Mechanical clocks became the technical thrill of the period. Mechanical timekeeping became "the single most important practical innovation of the entire Middle Ages, both scientifically and socially" (North 2005: 143). The mechanical clock became perceived as closely resembling the medieval world view with the heavens revolving around the earth, and turned into the metaphor that adequately depicted a world created by a divine clockmaker and – as a consequence – identified as a mechanism which can be investigated by reason.

The pressure for time signals was especially strong in cities that were engaged in textile manufacture such as Florence, the first and greatest of medieval industries.

This industry started to engage in large-scale production for export and hence became the first to overflow the traditional workshop and engage the dispersed workforce in the cities in larger manufacturers. The process of fulling by using a wooden roller with protruding blocks proved more effective than muscle power in kneading the cloth. The use of water power to turn the rollers led to the adoption of the term "mill" as an early form of a mechanized industry.[20] In some cities in the southern Low Countries, textile manufacturers also installed work bells to define the length of working time, which was crucial to the profitability of the enterprise and the prosperity of the commune. Textile artisans and spinners in Brussels had several *werckclocken* to tell when to start and stop work. Sometimes these bells were public, installed by the municipal authorities in a church tower or in a Belfry erected for that purpose.

In 1335, the mayor and council members of Amiens received from King Philip VI of France the right to issue an ordinance fixing the time when the workers had to stop their work and when to resume their work again. Henceforward, Amiens used the bell to signal when people were to eat and when they were to begin and finish working (Crosby 1997). Sometimes, workers did not rely on the public clocks but instead installed their own. This occurred, for example, in Ghent in 1324 when the Abbot of Saint Pierre authorized the fullers to install a bell. More often, these work bells were in private hands, which inevitably gave rise to conflict about the imposition of time discipline. The textile workers questioned whether the work bell was honest time. How could they trust the municipal bells when representatives of the employers or guild-masters dominated the town council? To prevent such kinds of conflicts and prevent a scandal, the city of Thérouanne decided in 1367 to silence the workers' bell. However, public clocks, especially those with dials, made the verification of time a continuous option for the inhabitants of the city.

Local city councils and government were inclined to prevent any kind of misunderstanding concerning the moment to start and stop work. For that reason Charles V of France decreed in 1370 that all clocks in the city of Paris should be regulated to the one he was installing in his palace. The King's instruction made the Horloge du Palais the "Zeitgeber" for the other clocks in the city.[21] It is questionable whether this actually happened as the decree itself has never been found (Dohrn-van Rossum 1992). It might have been the case that the clock keepers of the city simply imitated and copied "the audio stimulus coming from the king's palace" (Bluedorn 2002: 158).

With the appearance of mechanical clocks came the conception of the hour as a secular unit of time. Next to a daily bell ringing, clocks soon gained visual displays. Italian hours were often counted from the end of twilight after sunset. Others started at midnight. Although a feverish horological activity spread across Europe, the early mechanical clocks were not yet very reliable. And yet, ordinary people began to distinguish between the hour of the day in traditional ways and regular, mechanical clock time (Dohrn-van Rossum 1992). Public mechanical clocks began to dominating city activity, and the unequal, variable, or temporal hours gave way to equal or constant hours. In the towns, the guild statutes registered work hours and imposed a strict working time arrangement.

But who were the many inventors working independently on the same sort of mechanical devices leading to the perfection of the mechanical clock? Some suggest that the source of its invention must be found in monastic Christianity where the setting of prayer times by the clock was becoming important, thus the interest of monks in time measurement may have facilitated the invention of the mechanical clock. Whereas the city dweller like the countryman still had a natural sequence of tasks which gave rhythm to his day, certain uniformity in the pattern of activities of the monks made them time-conscious (Le Goff 1977).

Monasteries

Monasteries were oases of learning, but they also became structured centers of organized labor whose leaders maintained administrative skills and discipline which were copied from Roman officials. Whereas the majority of early monasteries and communities of monks across Europe avoided any knowledge not directly applicable to their faith, Flavius Magnus Aurelius Cassiodorus (*c.* 485 to *c.* 585) set up a scholars' monastery to preserve the general spirit of learning. In his monastery, he studied mathematics, time reckoning, and other elements of a classical curriculum. He drew a clear distinction between time reckoning and time measurement: *time reckoning* is a purely intellectual achievement. It recognizes the miracles of numbers and their usefulness in making calculations of time critical to a believer for planning when and how he will worship the deity. *Time measurement*, on the other hand, is "a matter of making observations of celestial bodies and jotting down numbers and using mechanical devices such as clocks that require technical skill to manufacture" (Duncan 1998: 92). Cassiodorus stressed that the *horologium* was the highest achievement of civilization. Abbott Benedict of Nursia in Umbria followed in the footsteps of Cassiodorus and was instructing his monks to make clocks and to use them to tell the time to the hour.

The first Benedictine monastery was established at Monte Cassino in Italy in 529 and was run with strict written guidelines governing times and types of work and prayer. Around 540, Benedict wrote down the Benedictine rule which included a table of hours setting out a strict list of duties, prayers, mealtimes, and ceremonies linked to a careful measuring of each hour of the day. He had copied the Roman army's system of daily watches and delineated canonical hours that did not have to be announced: dawn (*matutina*), sunrise (*prima hora*), sunset (*vespera*), and the coming of complete darkness at night (*completorium*). Unlike the Roman soldiers, the monks were not ordered by a temporal power or a political order to comply with the rules but based it on their own self-discipline.

In the rule of St Benedict, the new order of the offices found its first complete and detailed realization: Six (later seven) daytime services (lauds, prime, tierce, sext, none, vespers and compline) and one at night (vigils, later matins). Most of these services were set in terms of temporal hours and were henceforward called canonical hours. The monastic order enforced a regime of performing each prayer precisely at its prescribed time. The monks further filled their days with manual work while their minds were kept focused on spiritual matters. Idleness was seen as the enemy of the soul. Time discipline and punctuality were taken seriously. Benedict ordered that time be kept accurately and objectively using the best clocks available. Monastic rules institutionalized a regular application of the collective praying, working, eating, and sleeping day.

> Monastic space was closed space – areas and corridors of collective occupancy and movement – so arranged that everyone could be seen at all times. So with time: there was only one time signaled by the bell, measured and kept by the sacristan, excluding individual and autonomous time.
>
> (Landes 1983: 69)

Cistercians built their monasteries near rivers so as to have access to water power. They learned to use this in multifunctional, staged installations to exploit power capacity to the maximum. For such an undertaking, timekeepers and bells were an indispensable instrument of organization and control (ibid.). The Benedictine rules, the mechanical clock, and the self-discipline of the monks can be translated into the trinity of documents/devices/drilled men, a framework that has been used earlier in this book to describe the way the Portuguese navigators had been trained for the proper execution of their job.

This time discipline spilled over into the monastic estates and hospitals, which were run on the same precise temporal lines. Tower bells summoned people to mass and began to slowly govern people in everyday life in Europe. The idea of the monk's self-discipline, however, was essentially negative: It was based on the principle of non-idleness (Thrift 1988), while the time discipline in the towns was generally positive in the sense of being related to the structuring of work time, which stimulated the interest to value the use of time. The monasteries demanded a controlled and disciplined life that was still mainly based on variable or canonical hours. Yet the result was the invention of the mechanical clock, which came out of an interest in measuring time but was certainly not the origin of that interest.

The monastic clergy may have provided an initial market for timekeepers, which stimulated a principal drive to technical advances in this domain, but the church cannot have been alone in accounting for the increasing popularity and development of the new device. Numerous courts and expanding commercial urban centers with their wealthy bourgeoisie, especially in Italy, stimulated the future of the infant clock industry as well. Many of the early clockmakers were also gun-founders and borrowed heavily from military engineers (May 1986). Large tower clocks were installed in churches and town halls, and near city gates of major commercial towns. Royal rulers and the mighty residents of the cities were in the position "to build them as complements or successors to the cathedrals – a symbol of new secular dignity and power and a contribution to the general welfare" (Landes 1983: 71).[22]

Revolution in time

The actual inventor of the mechanical clock is not known. Neither is the place of origin, which may lie anywhere between northern Italy and England (North 2005). Traditionally, the hours both of religious services and of town activities such as curfew, market time, as well as the opening and closing of the gates had been struck by hand on special bells. If anyone needed a reliable clock, it was the tower warden who had the job of striking the hours. In the past, he had applied traditional devices such as sundials in daytime and water clocks at night. The limitations were made even more painfully evident in the cloudy skies and long nights in cold winters of northern Europe than they had been earlier in Mediterranean latitudes. The tower warden was probably the first user of the new mechanical clock to strike constant hours and was also the first to be replaced by it once the clock had advanced enough to strike the hours automatically.

The mechanical clock was a creative achievement of high intellectual rank. Nothing of comparable ingenuity had ever been invented before. The basic invention created an increasing interest in timekeepers. In general terms, the history of time measurement is the history of techniques for dividing time with greater accuracy into smaller divisions based on technical improvements of the clockwork. Due to this progress, clocks became transformed into symbols of regularity and order with which the secular society became structured and organized. According to May (1986), this development followed three stages: The mechanical, weight-driven clocks of the first phase (1300–1560); the second phase improving the reliability of the clock (1560–1657); and the third phase which led to the introduction of the chronometer (1657–1759). Here, we will only deal with the first two phases; in Chapter 4 the third phase will be dealt with.

The first phase of the mechanical clock (1300–1500)

The heart of the mechanical clock is its escapement, the device which regulated the running down of the motive power into repetitive mechanical motion. Its secret lies in its capacity to transform "the continuous accelerating, uni-directional motion of a falling weight into discrete, regular oscillating motion" (Rispens 2005: 118). The mechanism that managed to translate the force of gravity into equal jerks was called the verge-and-foliot escapement.[23] The first mechanical clocks fell into two categories: astronomical clocks that attempted to reproduce motions of the heavenly bodies; and figure clocks that imitated the actions of living beings and, as such, were instruments for entertainment and magic, giving pleasure to friends and guests. Clocks and automata were looked at as mediators between the worlds of magic and rationality. Clocks symbolized a representation of the known universe and helped to teach Europeans how to think mechanically.

The first mechanical clocks may have been made by wandering clockmakers, by practitioners of older crafts such as blacksmiths and locksmiths, and by scholarly amateurs such as theologians, astronomers, and medical doctors. Weight-driven mechanical clocks striking the hours became installed in full view of the public in all major cities and were rapidly spreading to the smaller towns. Some of these clocks were indicating the hours, too, following the practice for the sundial's shadow: the hands or dials moved in the same way, thereby reproducing and reaffirming a traditional practice. Remarkable achievements in this first stage have been the astronomical clock of Richard Wallingford, the Abbot of Saint Albans in Hertfordshire (England), and Giovanni de Dondi's astrarium in Padua in 1348–64.

The city of St Albans had grown quite naturally around the walls of the great Abbey of St Albans, a Benedictine monastery located between Oxford and London. It had a marketplace which added greatly to the town's commercial value and to the prosperity of the abbey. The Abbey's cellarer was controller of the market, and the abbot appointed the keeper of the town prison. The wealth of the abbey depended to a great extent on the monopolies it held. There were intensive contacts between the abbots and the townsmen concerning the use of water-powered mills for grinding grain, fulling cloth, sawing wood, crushing seeds for oil and crushing ore for melting.[24] Although Benedict's Rule could be applied in a flexible manner, it had an inflexible core which was related to its insistence on prayer, the reading of the scriptures and manual work. The bells were sounded to mark the hours of the day and to summon the brethren to their obligations.

Richard of Wallingford (1292–1336) became the abbot of St Albans in 1327. He had studied theology in his younger years and law at Oxford and had demonstrated a particular interest in astronomical instruments. He wrote several different treatises about astronomical issues and "was heir to a European tradition, which had benefited greatly from the astronomy of Muslim Spain. Richard not only had a lawyer's mind but a puritanical spirit to go along with it" (North 2005: 98). Although the mechanical clock might have been half a century old when Richard became abbot of St Albans, the instrument he developed is one of the earliest examples which survived (North 2005). The social time pattern that the monastery had introduced certainly became perfected with the introduction of the mechanical clock (Thrift 1996).

Independent from Wallingford, without any possibility to exchange information and knowledge but working in the same vein, Father Jacopo and his son Giovanni de Dondi, who was a physician and astronomer, published their *Tractatus astrarii* in 1364. This text contained one of the earliest known sketches of a mechanical clock. They began to build their astronomical clockwork in 1344 and, after almost 20 years, were able to expose it. The astrarium was more than a clock, it used a year wheel and a geared assembly to track the movements of the planets. In 1364, the new instrument was shown to the public at the Palazzo del Capitano of Padua. De Dondi's astronomical clock had several dials which showed the sun and the moon and the five spheres of Venus, Mercury, Saturn, Jupiter, and Mars (Uranus, Neptune, and Pluto were not yet discovered), and provided a perpetual calendar of all religious feasts. However, its complex movement still lacked reliability.

The first great clocks of medieval Europe were mainly designed as astronomical showpieces, but that situation quickly changed.[25] As the towns ordered public life according to the mechanical clock, individuals also welcomed the clock with constant hours into their private lives. As the clock extended its dominion, the concept of the *hour* acquired new meaning. Day and night had been independent quantities, each divided separately into temporal hours that varied with the seasons and with latitude. The mechanical clock, however, being oblivious of the cosmic rhythms, favored units of time that were rigidly constant. "The introduction of equal hours and the habituation of urban populations to public time announcements had profound consequences for the European mentality" (Landes 1983: 77).

The urban, commercial population began to learn a new language of time and its techniques, but its development took some time as misunderstanding continued about the meaning of the word *horologium*, a generic term that not only included mechanical clocks, water clocks, and sundials, but also astronomical instruments such as quadrants and astrolabes. Nevertheless, "equal hours announced the victory of a new cultural and economic order. Here indeed was an unintended consequence: the monks had wrought too well" (ibid.: 78). In other words, theological time transformed into technical time, and both clock and cannon became the product of a remarkable growth in the number and quality of metalworkers.

The mechanical clocks became symbols of power for the cities that owned one. Competing rich Italian city-states gave rise to a new style of public clocks: they became symbols of autonomy and independence, show cases of prestige, an edifying spectacle for residents and visitors alike. However, they also had an inverse effect.

> When Philip the Bold of Burgundy defeated the Flemish burghers at Rosebecke in 1382 and wanted to punish those proud and troublesome clothiers, he could do no worse than seize the Belfry clock at Courtrai and take it off to his capital at Dijon.
>
> (Ibid.: 79)

Additionally, the clock became an attribute of nobility and a symbol of aristocratic rank. Princes and their courts led the way in demanding punctuality from everyone. The habit of disciplining one's daily life became methodically inculcated into young princes. A particular appeal of the mechanical clock was rooted in the magic of the self-moving machine. But someone had to initiate and control it and keep it in balance. This purpose was symbolically expressed by temperance.

Temperance

When, by the mid-1250s, Aristotle's work was rediscovered, the interest in the meaning of cardinal virtues (prudence, courage, temperance, and justice) spread quickly. The Italian city-states became centers for a discourse about good governors, based on respecting these cardinal virtues. Initially, temperance had been primarily depicted as a female figure, either holding a jug from which she poured water into a cup to dilute its potency, or holding a pair of compasses. After the discovery of Aristotle's work, the popularity of temperance as a figure symbolizing a balanced execution of power increased (Skinner 1978). Henceforward, temperance became depicted with a sandglass/hourglass, which has an obscure origin but metaphorically symbolizes measure, the chief source of human goodness. Virtues and emotions such as honor, loyalty, humility, and patience became portrayed as mechanical counterparts of the clock. Authors such as the influential Belgian priest Jean Froissart spent his life at the courts of various princes and kings in northwestern Europe and compared courtly love with the working of the mechanical clock. He compared temperance with the escapement. Measure, or *temperantia* was the virtue that defined the perfect knight and the good governor.

Theological virtues (faith, hope, and charity) and cardinal virtues emerged as eminent, but the mechanical clock was not yet the dominant point of reference to illustrate the important role of temperance for the good ruler. While time measurement was still very inaccurate, temperance was seen as the symbol of measure regulating the mechanical clock. To illustrate this, a link was drawn between temperance and the hourglass. Our first reliable proof for the existence and application of the sandglass in relation to temperance can be found in a fresco by Ambrosio Lorenzetti (1290–1348). The painted allegory of 1338 is hanging in Sienna's town hall, the Salla Della Pace of the Palazzo Pubblico. The goddess of temperance presides with wisdom, peace, justice, faith, charity, magnanimity, and concord. While holding an hourglass, she is sitting above people going about their business. The good governor is surrounded by these seven female figures representing the cardinal and Christian virtues. Temperance symbolizes a virtuous attitude of the good governor to manage a beneficent city-state and making decisions based on good management.[26]

White (1969) supposes that the Italian idea of representing temperance with a timepiece was brought to northern Europe by Thomas de Pisan, a native of Bologna, who served King Charles V of France as his personal medical doctor and astrologer. The King behaved like a *magister temporis*, much like Julius Caesar, who used his civil authority to organize social time in the most practical way. Charles' royal palace gathered a remarkable group of intellectuals such as Nicole Oresme (1325–1382), who was the first to introduce the metaphor of the universe as a vast mechanical clock which functioned in a perfect, tempered, and harmonized way (Crosby 1997). Probably inspired by the mechanical clock, the prelate Oresme wrote in *Heaven and the World* that the universe was like a "horologe." Pisan's daughter, Christine de Pisan, lived at the same court in Paris and wrote in 1400 *The Epistle of Othea* (Pisan 1970), a book about manners and morals written for the benefit of adolescent aristocrats.

> Temperance should be called a goddess likewise. And because our human body is made of many parts and should be regulated by reason, it may be represented as a clock in which there are several wheels and measures. And just as the clock is worth nothing unless it is regulated, so our human body does not work unless Temperance orders it.
>
> (Pisan, quoted in White 1969: 209)

Her text belonged to a class of books with princely advices that enjoyed continuous demand.[27] Pisan's work was well known, found several translations throughout Europe, and was read at royal courts and in aristocratic circles. It belonged to a tradition of a vocabulary about virtues where the importance of and participation of citizens in the maintenance of the city-state was steered (Skinner 1978).

Although the clock began to stand for regularity, promptitude, and reliability, temperance was the metaphor that still corrected the clock for its mechanical crudities. Temperance became the noblest virtue, the pre-eminent virtue, and her essential character was Measure. Temperantia stayed a familiar figure in the paintings and sculptures of the fifteenth and early sixteenth century as long as the mechanical, but inaccurate, clock was not yet sharply distinguished from other timekeepers such as the sundial and the hourglass. The different time instruments were associated with Father Time or Chronos, and sometimes with the beating of the heart and with death in the sense that the clock suddenly could be "out of service." Timekeepers were employed indiscriminately to illustrate various general aspects of time in social life.

The Italian city-states were not only the kernel of a discourse about good governance but also a seedbed for a growing public interest in the functioning of mechanical clocks. The mechanical clock seized the imagination of the city-dwellers (White 1962). The overarching tone became one of a growing appreciation and unqualified approval for the clock. Temperance became associated with punctuality, discretion, circumspection, caution, and frugality. For a while, it became the ideal of the bourgeois townsman and a focus of behavior for the rising middle class (Mayr 1986) One might say, however, that with the continuous improvement and sophistication of the mechanical clock, and the rise of modern science the interest in temperance as a key symbol withered away.

The close of the first period of the rudimentary clock can be illustrated by an etching made by Pieter Bruegel the Elder in 1560, which shows how temperance still stood at the center of intellectual development (Figure 3.1). But on her head, she wears the mechanical clock: "the most distinctively Western of all contrivances for measuring quantity" (Crosby 1997: 10) Although she still held center stage, the fact that she wore a miniature mechanical clock on her head indicated that she had begun to lose her power to keep time in balance. Based on an innovation in the late sixteenth century, the mechanism of the clock improved, became more accurate and could regulate itself more precisely. Temperance disappeared slowly but surely into the background of the royal and bourgeois stage of western civilization.[28]

The second phase of the mechanical clock (1560–1650)

One can discern a second stage in the revolution of time due to the application of spring-power to clockwork. Once Jost Buergi had invented the cross-beat escapement in 1584 (Landes 1983), spring-powered clocks became compact, portable, and capable of storing their energy for weeks of continuous operation. In a century when paper milling, gunpowder production, cannon founding, brass making, sugar refining, alum and copper melting, and the mining of zinc and copper had all become organized industries, clock making and miniaturization became a specialty of full-time professionals. Clockmakers investigated the properties of the types of copper and steel used in their work. They studied the thermal expansion of metals and the elasticity and resilience of springs. These clocks were mostly made for

FIGURE 3.1 *Temperantia* (*Temperance*). Pieter Bruegel the Elder, 1560. (Courtesy Dr Harris Schrank.)

domestic and personal use and became more accurate. This development stimulated, according to Crosby (1997), a shift away from the *venerable* model – the period where measure was important but without quantification – to the modern model of time measurement. Time became isolated as a pure form exterior to life which led toward a culture-wide acceptance of homogeneous, empty time.

Clock-making began to flourish, amongst others, in German-speaking central Europe with its free cities such as Strasbourg, Ulm, Nuremburg, and Augsburg. These city-republics were dependent on trade and craft production for their prosperity and provided a favorable environment for clock-making. Productivity was enhanced by an initial phase of standardization and specialization and by extensive use of prefabricated parts which led to a successful refinement of clocks. Its development was further stimulated by the scientific revolution that was intimately connected with the invention of instruments such as the telescope, microscope, and air-pump.[29]

These instruments allowed scientists to investigate whole new fields of study that had previously been impractical, unthinkable, or even impossible. There has always been a debate about this relationship between the development of a mechanical clock and the appearance of modern science, i.e., the scientific revolution, which turned the conception of the cosmos from a closed, qualitative, and purposive world into an infinite quantitative universe.[30] An epochal shift from qualitative to quantitative perception took place in the seventeenth century and changed the view on measure. It was one of the West's distinctive intellectual accomplishments to bring mathematics and measurement together, and the scientific revolution introduced a conception of time as a scaled continuum of discrete moments leading to constant hours throughout the day (Dohrn-van Rossum 1992).

The clock as metaphor

In general terms then, the history of time measurement is the history of techniques for dividing time with ever greater accuracy into smaller divisions. (Especially in marine transport, as will be shown later with the development of the chronometer in Chapter 4, precise measurement was required to be sure that scheduled harbors would be reached safely.) The clock obtained a dominant metaphorical influence ever since Oresme used it to illustrate the functioning of the cosmos and Pisan compared it to the image of temperance. Ever since the mechanical clock as a metaphor described the regularities in the universe it has kept popular applicability. As such it has been used as a powerful way to produce insight, understanding, and meaning in the functioning of economic organizations too. It has become a template, a tacit imperative for managing and organizing economic activities (Bluedorn 2002; cf. Morgan 1986).

In the seventeenth century, the mechanical philosophy also introduced clockwork as the metaphor to describe the functioning of the body of an animal. The most famous philosopher of that time, René Descartes (1596–1650), had elaborated some general thoughts about the analogies between the movements of mechanical clocks and those of all natural bodies. Descartes served for a while in the army of the Dutch military leader Maurits, and developed a fascination for practical technology. In the 1630s, while staying in the Dutch city Deventer, Descartes worked on a text which became published as *Traité de l'Homme* (1664). Throughout his writing, he used words such as clock, automaton, and self-moving machine as interchangeable constructs. He postulated an account of the physical world that was thoroughly materialistic. His mechanical view of nature replaced the organism model which had been popular since the Renaissance. The structure and the features of the world were ultimately to be accounted for in terms of material bodies in motion. Even the working of the human body was seen as rigorously mechanical but, compared with the animal, it was endowed with a soul. This led to his dualistic doctrine of the two aspects of the human being: the physical aspect of the body functions like a clock, whereas the functioning of the soul, reason, emotions, and will is spiritual in kind. The soul as the highest in the hierarchy of organs of the human body was located in the "pineal gland." Human reasoning is the balanced interaction between the will, passions, and emotions of a conscious and moral subject.

Descartes' views provoked tremendous debates in circles of theologians and philosophers. Some theologians abhorred the idea that a human being could be compared to a clock. Others, however, were charmed by the idea that human beings are entities that fully operate mechanically. The German philosopher

C. Wolff explicitly defended this monistic view. He went even one step further and declared that the world did not function differently than a clock.

He was convinced that theological truths could be based on mathematical certain evidence. In 1721, he delivered a speech at the University of Halle "On the Practical Philosophy of the Chinese" in which he praised the purity of the moral precepts of Confucius, as evidence of the power of human reason to reach moral truth by its own efforts. This so upset the protestant community that Wolff was ousted in 1723 from his chair at Halle. The case became one of the most celebrated academic dramas of the eighteenth century. His enemies told King Friedrich Wilhelm I of Prussia that, if Wolff's determinism were recognized, no soldier who deserted could be punished, since he would only have acted as was necessarily predetermined. This so enraged the soldier-king that he immediately deprived Wolff of his office and commanded him to leave Prussian territory within 48 hours or be hanged. Wolff did stick to his monistic view, which was inspired by Jacques Vaucanson's (1709–1782) flute player and other objects which had been attracting much attention in Paris. They were seen as marvelous illustrations of the way any human being functions.[31] When Friedrich William died in 1740, his son and successor, Frederick II the Great (1712–1786) recalled Wolff from his exile.

It is commonly said that the French medical doctor Julien Offray de La Mettrie (1709–1751) took Wolff's determinism one step further. The young man from Saint Malo stayed for his medical training in Leiden (the Netherlands) where he studied the human body based on the laws of mechanics and hydraulics. He became a medical doctor in the French army and was involved in military operations in France, Flanders, Holland, and Austria. There he saw with his own eyes the cruelties of war and how wounded soldiers died in hospitals (Jauch 1998). In 1747, La Mettrie published an anonymous pamphlet *L'Homme Machine* (*The Man Machine*) (Clark 2006).[32] The book was read as if La Mettie denied the Cartesian view of a fundamental distinction between the physical (*bête machine*) and mental entities and that he defended a fundamental materialism of the total human being. He did compare the functioning of the human body with a clock. It was said that La Mettrie defended only a gradual distinction between the animal and human being and that he denied the existence of an immaterial soul. *The Man Machine* was thought to present a view that not only animals are just clockwork mechanisms but also that human beings are but machines which wind themselves up, a living picture of perpetual motion. His book provoked much disturbance for suggesting that monkeys and men can be seen as mere machines, comparable to clocks. The human body is like a large watch with its own skill and ingenuity.

La Mettrie's materialistic principles caused outrage even in the relatively tolerant Netherlands. In 1748, he was compelled to leave for Berlin where Frederick the Great took him into his Court as his personal doctor. The King was a well-known musician and lover of the arts, but he also meticulously orchestrated military actions and imposed tyrannical military operations onto his army. He was obsessed with the disciplining of his soldiers as repetitive-acting subjects, as if they were machines, an attitude which would not have been unfamiliar to Maurits.[33]

The King and La Mettrie became friends, and the King defended his "court atheist" against religious fanatics. At the court, La Mettrie wrote a hedonistic book *Discours sur le Bonheur* (1748) in which he stressed the unnaturalness of shame and guilt.[34] He was actually much more interested in the lust of the body and soul than in its physical sublimation and dressage. Negative views about La Mettrie's materialism, however, have been strongly influenced by Voltaire who had stayed at the same court since 1750. Voltaire (1694–1778), the ardent campaigner for human rights, had initially been a friend but finally wrote in his *Letters From Prussia*– which he later falsified – very negatively about La Mettrie and accused him of atheism.

Why should so many believe that *L'Homme Machine* is a simplified metaphor of the human being as a clock and a machine? According to Sutter (1988) and Jauch (1998), it has systematically been neglected that La Mettrie had written his books in an ironical way. He was himself a radical skeptic about any materialistic view on the human being. He had tried to demonstrate that *l'Homme Machine* is *l'homme sensible*, a sensible human being. La Mettrie used irony to demonstrate that the mechanical philosophers were boring slaves of order and methods. He was using a masquerade by saying that he did not mean what he did not say. The title of his book questioned the Cartesian hypothesis that if animals are merely clockwork mechanisms without a soul, then this hypothesis might as well be applied to men. However, reducing living beings to insentient automata means an immense loss of wisdom (Rispens 2005). The book is a kind of labyrinth and clearly indicated that the author was an agnostic who actually contested Descartes' body-automaton doctrine. The text was composed of all kinds of observations, playful polemical statements, and references to actual debates, frivolous descriptions, rhetorical maneuvering and remarkable philosophical explanations.[35] Nowhere did he state that a human being is identical to a machine.[36]

The Prussian King was so upset about the way Voltaire had treated La Mettrie that he ordered the public burning of Voltaire's *Diatribe du Docteur Akakia* where he had turned La Mettrie into an atheistic figure. Yet the monist image of the human being did stick to La Mettrie's materialistic views which became very dominant in the Western thinking about man and work. As such it also has been a source of inspiration for engineers, who began to apply this concept as an instance of the same phenomenon in nature and human life and so developed "time and motion" studies, as will be illustrated later.

With the further refinement of the mechanical clock, abstract time became dissociated from planetary rhythms and seasons, from change and aging, from experience and memory. It generated a concept of time, which became self-sufficient, empty of meaning and remarkably neutral. The mechanical clock seeped deeply into the fabric of social life. During the process of industrial globalization it spread "like a spider's web across the globe, leaving those who resisted the new time defined as backward and old-fashioned" (Adam 2004: 114). The clock became the standard instrument of co-ordination and control. Its tick-tock replaced fluctuating rhythms. The clock as machine-pacing replaced self-pacing. Clock time revolutionized the cadence of daily life.

4

THE AGE OF MERCHANT CAPITALISM

This chapter addresses the development of international trade and commerce from the sixteenth century forward as Europe's merchant capitalism grew into a significant economic force, and Europe introduced state-sponsored trading companies which became known as "proto-multinationals" (Jones 2007).[1] The period of merchant capitalism, which usually refers to the earliest phase in the development of capitalism with a particular economic and social system, was based on a growing influence of traders and merchants intensifying international contacts. During the last period of medieval Europe, merchant houses that were often backed by financiers began to act as intermediaries between simple commodity producers. The intensification of large-scale, long-distance, multilateral trade relations stimulated the appearance of a large merchant class, which required a new financing structure and stimulated new forms of labor relations (Jones 1981, 2003).

During the seventeenth century, the Dutch Republic became the main dominating country ruling the seas thanks to their strong joint-stock companies and their military power. The merchant marine power of Venice was replaced by Dutch ships which, since the 1590s, appeared in the Mediterranean Sea and operated more cheaply than Venetian vessels. Dutch, as well as English, ships were "faster and more heavily gunned than the Italian ships they displaced" (McNeill 1974: 142). The Dutch, however, needed to liberate themselves from Spanish occupation to become an independent Dutch Republic. As this chapter will demonstrate, during the time of merchant capitalism, the Dutch Republic developed two remarkable methods to improve the mastering of time.

International trading companies

As can be seen in Table 4.1, the early trading companies had a broad European background. Usually the King, as the head of state, granted a charter to a group of

TABLE 4.1 Some of the major international trading companies and their year of establishment

Company	*Established*
The Muscovy Company	1553
Compagnie van Moscovia	1577
The Levant Company	1581
The Africa Company	1588
The Eastland Company	1592[a]
The English East India Company (EIC)	1600
The Dutch East India Company (VOC)	1602
The Virginia Company	1606
The Danish East Indian Trading Company	1616
The Dutch West India Company (WIC)	1621
The Compagnia delle Indie Orientali	1647
The Royal African Company	1662
The French Compagnie des Indes	1664
The Hudson Bay Company	1670
The South Sea Company	1711
The Swedish East Indian Trading Company	1731

Notes: [a]Established in 1579 and became a joint stock company in 1592.

merchants to engage in trade dealing with certain commodities coming from particular regions.[2] The charter granted the company a monopoly in order to attract merchants into risky ventures. Early chartered joint-stock companies had specialized by shipping specific products such as sugar, grain, coal, or textiles. Some of them also specialized by region such as the Baltic area or the Levant and in handling multiple products. Generally, trading companies combined both regional and product specific features.

The Muscovy Company and the Levant Company were among the earliest general trading companies chartered by the British Crown and evolved into vast, and vastly profitable, enterprises. The Russia or Muscovy Company began as a joint-stock company, "a revolutionary new form of business organization" (Munro 2010: 128). Compared with a simple partnership, which was subject to unlimited liability for all partners and involved running the risk of partners being jointly liable for all debts and losses, the joint-stock company was based on limited liability including that every shareholder was only liable to the amount that one had agreed to pay in buying shares.

Charters had become common phenomena for medieval guilds and civic corporations which turned them into separate bodies with the status of a legal entity. After its third successful trip reaching Archangel, the Muscovy Company was granted a royal charter on February 26, 1555. The charter empowered the company to trade with any part of the world, but its purpose was an attempt to find the northeast passage to the Indies.

The Venetian captain Sebastian Cabot was highly involved in the first voyage to Russia in 1553 and was appointed governor of the Muscovy Company in 1555

(Willan 1953). It seems plausible that its joint-stock character was inspired by similar Italian practices (*commenda*), with the distinction that Italian *commenda* were only undertaken for one maritime venture.

Based on the success of the Muscovy Company, the Levant Company was formed in 1581 after a group of London merchants heard about the arrival of four Dutch ships in Amsterdam heavily loaded with spices. They petitioned for another English charter to begin trading in the Levant, a trade that had been lost in the previous decades. By 1588, the Levant Company was converted to a joint-stock company, and a new charter granted its role of monopolist in January 1592. It was England's first entry into in the lucrative Mediterranean trade and was established ten years after the Battle of Lepanto where European forces had crushed Ottoman naval supremacy (Munro 2010).

Soon, the Dutch and English merchant ships began to appear in the Mediterranean waters to bring grain from Danzig to Mediterranean cities. These ships were faster and more heavily gunned than Italian or Ottoman ships (McNeill 1974). Northern grain imports to southern Mediterranean ports provided substantial capital for the creation of the subsequent trade with Asia which was carried out by the use of a new type of ship, the carrack (Findlay and O'Rourke 2003: 28), which had more cargo capacity than the caravel. The improved sail rigging and also additional sails allowed a much improved sailing characteristic in the heavy winds and waves of the Atlantic Ocean. During the fifteenth century, the carrack had been used by the Portuguese when they gradually extended their explorations and trade ever further south along Africa's Atlantic coast and needed a larger and more advanced ship for their long oceanic adventures.

Some of the partners in the Levant Company took the initiative to create the East India Company (EIC) by using some of the capital that was earned by the Levant Company. It was not, however, a solid organization but only a conglomerate of merchants who bundled separately financed sets of ships to Asia. It was not until 1657 that the EIC was restructured into a company with shareholders providing permanent capital that facilitated long-term investments (Heijer 2005). The EIC soon obtained a monopoly for all of Asia.

Many of the early colonies were founded by these joint-stock companies seeking profits in the New World. The Massachusetts Bay Colony, as well as the Plymouth Colony, was organized as joint-stock companies to send all types of products back to England. Without the joint-stock companies, the history of the United States would probably have taken a different track.

The purpose of the EIC was to compete with the Dutch and establish a direct link with the East Indies spice trade (Munro 2010). Although the Dutch were not alone, nor indeed the first, in creating a joint-stock company, the Dutch United East India Company (Verenigde Oost-Indische Compagnie, VOC) was, unlike the EIC, the first publicly funded company from the outset, and became, as the world's first public company, the greatest success by far.

In order to clarify the reasoning of the English in their competition with the Dutch, an explanation is needed of the manner in which the Dutch Republic became such an economic and military stronghold.

Amsterdam as center of trade and commerce

For many years, the primary source of Dutch wealth and power had been the control of supplies of grain and naval stores from the Baltic Sea. Through the German Hansa, a federation of many Nordic cities, towns, and communities which lasted from 1250 to 1750 was established to protect against unfair taxation and obtain trading privileges in sea trade activities. It was an almost near-monopoly that facilitated and intensified international trade and commerce in northern Europe, thereby fostering regional economic development. Hanseatic merchants sailed with their cogs (small flat-bottomed vessels) and transported raw materials from the rural east of the Hansa to trade for manufactured and sophisticated goods in the hanseatic league of cities in and around the Baltic and North Sea.

Amsterdam successfully challenged this monopoly in the fourteenth century, and the city established a dominant commercial position through its mother trade (*moeder negotie*) with the Baltic Sea, its own course, and the connecting lines with southern Europe. The Dutch mercantile community was geographically well positioned to supply the increasing demand for supplies of grain, forest products, and metal goods as well as salt and herring (Hunt and Murray 1999). By establishing a tight control of the transfer of Baltic supplies through the Sont, the Dutch occupied a strategic market niche.

> After the capture and sack of Antwerp by Spanish troops in 1585, the displacement of Antwerp by Amsterdam as a central hub of world commerce and the transformation of the territories that were in the process of becoming the United Provinces into a secure refuge, jointly contributed to generating a massive migration of traders and artisans from the southern to the northern Netherlands. As a result, the population of Amsterdam grew from 40,000 in 1585 to 105,000 in 1622. The Antwerp textile industry was transplanted almost "en bloc" to Leiden.
>
> (Arrighi 2010: 203)

The Dutch became leaders in the commercial expansion of the entire European economy. There was a popular saying stating that Amsterdam turned into the "Venice of the North," and the city transformed into "the central entrepot of European and world commerce" (Arrighi 2010: 137).

Dutch surplus capital was not only invested in the development of internationally operating trading companies but also in land and in the development of commercial agriculture. Yet the frequently changing composition of partnerships prevented long-term investments in overseas trading companies (Heijer 2005). The Low Countries with their increasing prosperity became the jewels of the Habsburg Empire (North and Thomas 1973) but no longer accepted their subjugated position.

The Dutch revolt (1568–1648)

The Dutch Republic was born out of a Calvinist rebellion against the Spanish Habsburg authority of Philip II, who held title to all of the Netherlands and ruled them through a governor in Brussels, the earliest great power of modern Europe. As the rebellion allied with rising political elites – in particular, urban burgers – "the result was a profound transformation of social and institutional life" (Gorski 1993: 266). The revolt marked not only a religious conflict but also arose from deeply rooted political, economic, and social differences. While the Spanish kingdom was centrally ruled, the Netherlands had been governed by elites of regents in local cities. Each province elected its own *stadhouder* (city holder) who was the highest military official. The national military, political, and economic policies were set by the States General, an assembly which met in The Hague.

Nevertheless, the country was strictly local, i.e., cellular, in its economic and political organization. This can be illustrated by the functioning of a water board that was responsible for the maintenance of the infrastructure of the Low Lands. They operated as autonomous democratic institutions, as a network of interconnected cells rather than as a hierarchical structure and cooperated on the basis of pragmatic consensus (Sorge and Iterson 1995).

After a peace arrangement with France in 1559, Philip II saw his empire confronted with a growing Ottoman offensive in the Mediterranean. The Ottoman empire was seen "as the superpower of the early modern world, stretching for 2.3 million square kilometres across three continents with at least 22 million inhabitants, well over three times the number in the Habsburg monarchy" (Wilson 2009: 76). Driven by a religious zeal of intolerance, Philip II revived his galley fleet to organize a counter-offensive against the advance of Islam. He was convinced that "only by reasserting royal authority and husbanding resources in Spain, would he be able to rescue his empire as a whole, including the Netherlands, from imminent disintegration" (Israel 1995: 138).

To compensate his absence in the Low Countries, he affected the central government of the area to regent Margaret of Parma, an illegitimate Habsburg half-sister, who lacked experience and political skill and needed extensive advice and guidance. Resentment deepened against the Spanish demands to continue with high levels of taxation. Protestantism saturated the entire country, but the Spanish crown insisted on a more active persecution of heresy.

When Margaret introduced a policy of religious moderation in 1566, it was "misinterpreted" by the Calvinists as a proclamation of religious toleration. "The Calvinist movement which had grown rapidly during the previous decade burst into the open" (Gorski 1993: 275). Conflicts between Dutch magnates and the Spanish rulers surrounding the anti-heresy campaign rapidly developed and reflected the wider rift in Dutch society between Spanish power structures and Dutch patronage systems in dealing with political and religious issues. Insurrections and upheavals occurred everywhere in the provinces of Holland and Zeeland. Mobs of Calvinists raged through major cities, desecrating churches and smashing images and statues.

In 1567, Philip was forced to send an army of 10,000 men under the command of Alba to the Netherlands. Alba was a noble of Castile who detested Protestant heresy. William the Silent, Prince of Orange (a region in south eastern France), was a member of the ruling class who had fled to Germany and had become the leader of the rebellion and organized an armed invasion. However, he was no match for Alba as a military commander. In order to help and support Philip's deteriorating financial situation, Alva imposed new taxes such as the Tenth Penny, which became a symbol of unbridled central authority and aggravated the tension between the Netherlands and Spain. The execution of the Dutch Counts Egmont and Hoorne, both having been accused of leading the resistance in Brussels in 1568, provoked an outcry against the regime of Alva and plunged the Netherlands into 80 years of struggle for independence (Kamen 2002).

The Ottoman threat in the Mediterranean sphere was reduced once the Turks were defeated in the sea battle at Lepanto in 1571 and the Sultan began facing struggles with Persia. This enabled the Spaniards to switch their resources to the Netherlands. Spain's determination to conquer the Dutch led to the creation of Europe's largest army. The insurrection of the Protestants, a reformed population which adhered to different streams of Protestant churches[3] was triggered in 1572 by sea beggars who, since 1568, had led Alva to station substantial forces around the Scheldt and Maas estuaries.

The provinces of Holland and Zeeland formed a natural boundary surrounded by the sea, rivers, and low-lying land that could be flooded. The decentralized political structure of the provinces assisted "the rebels since each province had its own states or assembly composed of knights from the rural areas and representatives of the regions, or magistrates, who ran the towns" (Wilson 2009: 130). The rebels, who dominated the states of Holland, elected William as *stadhouder* and provincial captain in charge of the militia. Rebellion increased and once more, cities and provinces such as Zeeland and Utrecht were liberated. The majority of Holland towns had become Orangist, and the revolt against Philip II intensified (Israel 1995).

Philip II, contemplating his deteriorating financial situation and strategic difficulties, tried to end the revolt by negotiations but was without success. The close cooperation between Holland and Zeeland created a joint political, military, financial, and religious framework that led to the creation of the Dutch Republic in January 1579 when the rebels formed the Union of Utrecht. In 1581, this union agreed upon an act of abjuration repudiating Phillip II's authority. The revolt strengthened the political position of the provincial states and the city councils, which then became imbued with a fierce spirit of independence.

After the assassination of William of Orange in 1584, Queen Elisabeth of England was offered sovereignty over the Netherlands.[4] She accepted and nominated Robert Dudley, the Earl of Leicester, Chancellor of the University of Oxford and a known patron of Puritanism, to command the Dutch army. He commanded the army for two years (1585–1587) and appointed Thomas Digges as master-general.[5]

During the 80-year war, Dutch republicanism gradually took shape, especially through the writings of the political philosopher and lawyer Hugo Grotius (see page 111, this volume).

> Republican ideals rested on the myth of the Batavian freedom (Batavia being the Roman name for the Netherlands) and combined a utopian vision of biblical Judaea with that of ancient Athens to claim that liberty, stability, virtue, and prosperity were best guaranteed by a consultative government of wealthy, educated men with time to dedicate themselves to the public good.
> (Wilson 2009: 134)

The influence of Calvinism in the Netherlands also continuously increased. "The hegemony of the reformed Church in Dutch spiritual life was paralleled by (and closely connected to) the predominance of Calvinists within Dutch political institutions" (Gorski 1993: 280).

Around 1590, Calvinist merchants took the first initiatives in the Republic to consider trading activities beyond European waters. Twenty years earlier, individual Dutch navigators had already sailed to Cuba and Brazil. The African coast was also reached by 1590. The expansion of the trading network in Europe, however, reached its limits, and the remarkable work of cartographers such as Gerard Mercator, Abraham Ortelius, and Pertus Plancius provided new geographical knowledge about the globe.[6] The provinces of Holland and Zeeland met to discuss future cooperation in international trade. Between 1595 and the establishment of the VOC in 1602, several companies organized 15 fleets with in total 65 ships for new ventures. Soon the Republic would send out more ships to Asia than to Portugal, where they had traditionally collected Asian commodities to be distributed throughout Europe (Heijer 2005).

The VOC

The VOC was a part-governmental and part-business organization which was in need of long-term investment. This issue was resolved by the establishment of a joint-stock, limited-liability organizational structure. Previously, Genoese merchants had associations of individuals establishing a chartered company to earn profits and undertake war making and state making activities such as colonizing particular areas. The Amsterdam merchant elite was also eager to launch a large-scale joint-stock company to exercise exclusive trading and sovereignty rights dealing with overseas commercial spaces. By enabling many individuals to combine their assets and buy shares of a particular enterprise, the limited liability company, which protected them from losing everything if the project failed, turned into one of the greatest innovations in financial history.

The VOC, established in 1602 and based on a charter valid for a period of 20 years, was the first company to issue its own stocks and bonds through the Amsterdam Stock Exchange. In 1623, VOC ships evicted the English from Amboyna (currently Ambon) and gained almost complete control of trade in the region. The trade in pepper and other spices initially dominated the business, but from the mid seventeenth century onwards, textile imports increased in importance (Findlay and O'Rourke 2003: 18). The British EIC was only able to imitate that status in about 1680 when they became a fierce competitor, especially in cotton textiles which mainly came from India.[7]

These internationally operating trading companies have sometimes been perceived as the first multinational corporations operating from one nation-state. This label, however, is misleading as these new trading companies were established during an era when some of the nations, such as the Dutch Republic, were still in their infancy (Israel 1995). Besides, the VOC only became a multinational when it combined trade with manufacturing and opened a plant in 1641 in Bengal to refine saltpeter and engaged itself in other manufacturing activities outside of the Netherlands. By 1717, it employed 4,000 silk spinners in Kaimbazar (Bengal).

The Low Countries, however, were still occupied by the Spanish ruler and, therefore, one of the main purposes of the charter of the VOC, next to trade with Asia, was to attack the power, prestige, and revenues of Spanish and Portuguese vessels. Soon after its incorporation, the VOC succeeded in forcing back the Portuguese and Spanish in international trade. The first expedition set sail on December 18, 1603, supported by 12 armed ships. The trip from the Dutch island of Texel, which provided direct access to the North Sea, to Batavia in the Indies, took about eight months. The navigators were required to maintain a harsh discipline on board.

The Dutch had gained essential insights into the waters and routes of the Indies from Jan Huyghen van Linschoten, who had worked for Spanish and Portuguese merchants. In 1583, he had been appointed as secretary to the archbishop of the Portuguese colony of Goa where he had gained access to secret information from the nautical maps, which were well guarded. He meticulously copied them and, by 1592, took them to his home at Enkhuizen (Netherlands) where he wrote his *Travel Accounts of Portuguese Navigation in the Orient,* which was published in 1595. This work contained a large number of sailing directions, not only for shipping between Portugal and the East Indies colonies, but also between India, China, and Japan.[8] However, without solving the problem of determining longitude, the maps were of little importance. Therefore seafaring nations kept a strong interest in the improvements of the mechanical clock.

The Dutch States-General had once asked astronomer Galileo, working as professor of mathematics at the university of Padua (1592–1610), to come to the Republic to solve the problem of longitude. Though the Dutch had offered him a sum of 15,000 guilders, Galileo declined in order to continue his research at the court of the Medici in Florence where his telescopic discoveries were drawing some positive attention. During the summer of 1609, he had demonstrated the use of the telescope to the Venetian Senate. They were standing on top of the clock tower at the San Marco Square pointing at the fact that, with this instrument, ships of the enemy could be spotted much earlier than ever before.

With his telescope, which was originally intended to be handheld for maritime use, Galileo was able to compose tables of sightings, but the need for further refinements in the tables forced refinements in the telescope itself. This helped create an initial technological advancement in astronomy that provided new kinds of information based on discoveries and inventions.

Morosely, Italy was so much ingrained in Aristotelian science that Galilean discoveries were rejected and mathematical physics advanced elsewhere with people favoring more scientific tolerance. "By the 1620s and 1630s, intellectual leadership of Europe departed Italian soil" (McNeill 1974: 195). Northern Europe, especially the Dutch Republic, facilitated an institutional context where an interest in the application of clock time grew rapidly not only for seafaring interests, but also in two other particular areas where the mastering of time gained a prominent role: one had to do with the timely coordination of the transportation along canals, the other with improvements in the timely coordination of the army.

Temporal coordination of barges

During the Middle Ages, the construction of waterways for drainage had become necessary in the Low Countries when "the line of dunes that had protected the peaty marshes and the lagoons of the coastal region" (Vance 1986: 74) were breached and water regulation over the whole polder region had become essential for survival. The Low Countries had a ubiquity of waterways unique in Europe. To be effectively exploited for agriculture and pasturage, it was necessary to keep the land dry. The use of drainage canals was both widespread and obviously interconnected, but the canals were not intensively exploited for transportation facilities.

In Flanders, some waterways were created to link developing cities. In the twelfth century, Bruges, Ypres, and Saint Omer were connected with coastal ports. Rich businessmen became involved in the practice of land reclamation and waterway construction (Hunt and Murray 1999).[9] In Holland, the first cuts for navigation purposes were made by 1500. Beginning in the 1580s, Amsterdam had settled agreements with Utrecht, Dordrecht, Hoorn, Leiden, Haarlem, and other cities to maintain a network of intercity common carrier transportation systems. Skippers organized in guilds functioning as a sort of business enterprise were responsible for the execution of the transport of cargo, parcels, mail, and, later on, passengers as well (Dehé 2005). Three separate components of this network can be discerned.

First, there had been a form of routes which provided market boats opportunities to connect villages with market towns. Market boats navigated on the drainage or irrigation ditches. Second, a network of *beurtveren* (a strictly organized transport facility between two places) provided routes for regulated transport by boats carrying cargo. The *beurtveren* suited the requirements of the city markets in a systematic way. This network maintained intercity connections between cities and traders based on a complete arrangement of direct services. "The cities licensed skippers who were committed to depart on a regular schedule from a set place to a set destination and to provide their services at tariffs established by the cities" (Vries and Woude 1997: 180).

Between 1580 and the mid-seventeenth century, *beurtveer* agreements proliferated (Dehé 2005) and turned into "one of the most radical institutional innovations in the world of transport" (Lesger 2006: 97). The service between Leiden and Utrecht, for example, was maintained daily by two sailboats in each direction. One boat departed with the opening of the city gates each morning and was supposed to arrive at Utrecht the same evening. The second boat departed at noon and was to be in Utrecht when the city gates opened the following morning. This indicated a scheduled time of 14 to 18 hours for an average speed of around four kilometres per hour. Inherent in the use of *beurtveren* with sailing vessels was uncertainty about the length of the journey. Yet the Republic became blanketed by an infrastructure of dense, bilateral transportation routes.

The third form, the *trekvaart* network, was a specific form of transportation no longer relying on sails but instead on barges (*trekschuiten* or drawn boats) pulled by horses along a towpath next to the canal.[10] The new infrastructure provided year-round regularity and became an interesting manifestation of a growing interdependence between cities based on a practice of temporal coordination. Isolated communities suddenly became incorporated into an integrated transportation system. The *trekvaart* system led to the adoption of a single uniform standard of time and, as such, was the first transportation industry to standardize time across distance. The desire for a speedy, reliable means of year-round travel increased. To shorten the distance between Amsterdam and Utrecht, a *trekvaart* development began in 1626–1628 with the construction of towpaths along the river Vecht and the cutting of new canals.

In 1631, the first canal constructed especially for transport of persons between Amsterdam and Haarlem was finished (Dehé 2005).[11] This service provided "hourly departures and a fixed trip duration of two hours and fifteen minutes. With the striking of the church bells, the barges departed, and the ordinances required the skipper to turn an hourglass designed to run out when the barge was due at its destination" (Vries and Woude 1997: 183). In 1636, Leiden and Delft signed an agreement to convert a canal connecting their cities to a *trekvaart* to provide for regular and year-round mobility. Pamphlets – public information provision before newspapers – printed in Amsterdam took just half a day by drawn boats to be distributed in Leiden. Around the same time, decisions were taken to dig *trekvaarten* between Amsterdam and Weesp, as well as between Amsterdam and Naarden. There was a general belief that the new network would stimulate commerce and facilitate municipalities to promote their commercial interests. The VOC, in particular, was interested in a reliable and quick transmission of information between their chambers located in several different cities. In 1640, 50 chief merchants of Leiden made a formal request to the city council urging the construction of a canal to Haarlem. Their petition stressed that the time needed to travel should be reduced (Vries 1982: 21).

To advance the efficiency of a *trekvaart*, the canals were drawn generally as straight as possible across the landscape to connect cities with the shortest distance possible. "Because of the straightness and their indifference to intermediate villages, these trekvaarten stand out on the seventeenth and eighteenth century maps in the same way that limited access highways stand out on today's maps" (Vries 1978: 83). They became the highways of the age and increased the advancement of the Dutch Golden Age (Dehé 2005). Next to speed, military factors also played a role in the construction of this infrastructure of canals: cities often functioned not only as gateways to the eastern provinces but also as fortresses guarding against Spanish invasion threats.[12]

The water infrastructure was an organizational innovation, and the new canals increased the speed of transportation. The ordinances under which the boats operated stipulated the exact time of their departure and the maximum duration of the trips. Bells rang to announce departure time. To facilitate the maintenance of the schedules and routes, many barges were equipped with hourglasses to respect these requirements. "The skippers faced heavy penalties for failure to depart or arrive on schedule" (Vries 1982: 24). The social implications of this innovation were remarkable; it offered mobility to a large majority of the population. Between Amsterdam and Haarlem, the barges departed hourly from the time the city gates opened in the morning until 8 p.m. Around 1640, a unique high-volume passenger service was put in place using a dedicated right-of-way.

The years from 1659 to 1664 were prominent as the most intense period of *trekvaart* construction. Braudel (1975) confirms that the Dutch Republic became a world whose dimensions could best be expressed in time rather than distance.[13] Distance between cities was reckoned in hours. The speed of the barges did not vary appreciably with the seasons, and transportation could take place on the same basis year-round. After 1667, gaps between the infrastructure of canals and cities not connected to the system were solved by the construction of paved roads. This led to a national infrastructure with the Zuider Zee acting as a unifier.

The drawn-barge era endowed the Low Countries with a unique capacity for human interaction which profoundly affected economic and social life. Its pulled barge network was a geographical phenomenon, but it also constituted an economic innovation created as much on commercial as on religious interests in terms of not wasting one's time. An extensive waterway network was in place and served the population with fast, direct service. "Probably the directness was more important than the speed which was little advanced over the speed of sailing boats" (Vance 1986: 76).

Some travelers from England, however, complained at times about the system of the barges. They referred to the slow pace of the barge as an expression of the phlegmatic character of the Dutch. Even some Dutch themselves became critical about the effectiveness of the barge in terms of speed. And yet the waterway transportation system with passenger barges and strict time schedules remained unrivaled for a long time due to the cheap fares and its performance. Its success can be compared to that of the railroads that appeared nearly two centuries later. The barge system developed into an extensive system and operated efficiently until 1870, when trains took over (Filarski 1997).

Maurits' military revolution

Another development of timely coordinating the interaction between members of the Dutch society took place in the army and, subsequently, spilled over into the navy and the rest of Dutch society. The rebellion of the Dutch against the Spanish ruler shaped the core of modern militant armed resistance. The revolt that was sparked by ascetic religious movements such as Calvinism promoted discipline at an individual level (self-discipline) but also at an institutional level by developing a variety of policies to maintain collective discipline.

A famous propagator of this attitude was Justus Lipsius, who taught at the University of Leiden. Like many other sixteenth-century thinkers, he was fascinated by the belief that strict patterns could be detected in the natural world which would unlock the Divine secrets of the universe. Lipsius was horrified by sectarian violence caused by irrational passions which blinded humans to share common interests. He had studied classical literature and found a remedy in Neo-stoicism. Emotions should be suppressed by iron self-discipline. His philosophy turned into a plea for a firm, yet responsible, government where rulers had a duty to protect their subjects from harming themselves. Lipsius articulated a detailed concept of discipline for soldiers defending public interest:

- Military commanders had to maintain a strict hierarchical command structure in the army.
- Soldiers have to be drilled in order to accept subordination and respect order.
- Regular drill should be imposed to break the autonomous mercenary culture amongst the soldiers and encourage self-discipline.
- A system of rewards and punishments is required to institutionalize a strict military culture.

Lipsius' ideas were meant "to render war less violent and destructive," but they were also in line with the social disciplining agenda of church and state (Wilson 2009: 140). His concept of discipline was not unlike the monastic practice meant to strengthen devotion through exercise, order, force, and providing the right example of what could be expected of good soldiers (Ortmann 1995).

Maurits, the youngest son of William of Orange, attended Lipsius' lectures in Leiden and fell under the spell of his views. After the murder of William, Maurits became the leader of the Dutch Republic and introduced disciplinary rules in the army.[14] Maurits and his cousin Willem Lodewijk, *stadhouder* of the province Friesland, were striving for more and looked for a transition from obedience to discipline.[15] Maurits introduced ways to quickly and flexibly maneuver smaller army units to assert the Dutch Republic's independence against the military might of Catholic Spain. To increase combat skills, he began training young officers – whose span of control was decreased to enable them to improve their co-coordinating skills – by compelling them to master standardized commandos which had to be learned in Dutch, German, French, English, and Scottish.

Soldiers' training consisted of drilling to act efficiently during combat line-up. Therefore, he introduced the contra-march, which enabled soldiers to fire continuously.[16] For the fine tuning within the ranks, it was necessary for the soldiers to not only watch their own footsteps but also those of their colleagues. Drill was meant to instill among the soldiers the right attitude of acting. This practice was intended to provide continuous fire during an advance or retreat.

> Each rank fired in turn; those who had just discharged their weapons remained stationary to reload while the next line stepped through the gaps between each man to take its turn. By the time the last line had fired, those who had shot first would have reloaded and could move forward.[17] Arquebuses and lighter muskets took around a minute to load, requiring fewer ranks to maintain continuous fire than heavier muskets that needed up to three minutes to reload.[18]
>
> (Wilson 2009: 87)

Quick maneuvering power became foremost in military power and was used to outperform the opponent (Rosa 2005: 102).[19] For the execution of the choreographic movements, the soldiers were still dependent on their own alertness and mutual atonement: there was no clock yet to synchronize their movements.[20] Although the rhythm of one's own body movement was now disciplined, it was not yet synchronized. That was only achieved after the implementation of the mechanical clock to control the physical work of workers in the factories. This implementation of discipline and the concurrent use of the clock came about clearly as will be shown later – in the British textile mills where the aim was to economize time as much as possible. There Maurits' choreographic approach eventually fused with the synchronizing of labor in the factory system.

Along with these choreographically designed marching lessons, soldiers were taught to handle weapons efficiently. In order to visualize this, the Dutch illustrator Jacob de Gheyn designed *Wapenhandelinghe* (*Exercise of Arms*), a weaponry handbook which was published in 1607.[21] This was a manual of instructions for the handling of arms, analyzing the movements required, e.g., to load and fire a gun (Figure 4.1) and was probably the first motion study of its kind. Maurits also stimulated the standardization of guns and munitions. Complete standardization of armament and equipment, however, proved to be very difficult as the workshops where the craftsmen manufactured the weapons were not sufficiently equipped for it.[22]

Maurits would not have been able to achieve these innovations without his learned advisor Simon Stevin, the Netherlands' most famous engineer at the time. Warfare changed from a practice based on experience to a practice based on mathematical precision. Thanks to Stevin's mathematical insights, military machinery turned into a scientific organization. Not only did this concern the make-up of the encampments or the seizing of strongholds as Maurits' English military assistant Digges had proposed but also the military exercises. Stevin designed tools such as shovels with which the soldiers were able to do their digging as efficiently as possible. "Siege techniques were transformed to increase the efficiency of military labor power, to cut costs in terms of casualties, and to facilitate the maintenance of discipline in the army's ranks" (Arrighi 2010: 46).

FIGURE 4.1 Illustration by Jacob de Gheyn from *Exercise of Arms*, 1607.

Military historians have stressed how the Dutch became the progenitors of new ways of fighting based on gunpowder weaponry and military discipline. Maurits built an army which allowed a more or less intelligent response to unforeseen circumstances.

> Every movement attained a new level of exactitude and speed. The individual movements of soldiers when firing and marching, as well as the movements of battalions across the battlefield, could be controlled and predicted as never before. A well-drilled unit, by making every motion count, could increase the amount of lead projected against the enemy per minute of battle. The dexterity and resolution of individual infantrymen scarcely mattered anymore. Prowess and personal courage all but disappeared beneath an

> armour-plated routine. Yet troops drilled in the Mauritian fashion automatically exhibited superior effectiveness in battle.
>
> (McNeill 1984: 130)

To attain such a well-drilled army, Count Johann VII of Nassau-Siegen, nephew of William of Orange, established a military academy between 1616 and 1623 to train army commanders. He combined Dutch military ideas with German military practices to make it more applicable to the local circumstances (Wilson 2009). These technical, social, and organizational innovations resulted in a disciplined and regularly paid army, which boasted huge firepower. Due to these exercises, Maurits had created a maneuverable and quickly movable army that was logistically well supported. As with the Portuguese navigators, a trinity of new documents with instructions, new devices in the form of standardized arms, and drilled men provided an excellent base for Maurits to create a new form of warfare and a properly drilled habitus amongst the soldiers.[23] Together with Stevin, Maurits achieved for the Dutch army and navy what scientific management would achieve for American industry two centuries later (McNeill 1984).

The military revolution that carried through in the United Provinces was characterized by innovations in artillery, tactics, fortification, siege techniques, military transportation, and by a vast improvement in the discipline and orderliness of the military. Maurits gained fame with these military innovations. Order and discipline became trademarks which were not only introduced in the army but also in the navy (Black 1995).[24] From the 1580s up to the middle of the seventeenth century, the Dutch Republic became the principal school of warfare for all of Protestant and Catholic Europe. Whereas Philip II lacked a military vision, had no clear strategy, and was afraid to make mistakes (zero defects mentality) (Parker 1998), Prince Maurits created the first, well-trained, modern Dutch army during a period when the Dutch revolt raged on.

The Dutch introduced a more flexible military system which was improved by Swedish and German armies and was finally perfected by the French army in the subsequent century. Numerous volunteers flocked to learn about the military art under Maurits. Many German Protestant nobles also joined Dutch service along with the French philosopher Descartes and Gaspard de Coligny, who commanded two Huguenot regiments in Dutch service and brought Maurits' reforms to France. Famous army leaders such as the Swedish Gustaf Adolf received direct or indirect training in the Dutch manner (Roberts 1967).

Signing up for the Dutch State Army was recommended for the proper schooling of Protestant leaders. A case in point was Frederick William of Brandenburg who was inclined to improve the effectiveness of his army to protect Brandenburg (later to become Prussia) against Catholic invasions. He formed an alliance in 1605 with the Dutch Republic for military security.[25] These novelties were of great importance for the organization of large groups of people, spilled over in the organization of large enterprises, and constituted part of the foundations of management thinking.

Frederick William, who became the great elector during the period 1640–88, was sent to the Dutch Republic for a period of four years. His Dutch education, military training, and understanding of the necessity to subordinate the military to the authority and discipline of the state reinforced his sense of allegiance to the Calvinist course.[26] Frederick William, the soldier King, strove to remodel his own patrimony in the image of what he had observed in the Netherlands. "The training regime adopted by his army in 1654 was based on the drill book of Prince Maurits" (Clark 2006: 41). He introduced the parade drill to improve the maneuverability of his troops. Artillery calibers were standardized, and he founded a Cadet School for officer recruits.

> He became an institution-builder by merging financial, economic, and military administration and shaping an architectonic vision ofAdministration based on a hatred of corruption, wastage and inefficiency. The ministers were expected to convene at seven in the morning in summer and at eight in wintertime. They were expected to remain at their desks until the day's work was accounted for.[27]
>
> (Clark 2006: 89)

Frederick William's successor followed in his father's footsteps by further improving the military system and schooling. By 1740, Prussia's army had doubled in size to 80,000 men.[28] With the peace treaty of Munster in 1648, there came an end to the 80-year war, yet rivalry among several European nation-states continued. The system of sovereign states required new rules for interactions as much militarily and politically as economically (Jones 1981/2003).[29] The application of timetables for water transportation, as well as the clockwork-like marching of soldiers in battle illustrates how, slowly but surely, mechanical thinking became ingrained in the European mentality. Some of this thinking became applied in international trade as well.

Organization of the VOC

In the seventeenth century, the Dutch Republic became the European headquarters of globalizing trade and created a distinctive military culture with recognizable modern features such as the regular and systematic drilling of troops in battleground maneuvers and a disciplined, professional officers' corps. With these means at hand, the Dutch, as well as other European nation-states, began to further territorialize and colonize global space.

The establishment of chartered monopolies promoted mercantilist governmental policies which were geared to the active promotion of positive trade balances. The Dutch merchant class provided the House of Orange with liquidity, business knowledge, and connections that strengthened the establishment of a governmental organization, the United Provinces, which fused capitalist and territorial interests. The merchants of the Republic became the transporters of

Europe, symbolized by the foundation of the VOC in 1602, which itself had been an initiative of Maurits. The ships' inventory lists showed that mechanical clocks were one of the curiosities the merchants took with them to the East where some of the leaders began to appreciate timekeepers as a symbol of political power.

Chartered trading companies became the main actor in the extension of a terrestrial globalization by attempting to reduce transaction costs and improve the security and speed of maritime transportation. One such company, the VOC, was the result of an amalgamation of six small companies. The unification was enforced by the government to become a powerful military and economic weapon in the struggle against Spain and Portugal. Maurits had personally intervened to convince traders and merchants from the province of Zeeland to join. The VOC specialized territorially to the exclusion of other similar organizations. The merchant class directed its capital utilization not only in commercial agriculture but also in war-making and state-building activities. The VOC recruited salaried managers holding positions of responsibility. The company was allowed to build forts in Asia, employ soldiers, conclude treaties with Asian rulers, and appoint judges. Armaments were especially meant to inflict damage on the Portuguese and Spanish. The VOC produced its own protection costs that were "lower and more calculable than the costs charged to caravans and ships by local powers in the form of tribute, fees and outright extortions" (Arrighi 2010: 144).

The main advantage of the new form of enterprise was its permanent structure, which made it possible to dispose continuously of a large stock of fixed assets (Klein 1981). Its organization was composed of six chambers each providing a set of directors, who were supposed to own shares in the company: Amsterdam (20), Zeeland (12), Delft (7), Rotterdam (7), Hoorn (7), and Enkhuizen (7) (Figure 4.2). The 60 directors carried out the decisions taken by the board of the company. Each chamber had committees for equipage, thesauruses, and personnel. The directors were usually appointed for one season or one year to a committee for shipbuilding, victualling, ammunition, bookkeeping, or the sales of goods.[30] Several committees assisting the board were:

- the committee for drawing the annual balance sheet;
- the committee for supervising the auctions;
- the committee for inspection of the books of the chambers;
- the committee for reading the correspondence (it met in The Hague and was called "Haagse Besogne");
- the committee for issuing secret routes in times of war.

Along with the 60 directors, there were extraordinary directors representing other cities and provinces that were influential due to their support for the rebellion against the Spanish ruler. Through the States General, they had committed themselves to grant the VOC substantial subsidies and were required to monitor the manner in which the subsidies were used.

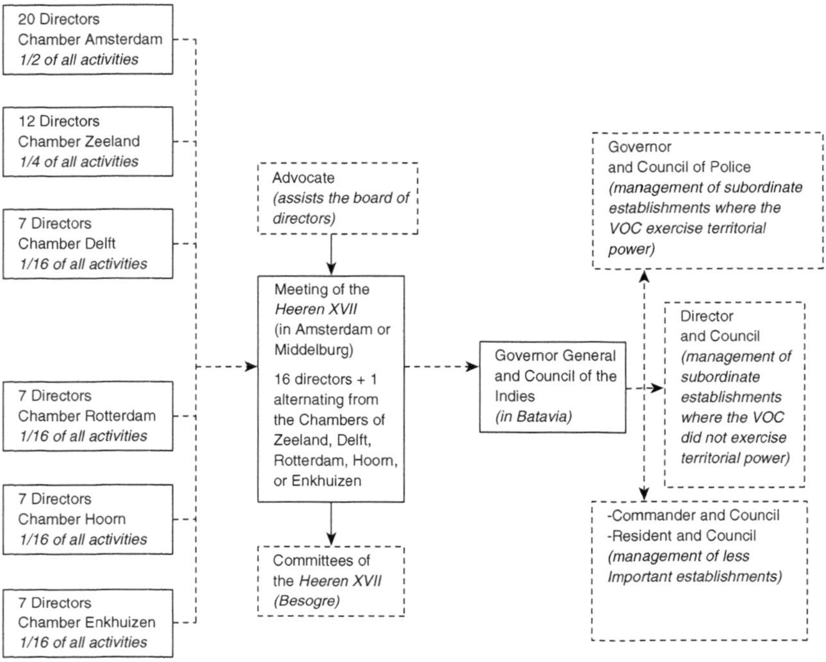

FIGURE 4.2 Organization of the VOC. (Reinterpretation of Gaastra 2003: 150.)

The board of the VOC consisted of the Heeren (Gentlemen) XVII and was composed of eight members from Amsterdam; four from Zeeland; the cities of Rotterdam, Enkhuizen, Delft, and Hoorn each had one seat only; the last seat was appointed in turn to one of the chambers other than Amsterdam. Membership was for life. These Heeren XVII met two to three times per year for two to three weeks. The composition of the board, however, was constantly changing, and the Heeren XVII lacked a supportive staff. Consequently, the administrative structure became extremely complicated and was struggling between administrative and market coordination.

The administrative practices of the Dutch Republic, with which the Heeren as regents were familiar, gradually seeped into the management of the VOC. The Amsterdam chamber wielded great influence in the central administration. Company advocates played an influential role as secretaries to the directors in gathering and distributing relevant information. They were the only permanent high-ranking officials who possessed a systematic influence within the central administration. The organizational structure was overcomplex, and it struggled with too many directors and too few merchants/traders. Eventually, the growth of its own company capital declined, and too much money went to stockholders.

In 1609, it was decided to set up one strong and powerful administrative center in Batavia on the island of Java (the East Indies). It was meant as a separate division under the nominal authority of a Director General. The organizational structure of chambers in the Netherlands and the division in Batavia reflected a cellular configuration of a central administrative office. When the city of Jakarta had been burned down, Batavia was founded in 1619 and its port became the transshipment of the VOC in Asia, the hub of flourishing intra-Asian trade (Gipouloux 2011), and the most important harbor of arrival and departure for ships to and from Europe. The city also provided the requisite infrastructure for the maintenance and repair of ships. A Director General, responsible for trade and shipping throughout the whole of the Asian branch of the enterprise, and a council were appointed to coordinate increasing inter-Asian trade activities. The Director General supervised the warehouses, both for merchandise and for provisions, and the pay office, and collected money.

Director General J. P. Coen turned Batavia into a stronghold which outstripped Goa and Malacca and attempted to take over the trading concerns of the Iberians. The inter-Asian trading network relied not only on European staff but also on the Asian workforce. In order to keep the geographical knowledge about intra-Asian trade to themselves, the company obtained a monopoly in 1619 on issuing maps dealing with the Indies. Gathering of knowledge about the local waters and ports in the Indies became a privileged activity of the VOC.

Although Batavia was a Dutch stronghold where almost 20,000 people lived by 1689, half of the population was slaves. The annual transport of slaves to Batavia is estimated to have been between 1,000 and 4,000 per annum. Although slave trading was not a primary activity of the VOC, its ships were incidentally sent from the Cape of Good Hope to Madagascar to collect slaves there.[31] "During the first 30 years of Batavia's existence, Indian and Arkanese slaves provided the main labor force of the company's Asian headquarters" (Vink 2003: 140). Slaves were sent to work in the company-operated goldmines along the west coast of Sumatra or toiled in Batavia. In 1689, the slave population there consisted of approximately 23,000 adults and 3,500 children (ibid.: 154).

The inner city of Batavia was dominated by Chinese traders and frequently referred to as the "Chinese colonial city." It became a trading place for spices such as cloves, nutmeg, mace, and cinnamon. Dutch ships were able to secure a monopoly for the cloves of the Spice Islands as well as cinnamon from Ceylon. Although the Portuguese had been the pioneers, and Spain secured control over territories in the new World, the Dutch developed commercial, organizational, and military capabilities which, until the 1740s, made it possible to fully exploit the new opportunities.

Yet compared with the much more centrally structured EIC, the overall management of the VOC remained a complicated construction. Whereas the EIC set up trading settlements in Batam (Java), Bombay, Madras, and Calcutta, its headquarters in London maintained control of the business.[32] Its Court of Directors was composed of 24 members. The secretary in London had its counterpart at each of the principal Asian settlements. The company developed a standardized method to deal with all financial transactions. "Because of the complicated time schedule for shipment and the difficulty of forecasting the quantities which were to be contracted for, there was ample room for a considerable discrepancy between planning and performance" (Chaudhuri 1981: 42). In contrast, the VOC kept two accounts, one in Amsterdam and one in Batavia. From Batavia, the VOC traded whatever was profitable, even British products for Asian ones. Yet the company was run as a business and not as an empire (Steengaard 1981).

The size of the VOC was quite impressive. Between 1602 and 1795, about 4,700 VOC ships set sail to Asia with 973,000 Europeans on board. The largest portion of passengers was composed of sailors (60 percent), while 30 percent were military people. Between 1613 and 1617 alone, the VOC sent 51 ships to Asia, while during the same time, the EIC only sent out 29. Whereas ships in the European trade could make up to four trips a year, the voyage to and from the Indies took two years (Wilson, 2009). Around 1615, the VOC employed about 3,500 sailors and 3,000 soldiers, the composition of which was very international, much more than within the EIC.[33] The VOC soldiers were operating under the banner of trust and honor (*Fidelitas et Honor*), which was carved in their tokens and served as a corporate logo.

Onboard the vessels of the VOC, a harsh discipline was maintained. Maurits' military rules were strictly respected: "Compared to the EIC, where more adventuresses reigned, the discipline at the VOC vessel was harsher" (Nagel 2007: 72). Dutch captains maintained a minimal social distance between crewmen and their officers, yet they developed an effective cooperation of all on board (McNeill 1974). As sailors earned wages for their labor, they have been seen as collective laborers, and their work on the ship required careful coordination and synchronization. In that sense, their work was a precursor of the factory system (Christopher 2006). Everyone was required to be on deck working all day from eight o'clock in the morning until six in the afternoon. Sailors were organized in night watches in a rhythm of four hours on and four hours off. The boatsman marked the changing of the hours by sounding the ship's bell.

The VOC self-financed the costly military operations necessary to win the world monopoly on fine spices. "The protection costs were brought within the range of rational calculation" (Steengaard 1981: 259). The monopoly on the trade in cinnamon, for example, was obtained by ousting the Portuguese from Ceylon. Trade and shipping between Europe and Asia increased tremendously. Textiles from India,[34] coffee from Arabia and Java, and tea from China captured the European markets. Between 1610 and 1730, the upper strata of the Dutch merchant class became the leaders and governors of a large European capitalist engine.

The creation of supply-and-demand conditions favorable in the East Indies involved a wide range of military actions and territorial conquests. The Dutch established monopolies not only for cinnamon but also for cloves, mace, and nutmeg by eliminating actual or potential competing centers of commodity exchange. Pepper, however, being harvested at so many different places, remained a product no one could monopolize. They secured a record of brutality in enslaving the indigenous peoples, depriving them of the means of livelihood, and using violence to break their resistance to the advantage of the company (Braudel 1984). The Dutch patriotic interests against the Spanish and Portuguese competitors were strongly supported by a specific interpretation of the open seas.

After the VOC confiscated the Portuguese ship Santa Catharina in 1604 in the Strait of Malacca, the young Dutch lawyer Grotius, or De Groot (1583–1645) had been asked by the VOC's order to produce a justification for seizure of hostile ships. On the basis of natural law, he produced a text that gave every human being the right to protect himself. This in itself justified the principal right of booty (*De Iure Praedae*). De Groot introduced the right to freely sail the oceans (*ius navigandi*), and his *De Mare Liberum* (1609), published anonymously, eloquently attacked the Iberian colonial monopolies. Advancing the view that, although other nations had their own territorial waters, they still had to respect the right to free trade on the high seas, which he defined as a free space (*res communis*) owned by no one. It became the document that broke open the route to the Indies and a global assault on the commercial and colonial practice of the King of Spain and Portugal (Blackburn 1997).

> The Dutch could not have imposed their trade supremacy in Asia, West Africa, and, more sporadically, in the Caribbean and Brazil had the States General not set up an armed politico-commercial organization of unprecedented scope and resources, not just with regard to the scale of their business operations but also in respect of the military and naval power.
>
> (Israel 1989: 411)

However, the company avoided military involvement and territorial acquisitions which did not directly maximize profits. "It cut down on involvements that did not yield satisfactory financial returns. The decisive weapon wielded by the Dutch in the struggle for wealth and power was exclusive control over regionally strategic supplies" (Arrighi 2010: 155).

The Dutch merchant class concentrated on the Indian Ocean rather than the Atlantic and, for commercial reasons, focused on the vitality of Asian markets. Despite its size and successes,[35] there was considerable discontent among the shareholders due to a lack of transparency in the management of the organization. The inventors asked for an accounting practice that would allow the shareholders to assess the stewardship of the Heeren XVII. The VOC applied a version of venture accounting as developed by the medieval Hanseatic merchants.

> The very success and self-sufficiency of the VOC increased the power of the managerial bureaucracy that was responsible for its day-to-day operations, and its greater power came to be exercised at the expense not so much of the board of directors of the company (de Heeren XVII) as of the VOC's shareholders.
>
> (Arrighi 2010: 157)

They felt that they themselves had very little influence on the management as such.

To counter this discontent, the shareholders were granted three committees in 1623 (the inspectors of accounts, posting of bills, and nine principal shareholders were delegated to meetings of the board) to gain some influence and control. Despite this criticism, there was an overall sense of commitment to the company. Compared with the Venetian oligarchy, the Dutch capitalist oligarchy demonstrated a joint involvement which was based on the communality of a long struggle of emancipation from Spanish imperial rule. In order to succeed:

> This oligarchy had to forge an alliance and share power with dynastic interests (the House of Orange) and had to ride the tiger of popular rebellion (Calvinism). As a consequence, the power of the capitalist oligarchy within the Dutch state was far less absolute than it had been within the Venetian state. For this very reason, the Dutch ruling group developed much greater capabilities than the Venetian rulers ever had to pose and solve the problems around which the European power struggle raged.
>
> (Arrighi 2010: 47)

Yet the business results in the period 1650–1780 turned out to be negative and a dividend was only paid 25 times (Zanden 1996). From 1710 onwards, the attitude of the initially dynamic merchants changed more into a rent-seeking behavior. The board became more interested in a wealthy lifestyle rather than active merchants.

The WIC

After the Twelve Years Truce which ended in 1621, military confrontations between Spain and the Republic resurfaced. In order to further crush the Iberian naval power overseas, The Dutch West India Company (WIC) was established in 1621 to attack the power, prestige, and revenues of Spain and Portugal.[36] The corporation was as much a Calvinist as an anti-Spanish institution and obtained a charter for 21 years operating in the Atlantic area and a large part of the Pacific Ocean (Emmer 1981). It was given a monopoly on trade in order to plunder the Spanish in the New World and initially turned its attention to sugar and tobacco. Dutch merchants were keenly aware of the value of the sugar trade. The sugar refineries in Amsterdam grew from three in 1598 to 29 in 1622 (Blackburn 1997: 189).

The WIC was to be both an instrument of state policy and a commercial organization. Like the VOC, the company was divided into a set of chambers (in this case five), each being responsible for a portion of the necessary capital. The Amsterdam chamber was the dominant one. It was commandeered by a board composed of the Heeren XIX with a chairman appointed by the States General. A portion of the capital came from public funds, and the government supplied soldiers and war material.[37]

The success of the WIC in terms of war and trade was also remarkable. On behalf of the WIC, Admiral Piet Heyn sailed to the Caribbean with a fleet of 32 ships and captured a large Spanish silver fleet of 15 vessels in 1628. Heyn's exploit caused anger in Madrid, but the fortunate shareholders of the WIC received a dividend of over 75 percent that year. Despite this and similar successes, the WIC faced possible bankruptcy. It was reorganized as a mere slave-trading enterprise with profitable side-lines and contraband trade with Spanish America and sugar production in Surinam.

After the conquest of several different regions such as the northern parts of Brazil, the company turned to intensified slave trading which further spurred naval efficiency and international regulation. Their sugar plantations required suitable labor, which was captured and transported from African estuaries. Slaves were taken from the Dutch castle Elmina on the West Coast of Africa. In 1634, the company occupied the island of Curacao, which became a station for the transport of black slaves. Under the governorship of Johan Maurits van Nassau-Siegen, the Dutch occupied northern Brazil between 1637 and 1644. Meanwhile, the WIC stayed heavily involved in the transatlantic transport of slaves and shipped approximately 80,000 West Africans between Africa and America.

This profitable activity was imitated by other Dutch private companies who traded slaves from Portuguese and French ships at Cape Town. In the period from 1625 to 1803, the Dutch transported more than 500,000 West Africans to the colonies, their share of overall slave trade across the Atlantic reaching some five percent (Priester 1987). Many other European nation-states, with the English taking the lead, also became heavily involved in this type of business which had carried a long tradition.

The slave trade

During the fifteenth century, Portuguese traders along the coast of West Africa had become aware of the profits they could gain by purchasing Africans and shipping them either to Lisbon for domestic services in Portuguese and Spanish households or to the newly settled Atlantic islands where sugar was being planted. With the extension of new occupied territories and more plantations realized, the lucrative slave trade spread across the Atlantic. When European settlers became eager to exploit the mineral and agricultural resources of their newly occupied lands, they required more slaves to replace the indigenous population, unsuited to performing the required labor. The outcome was the introduction of a large network of commerce in human chattels.

Yet the phenomenon was in itself not new. Slavery was, and is, the most uncivilized form of the human condition. With the replacement of egalitarian forms of human social organization by hierarchical ones, and with the drive for territorial expansion by warrior-led chieftains, belligerent conflicts between populations since ancient times had increased. With prisoners and "foreigners" as captives, slavery was not yet an institution of the early belligerent societies, but they remained societies with slaves. Throughout the Mediterranean Sea, ancient societies with slaves changed into slave societies when three conditions according to the historian Finley (1983) became respected:

- private ownership (property) of land requiring a permanent workforce;
- development of commodity production and free markets;
- scarcity of an alternative internal supply of labor.

The Mediterranean societies of Greece and Rome became the first slave societies where slavery as an institution became documented in some detail and fixed into written law.[38] Slaves were seen as unfree and had no civil or social status. In the Roman Empire, slaves worked as domestic servants, in mines, in public works, in brothels, on farms, and in cottage industries which made "the slave mode of production the decisive invention of the Greco-Roman world" (Anderson 1974: 21; Rotman 2009). The slave practice created an interest in the management of slaves. Columella was one of the first Roman authors who provided guidelines for their use. He enumerated all kinds of tasks slaves could perform full time even when there were no fields to be tilled or crops to be gathered. His *De Re Rustica* became a management textbook which propagated the stick and the carrot (Aubert 1994). He recommended teaching the slaves servile efficiency in terms of loyalty and obedience ("fides et obsequium") as the key to economic productivity.

During the medieval period, however, slavery in most of Europe declined and during feudalism transformed into serfdom. Yet with the rise of the Islam, slavery increased in the rapidly expanding Islamic world, and East Africa turned into an important source (Manning 2003). Before the coming of Islam, slavery in Africa had only existed on a small scale. "The most common manner in which slaves were taken was through villages or states raiding one another" (Nunn 2008: 4). Both the Arabs and the Byzantines enslaved prisoners of war and citizens captured during raids as well as persons kidnapped by pirates. Raids were primarily undertaken to make a profit from the sale of the captives. Yet when Arab civilization began to rule the East Coast of Africa during the seventh and eighth centuries, exchanging captives became a lucrative business (Brode 2000). Local African kings showed an increasing interest in Arabic horses and precious cloth, which were traded to them for slaves, thus, they began to capture more slaves on a larger scale.[39]

In the tenth century, Islamic Byzantium preserved a monopoly of the slave trade where Russians also figured in as suppliers for slaves from their own regions to Byzantine traders (Rotman 2009).[40] With the expansion of Muslim dominance in North Africa, Muslim slave traders also began to operate on the other side of Africa. Blackburn (1997: 80) estimates that between the tenth and the fifteenth century, "an annual average of some five thousand slaves a year was dispatched across the Sahara from West Africa." Historian Rediker (2007: 77) estimates that from the seventh to the nineteenth century, "more than nine million souls were carried northward in the trans-Saharan trade organized by Arab merchants." During the time of Muslim Spain, Muslim traders used the Iberian Peninsula as entrepot between Europe and North Africa. However, when their influence declined, the slave trade dried up there. During the Crusades – the Mediterranean religious wars (eleventh to fifteenth century) of the Middle Ages – Christians and Muslims enslaved each other in the Mediterranean, which at that time functioned as the "commercial intermediary between the European side and the Arab side as the arena in which and around which the movement of people and commodities occurred. The slave trade had a central role in that dynamic" (Rotman 2009: 58–59). The situation, however, completely changed with the exploitation of sugar plantations.

For the production of sugar, the Islamic world had increased the trade in slaves. Plantations initially utilized white slaves – Slavs and Bulgars – yet soon African slaves were imported in large quantities to satisfy a growing demand. Sugar was introduced into the Levant in the seventh century by the Arabs. Sugar plantations developed in the eleventh century at the Syrian coast. During the thirteenth century, Venetian merchants energetically developed private sugar enterprises on Sicily, Crete, and Cyprus.[41] Crete was, at that time, the most strategic site in the eastern Mediterranean.[42] With the introduction and extension of sugar plantations by the Portuguese on newly captured islands, a new wave of slave trading exploded. During the Crusades, Europeans had become familiar with sugar, which was shipped to all parts of Europe. The rapid growth of European demand led Portuguese merchants to extend sugar cultivation to the Iberian Peninsula. While sugar cane cultivation in the Algarve (Portugal) was not very promising, Prince Henry planted sugar cane successfully in Madeira. This practice was then transferred to other Iberian-owned Atlantic islands off the coast of Africa such as Sao Thomé and the Cape Verde where the use of slave labor expanded quickly (Blaut 1992).

When Portuguese raiding parties more intensively captured Africans at the African West coast and sold them at the Lisbon slave market, Prince Henry decreed the slave trade a state monopoly (Postma 1990).[43]

Beginning in 1486, the Portuguese began the settlement of Sao Tome by shipping Jews and black slaves to serve the sugar plantations and export the product back to their metropolis. Colonial plantations became outdoor manufacturers applying principles of industrial organization and production to tropical agriculture. Individual entrepreneurs were encouraged to establish sugar cane plantations on these islands to produce sugar for European markets. They were to man them with salaried specialists and African slaves (Blackburn 1997). Soon the first sugar mill as a type of commercial enterprise opened on Grand Canary. The sugar industry was characterized by the use of both free and enslaved labor. A substantial amount of the labor was actually done by free wage earners paid partly in kind (Mintz 1985).[44] Although Africans were still a minority of the overall slave population by 1500, "with the spread of the sugar cultivation to the Atlantic islands, Africans became the majority of the world's slave population and, therewith, a key characteristic of the modern world" (Manning 2003: 32). With the expansion of sugar, tobacco, and cotton plantations in the New World, the slave trade gained extraordinary proportions.

When the European demand for slaves increased due to the growing sugar plantations, the more prosperous areas with well-structured African communities and densely populated areas were preferably raided. "These 'razzias', as the odious practice of man-stealing was known, had been carried out throughout the Middle Ages in Spain and Africa by Muslim merchants, and their Christian equivalents had done the same" (Thomas 1997: 23). European traders learned that large numbers of slaves could easily be obtained if conflicts or civil wars were instigated (Nunn 2008). African merchants began to provide black slaves as captives of raids or wars "a euphemism for organized theft of human beings" (Rediker 2007: 99). Sometimes local African countrymen, as in Gabon, turned against the foreign intruders such as the Portuguese and limited the slave trade but, in general, the European slavers obtained more and more slaves via African dealers.

The Atlantic islands were the stepping stones by which the sugar industry moved from the Old to the New World (Mintz 1985).[45] Sugar cane was first carried to the New World by Columbus on his second voyage in 1493. It was first grown in Spanish Santo Domingo beginning around 1516, and the fields were worked by freemen as well as enslaved Native Americans, but a high mortality rate made the exploitation of the plantations less successful. To replace those who had died, African slaves were imported to America. "Hence, it was Spain that pioneered sugar cane, sugar making, African slave labor, and the plantation form in the Americas" (Mintz 1985: 32). The sugar cane plantations became colonial enterprises and were led by a new combination of the farmer-manufacturer roles.

Sugar mills in the Americas were mainly financed and run by the Genoese (Kamen 2002). Italian merchant families and military functionaries quickly turned to sugar production. The technical skills required for the plantations and production of sugar was imported from the Canaries. Specialization by skill, a division of labor by age and gender, shift work and gangs, together with the stress upon discipline, were soon introduced in this industry. Freemen including debt servants, petty criminals, political and religious non-conformists, labor organizers, and Irish revolutionaries worked together with slaves. Enslaved Africans were imported to help exploit the economic potentialities of the New World.

In that sense, the connection between the sugar trade and the slave trade became a direct one. The production process in the sugar mill with all its heat and noise was time-driven with synchronized labor activities from the moment the cane was ready for cutting until the semi-crystalline product was pouring into molds to drain and dry. "What made the early plantation system agro-industrial was a combination of agriculture and processing under one authority: discipline was probably its first essential feature" (Mintz 1985: 51). Indeed, a strict discipline was required from the slaves who had flocked in large quantities to the New World since the 1510s when King Ferdinand of Spain had approved the transport of 250 slaves to the New World.

> It was Europe's sweet tooth rather than its addiction to tobacco or its infatuation with cotton cloth that determined the extent of the Atlantic slave trade. Between 60 and 70 percent of all the Africans who survived the Atlantic voyages ended up in one or the other of Europe's sugar colonies.
>
> (Fogel and Engerman 1974: 16)

By 1526, Brazil was shipping sugar to Lisbon in commercial quantities. Once the Dutch WIC had established a stronghold in the area in the seventeenth century, the Dutch mercantile class and state became heavily involved in the Brazilian sugar enterprise. "Between 1700 and 1748, the number of refineries in Amsterdam rose from 20 to 95. Behind this expansion lay the Dutch attempt to conquer the Portuguese sugar empire in Brazil" (Berg 1994: 125). Dutch sugar refineries became the leading centralized manufacturers of the time. For all of these plantation activities, however, more and more labor was urgently needed. Elmina Castle, which had originally been built by the Portuguese in 1482, became the headquarters of the Dutch slave industry, a few kilometers along the coast from Cape Coast Castle (Ghana).

At Cape Coast Castle, a gun was fired at a set time every morning and evening so that the captains of the ships outside the ports could synchronize their watches. The castles functioned as centers for the gathering, collating, and the distribution of information concerning the Atlantic oceanic world which was vital to the trading companies. More ships meant more information carried more quickly, leading to better-informed commercial judgments by an increasing number of involved participants. Soon, British ships also transported slaves from Cape Coast Castle and became the headquarters in Africa of the entire British involvement in the transatlantic slave trade (Clair 2007).[46]

The transatlantic slave trade was a continuous business for the chartered companies of the European countries, who were all awarded with monopoly-trading privileges by their own home government. By 1650, the VOC, WIC, and EIC together controlled most of the international trade with Asia and the slave trade in Africa. A so-called triangle of trade was established linking Britain to Africa and to the New World: finished goods were sold to Africa, African slaves to the Americas, and American tropical commodities to the mother country and her importing neighbors (Mintz 1985). During the early 1700s, the sugar-producing Caribbean colonies created the most profitable economy, per capita, in the world. Their exports were worth two-and-a-half times those of the partly free-labor economies of North America (Drescher 2009).

By 1770, the British venture into sugar production which had begun in Barbados was leading in the international sugar trade, followed by the French and Portuguese, with the Dutch now in fourth position.[47] Sugar accounted for about 50 percent of Europe's imports from America (Findlay and O'Rourke 2003). With the booming sugar business, slave trading became an integral part of British commercial activities. British maritime vessels transported millions of slaves to every part of the New World with available capital heavily invested in it. "Among those who received dividends from the slave trade were the British Royal family, the British aristocracy, and the English Church" (Clair 2007: 4). In Spanish America, slave populations from Africa substantially contributed to the wealth of the Spanish empire and resembled those of Portugal in Brazil and the north European countries in the West Indies.

Between 1502 and the 1860s, about twelve million West Africans were transported to the other side of the Atlantic (Nunn 2008; Osterhammel and Petersson

2004). A large proportion of them went to North America to work not only on the sugar plantations but also in the cotton fields. While admitting the immorality of slavery, American economic historians Fogel and Engerman (1974) analyzed the antebellum (i.e., before the Civil War of 1861–1865) economy in the south of North America and found that blacks in the South did remarkably well under the institution of slavery. Their view undermines the deeply entrenched belief that slavery was backward, inefficient, and obsolete. However, they comment:

> The years of black enslavement and the Civil War in which it was terminated were our nation's time on the cross. Certainly if any aspect of the American past invokes a sense of shame, it is the system of slavery.
>
> (Ibid.: 3)

It has also been concluded that the notion that slave owners relied on the lash alone to promote discipline and efficiency has been a highly misleading myth. Instructions to overseers contained complete descriptions about the manner in which plantations should be organized. They were required to look after appropriate sheltering, feeding, and medical care of the slaves and their families. The slave family was seen as an important instrument to maintain labor discipline. Compared with the Roman *latifundia* where bailiffs were either slaves or freedmen, the managers-stewards, overseers, and drivers stemmed from the same classes as the landowners and received a salary.

The exploitative nature of slavery was most apparent in a system of punishment and rewards. Whipping was a very common punishment but reliable data on the frequency of whipping has been sparse. Plantation owners did not want to work with sullen and discontented slaves. So they tried to imbue slaves "with the Protestant work ethic and to transform that ethic from a state of mind into a high level of production" (Fogel and Engerman 1974: 147). Propelled by self-interest, some slaves emulated their masters' work ethic.[48]

Planters organized the slave workers into work gangs which were systematically monitored. Slave drivers set the pace of work, and their working hours lasted from dawn to dusk. Planters relied on direct disciplinary control to elicit greater work effort and tried to bend the lives of the slaves as much as possible. The planters shared ideas on how to manage slaves, entered essay contests on plantation management, and published scores of articles on the subject in agricultural journals. Some slave owners ensured gentle management to keep slaves productive. Others believed that blacks were lazy and dishonest and could not be trusted. Others again believed that "godlike power over other human beings can only be justified by godlike virtue," (Hoopes 2003: 12), an attitude the owners generally translated into strict disciplinary measures.

The 1850 US census reported that 18,859 Americans earned their living as slave overseers, probably the largest group of salaried managers in the world at that time. According to the American historian Hoopes (2003), the slave system anticipated the efficiency and productivity of the assembly lines that the car manufacturer Henry Ford would introduce in the early twentieth century. The slave practice was, so to say, one of the precursors to the stage of the modern factory or machine facture during the second phase of industrial globalization.

The European miracle in the time of terrestrial globalization and rising industrial globalization depended not only on the application of the mechanical clock and on the control of intercontinental commerce, but also on the profits of slavery made for the plantations (Blackburn 1997). The slaves were forced to work from dawn to dusk and overseers learned to design new requirements of coordinating large workforces on the plantations.

The expansion in the number and industriousness of the colonists, however, was continually limited by shortages of labor power which could not be satisfied by relying on the supplies engendered spontaneously from within the ranks of the settler populations or extracted forcibly from the indigenous populations. "This chronic labor shortage enhanced the profitability of capitalist enterprises engaged in the procurement (primarily in Africa), transport, and productive use (primarily in the Americas) of slave labor" (Arrighi 2010: 49). Slavery became partly a condition and partly a result of the success of settler colonialism. Slaves were traded for British manufactures such as cloth, guns, metalware, glass, and paper. The British slave trade reinforced at an unprecedented scale one of the greatest involuntary migrations in history. "British hegemony stimulated the capitalist world economy and moved it to dominion globally" (Hopkins 1990: 411).

Although the economic importance of slavery increased with the soaring global demand for consumer goods such as sugar, tobacco, and cotton textiles, the abolitionist movement gained strength in Great Britain and France.[49] In 1787, reformers in London founded the Society for Effecting the Abolition of the Slave Trade. During the time that the British slave system peaked in leading the world in plantation production, monopolizing the slave trade and gaining an overwhelming share of the growing world market for sugar and coffee, the reformers were able to convince parliament to intervene. The official cessation of the slave trade in both the United States and the British Empire in 1807 did not fully prevent the continuation of the atrocious slave trade and slave use on plantations. Although Britain took leadership in suppressing worldwide slave trade, between 1826 and 1850 the volume of Atlantic slave trade declined by only 5 percent (Drescher 2009).

By 1850, the British Royal Navy captured around 1600 illegal slave ships, the last ones – two American vessels – still crossing the Atlantic Ocean in 1858 and 1859. Notwithstanding this illegal trade practice, the rural labor of freed slaves and freemen no longer worked from dawn to dusk but became organized on a task work system, i.e., individualized and quantified norms of effort and output with a set number of hours. During a four-year period after abolition, the freed slaves in the West Indies – now called apprentices while still working for the former masters – were taught how to behave as responsible waged workers (Ackroyd 2007). Freed slaves working on the plantations of South America began to learn new practices concerning regulation of working hours and intensity of work (Genovese 1976).

Unlike the American slaves, free workers were hired in the marketplace, but if they demonstrated that they were lazy, indifferent, or otherwise shirking their duties, they could be fired and denied their food. This task labor introduced measurements of labor and quantities of products. Tasking rates were frequently discussed and articulated in the agricultural literature (Tyson *et al.* 2004). By the 1830s, tasking had become the prevalent form of the plantation work system. This method offered businessmen, factory owners, and managers a powerful new weapon and became common practice during the industrial era.[50] Supervision costs diminished and free time became an incentive for faster work. As a top-down approach, it certainly preceded the development of scientific management and became a direct precursor to modern managerialism (Tyson *et al.* 2004).

The demise of the VOC

For two centuries, the VOC was the biggest international trading company in the world and a powerful instrument in the on-going centralization in Amsterdam of world-embracing commerce and high finance.[51] It was a chartered monopoly backed with the power of the Dutch government. The same holds for the WIC. Both gained and maintained large territorial overseas areas during the seventeenth and eighteenth centuries and turned into extensions of the Dutch Republic. The involvement of the States General in the establishment of both companies demonstrated the symbiotic relationship between government and private trading companies and prepared the way for Dutch colonial power (Heijer 2005). Until 1800, the VOC remained the largest of the Asiatic companies which, together with other European joint-stock chartered companies, were the first remarkable actors in the process of terrestrial globalization. They had put a moderate form of managerial bureaucracy in place to organize their overseas activities and assure themselves of a reliable system of information sharing. Arrighi (2010) draws the conclusion that top managers of today's multinationals may find themselves more at home among the directors of these early joint-stock companies than in the family businesses that constituted the backbone of the first Industrial Revolution and concurrent British capitalism.

Although economically successful throughout the seventeenth and eighteenth centuries, the United Provinces, however, never became "a nation-state proper in terms of a territorialist organization" (Arrighi 2010: 77). Geographically, as well as demographically, with a population of around two million, the country was too small to build a territorialist empire. By the time England and France entered the business of territorial expansion in the extra-European world, "the spread of Mauritian 'scientific management' to the European armies had become a success due to which the Dutch lost their status and comparative advantage in this area, too" (ibid.: 48). The Dutch Republic had fused the advantages of commercial capitalism and territorialism "far more effectively than any of the northern Italian city-states had ever managed to do" (Arrighi 2010: 139). Yet the narrow territorial and demographic base of the Dutch nation-state limited the expansion of their fortunes.

During the Anglo-Dutch wars (1652–1654, 1664–1667, and 1672–1674), which frequently impeded a more intensified economic integration in history,

English aggression was triggered by resentment toward Dutch commercial prosperity. The English Navigation Act of 1651 had given English ships a monopoly on trade with the colonies, which resulted in wars with the Dutch. The Dutch capitalist class turned away from trade and focused more on high finance based on a carefully regulated banking and insurance practice.

While the Catholic Stuart monarchy lost popular support, the Protestant British made it possible for Dutch William III (1650–1702) of Orange to ascend the English throne in 1689. Dutch capital shifted from Dutch to British soil. Financial markets and financial institutions, which were new in seventeenth century London, were based on practices developed earlier in Amsterdam. The so-called Glorious Revolution in politics transformed England into a powerful and successful state, economy, and empire. The influence in the way the Dutch invested their capital, and applied modern organizational techniques in the army and navy eclipsed. William intensively relied on troops, artillery, equipment, money, and techniques to win the support of the English population. Surplus capital became not only invested in British stock and government securities, but the Dutch capitalist strata also invested in British enterprises which boosted British textile and metal industries. In this transition, the VOC gradually ceded its unique position.

Britain takes over

It is no surprise that after the religious wars (1618–1648), several different European countries fought several wars for the defense of their commercial and military interests. Between 1652, the year of the outbreak of the Anglo-Dutch Wars,[52] and the end of the Napoleonic War in 1815, the European interstate system was dominated by the struggle for world supremacy between Britain[53] and France. In France, the Bourbon Dynasty ascended, and French absolutism rapidly came of age. Administrative architects such as Sully, Richelieu, and Colbert structured the French state system. The country became a classical example of the sale of lucrative offices, creating an ever-expanding bureaucracy. The purchase of offices offered a profitable investment, but it resulted in less available capital being invested in manufacturing activities and restrained the development of a French bourgeoisie for many years.

The exception was the activities of merchants operating in international trade. With royal subventions, chartered companies were created to exploit the trade of the East and West Indies as well as colonial expansion overseas. The King of France, Louis XIV, attempted to dominate continental Europe, which brought European states into coalition against him. The ensuing series of interminable wars left France almost exhausted and nearly bankrupt. The French absolutist state in the late seventeenth and eighteenth centuries did not enjoy a hegemonic position for very long. Its expansion became effectively halted by the economic rise of English capitalism (Anderson 1974).

Since the late seventeenth century, England's economy and society had experienced a period of booming manufacturing and international commerce. A growing urbanity, together with an elaborated infrastructure of roads, ports, and navigable rivers, as well as an efficient, new postal system and new financial institutions, especially a national bank – the Bank of England (established 1694) – and a regular trade in government bonds turned the country into an energetic, thriving place. Its foreign policy was focused on expansion by naval power, not territorial armies. It attempted to internalize the network of trade and power of the United Provinces within its own domains and fought for control over the Atlantic. Around 1700, England had established a single internal market that offered advances for agrarian and mercantile capitalism. As the aristocracy had assumed local administrative functions for a long time, the monarchy was deprived of any professional regional apparatus. The state did not have to maintain an expensive standing army, and the peasantry was not subjected to a heavy tax burden (Anderson 1974). These circumstances disposed extra investments for international trade and commerce.

Next to its success as a commercial empire, productivity gains in agriculture made Great Britain become self-sufficient in food production. Better medical and hygienic provisions contributed to a combination of rising birth rates and falling death rates. The population of England and Wales more than doubled from seven million in 1770 to 17 million in 1851. Cultivated land increased from 10 million to 15 million acres. An increase in the mechanization of agriculture encouraged the rise of commercial agriculture. The enclosure movement permitted aristocratic landowners to take common land away from peasants in order to keep large flocks of sheep to provide the textile industry with raw material. This caused a tremendous shift away from the countryside to the cities where factories took in the free hands.

Great Britain was also endowed with other primary sources such as iron ore and coal. The products that the country could not provide for itself in large quantities were imported from America (cotton), and Scandinavia (iron ore). New technologies and new forms of industrial organization – above all in the textile factories – developed to a remarkable scale and power. With the expansion of its commercial position, opportunities to export its own products increased tremendously.[54]

The pattern of British merchant enterprise in Asia was initially conditioned by the existence of the EIC. It traded on the basis of its textile monopoly from India which still had the largest textile industry in the world. The import of oriental printed silks and Indian printed cottons to European markets was an innovation in pre-industrial long-distance trade. In Asia, British control was extended over Burma and Malaya.[55] The British West Indies, which provided sugar for European markets, remained a highly prized imperial possession. Britain believed in funding settlement colonies and not just ports-of-call en route to the Indies. Beginning in the 1750s, the EIC, along with its trading functions, additionally began to assume governing functions in parts of India. However, the exclusivity of commerce enjoyed by the joint-stock trading companies such as the EIC steadily declined.

> British private traders, who had often served with the EIC and who retained close collaborative links with it, developed a large clandestine

> trade within Asia and between Europe and India, trading in products such as indigo, sugar, cotton, spices, and opium.
>
> (Jones 2000: 29)

Exports of manufactured goods, as well as imports, increased tremendously. The country abandoned its protectionist policy with the repeal of the Navigation Acts and the Corn Laws. The EIC lost its monopoly; tariff reforms eroded the Navigation Acts with all its privileges for the traders and merchants. The repeal of the Corn Laws in 1846, which had been introduced to protect the landed classes against imports of cheap grain, marked the triumph of the ideology of free trade. British trade evolved into a decentralized web of small specialized firms.[56] More merchants were able to raise capital in partnership agreements primarily based on family ties.

> They made a profit by taking a commission and fees for acting as commission agents, or agents for shipping companies and insurance firms. There were always tensions between this ethos and making large capital investments necessary, in some cases, to facilitate trade.
>
> (Jones 2000: 27)

In some cases, the operations of merchants grew into manufacturing activities as well, which led to the establishment of early multinationals. At the same time, many of the wealthiest merchants began to act as bankers, in turn financing a group of middle-level merchants who acted as wholesalers. All these international trading and commercial activities, as well as improved communication, were accelerating at an unprecedented rate.[57]

Yet the British Empire's international expansion was not only based on international trade and industrial expansion. The nation became the primary slave trader of the world (Christopher 2006). Its military aggression also played a large part, and a wave of imperialism provided further opportunities for trading companies by extending the borders of Western influence. The wars against the French between 1793 and 1815 extensively increased British power and possessions. Although Britain lost its imperial position after the American Revolution in the west in 1776, it gained a second one in the east (Brendon 2009). After the collapse of the Mughal Empire, Britain was able to build a synthesis of capitalism and territorialism to dominate India and other colonies, a policy which, in 1858, made the British government assume control over most of India.

Although British drill-masters initially began to create miniature armies by recruiting local manpower for the protection of the British trading stations on the shores of the Indian Ocean, this approach was soon expanded to occupy larger areas. "Such forces, however miniscule, exhibited a clear superiority over the unwieldy armies that local rules were accustomed to bring into the field" (McNeill 1984: 135). As the EIC was mainly concerned with trade and not the running of an empire, it was less interested in the administrative burden for running the subcontinent. In 1773, the British government brought the company partly under governmental supervision to improve efficient control and raised taxation instead of increasing trade. "A third of the Indian government's expenditure was devoted to the military" (Brendon 2009: 55) which confirmed Britain's position as a global power. The Royal Navy dominated the seas and facilitated further international trade by British ships and British merchant companies. Its naval supremacy at sea culminated in the British victory at Trafalgar in 1805. At sea, trade followed the flag, even in the Mediterranean Sea.

The British Empire grew on the back of a commercially driven expansion of overseas trade. There was an incessant search for markets of manufactured goods, as well as supplies of commodities and raw materials. The British swept up parts of the Dutch empire in the Cape, Ceylon, and Java. It established close trading links with China to purchase tea, silk, and porcelain. The EIC still monopolized trade with China and traded luxurious objects such as highly ornamented musical clocks or "singsongs" to balance the otherwise one-sided trading relations.[58]

Eventually, England became the principle beneficiary of the system of rule in Europe and with a powerful army and navy was able to put a remarkable stamp on international trade as well as industrial activities. To demonstrate its increasing military and political superiority, it was, between 1662 and 1815, at war with France nine times, with Spain six times, with the Netherlands three times, and with the mainland American colonies twice. Britain fought the most wars, conquered the most colonies, imported the most slaves, and was the quickest in expanding the plantation system. Global space became territorialized and colonized. Within such a setting, there was an increasing demand in English markets for colonial products such as spices, silk, and porcelain, which were consumed by expanding strata of the population. Sugar became a highly attractive commercial commodity and a competitive industry where business ingenuity was at the forefront of success.

To be assured of successfully returning their ships to the mainland, British merchants and captains insisted on improvement of nautical knowledge and skills, especially the accurate determination of the position of the longitude.

The third phase of the mechanical clock: the marine chronometer (1657–1761)

Like the Spanish empire, other European nation-states held a growing interest for the precise determination of the position of longitude to determine an empire's size. England and France tried to beat each other in the race for the discovery of longitude and its use as a military and commercial weapon. For both countries, the eighteenth century was a period of rapid growth in trade and competition in what were known as colonial wares: sugar, coffee, tea, and tobacco, all commodities which were the source of fabulous fortunes and windfall revenues.[59] France was the more populous country and richer in aggregate, hence, the stronger land power. However, England had the advantage in ships and seamen and exploited this superiority to strip France of most of its colonial possessions.

Maritime troubles were linked to commercial rivalry, but there was one nagging problem. As more and more sailing vessels set out to conquer or explore new territories and bring back gold and commodities to their home country, the wealth of nations floated upon the oceans. As ships did not yet have the most reliable means of establishing their specific locations, untold numbers of sailors died when rock formations suddenly loomed out of the sea and took them by surprise.

Thus, for the precise location of ships, the determination of latitude and longitude were essential, and for longitude, the proper application of precise instruments was required. This led to the so-called horological revolution of the mechanical clock, i.e., its third phase of development (1657–1761) as a precision instrument. In 1657, the Dutchman Christiaan Huygens (1629–1695) achieved a major horological advance by substituting the pendulum for the foliot.[60] He built the new clock with the assistance of the Amsterdam clockmaker Salomon Coster. The process of finding solutions for mechanical problems was one that needed a continuous interaction between men of the hand and men of the head. The result was a process of mutual powering between the scholar and a craftsman with mechanical images, mechanical problems, and mechanical solutions.

Christiaan Huygens' invention achieved a sensational increase in accuracy. The precision of the pendulum clock led scientists and clockmakers to believe that the problem of longitude to which the seafaring, powerful, and prosperous Dutch nation-state had attached such great importance could now be solved. Huygens designed a spring balance for watches in 1674–75 as a suitable escapement for chronometers. It was, for the first time, possible to produce timekeepers which were accurate enough for any normal domestic purpose by an urbanized man (Macey 1980). However, the application of the spring balance experiments for navigational purposes failed. What was needed was a more precise and elaborate chronometer.

During the third quarter of the seventeenth century, Holland as well as England set this horological revolution into motion. Both countries developed the pendulum for standing clocks and the balance spring for movable or portable timepieces.[61] On a symbolic level, the chronometer introduced an improved measurement of time that reinforced the process of quantification of time and space (Mumford 1934; 1964).

London, which had become fast-growing due to its concentration of international commerce and industry and its huge urban population, developed into the new capital of clock making. The London Royal Society had regularly discussed Christiaan Huygens' invention of the pendulum clock.[62] In 1675, Charles II ordered the building of an observatory on Greenwich Hill. It should have as much scientific, astronomical, and commercial background as possible and should provide accurate celestial charts to facilitate navigation at sea. In 1676, two Tompion clocks were installed at Greenwich that had to be wound up once a year and were the most accurate clocks in the world at that time. Before 1657, clocks could not generally keep time more closely than to about 15 minutes per day.

Britain took over clock manufacturing and highly stimulated the development of the clock industry. Watches, which had previously been plump and ornate and worn mainly around the neck, became plain and flat, and disappeared into the vest pocket. The clock had turned into an unpretentious precision instrument with only one function: timekeeping.[63]

However, all the maps and precise tables that were produced could not prevent shipwrecks from taking place. In a single such accident on October 22, 1707, near the Scilly Isles off the coast of Cornwall in southwest England, four homebound British warships in the fleet of Admiral Sir Cloudisley Shovel, including the admiral's ship, ran aground and nearly 2,000 men lost their lives (Sobel and Andrewes 1998b). This incident catapulted the longitude question into the forefront of British national affairs. The tragedy led to the Merchants' and Seamens' petition offering a prize of 20,000 pounds for a method of determining longitude to an accuracy of half a degree of a great circle (equal to 60 nautical miles).

The accident and the petition precipitated the Longitude Act which was passed at the Westminster Palace in 1714. The Act intended to determine how, if a mariner had with him an accurate clock regulated to the time at Greenwich, England, and found that (in the Atlantic) his clock indicated 1:00 p.m. when the sun was directly to the north (local noon time), he then could know for sure that he was somewhere on the longitude 15° (360° divided by the difference in time) west of Greenwich.[64]

In 1730, British amateur clockmaker John Harrison (1693–1776) met Edmond Halley, the astronomer at the Royal Observatory at Greenwich, who stimulated him to work on a marine timekeeper. After five years, the first chronometer was finished and taken onboard the British vessel that sailed to Lisbon to be tested, but the ship never went to the West Indies to prove the accuracy of the clock. After a few new trials to improve the chronometer, Harrison completed a refined version in 1753 which looked like a pocket watch. His son, William, took the timekeeper onboard the ship Deptford at the harbor of Portsmouth, which set sail to Bridgetown (Barbados) in 1761. When the ship arrived in Jamaica after three months of sailing, it was found that the watch was only 5.1 seconds slow. The watch proved to be more accurate than the offered prize had required. It took, however, until 1773, another 12 years later, before Harrison received part of the reward from the Board of Longitude to recognize his achievement.

Thanks to the pioneering work of John Arnold (1736–1799), who was able to miniaturize the chronometer, the British turned the marine chronometer into an object of industrial manufacture and commercial use. By the middle of the eighteenth century, it had become possible to build chronometers for the first time that possessed the extremely high accuracy necessary for determining longitude at sea. The horological revolution created a new reality in economic and social terms: "The introduction of the balance spring revolutionized time discipline" (Landes 1983: 128). Punctuality became a new word.

John Harrison had solved the problem of the reliable clock in time for the first Industrial Revolution to take off.[65] It also ignited the design and application of machine tools with which craftsmen were able to manufacture precisely shaped and well-fitting parts. Precise fitting became a core goal of British craftsmanship (Friedel 2007).

In the same year Harrison died, Adam Smith published *The Wealth of Nations* (1776), a book that showed how the invisible hand as a metaphor can be compared to a spring balance, keeping markets in movement. He pointed out that better watches were being produced compared with a century earlier and for one-twentieth of the price. Clocks and watches were no longer expensive toys. The large-scale manufacture of watches – particularly driven through the work of Thomas Tompion (1639–1713) – brought industrial innovations such as batch production and the application of interchangeable parts, tooth- and wheel-cutting machines, requisite advances in technology and metalworking which became provided by sub-trades (Macey 1980).

However, the adoption of the chronometer in the maritime practice evolved slowly due to its high costs. Well into the nineteenth century, navigators continued to also use "sightings on the moon to place themselves in longitude" (Vance 1986: 429).[66] But it was also noted that a navy which became equipped with the new navigational chronometer would be in a better position to fight others.

In 1805, naval commander Horatio Nelson defeated the Franco-Spanish fleet off the Spanish coast near Cape Trafalgar which allowed Britain to enjoy the total mastery of the Atlantic Ocean. Nelson's victory conceived the Pax Britannica and turned the British Empire into an oceanic empire (Winchester 2010). The North Atlantic region's drive for political, military, and commercial hegemony realized a terrestrial globalization which the modern world increasingly drew into the ascendance of a market-oriented logic of profit seeking. Industrial developments would become the antecedents for the epoch of industrial globalization. After the abolition of slavery and the end of serfdom in Europe, one would expect that the value, dignity, and freedom of the human person would be fully respected. Yet the impact of new forms of transportation and industrialization on working conditions would demonstrate otherwise. Working and living within the rhythm of one's own local habitat or at other localities of the globe would become disturbed anew.

PART II
Industrial globalization

Great Britain not only assumed a new, leading, commercial role but also succeeded in combining international trade with industrial production, launching the epoch of industrial globalization, which introduced and expanded the trade of industrial products to new markets elsewhere in the world.

During the course of the eighteenth century, there was a growing interest in favor of the improvement of transportation. International trade relied on effective use of British harbors that could be served efficiently from within the country. The need for more and better roads and canals grew along with the growth of commerce and manufactures. To reduce the road transport bottlenecks, merchants and manufacturers took the initiative to create turnpike trusts. Between 1720 and 1749, approximately 130 turnpike companies were established, spurring the inland transportation of goods (Bagwell 1974).

One clear measure of the impact of infrastructural improvements was the reduction in the amount of travel time for journeys between the principle centers of population. When comparing the travel time along the turnpikes between 1770 and 1830, one could ascertain that "the times were halved" (Bagwell 1974: 29). It was not only the speed but also the amount of traveling that increased dramatically. More stagecoach services were developed. The Royal Mail, in service since 1516, fell behind the new stagecoach services and had to improve its performance by redesigning its mail coaches and material. Merchants and traders continued to be involved in projects which could reduce infrastructural limits. With an improved infrastructure, the export of manufactured goods such as metal products, cotton textiles, and woollen textiles to the colonies could be better facilitated, as well as imports of raw materials from the colonies.

Manufacturing for anonymous (inter)national markets facilitated the establishment of a factory system where operating according to strict working time schedules increased the scale of production.

> In 1750, more than 50 percent of the world's industrial output was produced in China and India, compared to some 18 percent in Western

> Europe. The following eighty years saw the Industrial Revolution, with Western Europe's industrial output more than doubling and that of the United Kingdom increasing by a factor of seven.
>
> (Crafts and Venables 2003: 35)

During the initial period of industrial globalization, Great Britain became the first society broadly ingrained by clock time patterns.

Next to the coordination of time in local places like factories, the ability to control time across geographical spaces with the building of railroads and steamships advanced in a spectacular way. After the first trains and telegraph systems introduced an increasing practice of adhering to clock time, this practice advanced in the sphere of distribution and communication as well as in the sphere of production. A remarkable interplay between industrial changes with wider British political, cultural, and social changes encouraged the appearance of an Industrial Revolution. The Industrial Revolution increased production output, steam engines driving the trains reduced inland transport costs, while improvements of ships reduced ocean transport costs.

At the end of the nineteenth century, however, Britain began to be displaced as the leading industrial power by the emergence of the United States as the main driver in industrial globalization. The advent of large-scale infrastructural provisions in transportation and communication facilities covering almost the entire American continent as well as the arrival of fully clock-time-driven machinefacture, intensified and expanded. The coevolution of a spatiotemporal infrastructure based on railroads connecting East and West America with the temporal infrastructure of assembly lines in manufacturing companies increasingly reinforced one another. Faster transport and transmission reduced the price of goods and information and further facilitated the growth of international trade and business. With the expansion of an American infrastructure of trains and telegraph systems, the practice of adhering to clock time increased tremendously, and this practice advanced in the sphere of distribution and communication as well.

In the period between 1870 and 1914, railways, telegraphs, steamships and automobiles facilitated a process of convergence among the American and Western European economies as well as with the temperate economies of European settlements like Canada, Australia, and New Zealand in terms of technologies, institutions, and productivity levels (Dowrick and DeLong 2003). Yet in 1955, North America was still the leading exporter and "accounted for over a quarter of manufactured exports" (Crafts and Venables 2003: 327). In the UK and the USA, this development concurred with the introduction of two new forms of applying mechanical clock time: factory time and railway time.

Whereas in the UK this development was only just beginning, the USA applied these new time practices on an unprecedented scale. Large American firms established bureaucratic organizational forms with steep hierarchies to safeguard the smooth transfer of information in order to make the appropriate strategic decisions. Speed of information provision as well as speed of transportation and production became key features in the first half of the twentieth century.

5

TIME IN BRITISH INDUSTRY AND RAILROAD TRANSPORT

Britain became the first nation in the world to industrialize and mark the development of the factory system in the textile industry. This system was characterized by a complex division of labor between and within work processes, the routinization of work tasks, as well as a complex interplay of technical and organizational changes. Industrialists replaced the merchants and traders as the dominant actor in the capitalist system. Substantial private capital seeking out profitable investments became available through the Bank of England. Merchants began to act as partners to industrialists who themselves affected the decline of the traditional handicraft skills of artisans, guilds, and journeymen.

Due to the improved developments in commercial agriculture, more hands became available for the textile industry. The Hungarian economist Karl Polanyi (1977), however, argued that capitalism not only emerged due to the progressive commodification of land and the concomitant technical improvements in agriculture, but that the same happened with the commodification of labor, which culminated in the establishment of a generalized labor market in Britain in the 1830s. Yet technical innovations alone would never have led to the Industrial Revolution if, concomitantly, some remarkable institutional changes had not been taking place (North 1990).

Evangelicalism became the force behind voluntary associations like the Society for Promoting the Religious Instruction of Youth (1800) and Society for the Suppression of Vice (1802). Protestants outside the Church of England (dissenters) and Catholics were granted civil rights. Institutional changes further led to a new pattern of inheriting wealth and the sharing of knowledge by associations like the Lunar Society. At the same time, technology facilitated the reduction of geographical distance. Railroads and steamships, while shrinking time over distance, made the exploitation of natural resources more feasible.

Mining

Due to commercial activities, as well as a growing population, British forests had become increasingly denuded to supply the navy and the charcoal industry. The rapidly industrializing British country had used up most of its firewood and become increasingly dependent on coal. Important industries like brewing, salt-, glass-, and brick-making began to use coal in the process of manufacturing. Prices for timber and firewood increased incessantly and ever growing numbers of households, workshops, and manufactures switched to coal for heating. This led to the invention of coke for the production of iron. The switch from charcoal to coke in smelting iron created pressures for innovations in the refining of iron. As the economist Werner Sombart (1916) clearly outlined, a transition took place from an organic economy, based on agriculture, to an inorganic economy based on mining activities. Yet the transition that took place in agriculture due to technical improvements mobilized a spirit of creation which turned England into an industrial nation. The transport of coal from the north of England to densely populated and booming industrial areas elsewhere in the country intensified and took place mainly via the business of coastal shipping.

Charcoal became used as fuel in iron and brass forging. In 1709, Abraham Darby introduced a more economical way of brass molding to cast-iron pots and designed a cheap and serviceable cooking pot. Cast iron became a very popular product and offered many opportunities. To convert ore into cast iron by smelting, Darby tried coal that had been turned by brewers into coke by heating it at high temperatures to change its structure.[1] Soon, blacksmiths were exploring the potential of this new product which would allow steam power to eventually come of age.

In order to provide enough coal, mineshafts needed to be dug significantly deeper. Mine work became a contracted-out job, and mining owners used middlemen to recruit mineworkers. Coal hewers, the workers actually cutting the coal, became the elite among the British workforce, earning high wages.[2] The productivity relied mainly on the strength and skill of the hewers, who were paid on a piece-rate basis. Supervision was required to assure the size and the quality of the output (Ackroyd 2007).

Unfortunately, new shafts were constantly flooding with groundwater. Many mines were installed with a crude horse-driven system with pumps, but it was slow, inefficient, and very expensive. Drainage became an important topic and, in 1712, Thomas Newcomen solved the issue when he invented the heat engine, a self-acting device that could pump water from flooded mineshafts in the Coneygree coal works in Staffordshire.[3] Annual British output of coal jumped from around three million tons in 1712 to nearly double that amount by 1750. Within a century of Newcomen's successful invention, the industrial world was being remade by coal energy.

The great abundance of coal, its energy density, and its more efficient use hastened the transition of production. Coal output and the abundance of iron accelerated the manufacture of machines including steam engines. The heat energy created by the coal was converted from combustion into a physical or mechanical energy of work. Factories of all kinds sprang up using steam engines to drive looms, lathes, and presses. By 1800, coal-fired steam locomotives began to carry cargo including coal. Iron foundries were relocated to coalfields in the north of England.[4] The advent of the steam railroads not only revolutionized the freight-hauling industry, but also made it possible for a growing number of passengers to undertake journeys over long distances. To transport coal to the cities and burgeoning industrial areas, an extensive network of canals and railroads was required.

Due to Britain's development as a commercial nation, it possessed a well-established system of wholesalers and retailers who were able to successfully sell the expanding output of the nation's workshops. However, British industrialists faced segmented markets with regional and overseas markets demanding large variety. Long production runs of identical goods were not an option.[5] By specializing in one particular part of the production chain, creating networks of close cooperation and operating as subcontractors for larger companies, small firms remained highly successful for a long period of time. British manufacturing ventures remained small family firms organized as a single proprietorship or partnership and kinship connections continued to be of utmost importance in the operation of the firms. Partnership based on ties of family and friendships made sense while risks were high, and the smaller family enterprises were personally responsible for any debt the company incurred. In these relatively small industrial firms, ownership and management remained united with business long remaining a personal and not a bureaucratic affair.[6]

Devoid of any common social circles to share particular experiences and exchange the latest insights, merchants, traders, and industrialists created their own in 1754, such as the Royal Society for the Encouragement of Arts, Manufacturers, and Commerce. These societies attempted to fill a gap in understanding the importance of the sciences to improve production methods. They also addressed the consequences of social changes arising from industrialization. This was particularly true in the case of pottery making.

Josiah Wedgwood

During the eighteenth century, the general rise in the British standard of living brought about an increase in tea and coffee drinking. These new, hot beverages took the country by storm and were strongly featured in the diet of the Britons. Tea was imported from China by the East India Company and sweetened by sugar cultivated in the slave economies of the Atlantic world. In 1730, Britons consumed around 1.4 million pounds (635,000 kg) of tea; by 1800, annual tea consumption had risen to over 12 million pounds (5,443,108 kg). This interest in drinking tea and coffee created a demand for inexpensive, good quality, earthenware pots and cups, which the existing potteries could hardly satisfy.

Josiah Wedgwood (1730–1795), after having begun as a potter's apprentice, observed these social changes that accompanied urban expansion in booming cities like Liverpool, Sheffield, Birmingham, and the mercantile center of Manchester.

He noticed that rich families required more exquisite creamware, decided to fill that gap, and began his first small operation in 1759 in Burslem (Staffordshire) making dishes, mugs, jugs, porringers, cups, and saucers. From borrowed books on chemistry, he learned how to combine metals. After much experimentation with mixing different types of ground metals and sponging them onto the piece to obtain a brilliant lead glaze, he produced an inexpensive, good, cream-colored earthenware product. With a new type of good-quality chinaware, he found himself forced to standardize the production process.

In his Brick House Works, he introduced a division of labor that was arranged sequentially to provide an efficient production line and to avoid unnecessary movement of workers and goods. He redesigned the production process and introduced a rudimentary form of bureaucracy involving task specialization, formalization of authority, standardization, and hierarchy. It was common practice in the workshop and manufacturing system for employers to collect workers in one spot, but the workers themselves controlled their own hours, work pace, and work patterns. Artisans usually worked long hours but took breaks whenever they wanted. They might work 12 or more hours per day for four or five days and then take several days off, including what was called Saint Monday. The skilled artisans and craftsmen with their long-established practices were accustomed to bouts of work, idleness, drinking, games, and gambling.

In many cases, workers did not have to produce any minimum output per week. Employers were forced to maintain extensive inventories to prevent interruption of the production process. Therefore, in order to reduce the costs involved and improve the level of output, coordination was required by imposing a quota of output per worker, but even that allowed workers freedom as to how they conducted themselves on the job.

Wedgwood intended to change the work pattern and regulate the working hours to at least guarantee a specific amount of output. As artisanal work was still very episodic, Wedgwood attempted to impose a strict discipline.[7] The work of craftsmen transformed into acts of labor no longer being performed for its own sake or for the sake of creating a lasting product but merely as a means to maintain the dynamics of the factory system. Labor became repetitive and endless and was redefined to accelerate the production process itself. The design of his new production method made it possible to operate a lesser-skilled labor force that could work for long hours under intense discipline for low wages.[8]

His factory rules dictated the conduct of the craftsmen in great detail. He demanded punctuality from the craftsmen "to make such machines of the Men as cannot err." The phrase summed up the efforts of many entrepreneurs at that time to rationalize the sequence of operations in factories and overcome human error like Prince Maurits had done with his soldiers in preparing them for the battlefield. To do so, he installed a factory clock on top of the roof to announce the start and the end of each working day and each break. The bell in the tower at top center summoned laborers to begin work after it rang at a quarter to six, which instigated the factory workers to soon rename the pottery factory "The Bell Works."

Installing the bell was Wedgwood's "first management innovation" (Dolan 2004: 127). Wedgwood imposed a schedule of regular working hours to distract dilatory workers from the allure of the alehouses. Foremen and managers like the Clerk of the Manufacturers and the Clerk of Weights and Measures vigilantly observed what was produced and how.

Employees were not immediately willing to cooperate; however, infringements of the rules were punishable. As semi-skilled workers rather than as master potters, they turned out sellable products, but they were forced either to work long hours in factories for low wages or to starve (Thompson 1967). They were required to appear every day at the same time and had to labor until the clock or a bell told them to stop. This application of abstract clock time and invariable working hours contrasted sharply with those imposed by the sun and seasons. To keep track of his workers' time, Wedgwood devised a primitive clocking-in system, "the first in the history of business" (Koehn 1997: 43).

The machines Wedgwood installed were tireless, regular, rapid, and precise. Industrial work became more uniform and more disciplined with the demands of machinery setting the pace. Wedgwood allowed workers less discretion over their workplace and methods. He imposed a strict 10-hour working day, six days a week with perhaps two holidays per year. Next to the overarching authority of the clock, factory work demanded a strict reduction of waste and obedience to the managerial hierarchy. The lodge keeper slammed the gate if workers arrived too late.[9] It cost Wedgwood a great deal of effort to recruit and keep staff. The workers compared factories to houses of correction with a harsh regime and little pay. He encountered stiff resistance while introducing factory obedience and new bureaucratic strictures on workers who were used to working in short intensive spurts and then slacking off for idleness.

To conform the workforce to factory rules on punctuality, precision, constant attendance, fixed hours, and avoidance of waste, he could not rely just on the proverbial stick and carrot but needed the theological aid of the Evangelical Movement. By happenstance, John Wesley began preaching in Burslem a year after Wedgwood opened the doors of his factory. This local preacher helped Wedgwood by preaching to the workers two alternatives; eternal damnation or subordination to authority by accepting the long hours of work under intense discipline.

Within the rural setting that Burslem was part of, the enclosure movement had allowed labor to become more mobile, with more freedom of movement. The large landholders no longer had to honor traditional norms nor the actual laws that bound communities together. The local poor labor force became sensitive to Wesley's preaching as local farming families became dispossessed, and the factory was the only place to find work. Wesley was one of those preachers who propagated the conviction – as formulated by the Reformation – that work was a way of serving God. The Reformation in the form of Puritanism, Calvinism, and Methodism attributed an intrinsic merit to work: work for the sake of work, work as an end and purpose in itself.[10]

The rise of Methodism in Staffordshire greatly aided Wedgwood in his efforts to organize his factory. With their churches and Sunday Schools, Methodists convinced the local people of the necessity to work in these new factories. Without a doubt, the local entrepreneurs financially supported this movement. Yet it took Wedgwood three decades to enforce obedience, discipline, and effective control of his workforce and convert his firm into a properly functioning bureaucracy.

Finally, the physical surroundings, Wedgwood's pervasive paternalism, and his style of management left the impression that his factory provided a sense of working in a family atmosphere. With the support of preachers, many more entrepreneurs were able to instill a time discipline in the factory population based on obedience to the clock and to the appointments made at times specified on it. Some critics denounced the dehumanizing effect of industrialism and would continue to do so when similar practices were later introduced in the nineteenth century by F. W. Taylor, who declared men and women mere appendages of machines.

The trinity for controlling a group of men, like the ones used for the training of Portuguese sailors and Maurits' soldiers, composed of new devices in the production process, documents concerning factory discipline in the form of printed rules, and drilling of the employees, was not sacrosanct. Many workers resented the enforced specialization of the factory system and felt the rules of behavior overly restrictive. The workers staged demonstrations and expressed grievances. The wage disputes of 1772 forced Wedgwood to examine the cost structure of his business in detail. The solution was to produce the greatest quality possible in a given time and spread the fixed costs over a larger number of articles.

Following the example of Matthew Boulton, who he knew quite well through membership at the Lunar Society, he set up a limited, but nonetheless valued, sickness club at his factories and contributed generously to the cost of establishing schools in the district. In this respect, virtues like prudence and humanity prompted Wedgwood to treat his workmen in an enlightened and liberal manner. Yet temperance was nowhere to be found.

The Lunar Society

As a successful entrepreneur, Wedgwood was drawn to scientific experiments, politics, literature, philosophy, and even botany. He had become a member of the Lunar Society, formed in 1775, combined with five friends who were practical men and extremely interested in electricity and chemistry, topics which had not yet received a commonly shared definition. The five friends involved with Wedgwood in the Society included Matthew Boulton (1728–1809),[11] the manufacturer of metal goods in his Soho works just outside Birmingham; Erasmus Darwin (1731–1802), the physician and grandfather of the author of the *Origin of Species*;[12] Joseph Priestley (1733–1804), the experimental chemist, political radical, and Unitarian preacher; and James Watt (1736–1819), the great advancer of the steam engine. The scene of their friendship was Birmingham, a fast expanding city of traders and manufacturers. As industrial activities were not dominated by craft guilds, the city breathed an atmosphere of free trade. The Lunar Society met every month on the Sunday nearest to the full moon so as to ensure enough light to return home safely.

The members of the society made little distinction between commercial ambition, freewheeling inventiveness, and the sheer rage of curiosity.[13] Boulton, who was busy creating an engine powered by steam, had heard how Watt was working on his own design for a steam engine in an iron works in Glasgow. They combined their efforts into the Soho works, a well-ordered, disciplined place with a highly skilled workforce. Together with Watt and Boulton, Wedgwood established a circle of new and powerful Midland industrial magnates. The Lunar Society became well known, ignited the Industrial Revolution, and was essential for Britain's industrial advancement (Uglow 2007). During the 1760s, Benjamin Franklin stayed in Britain and maintained close contact with the Society.[14] The scientist and deputy postmaster general for North America became famous for his aphorism "Remember that Time is Money."

The Lunar Society, and other learned societies and Royal Academies, became "the morally influential centers where scientists were organized to find ways and means by which nature could be trapped by experiments and instruments so that she would be forced to yield her secrets" (Arendt 1958: 278).

Within the circle of the Lunar Society, Wedgwood showed extreme interest in the opportunities offered by scientific progress and the economic advantages of new techniques. Greater mechanization would enable him to increase output by lowering price rates. He built a new factory adjacent to the new Trent-Mersey Canal and called it Etruria after the region in central Italy where ancient pottery was being excavated. As he saw this as a new business opportunity for his potteries, Wedgwood intensively emulated the excavated material.

To produce the new pottery, Wedgwood installed a steam engine in 1782 manufactured by the foundry of Boulton and Watt. Thanks to this engine, Wedgwood was able to mechanize the labor process and make it possible to calculate standard costs and determine the extent of the production per unit of time. The factory became the first standardized production system to be applied to pottery and turned it into a time-coordinating institution. The rapidly improving internal manufacturing with a growing accessible international market embodied new challenges. Wedgwood was able to realize high-volume production to meet the growing international demand, which his partner Bentley began to effectively serve.

Early forms of marketing

Wedgwood soon became one of the largest potters of earthenware and china in the country (Dolan 2004). Since the local roads were the main culprit in damaging the goods, more reliable and efficient transport of their goods was a necessity. British roads at that time were extremely crude, rugged, and often impassable in the winter even by packhorse. A large group of potters pressured Parliament to build roads for shipment to London, Liverpool, and other markets, indicating that a turnpike and canal would get their creamware transported in a fast and reliable way and move their raw materials by boat to decrease travel time. The canal nearest to Wedgwood's factory was 30 kilometers away. The British government provided the turnpike and canal that Wedgwood and others had pressed for.[15] The new canal linked the port cities of Liverpool and Hull with the Staffordshire potteries and improved the speed of transport. Even though Wedgwood included the cost of transporting bulky raw materials and fragile finished products in the final selling price of every article, the British government did not recover the costs in building the canal and turnpike as they did not charge the merchants a contribution for their use.

Thomas Bentley (1730–1780) began cooperating with Wedgwood in introducing his pottery to fine society. Bentley was the son of a successful farmer, had studied classical languages, learned French and Italian, and was trained in mathematics. Following his studies, he had been indentured to a wool and cotton manufacturer in Manchester, became a Liverpool merchant, and traveled widely in Europe to gain international commercial experience (Dolan 2004). It was Bentley who noticed that the upper classes had become extremely sensitive to a particular "vase mania" which had been sparked by Sir William Hamilton, ambassador to Naples, who owned a great collection of so-called Etruscan vases.

Around 1760, excavations began at Herculaneum and Pompeii near Naples.[16] Bronzes, statues, jewels, and paintings were unearthed, and the public imagination was enraptured. The ambassador promoted an interest in the study of the classical past and was responsible for the diffusion of the classic art of earthenware at Herculaneum and Pompeii. This growing interest in Roman, Greek, and Etruscan sculpture and design set Wedgwood on the path of duplicating vases and classical sculptures in his pottery.[17]

Bentley decided to reside in London where he set up a showroom which later became known as the Queen's Arms after Wedgwood had delivered a tea service to Queen Charlotte in 1765. Thanks to Bentley, ambassadors, envoys, and consuls introduced Wedgwood's earthenware to the courts of Russia (Catharine the Great), Poland, Spain, Portugal, Denmark, Sweden, the Netherlands, Turkey, Naples, and Turin. Unsolicited parcels of Wedgwood's pottery were sent to a thousand members of the German aristocracy and the nobility, a modern form of marketing to selected customers offering them the opportunity either to purchase the items at said prices or return them to the manufacturer at no cost to themselves. Bentley introduced a catalogue in 1774 of ornamental ware to show future clients what they could order for a stylish practice of tea drinking.

The general public began to imitate the new living styles developed by the aristocratic elite. Middle-class customers were won over by the quality and fashionable style of the products and became interested in imitating an aristocratic ideal of drinking tea. "Domestic servants were vital in this process of transmitting the attractions of fashion from the higher and middle classes to the rest of the social pyramid" (Voth 2000: 195).[18] Soon Wedgwood's main European competitors, Sèvres and Meissen, began to imitate his designs. The consumption of coffee and tea increased, which stimulated the purchase of sugar, fine breads, and easily replaceable plates. Coffee shops became a common phenomenon, as was having breakfast. The effect was that the caloric intake increased together with a new pattern for sociability in the household (Bayly 2004). New eating habits at regular times and an increase in industrious activities gave the whole of society a sense of a new time regime (Voth 2000).

Wedgwood's success can be related to his managerial efficiency, his avoidance of waste, and his use of cheap transport by turnpike and canal that enabled him to undercut his competitors.[19] Wedgwood and Bentley became successful in combining mass production with high quality in design and final work. The close personal involvement of both of them reduced the need for large numbers of office staff and their attendant paperwork.

Bentley and Wedgwood, however, also became sensitive to the misfortunes of merchant capitalism. They became active in the movement to abolish slavery in imperial Britain. The Society for Effecting the Abolition of Slave Trade which began a campaign against slavery in 1787 quickly turned into a popular movement. After having met the popular Frenchman Jean Jacques Rousseau, Bentley stimulated Wedgwood in his struggle against slavery. With his own strictly regulated workforce, he produced a medallion with a kneeling slave in chains. This depiction of slave emancipation was in support of the campaign for the abolition of slavery. The Slavery Abolition Act of 1807 ended the most extreme form of dehumanization, but in the British textile industry a new form of dehumanization took place.

Britain's industrious revolution

Some economic and social historians have been reluctant to call the overall changes that took place in Britain an industrial revolution. Although the development in scientific areas in the seventeenth century indeed provided new opportunities and applications, they alone did not lead to an industrial revolution. Next to new production techniques, there was also an increasing demand for new products. Economic historians like Vries (1994) argue that large structural changes in the economy actually occurred earlier in the late seventeenth and early eighteenth centuries and emphasize the importance of workshops and small units of production rather than the factory in industrial growth. In that sense, Vries (1994) prefers to talk about an *industrious* revolution as people began working longer hours and more intensively, which made labor input per member of the labor force increase. His concept of an industrious revolution highlights the savings in efficiency and the reorientation of consumption patterns well before the Industrial Revolution itself took off.

In his analysis of working hours and annual working time, the economic historian H. Voth has stressed that the two main factors underpinning this movement toward more working hours were "the disappearance of 'Saint Monday' and the reduced importance of holy days" (Voth 2000: 175). Although the practice of Saint Monday continued until the beginning of the nineteenth century as a day off (Thompson 1967), its frequency and commonality for many people diminished. Voth (2000) concluded that British people rose earlier from their beds, took breakfast around 8:00 at the workshops, and finished their work around 18:50. This working pattern suggested workdays of 12 hours including Saturdays. Symbolically illustrating this change was an increasing application of pocket watches for personal use.

While in 1675 not even 10 percent of English households had a clock, by 1800 more than a third of all British paupers had one. Shops closed late and offered the population more time to shop.[20] Yet seasonality still had an influence on the daily life of workers stopping earlier in winter than in summer time. Adam Smith was well aware of the practices of working longer hours and wrote in *the Wealth of Nations* (1776: 72–73):

> Where wages are high, accordingly, we shall always find the workmen more active, diligent, and expeditious than where they are low. [...] Some workmen, indeed, when they can earn in four days what will maintain them through the week, will be idle the other three. This, however, is by no means the case with the greater part. Workmen, on the contrary, when they are literally paid by the piece, are very apt to overwork themselves and to ruin their health and constitution in a few years.[21]

Whatever the explanation for the sudden rise in working hours between 1760 and 1830, it has become clear that before the Industrial Revolution took off, living patterns had already begun to change. Polanyi (1977), therefore, called the period of this fundamental transition "the great transformation" as it demonstrated the emergence of a market economy in tandem with the establishment of a modern nation-state and subsequent reforms that reshaped the relations between state and society, church and state. A mixture of political and economic reforms turned into forms of political conservatism and economic liberalism. Guild restrictions on work were removed in 1814; price regulation the following year.[22]

The percentage of the labor force employed in agriculture declined quickly.[23] At the same time, there was a rise in factory work as the result of a complex interplay between technological and economic factors. The factories began to introduce and maintain a strict labor discipline. The mastering of time by the factory owners became a key feature. Cultural and social changes in attitudes towards social time during the Industrial Revolution are best illustrated by Robert Owen, owner of a large textile factory.

The Industrial Revolution in the textile industry

International trade and industrialization created new challenges for business. British manufacturing ventures remained small family firms for a long period of time organized as a single proprietorship or partnership, and kinship connections continued to be of utmost importance in the operation of the firms. Larger firms, however, became a common phenomenon in the cotton textile industry, which exemplified the first Industrial Revolution in Britain.

From the 1660s onwards, Asian textiles rapidly penetrated European markets. Soon, some European governments took initiatives to regulate or ban these imports to support the development of their own infant textile industries. In 1721, the British government introduced legislation to ban Asian finished and unfinished textiles from the home market.

Initially, the EIC had transported calicoes from India to the West Indies, but soon merchants and traders began to transport cotton directly from India to Britain to make finished products at home. British manufacturers created added value in producing cloth, which was subsequently transported to the West Indies and the rest of the world. Due to the expansion of international markets, merchants and traders discovered opportunities to serve these markets more profitably by expanding the production capacities in the home textile industry. "The result was that British textile manufacturers moved into the production of mixed fibres and cloth, then into cotton" (Berg 1994: 126).

Prior to 1780, British cotton textiles were manufactured in small artisan shops and on family farms using spinning wheels to make the thread and hand-driven looms to weave the cloth. Traditionally, clothing made of wool, linen, and cotton were either produced within the guild system with its strict rules concerning production capacity and quality assurance or at home in the countryside. As these systems still relied on servicing local and regional markets with well-known clients, not being much interested in producing for anonymous markets elsewhere in the world, some merchants took the initiative to concentrate textile workers in their manufactures (places were weavers still worked manually) outside the cities where they imposed strict working habits.

In the mid-1700s, however, the textile sector was struggling with supplying cotton-cloth weavers with enough cotton yarn. Due to an invention by John Kay, the issue would soon be in the past. Traditionally, two weavers threw a shuttle carrying a thread (the weft) across a hand loom to each other; the weft passed between a set of tensioned lengthwise threads (the warp) to produce the fabric. John Kay, a clockmaker, patented a shuttle for the hand loom in 1733 (which became known as the "flying shuttle"), which ran on wheels and made it possible for one man to pull the weft through the warp. Workers' shuttles could now work so quickly that thread manufactures could not keep up with demand. As the local weavers of Colchester feared that the shuttle might take their jobs, Kay was chased away and had to flee the region. Yet his technique was put into general use and efficiency of cloth manufacturing increased. His invention was a landmark in textile mass production.

New spinning facilities were needed to help improve weaving output. Different technical improvements stood out. Carding machines prepared the cotton fibers, and a host of spinning machines replaced the spinning wheel in making thread (Berg 1994). In 1764, James Hargreaves perfected the spinning jenny, which reduced the gear ratio of the old spinning wheel.[24] The new machine could be equipped with up to hundreds of spindles but still relied on human power. To remedy that shortcoming, Richard Arkwright invented the spinning throttle in 1767, a device which needed to be activated by hydraulic energy or by a team of horses. This machine produced series of gear rollers at the top of the spinning frame which mechanically imitated the action of the hand spinner. In 1777, Arkwright appealed to James Watt to help improve the power capacity by building a steam engine to increase mechanical output. With the combination of the peculiarities of the jenny and the throttle, Samuel Crompton (of Lancashire) contrived the mule, which came into general use in the 1780s. The introduction of steam power allowed the mills to leave the streams of rural England in search of cheaper urban labor.

The next step was the improvement in the production process of weaving cloth. When, in 1785, the power loom of Edmund Cartwright was attached to a steam engine, production remarkably increased. With Arkwright's carding engine and preparatory frames, mechanical weaving became a challenging new development. Around 1804, Cartwright's power loom could compete with the hand weaver (Cohen 2003). The weaving industry took off, but textile mills still could not meet the new demands. Weaving prices increased as well as delivery delays. Yet, through a combination of factory organization and new technology, the average throughput per cotton textile firm rose by a factor of 13 between 1792 and 1850.

After the American Eli Whitney invented the cotton gin in 1793, a rapid expansion of the supply of raw cotton from the USA was assured. During the first three decades of the nineteenth century, real output in the industry grew by more than 5.5 percent annually. "Improvements in process technology in the cotton industry made possible the rapidly falling prices, which in turn provided the competitive strength for British exports to undercut Indian and other Asian textiles and indeed all other producers" (Freeman and Louca 2001: 155).

Although the cotton textile industry was one of the first British industries to industrialize, hand-loom weavers continued to be very important in this industry as late as the 1840s. The concentration of work in manufactures and later in factories – with steam-powered engines – still took decades to accomplish. The shift in Britain from the preindustrial system of workshop production to the factory system became most evident from 1750 onwards, but it was a gradual process, which meant that the factory system did not immediately displace the domestic production system.

The concentration of work in factories offered factory owners several advantages. The factories cut transportation costs associated with an earlier form of production, the putting-out system. In this system, merchants distributed or sent out raw materials to skilled workers, i.e., to local farmers with their rural families, who made them into finished items in their homes. This putting-out system of production – outsourcing in modern parlance – made it difficult for merchants to enforce control. Within the putting-out system with weavers working in their own cottages, achievement of control over production, more specifically its speed, regularity, and quality, as well as the prevention of embezzlement, were almost impossible (Pollard 1968). Under the putting-out system, merchants could manipulate only the prices they paid per piece. Effort was controlled by the worker who could take anywhere from two days to two weeks to turn in his goods. Workers' time-frames for completing a simple task tended to be dubious (Jacoby 2004). The merchants collected the finished goods and transported them to national and international markets.

Concentrating workers in factories operating a distinct division of labor allowed business people to better supervise employees, thus permitting improvements in scheduling work processes and the quality of goods. The textile trades learned much from the advances in horology. Charles Babbage used watch making as the best example for the introduction of the division of labor. He noticed that there were 102 distinct branches of the watch making trade, to each of which a boy might be put as an apprentice. The division of labor had become so specialized that "the watch finisher, who assembled the parts, was the only tradesman who could work in any other department than his own" (Macey 1980: 36). Adam Smith had been so surprised about this development, as he had recalled that the watch-making practice was generally an easy trade to learn. Not only did the transfer of this kind of practice from the clock-making sector to the textile industry improve production methods, the transfer of knowledge from a French weaving practice also had a similar impact.

In 1804–1805, French weaver J. M. Jacquard invented a silk-weaving loom with an information-processing device. The problem with weaving complex patterns like damask was to select the warp threads for each successive row of the material being woven. These threads need to be raised before the shuttle goes across the material to guide the warp between the weft. The loom works by tugging various colored threads into patterns by means of an array of rods. By inserting a card with punched holes, an operator can control the motion of the rods and can alter the pattern of the weave. The loom, equipped with a card-reading device, automatically produced complex weaving patterns by utilizing interchangeable punched cards that controlled the weaving of the cloth. The new machine transferred the design process from a labor-intensive weaving stage to a card-punching stage.[25] Literature usually projects Jacquard as a precursor of computer-supported automated production.

The need to simultaneously control machinery, materials, and men to ensure that business profits were maximized was the driving force behind the factory system. Factory owners increasingly opted to bring the labor management function in house by direct employment of a salaried manager and the people he managed. This new practice was still complicated because industrial supervision in the nineteenth century was mainly based on the practice of subcontracting. Internal subcontractors (skilled craftsmen) were integrated within the organization either as directly employed foremen or as semi-independent subcontractors answerable to the directly employed foremen (Dawson 1991). Inside contracting gave skilled workers the opportunity to continue arranging working patterns for themselves based on an agreement with the general superintendent or the owner of the company to make a part of the product and receive a certain price for each completed unit (Clawson 1980).

With the advent of industrial capitalism, the role of craft workers to control both the technical details of the work process and the social order of the workplace in terms of inside contracting and working time arrangements withered away. In

British textile factories, owners began to mark the beginning and close of the daily cycle of production by subjecting the workers to rigid controls of entry and exit from the factory. Whereas the factory owners tried to master time in order to prolong it as well as intensify it, labor tried to master time in order to shorten it (Starkey 1988). The doors were locked at the start of the workday, compelling latecomers to return to their homes. Occasionally, the manager came down to the gate to decide whether the latecomers could be let into the factory. Once the workers were inside the gates, they were paid only for what they made, whereas secretaries, clerical workers, and managers enjoyed a secure annual income. They could leave the factory unquestioned through the lodge, whereas a workman required a written pass from his foreman (Gater and Vincent 1988; Voth 2000).[26] The new practice of continuous employment in the factory became "one of the most hated aspects of factory work" (Pollard 1968: 196).

With the advent of technical improvements and new production practices in the textile industry, production capacity expanded incessantly and created the advent of the first Industrial Revolution.

> The share of cotton in total value that was added to industry grew from 2.6 percent in 1770 to 17 percent in 1801. It was on the basis of a whole series of inventions and improvements that big increases in productivity became possible based increasingly on their exploitation in the new system of factory based production.
>
> (Freeman and Louca 2001: 154–155)

A craze for calicoes was launched, but the same development also created discontent and resistance. English textile workers caught in the undertow of mechanization and standardization responded with the destruction of textile machinery. They were upset about the intrusion the machines made on their traditional work patterns and setting of prices. Traditionally, prices were negotiated and skilled workers' interests could be pursued by the benevolent master craftsmen based on relationship-based links. With the product being standardized and the process of production agonized, the price was set by the market and laborers were to bid against each other to get the work. They felt turned over to free-market forces and loss of control of their own working patterns as well their financial security and their pride. The Luddites resented the fact that mechanization and standardization of textile production were turning their jobs into contestable tasks and began to smash the machines. Similar developments took place in France where weavers used their wooden shoes (sabots) to smash the machines (this is where the term "sabotage" originated).

The textile business man Robert Owen became one of the first entrepreneurs to express worries about the effects of this development. He was the first capitalist to practice what is now called paternalism. He became the patriarch of personnel management (henceforth called the stakeholder approach within Human Resource Management) and the figurehead of an upcoming group of leading managers and managing directors who tried to improve, in a social way, the timely organization and coordination of the production processes in factories. He became seriously aware of the negative consequences of the slogan "Time is Money."

Robert Owen

Robert Owen (1771–1858) was born into a family of farmers, shopkeepers, and saddle makers in Newtown (Wales), a place of emergent flannel making. His father was postmaster and churchwarden. Robert had a certain enthusiasm for dancing, music, and sports. He began his career in the retail business of the Presbyterian James McGuffog who traded in linen and woolen products. In London, he worked in a department store and in a firm of silk merchants in Manchester. With easy credit conditions available, he set up with a partner in a machine workshop for spinning mules and roving frames. There, he intensively observed what the workmen were doing to understand the best way to maintain order and regularity.

Peter Drinkwater, owner of a large, modern spinning factory, recruited Owen as manager, and he soon began to improve overall quality control and enforced a strict discipline. By 1789, Drinkwater installed a steam engine supplied by Boulton and Watt, and the factory began to spin American cotton. Drinkwater asked Owen to reorganize his mill in Northwick and keep it under regular supervision.

In 1793, Owen joined the Manchester Literary and Philosophical Society where meetings were held about scientific developments, the impact of technology on social conditions, religious issues, and necessary medical reforms. He delivered a few papers about the influence of the environment on shaping the character of the workman.[27] Owen favored an educational system that would, under favorable circumstances, mold the character of the laboring class. The best way of eradicating the evils that beset society in terms of unhealthy living conditions and drinking habits was not the system of punishments and rewards as advocated by Jeremy Bentham, British moral philosopher, legal theorist, and author of *Punishments and Rewards* (1811), but by rational education and universal enlightenment.

During a business trip to Glasgow, Owen met Caroline Dale, who took him to the nearby cotton mills of her father. Owen was impressed by the remarkably successful textile merchant, manufacturer, banker, and humanitarian employer David Dale, who had built a cotton factory in 1785 near the Falls of Clyde in New Lanark in close cooperation with Richard Arkwright. The mills were built in a steep valley away from existing centers of population. Housing and other recreational facilities were provided in order to attract and retain a skilled and stable workforce.[28]

In September 1799, Owen married Caroline Dale and when her father retired, Owen took over New Lanark on the first of January 1800. He proceeded to put into practice his creed that a man's character is made for him and not by him.[29]

A proportion of the labor force in New Lanark was composed of orphanage pauper apprentices drawn from institutions in the vicinity or recruited from parish authorities in Glasgow and Edinburgh. Owen enforced tighter control with overseers who assessed the behavior and the performance of every worker and duly recorded it in his "Book of Character." For that purpose, he introduced a "silent monitor" which he borrowed from the Quaker educationalist Joseph Lancaster. This consisted of a four-sided rectangular block of wood suspended near every worker. Each side was painted a different color – white, yellow, blue, and black. Whatever color was turned to the front represented an assessment of the conduct of the individual on the previous day: white denoted super excellence in conduct, yellow equated to moderate goodness, blue represented a neutral state of morals, and black stood for excessive naughtiness. The previous day's conduct highlighted on the monitor was noted in the Book of Character.[30]

Owen kept production books in which he meticulously registered output, labor costs, hours of work, stocks of yarn, types of cotton, and their origin. All aspects of factory production were supervised. His prime concern was maintaining business efficiency and improving productivity at the time when the war with France, due to the French Revolution, continued to generate economic uncertainty. During a temporary embargo on cotton imports due to this war, he was forced to halt production temporarily until the embargo expired in March 1809, but he continued to pay wages to his workers.[31] To enhance the general level of skills, the company provided school facilities for the children and instructed the inhabitants of the village of the mill to observe proper cleaning, order, and good behavior. He provided decent accommodation and health care and initiated a sick fund, a savings bank, and a factory store where the factory workers could buy their goods based on labor notes.

Due to efficient factory operations, work discipline, a rigid adherence to timekeeping, and cheap US cotton, New Lanark turned into a profitable enterprise. From 1818 onwards, Owens practiced daylight saving by putting the clocks forward half an hour in summertime and then back in October (Donnachie 200: 160). In March 1816, a shorter working day was introduced, and work was cut by ten minutes. However, as the factory was wholly dependent on waterpower, a dry summer or cold winter could affect production in the mills more severely than the lack of daylight. Yet, his partners argued that there was scope to make New Lanark far more profitable. They intended to remove Owen from the mill management and decided to buy him out. After some internal dispute about the management of the mills and Owen's "progressive ideas", which had offended some partners, the partnership disintegrated in 1812. Owen was able to obtain the support of Quaker partners and Jeremy Bentham to run New Lanark autonomously.

By 1815, New Lanark employed 1,600 workers, whereby similar companies employed fewer.[32] The prime vehicle for social reform in his view was education. Along with dancing and music, the school curriculum introduced military-style exercises. Visitors, especially the upper-class dignitaries, looking at the youthful marches led by flute and drum, much approved of such discipline. In 1816, Owen developed an institute for evening classes, and when he reduced the working day to 10¾ hours, attendance rose rapidly.

Even though the factory system became hailed in industrial circles as the example of industrial progress, Owen noticed that mechanization had made things worse in the textile industry. Children were required to labor more than twelve hours a day, including the time allotted for meals. In 1813, he decided to introduce a ten-hour working day for women and children. During the period 1815–1819, he participated in the movement for factory reform, resenting the fact that the prosperity of the British country was based on a degradation of its workers. Morals had degenerated; therefore, to counter this development, a more humanitarian approach to the working people was urgently needed. Owen placed his ideas stressing the need for factory reform into a pamphlet that he distributed among members of Parliament.

Owen was convinced that a reduction of factory hours and other reforms were partly humanitarian and partly legitimated on the grounds of improved efficiency. He proposed the draft of a Parliamentary Bill which called for a limit of regular working hours to twelve per day for adults, including one-and-a-half hours for meals. The bill should also prevent the employment of children under ten and limit the working day for those under twelve to six hours, along with providing them a basic education.

Although Robert Peel[33] defended Owen's bill, it was not accepted as Parliament thought more evidence was needed. In 1818, Peel introduced a much more watered-down bill which applied only to children employed in cotton mills. This version was accepted in 1819.

Frustrated by the slow progress of parliamentary reforms and lack of support from other employers, Owen continued to plea for legislation to protect the work-force and to improve their health and psychological wellbeing. In that aspect, he began to agitate for the eight-hour working day as a justified balance between working time, sleeping time, and leisure time. Just as the author and political figure Saint Simon did in France, Owen pleaded for more acknowledgement of the social position of businessmen and entrepreneurs who could give substance to what today is called corporate social responsibility. Managers were not supposed to rule over people but instead govern matters efficiently. He established both consumer and producer cooperative movements and demonstrated that effective leadership is a matter of communicating a vision, not barking orders (O'Toole 1995).

Unlike the Luddites who resented the introduction of new production techniques and smashed the new machinery, he was not against technology and, unlike the radicals, neither was he against profit or private property. For Owen, it was quintessential that a proper management of time is determinative for an adequate functioning of organizations. He was remarkably successful in cutting out waste and raising productivity by dispensing with child labor and bringing the mills into full production.[34] Owen continued to stress his views on improved factory production practices in publications such as *A New View of Society*, in which he insisted on the humane treatment of employees. With this and similar activities, he stressed the overwhelming effect of the new mechanical power for the generation of new riches and satisfaction of the wants and desires of every human being. He was among the first of the new breed of entrepreneurial industrialists who created a rationale for a society that could banish social ills through proper education offered in a favorable environment.

During 1814–1824, the New Lanark mills remained a professionally managed and highly profitable enterprise. Owen enforced regulations concerning punctuality and conscientious performance that were closely supervised by overseers. However, the labor-intensive nature of cotton spinning created a management attitude among other factory owners built on the presupposition that problems could only be solved by resorting to a mixture of strong discipline and incentive.

> Mills relying on water to drive the machinery would sometimes use a two-faced clock. One phase would tell the real time and the second phase would tell the time for which an individual would be paid. This factory time was governed by the flow of the river that worked the machinery. If the flow slowed, so did production. The elongated hours measured by the clock geared to the water flow were used, therefore, to calculate the excess hours that employees would need to work to make up their pay.
>
> (Donkin 2001: 67–68)

Sometimes, factory masters would cheat their workers by putting the clocks forward in the morning and back at night.

Owen demonstrated a benevolent paternalism, but not everyone was impressed by his results. Robert Southey, the poet, wrote that Owen might as well be "the director of a plantation. Although the workers at New Lanark were white and could quit the service at any time, they were under the same 'absolute management' as so many negro slaves" (quoted in Donnachie 2000: 182). Despite Southey's misgivings, the New Lanark experiment was regarded by most onlookers, especially American visitors, as a success. American factory owners as well as politicians and medical doctors were touring Britain to learn about more humane labor conditions in factories.

Approximately 1,800 textile mills in New England had become America's first factories that supplanted home methods of manufacturing to factory production. These early mills shared a number of features that distinguished the factory system from other more traditional forms of production. The system relied on power-driven machinery, integrated different production processes at a single site, introduced an elaborate division of labor, and developed new methods of administration based on the role of the overseer or foreman. The textile factories required more effective labor supervision than was previously possible, and American visitors came to New Lanark to learn about Owens' approach.

However, as Owen did not obtain the general success he expected, he looked for another, international challenge. He left Britain in 1824 and sailed for the United States to establish a new settlement. He bought the community New Harmony in Indiana which had been founded by the German priest Frederik Rapp with other German Lutheran dissenters from Württemberg. It had been advertised for sale in the European press, thus the reason for Owen's interest. Due to the economic depression in 1819 and social conflicts between New Harmony and the local population, Rapp's disciples had grown restless, at which time he had decided to abandon it and build another settlement elsewhere.[35]

In March 1825, Owen got the chance to speak to several members of the American Cabinet, Supreme Court, and Congress assembled on Capitol Hill. He visited former presidents John Quincy Adams, Thomas Jefferson, and James Madison to talk about his project. However, deception soon followed. New Harmony dissolved amid bad planning and internal dissension. It became difficult to turn the broad range of cooperative societies into a more general movement and create a new way of producing and living in America. Disappointed by the results, Owen returned to England where he created "a movement, modest certainly, but nonetheless some of his most enthusiastic disciples had actually been inspired to put the precepts of the new system into practice" (Donnachie 2000: 227). One of his new initiatives was the establishment of a trade union in the textile industry and support for the Grand National Consolidated Trade Union in the 1830s which aspired to organize all workers and resist the pressure of increasing working hours.

Owen became the icon of a social business man introducing humane working conditions while at the same time improving the value of the company. During the British industrialization, however, the introduction of gas lighting facilitated the use of longer working hours and shift work in factories. British textile manufactures and factory owners in other industrializing countries rarely imitated Owens's policy and exploited the factory workers with long working hours for profit reasons. As will be shown in the next chapter, the international trade union movement fought throughout the second half of the nineteenth century for a legal limitation of factory working time to eight hours per day.

Trains

The economic success of the Industrial Revolution in Britain did not merely rely on the combined introduction of technical improvements and the factory system in the textile industry. Businessmen like Wedgwood had also pressed for improvements in the transportation sector to enhance a timely provision of raw materials and distribution of finished products at national and international markets.

Beginning in 1784, British mail coaches developed a certain level of synchronization by running a service of operations strictly according to timetables. However, punctuality of the delivery of mail remained a problem. In every stagecoach, a timepiece was set to Greenwich Mean Time, and British mail tried to synchronize every single town and village which still relied on their own local time. Step by step, the local towns did abide, and the system of coach time began to replace the local times.

Only with the construction of the railroads did the nation obtain an artificial space in which synchronization of different places really became possible. The idea of a railroad was not new. Wooden rails had been in common use in British mines with horse-drawn colliery wagons moving up and down the mines of Nottinghamshire to take the coal from the face to the shafts and from the pits to the canals or other nearby waterways. Richard Trevithick ran his first full-size locomotive at Camborne on Christmas Eve of 1801 using steam power as a means of propulsion. He later demonstrated to the public his locomotive called *Catch Me Who Can* on a circular track encased by high wooden walls on desolate land.[36]

In 1825, Britain experienced the first 40-kilometer railroad services between Stockton with its abundant Durham coalfields and the market town of Darlington in the northeast of England. The Stockton & Darlington Railway was constructed by subcontractors, who maintained the railroad's right-of-way, repaired its locomotives, and even wound its clocks.

The coming of the railroads created a new system of transportation. Some of the new railroad companies were created to break down a canal or navigation monopoly (Clapham 1950). The Birmingham Canals, on the other hand, became a railroad company just like the Trent and Mersey Canal. The success of the initial railroads was mainly due to the resolution of small groups of local businessmen and the conviction of a few engineers (ibid.). Liverpool businessmen were particularly eager to invest in the railroads and link together several different cities of importance.

In order to reach the goal of safe and quick journeys by improving railroad technology, a competition was organized between locomotive constructors. On Tuesday, October 6, 1829, a huge crowd gathered at a meeting ground in Rainhill, about 15 kilometers from Liverpool. The competition between eight locomotives was organized to see which one would be fast enough for regular railroad service.

English engineer George Stephenson, who was the chief engineer of the Stockton & Darlington Railway and an agile repairer of clocks and watches, had previously won a contest to design a locomotive for the new Liverpool & Manchester Railway (McGowan 2004). With his steam locomotive *Rocket*, built with the assistance of his son Robert in their factory in Newcastle, George Stephenson triumphed at the Rainhill locomotive competition. His *Rocket* traveled as fast as 32 kilometers an hour at a time when a horse-drawn wagon needed a whole day to travel the same distance. The 50 men who had traveled with the *Rocket* were astonished at the velocity: with their pocket watches, they had timed their journey at 55 minutes.

The railroad represented an almost blinding speed, and many people wondered if the human body could endure the strain. Soon steam locomotives pulled several different carriages for transport of goods and persons on different tracks.[37] The Rainhill trials and the creation of the Liverpool & Manchester Railway were "decisive events that demonstrated to the public at large [...] that steam locomotives were the power of the future" (Freeman and Louca 2001: 203–204).

In 1830, the first major railroad with a track of 48 kilometers between Liverpool and Manchester opened the modern transportation age. The two cities were primarily looking for a means to improve interurban transportation facilities in order to strengthen the link between the textile mills of Manchester and the docks of Liverpool. Along with freight haulage activities, the railroad made it possible for large numbers of passengers to undertake quick journeys over long distances. Although the line ran 22 coaches carrying more than 600 passengers and could operate incessantly all days of the week, the directors of the Liverpool & Manchester Railway held an energetic discussion about the morality of running their trains on Sundays (Garfield 2002).

The general consensus was that journeys still took a long time, possibly between three to four-and-a-half hours. However, the intrinsic advantages of a railroad system were widely discussed in the newspapers, explaining that it was expected that the transport would become cheaper, faster (around two hours), and independent of extreme weather (Garfield 2002). Other railroads quickly followed and established links to all large towns. Traveling by railroad at a common speed rather than by stagecoach became a new phenomenon. Though there was less personal danger associated with this new travel experience, the first serious accident took place in the early morning fog of November 23, 1832, when a delayed train at Rainhill station was hit by another incoming train. Discussions in the newspapers focused on the safety and reliability of the system, and for years, the news reported more disastrous railroad accidents.

In 1833, a group of Bristol financiers founded the Great Western Railway (GWR) which provided a straight connection with London. It extended its railroad transport in the West Country and developed links to the Midlands. The GWR owed its reputation as one of the big four railroad companies to the engineer I. K. Brunel, who became a famous shipbuilder. The GWR pioneered express travel with its *Flying Dutchman*, a service which at that time was the fastest in the world. Before long, greater value was placed on speed and directness of the railroads. The company created a centralized structure with a high degree of managerial control and recruited staff no longer based on patronage but on a modern career track. It developed an incremental wage system that rewarded long-serving workers.

In June 1836, the London & Greenwich Railway was opened. One of the assistants at Greenwich Observatory, John Henry Belville, convinced the company to implement Greenwich Mean Time and synchronize pocket chronometers by setting them to Greenwich Time. "The mail coach tradition of a person responsible for punctuality carrying an official timepiece was continued by the railways" (Howse 1980: 83). In November 1840, the GWR ordered that Greenwich Mean Time should be kept at all of its stations and in its timetables.[38] Many other railroads followed suit during the next few years.

In 1839, G. Bradshaw, the publisher of *Bradshaw's Maps of Inland Navigation*, introduced the world's first compilation of railroad timetables to help passengers synchronize their travel plans with the schedules of the trains. *Bradshaw's Railway Companion*, in a yellow wrapper, became synonymous with its publisher. For Victorians and Edwardians alike, a railroad timetable was a *Bradshaw*, no matter which railroad company issued it.

By 1840, approximately 2,253 kilometers of steam-powered track had been laid, operated by about 40 independent companies. The London & North Western Railway (LNWR), which was formed in 1846, alone exploited around 1,287 kilometers of track and employed 12,000 workers. Although other countries like Germany and the United States also embarked on railroad construction, the British lead remained substantial for some time (Freeman and Louca 2001: 190–191). When LNWR began to expand its operations, "it put in place a structure of three regional divisions, each of which had its own secretary, superintendent, locomotive manager, and goods manager" (Blackford 1998: 70). The divisional managers reported to a general manager who oversaw all the operations of the line. Precise record keeping systems, as well as accounting methods, were implemented to improve overall control.

By the mid-1840s, the predominance of passenger traffic was visible, and passengers began to appreciate the railroad's speed. However, the railroad companies were faced with serious concerns about the safety of the trains. As the lines grew longer and became more complex in their operations, railroads found it necessary to take over functions previously handled by their subcontractors, thus internalizing actions previously undertaken by independent small businesses and structuring the company into a solid bureaucracy. It was essential to get the trains to run between stations roughly on time. If a train did not arrive at a station and the cause of the delay was unknown, another train had to be sent down the line to search for it.

On September 22, 1847, the Railway Clearing House (a body that coordinated many aspects of railroad operation in Great Britain) recommended that each company adopt Greenwich Mean Time. On the night of Saturday, January 29, 1848, Edinburgh, Glasgow, Greenock, Stirling, and Perth all reset their public clocks on the same date. With the growing use of railroad transportation, many cities followed the railroad example and, by 1855, almost all public clocks in Britain were set to Greenwich Mean Time (Zerubavel 1981).[39]

By 1870, four substantial companies – the London & North Western, Midlands, Great Western, and North Eastern – dominated the British railroad industry. Railroad companies were joint-stock concerns with limited liability and a dispersed ownership. British rail was entirely financed by risk-bearing investors. Their organizational structures were mainly organized on a functional basis, and many newly recruited managers had a university degree. However, the new managerial practice of coordinating a large railroad company rarely spilled over to other large British industrial firms (Blackford 1998). Yet the railroads increased the influence of machine builders: the machinist/craftsman turned into the mechanical engineer who improved the building and construction of the railroads.

Although a reliable time system was necessary, it was not sufficient to guarantee the safety and the synchronization of the railroad system. The increasing alarm amongst the population concerning accidents moved Queen Victoria to ask railroad companies to provide safe conditions for journeys, but the British government found it difficult to establish an effective system of control (Cohen and Quinter 1996). Due to the spatial dispersal of their employees to a vast network of stations, direct supervision of all staff was impossible. Railroad companies turned their attention to the military to ensure discipline and control. They succeeded in securing special legislation to allow their employees to be treated in a military manner (Savage 1998). Every new employee was given a Rule Book which described the required behavior in detail. The Rule Book specified the necessary working practices and how to devote them to the company. As the companies were not sure

FIGURE 5.1 *Rain, Steam, and Speed – The Great Western Railway*. J. M. W. Turner, 1844.

about the way new staff would internalize the prescriptions from the Rule Book, traveling inspectors were introduced for visual inspection.

The general public, as well as artists like J. M. W. Turner (1775–1851), admired the impact the trains had on public life. Steam engines became an object of art, and railroad stations were built as the new cathedrals. Turner's painting *Rain, Steam, and Speed – The Great Western Railway* (1844), probably the first in its style, provides a perfect image of the coming age of speed (Figure 5.1). Details fade away when speed is represented, and yet one immediately has the sense of an object rushing towards the viewer. Here the atmosphere created by velocity is all that counts, rather than the precision of painted objects.

Telegraphy

An important adjunct to the railroad system was the development of the electric telegraph which became essential for the signaling and safe dispatch of trains. The telegraph was the brainchild of several people, and no one could lay claim to the title of "inventor" of the telegraph. Nevertheless, Frenchman and amateur physicist Claude Chappe certainly played a prominent role by inventing the optical telegraph during the French Revolution. His invention provided a device that would certainly defend the revolutionary government and army against the sudden appearance of foreign enemies.[40]

With the application of electricity, the transmission of information fundamentally changed. Several different inventors had been involved in the design of the electric telegraph including two British men, William F. Cooke, who studied at the University of Heidelberg (Germany), and Charles Wheatstone, who studied physics at King's College in London, as well as the American portrait painter Samuel F. B. Morse, who designed an efficient code.

Remarkably, there was no initial particular or political necessity that lay waiting for the electrical telegraph (Friedel 2007). Wheatstone and Cooke developed a viable system which was first demonstrated on July 24, 1837, when they ran a telegraph line along a railroad track and successfully transmitted and received a message. The London & Birmingham Railway showed interest in the experimental telegraph link between Euston and Camden Town stations, which Cooke and Wheatstone had built over a distance of one and a quarter miles (about 2 km), and introduced Morse's telegraphic codes to regulate the departure and arrival of trains on the basis of standard times. Yet, for whatever reason, the company decided not to proceed further. When Cooke turned to the GWR, he encountered more sympathy, and 2,000 kilometers of telegraph link between Paddington and West Drayton was soon installed (Standage 1998: 45). When other railroad companies began to discover the advantages of quickly sending messages across the line, telegraph facilities increased in popularity.

In June 1843, there were about 3,057 kilometers of line opened in Great Britain (Clapham 1950). The same year, the first public telegraph was in operation alongside the GWR line. By 1846, the privately owned Electric Telegraph Company was formed to exploit a large telegraph network. The railroad companies decided to operate at a national level with standard time throughout their system. The company bought up most of Wheatstone and Cooke's patents together with those of other inventors and pressed telegraphy on the railroads and on the public.

The transmission of messages to synchronize trains made the public aware of the existence of a new time infrastructure that also began to determine their pattern of mobility. The efficiency of the organization of railroads, however, did not solely rely on technological improvements but also on administrative control. Yet management level of the railroad companies had difficulty with the surveillance of their own personnel in respecting and imposing strict time schedules. The surveillance practice by inspectors turned out to be less effective as visual supervision was difficult to enforce. Railroad companies needed to structure their organizations in more hierarchical and bureaucratic form to ensure the effectiveness of their personnel.

An essential by-product of modern bureaucracy is that it secures efficiency and organizational cohesion. By 1870, the GWR replaced the military-like practice of punitive discipline (the use of demotion, suspension, and reductions in pay) by a designed form of discipline (dressage) based on a policy of modern career progress and the systematic construction of career routes which encouraged workers to monitor and regulate their own actions.[41] The new approach worked into the soul of the employee and marked a breakthrough in the management approach in the railroads. The bureaucratic career culture which internalized ethical concerns was, so to say, first developed as a disciplinary device in the railroad companies. Normative control seeped into the unspoken codes of organizational life based on respect for the application of a strict time regime as defined by abstract clock time.

By the mid-nineteenth century, the post office, railroads, and telegraph companies were all keeping Greenwich Mean Time. It had become essential to these organizations to have a standard time throughout the system. Factory owners, however, were less concerned whether their factory clocks were all synchronized to Greenwich Time as long as they could maintain the control over their clocks.[42]

Steamships

The modernization of Britain not only relied on the coming of the railroads. The quickly expanding scale of human activity was also hastened by the introduction of steamships which eventually eclipsed sailing craft. Crucial new inventions were introduced which Britain quickly took advantage of to produce cheaper steel and develop less expensive shipbuilding. The mass-scale production of steel for the building of ships was largely made possible by the invention of the converter by Englishman Henry Bessemer and the open hearth furnace by the German Siemens brothers.[43] New battleships, field guns, and armed steamboats facilitated access into protected markets in the Ottoman Empire, China, and South America (Bayly, 2004). The new ship-building technology enhanced British power to dominate huge regions of the globe at a time when other nations began to struggle with internal affairs, such as Germany with unification and America with civil war. Although by 1820 "world trade accounted for only about two percent of all combinations," progress of international trade was on its way (Lang 2006: 920).

While steel took the place of wood, steam replaced sails. The appearance of iron steamships maintained the British competitive position.[44] Brunel was not only involved in railroad building but also moved into ship design and built the first transatlantic steamer *Great Western* which, from the 1840s onwards, established a transatlantic steam service to New York. On its maiden trip it "smashed the transatlantic speed record, crossing in 15 days and 5 hours" (Gordon 2003: 152). Brunel's genius was to design a ship fully five times the size of any ever built before. The next ship he built was the *Great Britain* which was the first large iron steamer and the first to rely on screw propulsion. His *Great Eastern* became the largest ship built in the nineteenth century.[45]

The Great Exhibition

Britain felt the time was right to demonstrate its industrial advances in production and transportation to the general public in a more peaceful manner. On May 1, 1851, trains brought passengers in large quantities to London, where they could visit the first Great Exhibition of the Works of Industry of all Nations in Hyde Park, usually referred to as the Crystal Palace Exhibition. It had been constructed from cast-iron frame components and glass and with its transparency was an instant success and announced the coming of a new dynamic industrial phase.[46] Queen Victoria opened the Exhibition that same day and acquired a key-wound timepiece there in a pendant style that was manufactured by the Geneva watch-making company Patek Philippe.[47] The exhibition received more than six million visitors and displayed a remarkable self-confident technological progress.

By the time of the Exhibition, Britain had become one of the largest empires in history, and the Exhibition demonstrated Britain's economic and technological supremacy to the world. Throughout the nineteenth century, Britain developed into an empire where industrialism merged with imperialism. An unprecedented territorial expansion across Asia, Africa, and the Pacific took place. British hegemony became based as much on its position of a world power as on its size as a world economy. This material expansion of the capitalist economy led to industrial globalization (Arrighi 2010).

Yet the upper classes and elites began to cling less to the British promotion of international trade and commerce[48] or the British industrial lead against the competition from Germany and the United States. Instead, they turned their interest to the running of the British Empire and focused more and more on jobs in the public sector where they aspired to obtain lucrative civil servants' jobs to run the Empire (Brendon 2009). British elites used their surplus capital less in the running of the industrial enterprises and more in the bureaucracy of the Empire.

For the British bourgeoisie, the period from 1896 to the outbreak of the First World War was one of the *belle époque* (Hobsbawm 1995). The cosmocratic network of high finance which had initially invested in industrial Britain moved into banking and financing of imperial Britain. Although the Great Depression of 1873–96 ended the tremendous expansion of the Industrial Revolution in Britain, its overabundance of money capital became primarily invested in the United States. The third systemic cycle of capital accumulation with British high finance in the lead came to a close. Eventually, even Britain's watch-making industry fell back technologically, while "both entrepreneurs and labor had resisted innovations" (Landes 1983: 300–301). It lost its position to the American watch-making industry.

6

POLITICS AND TIME

Industrial globalization initially developed in Great Britain but soon spilled over to other European countries. The social disruptions that occurred due to this development led to political clashes. One of the main issues that arose was the organization of factory workers in labor unions struggling to improve labor conditions and create shorter working hours. Hence, Owenite Socialism became an important influence in the British working-class movement. However, most entrepreneurs cared little about what Robert Owen proclaimed and wanted to stay masters over time and maintain long working days. The monotonous labor in textile factories and the lack of variability in labor patterns, nonetheless, evoked resistance. Factory workers defended the Blue Monday (a free Monday to resist the increasing discipline of the factory bell that was imposed on weavers and cobblers during the six working days of the week, cf. Thompson 1963), regularly changed jobs, and sabotaged the homogeneous time structure that had become imposed within the textile factories.

To guide this resistance in a more structured way, Owen, at the age of 63, initiated and supported the creation of the Grand National Consolidated Trades Union and briefly chaired the forerunner of the modern British National Trade Union movement. Resistance against the long working hours gained voice. Within the English labor unions, the slogan "the Master's Right in the Master's Time and the Workman's Right in his own Time" could soon be heard everywhere.

The French Revolution and a new rational order

During the seventeenth century, European wars arose mainly due to clashes between religious beliefs and the power of the state to suppress dissidents and heretics. During the eighteenth century, however, wars focused primarily on dynastic resources and the control of towns, trade, and customs. The status of monopolist corporations such as the Dutch, English, and French East India Companies which had stimulated overseas commerce and had gained increasing political influence while providing credit facilities, were eventually contested. With the support of the EIC, the British commercial empire had successfully expanded its network of forts and trades and advantageously bargained with continental powers.

Meanwhile, as European governments could hardly cut into the earnings of the aristocracy to finance their military and naval operations, taxes needed to be continuously increased. Riots over taxation and excessive duties infected continental Europe. The historical collapse of French absolutism was the outcome of the state's inability to tax the class which it represented. "The very rigidity of the nexus between State and nobility ultimately precipitated their common downfall" (Anderson 1974: 113). However, France was not the only country confronted with this collapse. "The world crisis of 1780–1820 had its origins in a growing imbalance between the perceived military needs of states and their financial capacity" (Bayly 2004: 100).

Americans resisted British dominance in the colonies by crying out in 1776 "no taxation without representation" and fought for their independence. The American example brought ideas of reform to Europe that rapidly germinated in France. Peasant protests, anticlerical explosions, and constant outbreaks of popular disorder against the aristocracy and the Church fueled the French Revolution, which burst into rebellion in 1789. It was the first movement of ideas in Europe which did not appear Christian and quite often presented itself as anti-Christian. It not only overthrew the monarchy, established a republic, and plunged the nation into a terrible civil war, but it also caused more than 20 years of European warfare.

The revolutionary French government wanted to present an example to the world of the clearing up of ancient muddles and disenchant the world by destroying the long-lasting influence of the old orders of Church, royal families, and trade guilds. It instructed the Académie des Sciences (Academy of Sciences) to devise a new, rational, unit of measurement which should regulate the society in a modern manner. Thereupon, the Academy introduced the "meter" as the new unit of length, with divisions and multiples based on the decimal system. This principle was extended to other units, becoming the metric system, which was seen as a scientific incarnation of the values of standardization, universality, and equality with science being the rational mediator between humanity and nature.

To lend authority to the new unit, it was determined that it should be related to some feature of the natural world. Scientists insisted upon a natural magnitude determined by scientific procedure as standard, be it a new unit of time or a unit of distance, whereas some revolutionaries stressed that it should be based on a mere convention. To solve this issue, the Academy had discerned two possibilities. The first was to take the length of a pendulum which would, at a given latitude and at a given temperature, mark out seconds of time.[1] The second option was to take a fraction of the length of the meridian, i.e., the distance on the earth's surface between the equator and the North or South Pole (the quadrant of the meridian). This proposal was inspired by some geodesists who, around 1760, had already suggested that for an accurate measurement of distance, a fraction of the length of the meridian would provide a convenient measure.

The revolutionary French government struggled with the two options and installed two committees to deal with the two propositions. One was instructed to verify the length of the seconds pendulum. The technique consisted of comparing "the oscillation of a bob suspended on the 12 foot wire with those of a pendulum clock accurately beating seconds" (Gillispie 2004: 252). The other committee, presided by the Bureau de Cadastre, had to determine the exact length of the meridian from Dunkirk to Barcelona.

The government decided to investigate the second option. Therefore, scientists Jean Baptiste Delambre and Andre Méchain were dispatched to survey, by triangulation (a technique developed by Gemmo Frisius in the sixteenth century) the meridian from Dunkirk via Prats-de-Mollo-la-Peste in the Pyrenees to Barcelona.[2] Mechain, who was in charge of the southern part of the route, suffered an accident which incapacitated him for six months; later, he discovered an inconsistency in the calculations he had already reported. The project took six years and the decision was ultimately made not to rely on a fraction of the length of a meridian but instead on the standard of a meter as a manmade convention.

The French mathematician and Marquis De Laplace proposed that a linear standard should be called the meter (*le mètre*). By 1792, the Academy accepted his proposal. The "Agence Temporaire", later to be called Bureau Consultative des Poids et Mesures, placed orders for the manufacture of one hundred standard meters in copper. The general principles of scientific nomenclature were further embraced by introducing the decimeter, centimeter, and millimeter as well. All units were to be calibrated against the original standard. The purpose of decimalization was to have a standard subdivision of all units, whether it be for money, length, surface, volume, or weight.

In 1798, the Minister of Foreign Affairs, Talleyrand, organized an international conference to get the metric system more generally adopted. Although no representative from England, the United States, Austria, Prussia, or Russia was present, the conference adopted the introduction of the metric system based on a standard meter of platinum, which had been manufactured by Marc Etienne Janety and his son. Janety had been working as a goldsmith for the royal family but fled to Marseille during the Revolution where he settled as a manufacturer of base-metal clock parts. He was sought out by the new French government because of his knowledge of the working of platinum and was asked to come back to Paris and prepare the standards of length and weight for the newly established metric system. Between 1795 and 1802, he delivered four standard meters.

One month after Napoleon seized power on November 9, 1799, legislation enacted the definite standards of the metric system.[3]

The French calendar

On December 20, 1792, the National Convention, a new assembly ruling France, authorized the Committee of Public Instruction to consider a general reform of the existing calendar and recommended to establish a new calendar based on a set of 30-day months, each of which would be divided into precisely 10-day weekly cycles called decades (Zerubavel 1981). The new calendar had to respect the division of the year into 12 months to remain in harmony with the annual frequency of the lunation.[4] Revolutionaries felt that the traditional Gregorian calendar should suit their revolutionary political system.[5] The French Republican calendar had no place for a Christian pattern of days and weeks. They insisted upon the decimal principle and insisted that the months should be divided into three ten-day periods, each ending with a free day dedicated to a popular French person.

Charles Gilbert Romme designed a calendar for the Republican era which would begin with the abolition of the monarchy on September 22, 1792. On that very day, at 18 minutes and 30 seconds after nine o'clock in the morning, the sun crossed the equinox. In accordance with this event, the opening of all successive years should coincide with the equinox (Gillespie 2004). The convention ordered that, henceforward, all public clocks were to be dated starting from the first year of the Republic. Romme applied the decimal system, and every month following would be divided into three parts of ten days. The year will then consist of 36½ decades, i.e., 73 demi-decades; each day of the demi-decade could then be identified on one of the fingers of the hand. As the sexagesimal subdivision of the hour and the minute would be extremely inconvenient in this new scheme of time, he proposed to partition the day into ten equal intervals. The day of rest would come at the end of every decade (ibid.). The declaration of war on Great Britain by the French National Convention on February 1, 1793, initially prevented a general implementation of this new decimal system.

On October 5, 1793, the National Convention however, adopted the new chronological framework of Romme. Inspectors became heavily involved in the enforcement of the new practice. The revolutionaries commanded the production of new watches that displayed two sets of times: an inner circle with ten hours and an outer circle displaying a twelve-hour frame. But the nomenclature to indicate the months and the days Romme had designed was another matter. He had proposed an inconsistent labeling of the days which was not well accepted. As chairman of a new committee entrusted with the creation of a natural vocabulary for the days and the months, Fabre d'Eglantine coined the months according to the seasons. This new calendar featured ten-day weeks so that Sunday would be skipped as a religious day.[6] The great Terror of Robespierre further delayed its full implementation. Romme was arrested and in 1795 committed suicide.

In 1798, however, the new calendar was made compulsory, and any other Christian observance of the week was forbidden. The general public perceived the Gregorian system as immutable, and the new approach of time in the form of the calendar with a decimal time frame was greatly resented. As there was still so much unfamiliarity with the nomenclature, a new decree was issued on November 4, 1800, stating that the new system would come into effect throughout the Republic on September 23, 1801. De Laplace noticed that resentment against the calendar continued and recommended Napoleon to return to the Gregorian calendar as of

January 1, 1806. Napoleon accepted and ordered the return to the Gregorian calendar which, however, was not applied everywhere in continental Europe.

On December 2, 1805, Napoleon landed a tremendous military success during the battle of Austerlitz against the Austrians and Russians.[7] The Austrians had been promised that Russian troops would arrive before the end of October, but the Russians still used the Julian calendar and arrived too late. The Austrians at the garrison at Ulm had to surrender due to lack of Russian support, which contributed to the final victory of Napoleon. As the Gregorian calendar was ten days ahead of the Julian system, the Russians could hardly have arrived on the scheduled Austrian date, and "this miscalculation was to ruin the balance of the entire Allied scheme" (Chandler 1967: 383).

Other nations including Britain had also struggled to adopt the Gregorian calendar. Although the Royal Society had paid heed several times to necessary calendar reforms, it would take another 170 years before Puritan Britain finally adopted the Gregorian calendar in 1752. The population revolted for a long time and finally came trailing in amid a clamor of "give us back our 11 days." Britain, however, was not the last country to change. Sweden changed in 1753; Greece did not reform its civil calendar until 1924. But some institutions like the Russian Orthodox Church continued to use the Julian calendar for the immovable feasts, even though the civil dates are 13 days later. [8]

Standardization

The introduction of the new metric system in France eventually succeeded in 1837 when the National Assembly carried the new practice. What finally made it succeed was not persuasion, reason, or authority but time itself: It had taken 40 years for the educational system to teach all French children meters, grams, and decimals, and this eventually weakened the resistance.[9] Yet elsewhere, the Julian calendar remained operational. The differences in calendar practices remained a nuisance for businessmen dealing with international trade. Slowly the Gregorian calendar, however, became more and more the dominant standard for reckoning time.

What the French revolutionaries had attempted to realize, a modern standard calendar with a new terminology, failed. Yet the metric system of weights and measures based on standardization, decimalization, and a particular definition of the meter remained the most pervasive legacy of the French Revolution. The metric system replaced traditional sizes of a unit like the measure of a man, the foot, the ell for cloth, and bushel for grain, which all varied according to localities. Standardization would continue to be hailed as the best way to improve industrial production, but standardization of labor time in terms of limiting the working hours to a certain standard was a different matter.

In the aftermath of the French Revolution and its overthrow of the existing social structure, political discussions about the position of the working class in Western society continued. The French Comte de Saint Simon, an aristocrat by birth, proposed to reorganize society on the basis of those who were mainly involved in productive activities. He intended to divest his own elite power and prestige by placing business and scientific-technological elites at the apex of the new social pyramid of society. Productive individuals were to replace the nonproducing aristocracy. He believed that this would lead to a new society with more equality. Much like Owen, Saint Simon became an essential figure in the struggle for a just combination of the drive for efficiency in the factory system with humanity at shop-floor level.

However, in Britain, Owen's views ran counter to the accepted beliefs and ideology of the employing class. Only a few would attempt to improve working conditions to prevent confrontations with unions.[10] In 1835, the first guide book to cotton factory management that included technical and managerial advice appeared, but it hardly changed the prevailing practice of excessive working hours (Pollard 1968).

Labor's protest

In 1851, Karl Marx listened to an address by Owen who, by then, was 80 years old. He was impressed and wrote about Owen's progressive views in a favorable manner to his friend and supporter, the industrialist Friedrich Engels. Still, in his *Condition of the Working Class*, Engels accused Owen of expropriating the profits at New Lanark from the working people. He was convinced that the factory system was detrimental to workers even though Owen tried to compensate for the worst effects. In factories, the new technologies, i.e., the increased fixed capital per worker, led to a greater division of labor which made it imperative for employers to coerce the workers into a stricter discipline and realize synchronization. British historian E. P. Thompson (1963) concluded that as long as the manufacturing industry had been run on a domestic or small workshop scale, little synchronization was demanded. Instead, task orientation prevailed.

With the introduction of the factory system and technical progress, however, working conditions deteriorated. As the factory system demanded the synchronization of labor, working time had become a key element of industrial control for the factory owner. With the introduction of water – and later steam – power in the textile industry, this need for synchronization had become paramount. Factory owners employed supervisors on the floor, levied fines, and dismissed their workers in order to instill obedience and discipline. "What was needed was regularity and steady intensity in place of irregular spurts of work; accuracy and standardization in place of individual design; and care of equipment and material in place of pride in one's tools" (Pollard 1968: 213).

Although factory owners were able to buy the laboring time of the workers, their ability to dispose of it in the most productive way was limited by the monopoly of industrial knowhow in the hands of the skilled workers. Marx insisted that employers would do anything to exploit any insights of the factory workers based on their own skills, dexterity, and experience. Although science could be an instrument for improving production methods and labor conditions, as Wedgwood and

Owen had illustrated, the factory system prevented this as the position of the factory workers was reduced to one of subordination. It was the master/owner who designed and controlled their work as well as their working time.

The factory clock became the "new kind of power machine" (Mumford 1934), a symbol of the discipline, regimentation, and synchronization of industrial production. By the mid-nineteenth century, discipline triumphed in the textile industry, with workers required to attend fixed hours and obey strict codes of conduct: in return for their wages, they surrendered complete command of their labor to the employer for a fixed period each day (Clark 1994). Cotton weavers were fined if they arrived only minutes late to work. A fine could rise to two hours of wages for tardiness of ten minutes. The factory meant economy of time and, as such, introduced the notion of time thrift.

With the application of technology, employers were able to introduce piece-rate approaches to stimulate workers even more for active involvement. They coerced the workers into doing more than they would have freely chosen to do. The overall purpose was to extract more work per unit of pay. The puzzling aspect was that employers in the textile industry began to pay their workers on the basis of the inputs, like hours of attendance, sobriety, and demeanor and not on the output which in the textile industry was easily measurable.

Nevertheless, factory workers were paid much more per week than hand workers working from home and in undisciplined shops. Workers accepted the imposition of discipline by being compensated with the premiums they received but also due to the fact that imperfect labor markets prevented textile workers from looking for other jobs. They stayed in the factory because, at the end of the week, their wages were percentages higher than they could have received without the discipline of the factory. Workers who retained their independence received on average about 20 percent less per piece of the same type of cloth in wage payments (Clark 1994).

Thompson (1967), who chronicled the deliberate efforts to discipline the British workforce, quoted the record of two factory men who testified that workers were not allowed to have their own clocks and watches within the company grounds:

> In reality, there were no regular hours: masters and managers did with us as they liked. The clocks at the factories were often put forward in the morning and back at night, and instead of being instruments for the measurement of time, they were used as cloaks for cheating and oppression. Though this was known amongst the hands, all were afraid to speak, and the workman then was afraid to carry a watch, as it was no uncommon event to dismiss anyone who presumed to know too much about the science of horology.
>
> (Quoted in Thompson 1967: 86)

The factories became "the familial landscape of disciplined industrial capitalism" (ibid.: 82). The introduction of the industrial clock and discipline was a dramatic and traumatic watershed in the history of Western society.

The progressive temporal commodification of labor was accompanied by an increase of deskilling of work itself. Yet, large factories did not sweep quickly across Great Britain. "Two thirds of the nation's cotton textile mills in the mid-nineteenth century still employed 50 workers or less" (Blackford 1998: 45). Out of a population of 17 million, British cotton textiles in 1851 employed no more than approximately 300,000 people. Although textiles dominated British exports for a long time, in terms of percentage of total export it slowly but definitely began to fall back from around 60 percent of total exports in 1794–1796 to almost 48.5 percent in 1834–1836.[11] Nevertheless, it remained one of the most important export products.

It took the entire nineteenth century before the factory system in the cotton textile industry became the dominant form of organization. In many other industries, the rise of the factory system was a "long drawn-out affair" (Berg 1994: 206). The British cotton industry remained vertically separated: Spinning and weaving remained separate activities and marketing was controlled by separate groups of merchants. Nonetheless, the impact of the factory system was clearly felt everywhere. The loss of independence frustrated many artisans. The organization of work processes and the structure of authority generated a host of grievances against increasing discipline, harsh foremen, and distress. Printers, engineering workers, and other skilled craftsmen began to form the rank-and-file of union organization in Britain as well as in France, Germany, and elsewhere in Europe.

> In England wages for handloom weavers had dropped from 30 shillings a week in 1800 to 5 shillings in the early 1830s. Aggregate calculations of real wages on the Continent suggest an even more general decline in living standards. In Germany one calculation suggests that average real wages declined by about 26 percent between 1800 and 1848, whilst average real wages in France also fell significantly between 1817 and 1848.
>
> (Geary 1981: 33)

The drive for synchronization of time in factories continued to be a source of serious conflict which created waves of strikes. To resist the exploitation of the workers in the factory system and safeguard some autonomy and human treatment for the worker, the labor union movement – with the intellectual and organizational support of Marx and Engels – continued to claim temporal autonomy by reducing working hours. The issue of shorter hours awakened a revolutionary spirit amongst the working class in Europe, America, and Australia.

Australia and the eight-hour day

From 1810 forward, the labor movement in Great Britain was organized around the following main issues: the right to vote, maintaining independence and autonomy for the workmen, and increasing their salaries and reduction of the working time. Inspired by the French Revolution and by the negative consequences to the factory system, a new popular radicalism and more militant trade-unionism appeared simultaneously.

Although the British government had introduced a Reform Bill in 1832, many British citizens were still exempt from voting. Disappointed by the outcome, radicals organized a movement around a Charter which asked for a general right to vote, secret ballot, annual elections, and salaries for members of parliament. The Charter became the symbol for a movement called Chartism. Chartists pointed out that low wages and long, exhausting working days simply meant that most of the skilled as well as the unskilled had neither the time, energy, nor the money to participate in economic and political activities. Although they initiated a petition a few times to ask for the implementation of the Charter, the British government responded reluctantly. The influence of the Chartists and their rallies for the general introduction of a ten hour working day remained weak and could only moderately improve the lot of the laboring class. "In half a century of the fullest development of industrialism, the standard of living remained – for very large but indeterminate groups – at the point of subsistence" (Thompson 1963: 226).

The increasing grievances of the laborers, the impact of the poor working conditions in the factory system, and the frustrations about the minimal political response to the demands of the charter, however, drove many Chartists to the decision to leave the country hoping for better living conditions in America. John Cluer, a Chartist and immigrant weaver, began to organize the textile workers in mills in Pittsburgh as well as in other mills in New England. He became spokesman for the New England Workingmen's Association and, as a ten-hour advocate, applied the tactics learned in the English Charter and Ten Hour Movement. He organized strikes for a shorter working time which would be in line with the strive for the *pursuit of happiness* as stated in the American declaration of independence. Although the eight-hour day remained out of reach and the first strike for an eight-hour day of house carpenters in October 1850 failed, the Ten Hour Movement remained strongly anchored in the working population (Roediger and Foner 1989). Many other Chartists flocked to America but some also went to other colonies.

In 1851, when gold was found in the Australian Blue Mountains, a gold rush ensued. In ten years' time, the Australian population tripled. Sydney and Melbourne became booming cities and attractive places for construction workers to work and live. They soon organized themselves into trade unions and, inspired by the Chartists, asked for a reduction of the working time to eight hours per day for a six-day working week. Although the masons in Sydney obtained the eight-hour day in 1856, they were also expected to accept a reduction of their wages. The construction unions in Melbourne did not accept this result as a fair deal.

Chartist James Stephens initiated a campaign for the eight-hour day in February 1856 without a reduction in salaries. On April 21, Australian stonemasons and other building workers marched through Melbourne to proclaim victory in obtaining the eight-hour work day that they had negotiated during the preceding months with the construction employers (Love 2006). A month later, they moved through the city in a triumphant march behind the banner "Eight Hours Labor, Eight Hours Recreation, Eight Hours Rest". The victory was one of the earliest officially sanctioned eight-hour working days in history.

The purpose of the eight-hour day was social and intellectual improvement. In line with what the Chartists had insisted upon in Great Britain, the Australian workers had a desire for self-improvement through education. The Chartists ascribed high moral values to knowledge and the process of learning, yet a general implementation of the eight-hour work day still encountered many barriers by the resistance of the employers and a reluctant Australian government to introduce factory acts which would limit the working time. Elsewhere, the movement for an eight-hour day gained more success.

During the American Civil War (1861–1865), eight-hour leagues, who were fighting for more freedom of the American workers, supported the Civil War as it would also liberate the slaves in the South. In 1868, a campaign for eight-hour legislation for American federal employees proved successful (Roediger and Foner 1989). The Civil War inspired labor unions and socialist political parties to take initiatives for shorter working hours.

The international labor movement

In 1864, the First International Workingmen's Association (IWA), also sometimes known as the First International, was founded in London aimed at uniting a variety of different left-wing political groups and trade union organizations in Europe, America (but not until 1869), and Australia. The movement's slogan was identical to the one put on a banner during the demonstration in Melbourne, and received widespread support.[12] In America, central eight-hour leagues were established in several big cities and, in August 1866, a convention of union representatives decided to establish a confederation which, in order to deliver the labor from thralldom, has as its main desideratum "to constitute a legal day's work in every state of the American union" (quoted in Cahill 1932: 35).

Construction workers in New York participated in a strike for an eight-hour day in 1871 and, as they won their case, organized a great parade with 150,000 people in New York a year later on the June 10. The parade was one of the unique events held in support of a shorter working time which many factory workers felt still to be utopian. To reach that goal, organization of the workers in different industries would be essential. The labor force within the British textile industry was able to form an effective industry-wide union which carried out successful strikes between 1869–1875 which forced employers to accept major concessions like shorter working hours.

The Amalgamated Association of Operative Cotton Spinners and Twiners, formed in 1870, became one of the strongest worker's organizations in Britain. The escalation of interstate rivalries facilitated a major sphere for profitable private investment in industrial activities. Yet the Franco-Prussian War of 1871 scattered solidarity amongst the European working class, and the First International was dissolved. Nation-states were promoting nationalism and patriotism as the new religion which might make it easier to mobilize soldiers as concerned citizens (Silver 2003).

Although, with the onset of the Great Depression of 1873–1896, union strength would be decimated, trade unions and socialist parties established a Second International in 1889 to reinforce a heartfelt solidarity amongst the working class. An expanding industrialism with increasingly homogeneous working conditions helped to stamp a shared consciousness about common social life problems and working conditions within the working class amongst the skilled and unskilled workers.

In Britain, it was the great dock strike of 1889 which heralded the solidarity among unskilled and skilled workers to protest against invariable economic hardship. Many felt that the eight-hour day as a legal imposition might compensate for the consequences of large-scale manufacturing practices. Universal manhood suffrage and parliamentary sovereignty would help render the state more sympathetic to the demands of the emerging working class (Geary 1981). To attain that goal, it became essential to gain the support of socialist labor parties as will be demonstrated in the next chapter.

7

SPEED IN TRANSPORT AND INDUSTRY

The opening of the Suez Canal in November, 1869, dramatically reduced the time needed to travel between West and East by sea from London to Bombay (by about 40 percent) and from London to Singapore (by about 30 percent) (Karabell 2004).[1] The shortening of distance achieved by building this canal was crucial to the integration of the world economy. Steamships, much like airplanes in the twentieth century, became the most important nineteenth century innovation in shipping technology. A regular transatlantic steam service carrying things of high value such as passengers and mail became fashionable from the 1860s forward. The screw propeller, the compound engine, and steel hulls allowed for bigger ships and shorter turnaround time in the ports, all of which made sea transport more efficient (Findlay and O'Rourke 2003).

Instead of expanding international trade and commerce over sea as Britain did, the American market developments mainly took place over land. The American government became involved in financing transportation improvements and in the building of the national road across the Appalachian Mountains, funded the construction of several canals, and provided much of the financing for some path-breaking railroads.[2] By the mid-1850s, four railroads – the Pennsylvania, the Baltimore & Ohio (B&O), the New York Central, and the New York & Erie – controlled an extensive network of trackage from the East Coast to the Midwest.

The increase in the pace of technical change required new organizational forms in regards to organizing the production processes that led to the modern factory system. The new factory system then became characterized by faster growth in manufacturing productivity as measured by output per hour. The economies of scale in manufacturing between 1870 and 1914 were attained through labor-saving techniques and materials using opportunities in technological improvements by which firms were able to respond to a rapidly expanding domestic market.

American capitalism extended its influence not by territorialist expansion as Britain had done, but by technologically driven industrial expansion. The first

Industrial Revolution had centered on improvement in coal, iron, and steam technologies, as well as the introduction of the factory system, which mainly took place in Great Britain. The second Industrial Revolution announced a new phase in the development of industrial capitalism. It began mainly with steel, electricity, chemicals, and the development of the modern factory system in the United States. It set off in the middle of the nineteenth century with the growth of railroads.

Speed became the main driver. To realize that goal, two new time practices were further developed and implemented at an unprecedented scale: railroad time zones and time and motion studies. Whereas in Britain railroad time developed after factory time had already found its implementation, it was the other way around in the American situation: railroad time zones were implemented in the American transportation era (1860–1890) and found widespread use throughout the industrializing world before time-and-motion studies and rapid developments in new manufacturing technologies based on electricity became fully applied in industry (1890–1930) which, subsequently, became implemented in other parts of the industrializing world. Eventually, it took the place of nineteenth-century Britain as the chief supplier of foreign investment funds. The dollar would soon obtain the status of a hegemonic currency as part of a new imperial system of political and industrial relations. Where do its origins lie?

The American system of manufacture

The Great Exhibition in London's Hyde Park, held in the Crystal Palace in 1851, had electrified the imagination of more than six million visitors. The building itself with its prefabricated parts of iron and glass, which was constructed in just nine months, symbolized the scale and scope of a growing industrial dynamism. Within the building, an American riflemaker Samuel Colt from Hartford (Connecticut) drew a lot of attention. He demonstrated his rifles that were composed of interchangeable parts. The visitors were surprised by the practice of standardization and uniformity.[3] A British military commission investigated the machinery of the United States as displayed in Hyde Park. Their report of 1854 coined the term "the American system of manufacture" while pointing at its remarkable division of labor and application of machinery in the production of firearms with interchangeable parts (Shenhav 1995).[4] Uniformity and mechanization became the new catchwords to describe the qualities of the American system of manufacturing.

The Colt Patent Firearms Manufacturing Company in Hartford played a significant role in enhancing the process of synchronization within the factory system by refining and improving the production process. The gun factory was fitted with 400 machines producing around 24,000 pistols per year. Its approach, however, was inspired by the work of the American watchmaker Eli Terry, one of the first to initiate a mass production process for making timepieces. The wooden clocks that came from his factory in 1816 were the world's first mass-manufactured clocks with interchangeable moving parts. The finished products were no longer the result of manufacture by skilled workmen: they were objects assembled from machine-made, standardized, and interchangeable parts. The working practices within the factory had also changed. "The technologies of machine manufacture, which made possible the mass production of inexpensive clocks and watches from standardized and interchangeable parts, brought with them a reconceptualization of the labor process itself" (Dan 1992: 37).[5]

The development and popularity of the American system of manufacture had been greatly enhanced by the innovative role of the US Army Ordnance Department. In 1815, a law had been passed which placed the National armories at Springfield and Harpers Ferry (Virginia) under its immediate command. The Chief of the Ordnance was empowered to impose a system of regulations for uniformity of manufactures of all arms. This policy had emanated from sources deeply rooted in the French military tradition.

After the Declaration of Independence in 1776, America continued to struggle with Britain until 1783 but, due to the diplomacy of Benjamin Franklin, France had supported the struggle for independence. During that time, military engineering treatises, testing procedures, arms design, and educational techniques had made their way to America and were assimilated by native officers (Smith 1985).[6]

The American Ordnance decided to improve its military program by imitating the French approach for appropriate models of military equipment and organizational structures. General Jean-Baptiste Vaquette de Gribeauval, previously an artillery officer, was chosen to lead the reforms in the organization and production of the French Ordnance and had developed various labor-intensive methods for manufacturing muskets with uniform parts. Thomas Jefferson, while serving as ambassador in France, had witnessed the way in which the components of muskets could be randomly selected and interchanged without any fitting or filing.[7]

One person in particular profited from the exchanges with the French concerning the sharing of this kind of military information: mechanic Eli Whitney from Connecticut. "He became a zealous advocate of the uniformity principle, popularized the concept, and persuaded many politicians to support policies aimed at standardizing the manufacture of military arms" (Smith 1985: 49). In June 1815, the superintendents of Springfield Armory, Harpers Ferry Armory, and Eli Whitney gathered with the board of the Ordnance Department to decide how the uniformity should be applied to the manufacture of muskets as the manufacturing of guns had yet to be streamlined.[8] Eli Whitney had become famous for his design of the cotton gin which was able to speed up the process of cleaning seeds from cotton, however, he was less successful in establishing an arms factory based on a mechanized production process with interchangeable parts.

Daniel Tyler, an engineer from West Point Military Academy was sent to Springfield Armory where he spent six months with a stopwatch in hand to examine every aspect of production. While there, he would calculate norms for all production activities in order to mechanize them. At the armory, he mathematized time and action through writing, grading, and examining production activities and, consequently, designed standard times. This intervention had taken place before the technique of the interchangeable part manufacture had been introduced (Hoskin

1998: 104). From 1815 onwards, Roswell Lee took over and, under his supervision, the company championed the policy of uniformity during the next 18 years.

Lt Colonel George Bomford, who was head of the Ordnance Department during the period 1821–1842, put forth tremendous effort to stimulate the production of the required weaponry according to set criteria. The armories imposed a specific regulation of the total production process from the initial distributional stock to the final accounting of costs. This required strict coordination and control, which eventually helped to enforce the American system of manufacture. Accurate bookkeeping enabled the Ordnance Department to survey the general spectrum of the production of the arms.

The rifles and muskets became the first fully interchangeable firearms to be made in large numbers but problems still occurred with complying to uniformity. Firearms producers such as Colt and E. Remington & Sons were allowed to acquire the latest armory knowhow to improve their own production techniques.

In 1850, David Davis, Edward Howard, and Aaron Lufkin Dennison formed a company together to produce watches mechanically based on the same principle of uniformity. They designed a revolutionary production process by manufacturing the movement parts of watches so precisely that they would become fully interchangeable. This mechanization of the manufacturing made watches so inexpensive that, in 1852, they became an immediate success. Their precision watch-making machines created the *American System of Watch Manufacturing*. In September 1853, a new factory built in Waltham (Massachusetts),[9] The Waltham Watch Company, fathered most other American factories developing mass-production systems. "Armory practice spread to technically related industries and by the late 1850s could be found in factories making sewing machines, pocket watches, railroad equipment, wagons, and hand tools" (Smith 1985: 78).

As conformity supplanted individuality in the workplace, craft skills and other preindustrial traditions became detrimental to production. Workers would need to rearrange their lives around values of diligence, thrift, and responsibility. Self-acting machinery enabled arms-makers who were paid by the piece to complete more work in less time. Superintendents, however, noticed a lack of internal discipline by the craftsmen and urged for reforms. The Ordnance Department replaced the civilian superintendents with military superintendents who introduced work rules, clocked the working days, and regularized the operating procedures in the workplace. The pieceworkers and inspectors complained about the regular hours and protested against the installation of time clocks while attacking the despotism and oppression of military rules.

In the end, however, workers gradually but reluctantly accepted the time-oriented world of a dawning new industrial era. Regardless of the resistance, more factory owners became interested in this systematic control of work processes. "They firmly believed that an orderly and well-regulated work environment would not only promote efficiency but also instill values conducive to the moral growth and wellbeing of the country" (Smith 1985: 85).

Uniformity and interchangeability became central drivers of the American system of manufacture. Until the American Civil War (1861–1865), the American government created a bureaucratic setting which favored the installation of new technologies of production. After the end of the American Civil War, accurate and dependable watches became mass produced and permitted "the synchronization of movement in the army company" (Landes 1983: 96). Yet after the war, government contracts for the production of new guns slowly evaporated. Companies such as arms producer Remington & Rand moved into the burgeoning business of making writing machines to facilitate the sending of telegraphic messages to serve the American railroads (Wershler-Henry 2007). Mechanical ingenuity turned into a "special American value and capability" (Fiedler 2007: 328). Engineers from West Point Academy not only played a crucial role in the structuring of arms factories to implement uniformity of arms based on interchangeable parts but also played a similar role in the development of the organizational structure of the American railroads.

Railroads and military practice

The construction of railroads in the "empty" landscape of the United States took place on an unprecedented scale and provided enormous challenges. Trains played a crucial role in facilitating the coming about of a national American market that reduced the price gap between producers and consumers and converged regional price differences. Initially, Britain had been the source of much of the capital invested in the American railroads, the iron rails, and most of the major inventions. The first steam locomotive on an American railroad began operation on August 8, 1829, and carried anthracite from the mines of Carbondale to the Delaware and Hudson Canal.[10]

The U.S. War Department allowed its engineers to work within private or locally sponsored transportation projects. A significant number of entrepreneurs who saw their business expanding turned to the army for advice. Military engineers were assigned to the service of the newly organized Baltimore and Ohio (B&O) Railroad Company. As a result, army engineers of B&O managed the company according to military standards and applied a hierarchy and techniques of accountability as fundamental organizational principles. Like B&O, other railroads also frequently turned to the military for the improvement of management procedures. J. C. Calhoun, the Secretary of War (1817–1825), laid out the first essential elements of the military system[11] and implemented a system of accountability and bureaucratic means by which this accountability was enforced.[12]

The General Regulations for the Army was the first comprehensive management manual published in the United States and described in great detail the rules and procedures that governed military life. As the army developed and refined its administrative techniques, it created a bureaucratic management structure. Line and staff officers reported where they were, what they were doing, and how they were expending the funds allotted to them. Calhoun's military model stressed a hierarchical, authoritarian organization that demanded strict fiscal and operational accountability and provided detailed written rules and regulations to guide organizational behavior. With the corps of engineers, the army was able to demonstrate its administrative competence to a wider civilian audience.

West Point also became essential for the improvements of the management of railroads. The military academy at West Point was founded in 1801 when President Jefferson signed its charter. Jefferson believed that civilian militias would protect national security better than a permanent bureaucratic army.[13] West Point cadets would be well equipped and instructed to run these militias. The academy designed a curriculum modeled after the French École Polytechnique.[14] Jonathan Williams, a mathematician Jefferson had met in Paris and who was acting as secretary to Benjamin Franklin, became the first director and implemented strict organizational procedures for West Point.

When Sylvanus Thayer took over as director in 1817, he continued to borrow the syllabus, textbooks, teachers, and grading system directly from the École Polytechnique. Thayer forced the academy into a house of scientific and technical distinction. "Between reveille and dusk, between the first drill parade and lights out, the days of the cadets were remorselessly filled with instruction on mathematics, chemistry, engineering, mechanical drawing, and even a little geology and history. French was mandatory" (Schama 2008: 40). He restructured the Academy by developing a human accountability system that reflected the academic and behavioral performance of the cadets.

Thayer knew the cadets principally through files kept meticulously by his personal clerk. He designed files for all sorts of activities the cadets were required to perform and kept track of their performance. Their activities were constantly measured, evaluated, and recorded in a numerical-linguistic language. He instituted a shared vocabulary, and the organization was structured according to this new vocabulary. His "bio-administration" was based on logic of classifying, qualifying, categorizing, and ranking the lives of the soldiers. The school was structured in a line-staff system with two divisions. The Central Staff Office kept track of all the performances of the cadets stored in daily, weekly, and monthly written reports. West Point introduced a system of accounting of the cadets based on writing, examining, and grading, i.e., practices of administrative coordination which institutionalized a new form of human accounting. Thayer also recruited French officers who

> provided manuals with explanations about the principle of uniformity. Not only were standard weapons and standardized units essential, but even more so the soldiers training on the basis of uniform regulations, standard procedures of grading, and military exercise. Thayer quickly developed an institution *sui generis*.
>
> (Hoskin and Macve 1986: 131)

Thayer's cadets were transformed into calculable men, and he perfected the Academy into a "disciplinary revolution."

Management of railroads

Though some railroads began to mimic military approaches to improve their managerial railroad practices, not all maintained the presence of the military engineers.

After some internal conflicts in 1830, military engineers were removed from the board of B&O, though their managerial legacy remained. Other railroad companies reflected similar managerial developments which they borrowed from B&O. When the first American railroads quickly expanded into multi-unit organizations, they required control mechanisms other than visual observation and inspection. West Point could still be of assistance in this area.

When the former cadet, teacher, and civil engineer G.W. Whistler became Chief Engineer of the Western Railroad in 1839, he introduced the line and staff structure and implemented procedures similar to those of the American army. He played a crucial role as intermediary in the genesis of accountability in the railroads. After a serious accident on October 5, 1841, with a head-on collision of two passenger trains on a single track between Boston and Albany, the Massachusetts government forced the Western Railroad to improve supervision on the exploitation of the railroad. The company was instructed to promote the safety of the traveler and the security of the trains but lacked a centralized system of control, efficient telecommunication, and formalized procedures to adequately process information concerning train movements.

Within six weeks, Whistler managed to implement a new organizational structure with geographical divisions operating with a sound timetable. He had devised a new system that introduced the first principle of a new management: the clearly defined relays of command and information. This system made every individual in the organization constantly visible and accountable for his behavior. Whistler's solid training from the West Point Academy was a significant factor in this achievement.

> While neither the military's organizational theory nor its administrative procedures offered the railroad precise guidelines for responding to the series of events that propelled reform, Whistler drew upon his military experience and his work with other systems based on the military's management model for a pattern of organizational and administrative procedures of proven utility to railroad.
>
> (O'Connell 1985: 110)

West Point had supplied the first generation of railroad managers, and Whistler played a key role as intermediary in the genesis of accountability practices between the West Point system and the railroads. Like Robert Owen, Whistler became, so to speak, a pioneer of modern management. West Point had provided the foundations for a professional management practice that were based on controlling man's time, as much as that of machines and organizations.[15]

In 1847, the Pennsylvania Railroad hired army officer Herman Haupt, who graduated from West Point in 1835, to improve its organizational structure. He divided responsibility for the operation of the railroad into four departments: transportation, maintenance of the way, motive power, and maintenance of cars. He also improved procedures in use within the railroads. Line officers directed the day-to-day operations while staff officers at the general transportation office concentrated on the proper problems of cost determination, competitive rate making, and strategic expansion. The key contribution of military men to the railroads was the ability of former officers "to translate the techniques used to administer engineering projects into operational procedures for managing the complex daily business of a railroad" (O'Connell 1985: 111).

Although the American Army had recognized that the division of management functions between line and staff officers yielded some improved efficiency, the Pennsylvania Railroad organized the first unique civilian application in the business community between line and staff officers and the separation of operations from decision making, resulting in American railroads emerging as the nation's first big business. By the mid-1850s, four railroads – B&O, the Pennsylvania, the New York Central, and Erie – controlled a large trackage from the East Coast to the Midwest. As they grew from small local carriers to interregional lines, these railroads faced unprecedented operating difficulties such as the best way to schedule the large numbers of trains and run them safely on time, as well as how to raise enormous amounts of capital needed for expansion. One solution to the need for expansion was the introduction of standard gauges to run trains from one company on the track of another (Beniger 1986).

The Academy had initiated a new form of hegemonic control. The creation of calculable subjects for capitalistic production processes had begun in this educational field (Hoskin and Macve 1986: 1988). Careful scheduling gave a moral value to time itself as a structural deployment of supervised activity. The cadets internalized this pedagogic approach with its exhortations to punctuality and regularity, and after their graduation, applied it as managers to the American railroads.

The time study of Tyler at Springfield, the line and staff structure of Whistler, as well as the management practices of Haupt became aspects of the same disciplinary transformations that had been initiated at West Point. These military officers subsequently had a great influence on the development of the art of management in the United States.[16] "The West Point Academy provided the best formal training in civil engineering until the 1860s" (Chandler 1977: 95). The trinity of clocking people and documents prescribing, instructing, and checking human behavior, as well as drilling men, provided a sound base for the improvement of the coordination of railroad companies. Military rationalism created a cadre of calculable men which was similar to the initiatives Dutch military leader Maurits had undertaken in the seventeenth century. His discourse on accountable soldiers found an exponent in the military training at West Point. It became the first institute where future engineers and managers were trained in such a disciplinary way that it became part of their management style.

> The West Point System produced a managerial accountability structure for the railroads by articulating the role of the chief executive and his senior functional officers, their lines of responsibility, and the detailed reporting required. The managers of the railroads themselves became accountable in

> a system of mutual surveillance, i.e. a reciprocal hierarchical observation and normalizing judgment.
>
> (Ezzamel *et al.* 1990: 161)

The impact on the role of engineers was not only felt in railroad companies but also in the wider American business community. Between 1802 and 1866, 1058 military officers graduated from West Point. Thirty-five became corporate presidents, 48 became chief engineers, 41 became superintendents and general managers, and eight became corporate treasurers of various kinds of businesses (O'Connell 1985: 114). Both civilian and military managers adopted military practices and adapted elements of these practices as needed. In the initial stages, military officers had been the catalysts of this adaptive process. The military model had provided a conceptual and procedural framework the officers could implement as no other suitable model was available. While further expanding, the railroads erected new and complicated management hierarchies to coordinate and control a variety of functionally diverse, geographically separated corporate activities. The civil engineer D. C. McCallum, superintendent of the Erie Railroad, noticed that management problems were not related to increasing scale but rather arose from a decreasing ability to efficiently control operations. He therefore introduced a system of greater regularity and speed of communication in order to increase control over the railroad. He implemented general principles of administration by placing major emphasis on intelligence gathering, hierarchical communication, feedback, and error detection (Beniger 1986).

Railroads pioneered modern management through bills of lading, intercompany billing, equipment identification, cost accounting, pricing, and management rules (Chandler 1977). Because of their unique size[17] and the complexity of their operations, railroads developed new administrative practices, quite independent of other institutions in American society. General personnel were replaced by bureaucratic management which soon spread to the big industrial enterprises arising in the 1880s. Andrew Carnegie, for example, who served as superintendent of the Western Division of the Pennsylvania Railroad, applied the lessons learned in the railroad company to run his large steel company. He delegated authority over operations to steel mill managers and rarely interfered with the decisions they made. Inspired by his railroad experiences, he took an abiding concern for controlling costs and promoting high volume production in his steel mills.

Railroad time zones

In 1862, Congress passed the funding for a transcontinental railroad, and the work began in 1863. The Union Pacific, which moved to the West, was built mainly by Irish labor. The Central Pacific which moved to the East of the country was made possible by a consortium of four major investors, one being Leland Stanford. Although Stanford, as governor of California, had spoken out against Asian immigrants, the other investors supported the idea of employing Chinese workers.

By 1867, about 75 percent of the Central Pacific workforce was composed of Chinese workmen. In 1867, working conditions were so terrible that around 2,000 Chinese workmen went on strike for a ten-hour day, an increase in monthly wage, the elimination of the degrading power to be whipped or confined to prevent them walking off the job and freedom for any worker to quit, if and when he chose.

One Saturday, a document listing demands circulated through the camps, which housed almost 10,000 Chinese workers (Montgomery 1987). On Monday morning, no Chinese went to work. Throughout the week that followed, they remained in their camps.

> But by the following Monday the strike was broken as all food trains headed for the camps had been stopped [...] Not only was the strike over, but the very existence of the Chinese who had built the railroad was also soon obliterated from the American consciousness.
>
> (Montgomery 1987: 67–68)

Nevertheless, the Chinese strike was one of the large-scale strikes of the century.

The Chinese may not have discovered America in 1421 as Menzies suggested, but they certainly helped build a continental America from coast to coast.[18] The meeting of the Union Pacific and the Central Pacific in May 1869 definitively turned America into a continental nation. While the politics of slavery had been tearing America apart during the Civil War, the railroads "knit the country together again" (Schama 2008: 272).

The railroad companies extended their track systems to the west and intensified the exploitation with ever faster running steam locomotives. New rail lines facilitated the growth of large cities which stimulated industrial development. Manufacturing activities became increasingly spatially concentrated. The American railroads posed new challenges in the design and construction of locomotives as well as building tracks across thousands of kilometers of barely charted territory, which greatly improved the identity of the engineering profession (Friedel 2007).

After the end of the Civil War, each American railroad company kept its own railroad time – as did each town and city along the way, which was a great inconvenience for the traveling public. These trains covered such distances that several time zones were necessary to secure the efficiency of the timetables. The difference between East and West amounted to more than three-and-a-half hours. Astronomers sent forth a strong plea for a uniform standard of time. In 1870, C. F. Dowd published *A System of National Time for Railroads* which was the result of discussions at the convention of railroad trunk lines in New York City in October of the previous year.

To obviate the inconvenience of having some 80 different time standards on the American railroads, Dowd proposed to divide the United States and Canada into four time belts with each 15 degrees of longitude wide (Zerubavel 1981). The eastern one would be the Washington Meridian passing the transit instrument at the US naval Observatory. While the Washington Meridian projected the other meridians too far west to suit the needs of the eastern and central sections, discussions about the precise location continued for years.

Yet an effective use of a complex railroad network and prevention of accidents forced the companies to consider a standardization of time reckoning at a level higher than that of the individual railroad company. An ambitious project began to settle an arrangement with standardized time reckoning on the entire North American continent: the Standard Time Zone. However, every region continued to maintain its own local time determined by the position of the sun. With the extension of the railroads through several regions, it became difficult to schedule the different local train times in a proper manner. With the introduction of the telegraph, the problem was aggravated even more. The national telegraph needed a single standard to which all local times could be referred (Kern 1983). Pressure for a general arrangement increased.

In August 1871, the first International Geology Conference took place in Antwerp (Europe) in order to agree upon a resolution to adopt the Greenwich meridian as the common zero for longitude. Since the increasing internationalization of trade and transport had made it convenient that all maps and charts should indicate the same longitude, this prime meridian would also yield a standard time (Holford-Strevens 2005). Sanford Fleming, director of the Canadian Pacific Railway, defended the interest of a uniform time for the whole world and talked about terrestrial or cosmopolitan time.

Unfortunately, the debate turned into a political clash about a universally applicable standard time concept. Some participants were of the opinion that the choice should be decided by Paris as it had previously introduced other standards such as the meter and not by London with its Greenwich Observatory. Other astronomers did not agree with either proposition and questioned the usefulness of adopting a universal time for railroads, shipping lines, telegraphs, and postal services coinciding with local or national time. The debates continued for several years and other conferences did not bring the discussion to an end (Bartky 2007).

On August 3, 1882, American Congress authorized the president to call an international conference to fix the common prime meridian and standard time throughout the world. The American railroad companies, however, felt a more urgent need to solve the problem and could not wait for the geologists or politicians to settle the matter. Railroad journal editor and secretary of the railroad industry's General Time Convention W. F. Allen pressed for harmonization. After a lengthy and complex process, 56 railroads inaugurated a uniform railroad time with four time zones in April 1883. On Sunday, November 18, 1883, the Sunday of the two noons, the railroads introduced a single uniform railroad time: Standard Railroad Time. All local solar times were changed into one common railroad time as a new application of abstract time. Most Americans set their clocks and watches to this new standard time.

From that moment forward, the United States obtained one synchronized time pattern. It was the most momentous development in the history of uniform public time since the invention of the mechanical clock. Despite all the sound scientific and military arguments for world time, it was the railroad companies and not the

FIGURE 7.1 Untitled. (Photograph by Chema Madoz, 2000.)

governments that were the first to institute it. Railroad time became firmly established as a regulator of public life. The railroads became, at one and the same time, an economic activity and an essential public good with a natural monopoly.[19]

The horometric harmony once firmly established was reported by many newspapers. The interventions of the civil engineers of the railroads gave the push to a new definer of abstract time. People began rewriting embedded social times into a new global time. While Philip II still needed the sun as reference point for the omnipotence of his empire, an empire of chrono-technology dawned. However, not everyone greeted the new time with open arms. Some people criticized this act of standardization, others pointed at the rigidity of the clock (O'Malley 1990). The editor Ch. D. Warner objected to the rigidity and invariability of time lived by the clock: "the chopping up of time into rigid periods is an invasion of individual freedom and makes no allowances for differences in temperament and feeling" (quoted in Rifkin 1987: 145).

On October 1, 1884, an international meridian conference with approximately 41 delegates from 25 countries assembled in Washington DC. After 12 days, the conference recommended Greenwich as prime meridian of the world. They had adopted the American proposal that the prime meridian should be the one passing through the center of the transit instrument at Greenwich Observatory. After discussion about longitude, it took a few more days before a similar recommendation about the instauration of universal time was adopted, starting the universal day at Greenwich noon and reconciling the new universal day which starts every night at twelve and the astronomical day which started at noon. Representatives of American railroad companies extolled the virtues of a standard time system for use in ordinary life, leaving Universal Time for science and international telegraphs. The conference recommended that Universal Time should be Greenwich Time, but France did not yet conform.

The coordination of time that the American railroads established also became the basis for an improvement of control of space (Giddens 1979). The American railroad engineers were the architects of a new time–space configuration. Their managerial approach led to the standardization of railroad times and of civic time in the United States as a whole. Railroads and telegraphy created competitive advantages "to many sectors of American industry and agriculture, notably in speed and reliability of transport and communication" (Freeman and Louca 2005: 211).

The world, however, was slow in adopting the system. In 1888, Japan coordinated railroads and telegraphic services nine hours ahead of Greenwich; Belgium and Holland implemented Greenwich Mean Time in 1892; Germany, Austria-Hungary, and Italy in 1893. Still, there continued to be a significant amount of confusion. Telegraph companies in China used a time that was approximately the same as in Shanghai; in Russia there were odd local times, such as the one of St Petersburg which was two hours, one minute and 18.7 seconds ahead of Greenwich. France still used four different times, none of which had a simple conversion to Greenwich Mean Time. Big cities had a local time taken from solar readings. The French railroad used Paris time which was nine minutes and 21 seconds ahead of Greenwich:

> A law of 1891 made it the legal time of France, but the railroads actually ran five minutes behind it in order to give passengers extra time to board: thus, the clocks inside railway stations were five minutes ahead of those on the tracks.
>
> (Kern 2001: 13)

Eventually, the entire world became organized according to a standard railroad time, even more so than the organization attained by the Gregorian calendar.[20]

The modern factory system

Traditionally, the first American management practices originated in the cotton plantations where slave overseers, and slave drivers became "the first large group of managers in American private enterprises. Cotton plantations presented a managerial challenge and brought into being America's first significant body of management writing, mainly by the slave owners" (Hoopes 2003: 10). Plantation managers regulated and coordinated the working time of slaves by the application of sounding bells which were attuned to the time of the master's watch (Bayly 2004). Thomas R. Dew, an American professor of political economy, defended the existence of slavery in 1832 by stating: "we take it for granted, that the right of the owner, holder to his slave, is to be respected, and, consequently, that he is not required to emancipate him, unless his full value is paid by the State." Dew was convinced that American slave owners could not be blamed for the practical treatment of slaves. The original sin of the introduction of slavery "rests not on our heads, and we shall soon see that all those dreadful calamities which the false prophets of our day are pointing to, will never, in all probability occur" (Dew 1832: 22).

The Civil War, which had caused a cotton famine in Britain and had thrown hundreds of thousands of textile workers and their families out of work, finally abolished slavery and was the only achievement that justified all the death and destruction. With the slaves being liberated and industrial progress on the horizon, a modern American industrial society was created where many large-scale companies would flourish.[21]

Next to the former slavery practices on the large cotton plantations and the influence of engineers from West Point in the exploitation of railroad companies, there was another source to the development of modern American management practices. Traditional metalworking industries had created a systematic management movement to minimize the hodgepodge of remedies for the waste and confusion found in machine shops. Articles on production management drew analogies to the military hierarchical system and to the human body with a firm's top managers representing the brain (Jacoby 2004).

It is noteworthy that the American system of management did not develop in the new science-based industries but was instead pioneered by mechanical engineers in these traditional metalworking industries where a highly developed division of labor necessitated greater coordination of operations. To effectively manage the increasing size of these companies, the owners turned to railroads' managerial practices as another source of inspiration where the influences from West Point were clearly noticeable.

Henry Metcalfe, the superintendent of several large arsenals run by the Army's Ordnance Department, also introduced the use of routing slips to streamline processes. These slips were prepared by the head office and indicated the sequence of the departments an order would pass through and the operations to be performed in each department. "The use of standard forms and instructions spread throughout industry and eventually came to be considered essential features of good management" (ibid.: 32).

The dramatic growth of the factory, however, did not immediately displace other organizational forms. For instance, rural forges and small foundries still coexisted during the 1860s and 1870s along with giant rail mills, employing more than a thousand workers. Also, although steam-powered machinery helped to establish large shoe factories in the 1850s, "certain types of women's shoes and slippers were manufactured on a putting-out basis until the end of the nineteenth century" (ibid.: 11).

By the 1880s, however, winds of change were beginning to erode the power of artisans and skilled workers in production practices. New industries such as electrical machinery and chemicals were based on a technology that could not rely on artisanal techniques. A spurt of innovations such as the introduction of the dynamo and the carbon filament lamp made electrical power available to industry and households. Based on the work of entrepreneurs such as Siemens, Halske, and Edison, electric power generation introduced tramways and electric railroads in the larger cities. By 1890, 15 percent of urban transit lines were using electric-powered streetcars (Chandler 1977). Concurrently, with the expansion of the American economy around 1900, a body of management thinking became diffused which would shape the foundations of modern Western management thinking (Gillispie 2004). The new electrical technology brought with it a shift in management style in industry. After the introduction of continuous flow methods in steel manufacturing, the foreman lost a lot of authority. Electrical machinery and motor vehicle manufacturing soon rivaled steel and textiles in size. Manufacturing employment nearly doubled during 1880–1900 and nearly doubled again during 1900–1920 (Jacoby 2004). A transformative change took place, a transition from industrial size based on scope to a size based on scale (Shenhav 1995).

> Not only the scale but also the speed of manufacturing operations increased as human labor was replaced by power machinery and handling equipment. This reduction in the amount of time required to process each unit of output, together with the growth in the volume of production, meant that unit costs were steadily reduced.
>
> (Jacoby 2004: 30)

The unprecedented size and complexity of the manufacturing firm created many administrative difficulties. Lack of coordination between various departments of a firm hampered an effective flow of production. If production speed was to be increased, greater coordination and systematization were required. Engineers and metallurgists began to make decisions about new production methods and more effective organizational forms.

The shift to large-scale factory production upset workers who were accustomed to traditional ways of producing goods. Industrial development was often accompanied by dangerous working conditions, longer hours of work, and low wages. Company owners, managers, and workers struggled about the issue of who would control the work process, how the work was performed, at what speed, and on what type of system work rewards would be based. In this struggle for control of the workplace, superintends of large American factories increasingly employed the "drive" system of managing labor, i.e., extracting as much work as possible in a given period of time, usually a ten- or twelve-hour working day.

Traditionally, foremen had been in charge of hiring and firing workers and had considerable say, along with skilled workers, in determining just how tasks were set up. These foremen, however, were becoming increasingly replaced by engineers who introduced managerial practices to make tasks and flow of work in factories more efficient. After 1880, incentive wage schemes, such as piece rate wages, became increasingly popular in American industry:

> [B]ut engineers were skeptical that productivity gains could be achieved by piecework, pointing to the worker's refusal to exceed traditional output norms and to the growing disparity in labor costs between United States and its competitors in England and Germany in the 1890s.
>
> (Jacoby 2004: 33)

Between 1880 and 1920, the number of engineers grew from 7,000 to 135,000. They became highly involved in the development of new methods of organization and management. In that sense, the American historian J. Hoopes provocatively stated that "modern business management is an American invention" (Hoopes 2003: 5). Yet, he also indicates that the increase of top-down management power created tensions in relation to bottom-up American democracy which were the sources of American independence as President Jefferson had clearly indicated. Management power might be a vital necessity of economic wellbeing yet it stands in an obvious contradiction vis-à-vis American democratic values which Hoopes (2003) labels as a paradox.

Labor resistance

With the increasing pressure to improve working efforts, workers wanted to be compensated with higher wages, shorter working hours, and better working conditions. In order to accomplish their goals, they formed nationwide unions to resist the growing power and influence of owners and managers in large-scale organizations. "The United States experienced, between 1876 and 1896, more strikes than any other industrial nation" (Blackford 1998: 99). The Knights of Labor, established in 1869, organized workers into local "assemblies" across the United States and questioned the existence of companies being owned by capitalists instead of the workers themselves. Initially, they achieved some success, but large railroad companies contested their position.

Changes in the exploitation of the railroads and application of strict schedules of railroad times caused resentment. In 1877, B&O ran freight trains on 100 miles (about 169 kilometers) of track, which acquired a travel time of six hours. The crew was only paid for the six hours and did not earn anything when there were delays along the route. The moment that management introduced a cut in pay rates, the engine's firemen stepped down and initiated a strike. Within a few days, the railroad network was paralyzed, and B&O demanded military intervention to break the strike (Montgomery 1987). Over the next ten days, disturbances cost the lives of more than a hundred people. Eventually, the powerful position of the Knights of Labor waned after they lost a nationwide railroad strike in 1877. The introduction of strict railroad time schedules did not translate to the implementation of a similar policy for the standardization of working time limiting them to eight hours per day.

In the 1880s, a group of trade unions amalgamated into the American Federation of Labor (AFL), which soon gained increasing influence and power. It was composed mainly of skilled workers in workshops and factories and accepted the need to bargain with business owners to secure higher wages, shorter working hours, and better working conditions.

The American system of manufacture propagated standardization of work processes based on a division of labor elaborated by a growing body of engineers. However, there was no concomitant standardization of working time for all workers involved in factories and transportation companies. After the implementation of rather effective ten-hour legislation by the American government, some companies began to introduce speed-up processes which, together with an increasing wave of immigrants, increased tensions in the labor market. When Pullman introduced a piece rate system in 1886, the workers went on strike. Although the strike was short lived, the AFL associated it with a claim for the introduction of an eight-hour working day.[22] Both the Knights of Labor and the Federation of Organized Trades and Labor Unions (later known as the American Federation of Labor) saw the shorter day as a cardinal issue in their activities. Samuel Gompers, who became the chairman of the AFL, insisted that the eight-hour day could generate working-class unity.

Unions and socialist parties supported the idea that May 1, 1886 – traditionally a day when labor contracts were renewed – should be used to organize local protests and strikes. A famous strike for the eight-hour day took place at the McCormick Harvester Works in Chicago. The strike led to the Haymarket riot where a bomb exploded in front of the police. Subsequently, they fired at the protesters which resulted in 13 deaths (Roediger and Rosemont 1986). Hysteria against the socialists, eight-hour advocates, and immigrants fed the demand for a court trial, yet the subsequent execution of four German-American anarchists on November 11, 1887, led to general protest in America as well as in Europe and became annually commemorated for a time. The eight-hour movement appeared to be devastated in America (Roediger and Foner 1989).

In order to restore some social wellbeing of the working class, national European confederations gathered during the Exposition Universelle in Paris in 1889. Earlier, they had decided that a new international labor organization should mobilize workers worldwide to restore that wellbeing. In 1886, the French political party Possibilistes had taken the initiative to organize an International Labor Conference in Paris to strengthen international cooperation between confederations.

During the conference, which gathered representatives from many European countries as well as Australia, it was decided to strive for an international implementation of a legal eight-hour day. The representative of the Belgian Labor Party, Edward Anseele, proposed that all members should reunite during the international exposition of 1889 in Paris and further elaborate the plan. The Possibilistes were asked to organize the conference in Paris, which at this time was very busy with many fairgoers roaming about town.

Like the 1851 Crystal Palace Exhibition in London, the Paris Exposition drew tremendous attention. As part of the Exposition festivities, which at the same time would be the commemoration of the centenary of the French Revolution of 1789, the French government had conducted a contest for the creation of an object to symbolize the advancements of industrial capitalism. The engineer Gustave Eiffel won the contest and built a magnificent tower of iron that would become at that time the tallest building in the world at 312 meters.

The American Edison Company was well represented at the Exposition with displays of hundreds of electric-light installations with incandescent bulbs of every size, color, and shape. In one pavilion, visitors were gasping at modern telegraph and telephone equipment and saw the foreshadowing of a future society transformed by electric lights for the interiors of houses and workplaces, as well as by quick communication via telephones. Inventor Thomas Edison visited his own site and, being fascinated by aerial navigation, he was surrounded by young bright designers with different models of flying machines for which he showed a special interest.

Within the Exposition, another pavilion was installed with Javanese dancers to demonstrate how far the industrial world had separated itself, in the meantime, from an innocent paradise like the Javanese society, where Arcadian circumstances facilitated a natural balance between work and life. The French painter Paul Gauguin, who had grown up in Peru and befriended Vincent van Gogh when they stayed in Arles (Jonnes 2009), was so impressed by this idyllic social life where time was still dictated by the rhythm of nature and not by clock time which fixed Western society more and more, that he soon, thereafter, left France to live in the Arcadian society of the Tahitians, from where he sent his idyllic paintings back to Paris.

Amidst the symbols of industrial progress at the Exposition on one hand, and the performance of Javanese dancers reflecting a more authentic natural life on the other, the Socialists organized two conferences about the future of the working class. The French labor party (POF) organized a similar conference in the Salle Pétrelle, where representatives of the SPD (German Socialist Party), J. Guesde and P. Lafargue, the leader of the PDF and the son-in-law of Marx, were present. The Possibilistes invited labor unions as well as socialist political parties to participate in the conference in the Rue de Lancry. Here, the representatives of the European Labor movement founded a new Second International.

In line with what Marx had pointed out, the reduction of work time became the key to the struggle between employer and worker over the extraction of surplus value.[23] A normal workday of eight hours would oblige employers to alter the structure of their business and distribute work evenly over the year. The desire for an eight-hour day reflected the quest for a predictable allocation of personal time and work time (Cross 1989). It would allow the factory workers some leisure time for reading, learning political activities, as well as enjoying social life.

The French delegate, Raymond Lavigne, proposed that a great international demonstration should be organized on a fixed date so that simultaneously, in all countries on the same day, the workers would demand public authorities to legally reduce the working day to a standard eight hours. When the delegate of the United States, Hugh McGregor representing the AFL, heard about this proposal he suggested the date of May 1, 1890, as it was already adopted by his organization. The conference accepted this date for the international manifestation and international socialism, henceforward, linked the eight-hour demand to an international May Day manifestation.[24]

Trade unions of Europe, America, and Australia adopted resolutions to legally impose eight hours for all workers. At the other conference in Rue de Lancry, the anarchists insisted on the maintenance of the celebration of November 11 in memory of the Chicago martyrs but were unable to distantiate themselves from the

May Day proposal for a legally implemented eight-hour day because of its appeal for the proletariat and its potential for a more general struggle (Panaccione 1989).

The goal of the May 1 demonstrations and manifestations was that they should lead to the introduction of a standard working time for all workers of the world and recreate a society with idyllic connotations, such as the one Gauguin began to paint in Tahiti. Workers living and working in major cities became too detached from millennia-old links between seasonal agricultural work and urban labor. Uniformity and similarity in the execution of factory work in the cities did not mean, however, that they would all have to think and act in the same way. Some American engineers in some industries soon began to think otherwise, as the next chapter will illustrate.[25]

The Greenwich Observatory bomb

The great majority of trade unions in England, France, Austria, Germany, the Netherlands, Italy, and Spain remained craft based and continued to pursue highly restrictive practices (Geary 1981). In some countries, however, some confederations did not limit their activities to the pursuit of immediate political and material gains but committed themselves to the revolutionary doctrine of anarcho-syndicalism. They condemned political rallies leading to participation in parliamentary elections and despised cooperation with middle-class elements or even the state. They fervently believed that proletarian emancipation would come from the workers themselves. The syndicates were seen as the organ for struggle and a source of constant resistance against capitalism, which had to be overthrown. France, in particular, possessed a strong anarchist tradition, and the following decades made anarchism synonymous with bloodshed across Europe.

In these circles, Greenwich was seen as a symbol of oppression. In 1884, Greenwich had obtained its zero meridian as the centerpiece of an undisputable framework of time and space which henceforward would coordinate trade, communication, and transport. Through a refined network of telegraphically linked and synchronized public clocks, it was able to transmit precise time signals that reinforced national societal punctuality (Aslet 1999). Chronometers were sent across the Atlantic many times to determine the longitudinal differences between the American and British observatories from which the difference of longitude between Harvard and Greenwich was accurately determined.[26]

Although precise measurement of time, disciplining, and synchronization of labor activities became symbols of Western industrial capitalism, anarchists resented the consequences of these developments within Western society. Powerful people such as kings, aristocrats, and businessmen were blamed for the rigorous exploitation of the population.

In the last two decades of the nineteenth century, a series of anarchist-inspired terrorist attacks hit many European countries. One of the earliest and most spectacular attacks was the bomb assassination of the Russian tsar Alexander which inspired anarchists to organize many other similar attacks on rulers and aristocrats

(Burleigh 2008). By late 1893, anarchist terrorists were particularly active in France targeting crowded opera houses and cafés, which culminated in the bombing of the Chamber of Deputies in Paris in December of that year. For this crime, anarchist Auguste Vaillant was convicted and executed.

Out of protest against the verdict, a bomb exploded in a Paris café on February 12, 1894. Three days later, another catastrophe took place on the afternoon of February 15, 1894, when two members of the Greenwich Observatory staff were still in the building. Around 4:45 p.m., they heard a sharp and clear detonation followed by a noise like a shell going through the air. They went outside to the courtyard where the park-warden was holding a man who, despite massive injuries, was still alive. Without having said anything, the Frenchman, Martial Bourdin, died about 30 minutes later. The explosion caused quite a disturbance, but no one knew precisely what the bomber's intention had been. He carried with him only a small bomb which could hardly have caused any serious damage to the Observatory. After extensive police investigation, it turned out that Bourdin was a French anarchist who had been member of the anarchist club Autonomie in London.

Although Britain had been unaffected by anarchistic campaigns so far, the Greenwich bomb changed everything. The funeral of Bourdin became a rallying point for anarchist sympathizers in London and attracted a huge crowd. People speculated about the intentions of the bomber. Some believed that he intended to blow up the Chronometer Chamber of the Observatory as he despised the dictatorship of universal time as expressed by Greenwich. Greenwich symbolized the modern society with its increased mobility, and anarchists feared that abstract quantifiable time had infiltrated into their personal, natural existence.

Others believed that the secular belief of the anarchists in promoting a more progressive society by bombing its symbol was a strange oxymoron. Violence itself could never lead to a better society. Whatever was the case, Bourdin must have made some miscalculation, and the bomb exploded in his hand. As he was carrying a considerable sum of money, it was supposed that he intended to leave for France immediately after the attack. If returning by boat, he, paradoxically, might have sailed using Greenwich Mean Time coordinates to arrive safely home in France. Nevertheless, the first international terrorist incident in Britain against the increasing dominance of abstract time practices fascinated the public.[27]

Despite this anarchistic attack on universal time arrangements, the period between 1870 and 1914 demonstrated an increasing industrial globalization of the world economy due to a rapid adoption and adaptation of industrial-core technologies and shifts in economic structures. Due to the impact of the two Industrial Revolutions, a pattern of convergence among West European economies, along with the temperate economies of the European settlements of Canada, Australia, and New Zealand was noticeable. Convergence means "the assimilation of countries outside northwest Europe of the institutions, technologies, and productivity levels currently in use in North West Europe and in the rest of the industrial core."

Yet, beyond this "charmed circle," it was, according to Dowrick and DeLong (2003: 192), not bringing convergence at all. "The integrating world economy was powerful enough to move tens of millions of people across oceans and shape crop and livestock patterns in Java, central Brazil, and Ceylon as well as on the pampas and in the outback." Although the convergence club remained of limited size, specific processes leading to common institutions and practices such as scientific management in factories, transportation, and information provision would soon be shared amongst its members.[28]

> Suburbanization, progressive ideas regarding leisure, the early growth of scientific management, and the development of labor statistics and labor economics as professions contributed to an intellectual climate in which shortening the working day could be approached not as a class issue but as a reform worthy of consideration on the grounds of efficiency, uplift, and safety.
> (Roediger and Foner 1989: 14)

However, it would need the First World War to function as a catalyst before these changes were fully implemented in the Western world.

8

TIME AND MOTION

During the first Industrial Revolution, factories developed into a system where workers were less and less forced by an external, i.e., religiously imposed, discipline and direction but instead became more and more internally regulated bodies which were ancillary to the machine. This system became more profound as artisans and skilled craftsmen lost their positions, and unskilled factory workers executed simple work in a monotonous way and operated mechanically. Human labor was viewed in a mechanical way, an approach which was rooted in a fascination for the clock which was glorified in dominant Western philosophy.

The image of the clock was linked with concepts that people held in high regard: the clock illustrated the attributes of a deity. The analogy drawn between the clock and the world in terms of both its creation and its normal functioning came to play the role of a master metaphor from which most of the other clock metaphors were derived. The clock idealized the quality of regularity, order, and harmony. It was identified with the mechanical character of the physical world, and the protagonists of this view tried to discredit magic and advance rationality. It symbolized as much the harmony of the universe as the effectiveness of an absolute monarchy, but it became more and more detached from temperance, which had been seen as the highest of the seven virtues since the late medieval period.

Thus, the mechanical clock stood as a physical illustration of the advantages of authoritarian, centralist command structures, be they in the body, in society, or in the universe. It was also eventually seen as proof of the success of modern science.

> From the conviction that the world had the character of clockwork followed the belief that nature obeyed the laws of mechanics. This belief became one of the constituting elements of the new science and found expression in the often repeated call that philosophers should approach the secrets of nature in the manner of the clockmaker who uncovers the causes

> of trouble by taking the malfunctioning mechanism apart. The new science insisted that all of nature was subject to the same laws.
>
> (Mayr 1986: 116)

The mechanical clock seeped deeply into the fabric of social life where abstract time became dissociated from planetary rhythms and seasons, from change and aging, from experience and memory. It became an abstract concept of time, self-sufficient, empty of meaning, and remarkably neutral. During the process of industrial globalization, it spread "like a spider's web across the globe, leaving those who resisted the new time defined as backward and old fashioned" (Adam 2004: 114). The clock became the standard instrument of coordination and control. Its tick-tock replaced fluctuating rhythms. The clock as machine-pacing replaced self-pacing. Clock time revolutionized the cadence of daily life.

This perception of the clock was the key background for the way "time-and-motion" studies were undertaken. These studies attempted to develop a modern American factory system during the second Industrial Revolution in order to improve working patterns in American factories. However, it is remarkable that the building blocks for time-and-motion studies have a European origin, in particular based on the research of a French scientist Etienne Jules Marey and a European Science of Work. Before we can elaborate the development of time-and-motion studies, we must, therefore, first clarify the way the mechanical approach changed some common views on work.

Work

From the eighteenth century onwards, the clock was used as the metaphor to describe the human being in mechanical terms as in La Mettrie's work of 1747, *L'Homme Machine* (*Machine Man*), discussed in Chapter 3. At the end of the nineteenth century, however, the view of the human as a machine became compared less to the clock and more to a steam engine which needed to be regulated in order to prevent wasteful activities which would require too much energy, a metaphor introduced by physicists. Whereas the clock had been perceived mainly as being capable of work when powered by some external sources, the metaphor of the steam engine was introduced to identify the main characteristics of work which, like the steam engine, are regulated by internal dynamic principles, converting fuel into heat and heat into mechanical work. The steam engine became the metaphor conceptualizing auto-regulation. A resemblance was seen between the human body and the steam engine as a motor in motion that was kept in movement by internal dynamic principles and by its capacity for self-regulation.

Benjamin Thompson, Count Rumford (1753–1814), who experimented in a Munich laboratory, began to interpret the nature of heat as a form of motion. Physicists began to look at energy as a universal force presented in all matter capable of converting itself into innumerable forms. It was the work of the German scientist Hermann von Helmholtz (1821–1894) with his principle of energy conservation that redirected science toward new fields of interest. His treatise on the conservation of energy (1847) was written in the context of his medical studies and his interest in philosophy. The discovery of energy conservation became the basis of all matter and motion in the physical world. This had a fundamental impact on the way labor was perceived once Helmholtz had identified mechanical work with labor power. This approach would boil down to "time-and-motion" studies, as will be illustrated later.

The language of labor power eliminated the need for a spiritual understanding of labor. It looked at labor power as the mere quantitative equivalent of work produced, regardless of the source of the energy transformed. This new focus had serious consequences for the way ethics could deal with labor. There was one stream of thinking which stressed a different view about the human being and labor which led to a major reevaluation of work in the West. German sociologist Max Weber, together with British historian Richard Tawny (1880–1962), paid particular attention to the influence of Protestantism, especially Calvinism, on work attitudes that developed during the period 1640–1740. The doctrine of predestination, the "calling," and worldly asceticism made them apt to reinvest their profits in businesses and not to use the results of their work for conspicuous consumption (Munro 2000). They did not say that Calvinism was responsible for the birth of European capitalism but that their religion certainly stimulated the development of the spirit and work ethos of capitalism.

Adam Smith also viewed work as the source of all wealth which would be further increased by an appropriate division of labor. French Enlightenment philosophers such as Diderot and Rousseau publicized the virtues of work. Labor was seen as the beginning point of human self-consciousness and autonomy (Rabinbach, 1990). Labor and selfhood being productive, rational, and moral became crucial labels in an extensive campaign against the unproductive position of the nobility and the idle poor. This view culminated in Marx' studies, to whom labor was not only the source of all productivity but the expression of man's humanity. He condemned situations in the factory where the human being was reduced to an "animal laborans." Marx considered labor itself as the quintessential human activity defining social being and offering humanity a way to re-appropriate and regain its essential attributes (Heller 1974). His key perspective had been the emancipation of mankind through labor.

The work ethic in the factories, however, seemed to eclipse into a "quantitative economy of energy" (Rabinbach 1990: 59). Marx feared that the value of human labor would be more determined by the expended force than by the skill involved, which became seen as a product of "time and trouble." Henceforward, the output of a worker and a machine became both measured and evaluated in terms of cost and yield of production. New technologies could stimulate ever increasing productivity. Yet he noticed a sense of alienation among the workforce in the factories which could only be countered by the reduction of working time and a simultaneous expansion of technology which could eliminate wasteful energy at work, abolish fatigue, and release human beings from the burden of labor by the productive forces of improved technology. The more the working day would be shortened, the more the intensity of labor may increase which implies that the factory owners will be economizing on the means of production by investing in ever better production technologies (Rabinbach 1990).[1]

The first signs of worries about the consequences of fatigue and wasteful energy expenditure among factory workers appeared in the European medical literature of the late 1870s. Physiologists and medical doctors began to discover the consequences of overwork, overexertion, and fatigue as forms of modern disorder. They began to focus on phenomena such as fatigue, adaptation, self-regulation, and interaction. Work hygienists considered the physiological and moral qualities of labor as complementary. Social hygienists, psychologists, and social reformers looked at fatigue as the threshold of human limitation and the barrier that society should strive to bring improved technology under control by the use of medicine and the support of politics by intervening with legal measures.

Their new insights strengthened the view that the human body can be seen as a motor, a thermodynamic machine. Work was interpreted as the conversion of energy into use and fatigue interpreted as the opposing concept of energy. "With this semantic shift in the meaning of work, all labor was reduced to its physical properties, devoid of context and inherent purpose. Work was universalized" (Rabinbach 1990: 47). Through mechanization, work might become more intense and increase fatigue. Long hours, however, would reduce the time available for bodily restoration and family life. Some social scientists, therefore, defended the idea that working hours had to be regulated in order to balance the "energy household" of human beings. Physiologists and medical doctors in general began to apply views of physics to explain the wrong energy household of living entities in terms of fatigue and the consequences this might have for their proper functioning. Marey was one of these professionals.

Etienne-Jules Marey

The French scientist, Etienne-Jules Marey (1830–1904) – physician, physiologist, pioneer of medical measurement, cardiologist, aviation pioneer, student of hydraulics, and pioneer of photography and cinema – introduced the new insights of the natural sciences to the domain of medicine. Marey, who was inspired by the work of Helmholtz, saw movement as the central fact of life. He was convinced that the new discoveries about nature can be stated in absolutely valid laws of nature. He felt that there was no difference between the scientific methods for analyzing inanimate nature and the work of physiologists who studied animate nature. Marey believed that the science of life could be explained by the laws of physics and chemistry. The concept of energy reintegrated the organic and inorganic order of things: the unity of all material being in energy conservation. Motion had to be understood not only in terms of the amount of energy used but also in terms of how the energy is deployed in space and in time. The time expended in any activity or reaction time is a function of the body's internal laws of energy and motion.

Physiology could become an exact science as soon as precise measuring instruments were available. Marey, therefore, devoted his career to charting and representing the movements of the animal machine including the human being.[2] He insisted that the study of the animal machine demanded both new instruments and new techniques whose application was firmly rooted in a mechanistic and materialistic theory. Marey translated his interest in the movement of bodies into a "dynamic physiology" and developed a technique of graphic inscriptions which permitted him to better understand human and equestrian gait, as well as aquatic locomotion. He studied the flight of birds and experimented with instruments for measuring the frequency of the movement of their wings.[3] For this purpose, he designed chrono-photographic techniques by using a gun with which he "shot" pictures. He also designed a new apparatus for measuring human walking and running consisting of a portable inscripture and an experimental shoe to reduce fatigue and suffering.

With his instruments, Marey created a new language to describe the body at work, "a language of time and motion" and developed the first complex, sophisticated, and practicable instruments for measuring bodily activities. "As Greenwich Time established an official standard of measurement for time around the world in 1884, Marey developed his techniques of physiological measurement in order to develop a standard for physiological time" (Rabinbach 1990: 94). He studied human and animal locomotion – the description of the moving parts of the animate motor and of the accumulated forces that propel them through space and time. With his "cinematographic" method, he was able to cut time into homogeneous, isolated parts, fully separated from each other. He reported his experiments in 1873 in *La Machine Animale: Locomotion Terrestre et Arienne*, part of which was later published as *Animal Mechanisms*.

Thus, it was no surprise that Marey was interested in the movement of horses. To improve the analysis of equestrian gait, he attached inscriptors to the horse's hooves while the rider held a portable inscriptive apparatus. These hippological investigations brought him into contact with the Anglo-American photographer Eadweard J. Muybridge who had made his reputation in the American West by trying to determine the precise gait of the horse. Muybridge's photography of horses began in 1872 when Leland Stanford, governor of California, invited him to make some photographic experiments on Stanford's horses that could help in the development of "the greatest racing stable in the West" (Braun 1992: 45).

In 1874, Stanford and Muybridge read Marey's *Animal Mechanisms* and set up a technique to take a series of photographs to end the debate about whether there were moments when a trotting or galloping horse had all four hooves free of the ground. Marey was struck by Muybridge's results, which were published in 1878 in the magazine *La Nature* and proved that a galloping horse is indeed free from the ground for a moment. The camera provided the visual solution to the problem.

Muybridge went to Paris in 1881 where Marey invited him to give the first European demonstration of his findings.[4] But Marey was not satisfied with what he saw. The pictures Muybridge had taken of horses with a battery of cameras showed the successive positions of the horse, but his technique could not successfully capture the movements of smaller animals like birds. While Muybridge's interest centered primarily on the decomposition of movement into phases, Marey wanted to determine the precise relationship between time and motion in the sequences. Muybridge could freeze the individual moments in series, but he could not integrate the crucial variable of time. His images were spatially distinct but temporally blurred. Marey, therefore, developed and experimented with a camera which made a picture of changes that occurred in instances of time.

Marey's graphic method had given movement as continuity, moving with the movement, but it did so at the cost of concrete detail. The camera provided pictures with as much concrete detail as one could wish, but it did so at the cost of continuity and clear expression (Braun 1992: 62).

Marey appealed to the French government for support, stressing that motion studies could have potential benefits for military training. Together with his assistant Georges Demeny and the photographer Albert Londe, Marey developed a technique of geometric chrono-photography, "an absolute original, indeed revolutionary, method of photographing movement by decomposing it and registering its segments on a single readable plate" (ibid.: 82). He analyzed motion by means of a series of instantaneous photographs taken at very short and equal intervals of time.

The government showed interest and, in 1882, a research laboratory, a *Station Physiologique*, was built in the Bois de Bologne near Paris, where Marey continued his experiments.[5] He began to analyze crippled soldiers and patients to improve their mobility and to reduce fatigue.[6] He attached "tiny electric bulbs connected to a generator to the patients' joints and then followed their light with his camera as each lamp unmistakably revealed the otherwise invisible consequences of accidents, war, old age, and disease" (ibid.: 102). Marey's new insights about the movements of the human body and their energy household led to improvements in military training programs and gymnastic exercises at schools which aimed at giving people better control over their bodies.

Yet, after 1888, Marey's complicated camera technique became obsolete as sensitized photographic paper film reached the French market. The film, invented by Thomas Edison, became the key to Marey's technological progress in deciphering human motion. Edison once met Marey, and this led to the Kinetoscope, the first commercially viable motion picture viewer.[7] Marey used the new cinematographic technique to reveal the essence of the movement of animate motors that could not be seen by the naked eye.

Marey's optimistic faith in objective quantification made him a firm believer in the objective treatment of the "human machine" as his countryman and medical doctor La Mettrie had described earlier. His techniques were the only way and the sole determinant of our knowledge of reality. He was also convinced that his research could help to improve the existence of human beings by reducing the waste of energy in their daily activities. In order to extend his research capacities, his laboratory was restructured in 1901 and became the Institut Marey for advance research in physiology.

The European Science of Work

Marey's insights were of interest to those who wanted to improve the human lot. Motion studies and studies of fatigue spawned a new generation of social scientists in France and Germany who focused on the problem of industrial work. Charles Fremont, a French engineer who was interested in the optimum work performance that workers could attain while working with specific tools, was one of the first to apply chrono-photography to the study of physical labor in Marey's laboratory in 1894. His chrono-photographs initiated several different studies of industrial occupations with the goal of discovering the optimum expenditure of energy. The interest in the Science of Work (*Science du Travail, Arbeitswissenschaft*) deepened.

With the French in the forefront, the new field was highly promoted among physiologists, hygienists, and engineers. The French advocated political and social reforms to improve labor conditions. Regulation of labor power and conservation of energy became synonymous labels. Studies appeared concerning the economy of the laboring body. Bodily movements were subjected to detailed investigations. It was generally believed that with a reduction of fatigue, both the performance in an industrial workplace and social harmony could be promoted.

Not everywhere were the links between science and business practice so clearly established as in France. Research at German university laboratories remained isolated from the business community and had little outside impact. Max Weber criticized the work of those "energeticists" and psychologists who were reluctant to promote interventions in industry. Weber and others developed an alternative approach to the problem of labor power by promoting sociological survey projects under the aegis of the "Verein für Sozialpolitik" (Social Policy Association). Their studies emphasized the importance of studying performance in actual work. The researchers needed to understand the tendency of workers to "regulate" their own productivity and respond to economic as well as physiological factors in real life situations. The Verein was not interested in the energy expenditure or performance of the individual worker per se, but in the way the social body of workers was conceived in terms of conditions of recruitment, structural components in the work environment, optimal deployment, and strategic mobilization of forces.[8] According to Rabinbach (1990), German sociologists were more oriented toward reform but focused less on the need for state intervention than the French.[9]

In many European countries, the Science of Work was based on the premise that greater productivity would lead to social happiness.[10] It shared the idea that "the elimination of fatigue, overwork, and excessive motion" could be achieved through reducing the acceleration of the work tempo, introduction of rest pauses, and a shorter working day (Rabinbach 1990). The Science of Work community began to worry about the consequences of the new factory system that was born in the heyday of the second Industrial Revolution. International congresses of hygiene and demography brought together all sorts of experts and reformers to deal with the risks of epidemic diseases, poverty, as well as industrial health and safety. Fatigue in industrial work became a key topic and prevention of it could become a promising means of resolving the social issues of the epoch.

Nonetheless, some physiologists and hygienists believed that an optimal performance could only be attained by achieving a proper balance between a shorter working day and the intensity of work. Concerns about fatigue and the quality of the work environment reflected the consequences of the factory system that began to rely on cheap, unskilled, and inexperienced labor, replacing the skilled, self-disciplined, and proud craftsmen. Physiologists believed that a rational solution to the labor problem could be found, but the moral discipline of work, which middle-class reformists had attempted to indoctrinate among earlier generations of industrial workers, had given way to debates about the discipline of the workplace and to conflicts over the tempo of work and the risks of mechanized industries.

Labor unions and socialist parties, however, resisted the efforts of members of the Science of Work to simply pursue a policy of labor time reduction in order to reduce fatigue without taking into consideration the negative consequences for their wages. They mobilized more frequent rallies for the eight-hour day without reduction in wages and organized strikes when employers suggested that the loss of income was inevitable (Cross 1989). As director of the Société des Produits Chimiques in Engis (Belgium), industrialist and engineer L. G. Fromont announced in 1897 that his factory would start to operate a system of three eight-hour shifts. Once resistance had been exhausted and wages increased, he noticed that productivity had risen and that labor relations had improved considerably.

Around 1900, German firms introduced the time clock and the punch-card system which exemplified a tighter restraint over work time, as did increased shop-floor hierarchy and surveillance. German labor unions organized protests to resist the increasing hegemony of business owners and managers over working time and refused to bend to the imposed discipline of the factory system. In 1904, the French military experimented with a reduced working day in its arsenals in Tarbes. Once positive results were reported, a ten-hour working day was extended to all war departments' industrial workshops. In 1906, the German industrial firm Bosch introduced an eight-hour working day with similar results. These local efforts to regulate the tempo of labor by socially concerned entrepreneurs and businessmen coincided with the first attempts by governments to regulate labor conflicts by initiating legally implemented shorter working days and labor policies (Rabinbach 1990). These initiatives, however, could not prevent a crisis of discipline which would burst at the end of the First World War.

In the United States, engineers in particular became highly involved in detailed studies of work practices in several American factories. Their practice-oriented approach provided industrialists and managers with arguments against unions who, in their eyes, resisted "technological progress." Engineers became involved in training unskilled laborers for work in modern industrial plants, taking control of the factory out of the hands of paternalistic and conservative business owners and putting management into their own hands as competent, trained professionals (Rabinbach 1990). One of the most famous of these persons was F. W. Taylor.

Frederick Winslow Taylor

The construction of American railroads had created an internal market which was constantly expanding. This caused a revolution in American industry. Opportunities to sell products at distances far away in booming places through department stores increased substantially. Large-scale production methods became necessary to manufacture the required goods and cut production costs. Production processes were increasingly standardized and mechanized as previously noted with the American Springfield Armory who had been among the first to gain experience with this by changing the method of the production of weapons from skilled, manual labor to factory production. With this mechanization and interchangeability of parts, the company was able to produce 30,000 muskets a year.[11] The same method was also attempted in other non-military factories, but in most cases the result was modest due to the lack of proper supervision during the production process. The work of Frederick Winslow Taylor (1856–1915) changed all that as chrono-technology took hold in the factories.

Taylor came from a well-to-do family. For three years, young Taylor traveled with his family in Europe where he had the habit of listing the arrival and departure times of trains in a diary. After his return to America, he attended Phillips Exeter Academy in Exeter (New Hampshire) where he experienced a remarkable behavior of his mathematics instructor, Mr Wentworth. In his diary, he wondered why they always had precisely two hours of math class. In time, he realized why.

> Wentworth would sit with his watch always hid behind a ledge on the desk and while it was there we did not know what the darn thing was used for. About once a week or sometimes twice a week he went through the same kind of exercise with the class. He would give out a series of problems and insist that the first boy who had them done should raise his hand and snap his fingers. Then he would call his name. He went right through the class until just one half of the class had held up their hands. We always noted when he got half way through the class and the middle boy would snap his fingers he would say: "That is enough, that will do". What he wanted was to find out just how many minutes it took the average boy in the class to do the examples which he gave. Then we found that Wentworth timed himself when he first tackled those problems. He calculated his own time for doing those five examples, and the ratio between his time to do the examples and the time of the middle boy of the class enabled him to fix the exact time for us right along. That was the first instance of a time study of mental operations which I had ever seen.
>
> (Taylor quoted in Kakar 1970: 120)

In January 1874, Taylor left school and began a career in practice as an apprentice for four years in a small Philadelphia enterprise in a hydraulic works company, Ferrell & Jones. At that time, the city was well known for the manufacturing of ships, tools, textiles, and locomotives.[12] At the age of 23, Taylor took a job at Midvale Steel where he became a foreman or "gang boss" of machinists in the tire department. This group formed the industrial world's broadest class of skilled workers producing the metal band on the outside of a wagon wheel. He registered observations in his notebook on the time it took workers to execute certain tasks. The idea of recording work times in a far more detailed way grew as he began to transfer the principles of measuring a machine's capabilities to measuring human capabilities.[13]

As the Navy asked Midvale to produce howitzers used for throwing lines to ships in distress, Taylor began to conduct experiments that would determine the maximum capacity of the firm's metal-cutting machines needed for the task. He rationalized Midvale's operations by replacing the general foremen, who operated too much as sergeants, with functional foremen who would primarily focus on efficiency. He designed a detailed approach to time specific elements of the task machinists executed.[14] He understood that the component times could be added together to fix piece rates for many different jobs (Donkin 2001). He substituted unskilled and inexpensive workers for skilled workers and improved productivity five to six times, raised wages, and increased profits. Traditional craft knowhow was replaced by instruction cards which were passed from the workmen to managers, from the shop floor to a front office. These detailed instruction cards deprived the machinist of his skill but improved the production process (Hoopes 2003).[15] In the big mills of the Midvale shop, Taylor created a new world of work.

In 1887, Taylor became chief engineer at Midvale[16] and began experiments with metal cutting for which he used a stopwatch to time the work that the workmen were required to perform. Out of the times that each of the elements of the job took, he established an overall time for a particular activity by summing up the total times of its component parts. To make the job easier, every required work load was described on the instruction card which became "to the art of management what the drawing is to engineering" (Kanigel 1997: 208). His aim was to move knowledge away from workers by breaking the job into component parts. By doing so, he could improve the speed of production and raise productivity. However, he frequently complained about the working attitude among the workmen, whom he accused of systematic soldiering, which he referred to as "goofing off," a behavior of slowing down during work time (ibid.).

After three years, Taylor left Midvale and became investor and manager of a paper-making company in Maine. However, the economic panic of 1893 ruined the company. Taylor lost a significant amount of money and developed a dislike for financiers and moneymakers. For the next six years, he operated as a consultant, one of the first in America.

In 1898, Bethlehem Steel asked him to improve the efficiency of their production process. Bethlehem steel was under attack as it was producing steel which was too expensive in the eyes of the public. This complaint became all the more serious during the interlude prior to the Spanish-American War. The *USS Maine* had been blown up in the harbor of Havana, killing hundreds of American sailors. American navy ships steamed up to Cuba. As it was still a colony of Spain, America declared war on Spain, ousted her from Cuba and Puerto Rico, and converted the Philippines into an American colony. The United States had, in that sense, become, an empire.[17]

Due to this incident, Bethlehem received an order from the American government to provide steel for the shipbuilding industry. It still had a large stock of around 80,000 tons available which was stockpiled in the yard at the company's site. In the spring of 1899, the company began to sell its surplus iron, and laborers were asked to carry and load the bars of heavy pig iron onto railroad cars. Taylor was asked to analyze the best way of completing this task. Stopwatch timing was carried out on a group of 12 workmen. Sometimes the men were asked to work steadily, and sometimes they were urged to work flat out for short periods.[18]

Taylor analyzed the loading of pig iron onto railroad wagons by introducing Henry Noll, who became the living emblem of Scientific Management.[19] Experiments with Noll loading pig iron on rail wagons showed that of the ten recruits, only Noll managed to attain the 45 ton requirement.[20] All the other nine laborers quit rather than put up with Taylor's backbreaking regime. The individual who was willing to follow Taylor's instructions had to suppress his normal working rhythms and adopt the rigid mechanical practices laid down by Taylor's principles.[21]

> His goal was the maximization of output productivity irrespective of the physiological cost to the worker. As an engineer, he considered the body as a "machine" which either operated efficiently, or it did not. He was not concerned, as was the physiologist, about the functioning of the "human motor," or how energy and fatigue might be optimally calculated for long term use rather than productivity.
>
> (Rabinbach 1990: 117)

By decomposing each task into a series of precise activities which were calculable in terms of time and units of work, Taylor achieved a higher level of work performance and by way of routinization adapted the body to technical demands. He reduced work to the realm of a techno-social order where roles and resources were expressed in functional relationships. By counting and calculating, he transformed management in this sector, which was traditionally oral, into a set of calculated and written techniques (Gowler and Legge 1983).

The productivity of pig iron loaders improved, but this was not enough to satisfy an increasingly impatient management. The technology-oriented supervisors at Bethlehem needed instructions about the effective running and maintenance of large-scale operations. Taylor had introduced the concept of the functional foreman and emphasized the importance of knowing exactly what foremen were to do and then do it in the cheapest possible way. The people-oriented supervisors were instructed by managers how to control the activities of laborers by close contact. Their main function was regulating and directing the activities of workers, monitoring and evaluating their performance, and disciplining.

In order to achieve this, he designed instruction cards, introduced a new device (stopwatch) and drilled the men according to his requirements. The trinity of initiatives Taylor introduced demonstrated remarkable similarities with the way Portuguese captains and Maurits' soldiers had been instructed (see Chapter 3). He assumed that by offering higher wages as bonuses for productivity, the "natural" unhappiness of workers could be compensated by non-work-related, material rewards through which workers could enjoy their free time. His approach, however, generated resentment among the machinists who pointed at the dubious character of his analyses.

Gilbreth's motion studies

Taylor's work was incomparable with the scientific approach Marey had developed. Marey's science of movement was deduced from general theoretical principles of mechanics and physics on the premise that work is the product of force or energy. Taylor's work was the systematization and classification of practical knowhow. It was induced from observation, comparison, and the measurement of numerous ways of doing things (Andrew 1981). Taylor believed he had found an objective measure, a formal research method, but its subjectivity still pervaded his time studies and task setting. He still had to rely on his own eyesight while watching the work being performed by the worker. He could only take imprecise notes and rely on his own interpretation of what a worker actually accomplished (Wrege and Perroni 1974). What Taylor did not have was a sound scientific analysis whereby human activities could be broken down into basic spatio-temporal units to be recomposed thereafter in such a way that it would improve output.

The studies of F. B. Gilbreth (1868–1924) closely paralleled what Taylor was doing. He began as an apprentice bricklayer and worked his way up to become an independent contractor. Thereafter, he established his own management consulting company and met Taylor for the first time in the New York Engineering Society's Building in 1907 (Spriegel and Myers 1953; Yost 1949). In his *Bricklaying System* (1909), he reported how a motion study of bricklaying enabled him to devise an adjustable scaffold for piling up bricks, a method which tripled workers' output. The higher rate was possible not because of harder work but because his method eliminated certain unnecessary movements. He became convinced that his motion study revealed the best way of laying bricks and could teach apprentices the right way to maximize output.

Gilbreth had to admit that it was difficult to transfer the knowledge he obtained from his micro-motion records as he could not make others visualize what he had noticed. Nevertheless, he extended his motion studies beyond bricklaying and advised in the general field of manufacturing. One of Gilbreth's first major clients was the Remington Typewriter Company, which had been split off from the Remington Arms Company in 1886. "This company did much to cement the typewriter into the daily operations of modern business" (Wershler-Henry 2007: 145).[22]

Fatigue became his major focus of attention, and with the help of his wife, Lillian Moeller (1878–1972), who was a trained psychologist, he attempted to detect its causes (Lancaster 2004). The prevailing approach was that "fatigue was caused by a toxin generated by physical exertion and released in the blood. Every motion caused fatigue; hence, any elimination of motions reduced fatigue" (Wren 1987: 144). After a number of experiments, they jointly developed a list of micro motions of the most elementary gestures. The list contained 17 basic motions – each called a Therblig (Gilbreth spelled backwards with the "th" transposed) – such as "search," "select," "transport loaded," "position," "hold," and so on. The list gave the Gilbreths a precise way of analyzing the exact elements of any worker movement.

The human eye, however, was ineffective when it came to following motions. There was no way of recording and measuring motions accurately, neither were there any writings available that could be called scientific. Gilbreth looked for a new frame and made use of the infant technology of the motion picture film. In 1910, he met Muybridge and became fascinated by its applications. In his *Motion Study and Time Study*, he wrote that "we are now able to use the motion picture camera with our speed clock, and other accessories, as a device for recording elements of motion and their corresponding times, simultaneously" (Gilbreth 1953: 230).

Gilbreth applied the film technique to analyze alternatives for human movements that were better suited to specific tasks. He began with a cycle graph method whereby the cycle was the unit of measurement to be recorded. He put a worker in a blacked-out room and tracked a light bulb attached to the man's hand with a still camera. He made photographic time exposure of motions that appeared on continuous white lines. These made it possible to see the path of a motion which was an exact emulation of the system Marey had devised in 1887.

The Gilbreths carried out studies in a laboratory to visualize each part of operations on a shop floor in relation to the whole. They adapted a motion picture camera to take chrono-cycle graphs "that would show the paths of each of several motions made by various parts of the body and their exact distances, exact times, relative times, exact speeds, relative speeds, and directions" (Kern 1983: 116). Thereupon, he constructed wire models of the motions which represented the path, speed, and direction of the motion they had registered. The wire model illustrated the movements of a skilled worker even better than writing could do. He also discovered that highly skilled workers made more rhythmic paths of motions than the less highly skilled.

Gilbreth has always been presented as the man who invented this new research method, creating a technical language which endowed the analysis of the labor processes with a new meaning. He knew that Taylor's work was lacking in scientific quality because it had not been supported by an adequate analytical study of motions preceding his time studies. Gilbreth made use of scientific insights, but he did not develop a theory about the human motor. His source of inspiration lay in Europe, and between 1910 and 1914, the Gilbreths often traveled to Europe to visit the Institut Marey. Frank was familiar with what had occurred in Paris but did not like to acknowledge it. He had a strong desire to be considered original in his field. In an unpublished typescript of a management seminar, Gilbreth declared that Marey was "the man I wish had never been born" (quoted in Braun 1992: 347; Yost 1949: 246). In his published work, Gilbreth mentioned Marey only once:

> As for motion study, Marey, with no thought of motion study in our present use of the term in his mind, developed, as one line of his multitudinous activities, a method of recording paths of motions, but never succeeded in his effort to record direction of motions photographically.
>
> (Gilbreth 1953: 229)

It is clear that this description is a biased one. Marey had not taken out any patent in France or America for his system, therefore, Gilbreth had free rein to do so.

Nevertheless, there is a clear difference between the systems of the two men. Marey was accurately charting the basic movements the human body could make, whereas Gilbreth intended his analysis of motion to be the key factor in improving the mechanical efficiency of a basic industrial element: the human being. Gilbreth initially wanted to combat symptoms of fatigue with which the workers had to cope. His research techniques could have been used to communicate with employees regarding the best way to improve their work habits, but it has hardly been used in that way.

Gilbreth mainly produced pictorial representations that decomposed human motion into its constitutive elements. At a meeting of the Taylor Society in New York in 1920, Gilbreth condemned Taylor's method as "unethical, wasteful, and inaccurate" (Nadwordny 1957: 31). Taylor considered time study to be the foundation of his work and saw motion studies as the development of a fine system of time study. But the Gilbreths argued that micro-motion studies were more accurate than simple time studies (Lancaster 2004).

Gilbreth's attitude toward workers also demonstrated a fierce difference of opinion with Taylor. He intended to reduce fatigue among workers in modern factories with his studies, but they were actually a very adequate technique to speed up working. In other words, both Taylor's and Gilbreth's approaches might have been justified within a techno-social order, but they were not within a moral aesthetic order where subjects are treated as human beings possessing consciousness and understanding. That conflict became apparent in the Hawthorne experiments which will be described in the next chapter.

Scientific Management popularized

Taylor's approach came to attention nationwide in the USA in 1910 when the Pennsylvania and other railroads petitioned the Interstate Commerce Commission for higher freight rates to cover a pay increase for workers. The national railroads had grown into interstate business giants. Several of them employed more than 100,000 people, and the industry as a whole was widely regarded as unduly wealthy and powerful. Shippers and consumers opposed an increase of rates, and the shippers hired lawyer L. D. Brandeis to oppose the railroads' petition. Brandeis (1856–1941) was a graduate from Harvard Law School and believed that, as a lawyer and as a citizen, public service is a duty. He embodied the popular revolt against big business dominating American economic life as they were monopolizing power and unduly influencing government (McCraw 1984).

Brandeis resented the fact that government had taken the side of capital by turning corporations into constitutionally protected "persons" while simultaneously denying legal protection to the activities of labor unions. He favored industrial democracy and urged unions to fight for reasonable working hours because, as members of a democracy, workers needed leisure to improve their wellbeing (Strum 1993). Brandeis saw it as the government's responsibility to alleviate the situation of the labor force by enacting minimum wage and maximum hour laws. He became one of the nation's best known lawyers in *pro bono* cases[23] and pleaded for the awakening of a social sense of community that would enable the citizenry to regain its rightful place in public affairs.[24]

After having read articles by Taylor and other efficiency engineers like H. Gantt, Gilbreth, and H. Emerson, Brandeis became enthusiastic about time-and-motion studies. He was especially under the spell of Emerson, who had been trained at a Polytechnic in Munich and was consultant for the Burlington Railroad and Santa Fe Railroad. The time-and-motion studies became "a revolution in industry comparable only to that effected in the transition from hand labor to machinery" (Brandeis quoted in Strum 1993: 43).[25]

During the hearings about the freight rates, Brandeis pointed out that the proper source of the financial problems with which the railroads were confronted should not be compensated by demanding higher rates for the shippers. Instead, the railroads should lower their own internal operating costs. Brandeis impressed the commissioners of the hearings by stressing that the railroads were operating inefficiently and that railroad magnates were gouging the public in order to make up for their own wastefulness (McCraw 1984). Brandeis called several efficiency experts like Emerson to the witness stand to show that the railroads could reduce the terrible waste of delays and avoid accidents by applying calculation techniques instead of guesses. During his witness testimony, Emerson had stated that the railroads could save 300 million dollars a year if they improved efficiency in the way the Prussian army had operated the German railroads during the war with France in 1871 (Martin 1971).

Brandeis referred to the work of Taylor and other efficiency experts, labeling it "Scientific Management." As the hearings had gained widespread interest, the general public suddenly became aware of this new label, which was taken up by newspapers and broadcasted throughout the country. Cited again and again during the rate hearings, it took on a life of its own. "Ultimately, it seemed to be made true by virtue of endless repetition" (McCraw 1984: 93). Subsequent to the hearings, Emerson published *Twelve Principles of Efficiency* (1912) which became a landmark in the history of management thought and turned him into a pioneer spreading the gospel of efficiency.

Brandeis won his case and Scientific Management obtained national popularity and created a national craze for efficiency. According to Martin (1971: 361), the impact of the hearings was that the railroads came under federal control and regulated. "What was lost was the spirit of enterprise which had produced such remarkable results." Yet with the nationalization of the railroads, the unions achieved a high degree of unity, and with the intervention of President Woodrow Wilson, obtained the eight-hour day in 1916.

Eventually, railroad transport became incapable of competing with the increasing popularity of trailer trucks which had more flexibility in transporting their cargo. The passenger train would lose its attractiveness as well once the American public began to show an increasing preference for their own cars, and a growing set of interstate highways was constructed. Intercity travel was also preferably done in a private automobile. The American rail industry would face serious problems in the 1970s when Penn Central went bankrupt and others abandoned their passenger service.[26] The public, however, credited Taylor with the victory of Scientific Management, and it became "the first of many twentieth century management fads to spill over into the general culture" (Hoopes 2003: 59).

In 1911, Taylor summarized the basic precepts of industrial mechanization in *The Principles of Scientific Management*. All the brainwork of the workmen should be removed from the shop floor and centered in a planning department where managers and engineers gathered all the explicit knowledge together in order to classify, tabulate, and reduce it to rules, laws, and formulae. The success of the modern factory system relied not only on the establishment of a full-swing time-coordinating production process but also depended on the possibilities to subsequently recruit, select, and employ men of "smaller caliber" who were cheaper than those required under the previous factory system still operating with skilled workmen. Taylor was clear about the policing of the workmen:

> It is only through enforced standardization of methods, enforced adoption of the best implements and working conditions, and enforced cooperation that this faster work can be assured. And the duty of enforcing the adoption of standards and of enforcing this cooperation rests with the management alone.
>
> (Taylor 1911: 83)

Taylor was convinced that with this discipline-based managerial approach, an economically effective work attitude among the workers would be constructed. He assumed that labor unions would not accept his new approach and that management would have to enforce the standardization of methods, the adoption of the best tools and equipment, and working conditions.

Timekeepers became the new regulators and controllers of the tasks of the workmen. They were required to measure productivity and efficiency by analyzing the operations involved in each job by purely calculating the efforts. They looked for the maximization of productivity time either by lengthening of the working day or, when this became impossible, by intensifying the work done in the maximum time available. Taylor mainly focused on physical labor and material efficiency to maximize productivity, irrespective of the physiological cost to the worker. "As an engineer, he considered the body as a machine, which either operated efficiently or it did not" (Schor 1992: 58).

After Brandeis' pleas and press coverage of Taylor's work, a national efficiency craze boomed. Taylor had shown the stopwatch to be a symbol of heightened "time thrift" and a greater intensity of production.[27] However, strikes at the US Arsenal in Watertown (Massachusetts) in 1911 against the activities of time managers dimmed the enthusiasm for Taylor's work. At the governmental hearings about the cause of the strike, Taylor gave a poor presentation when he tried to justify the application of the stopwatch to improve performance of the arsenals. Finally, the US Government banned Taylorism in government-owned organizations until 1948.[28]

The American business community, however, showed an increasing interest in Scientific Management, and Gilbreth created the Society for the Promotion of the Science of Management (SPSM) to respond adequately. As consultant, he advised the New England Butt Company in 1912, which produced braiding machines by applying micro-motion studies and filming workers' operations. His films demonstrated that motion studies increased output through a more effective use of time rather than through faster speed. During 1913–14, Gilbreth undertook a major reorganization of the Herrmann-Aukam Company, a manufacturer of handkerchiefs, by applying chrono-cycle graphs and stereo chrono-cycle graphs. He filmed many workers in order to arrive at an ideal pattern of movements. He translated his analyses into cords with complicated twists which illustrated the complicated hand movements of workers. These cord constructions showed the precise pores of empty time and were used as instruction material for other workers to imitate. However, Gilbreth became irritated frequently by the behavior of foremen, superintendents, and managers who were even more recalcitrant than workers in applying his techniques. During his work as adviser at the American truck producer Pierce-Arrow, Gilbreth failed to respond to the directives of management and resigned prematurely (Nelson 1992).

During the First World War, Gilbreth made a film for the American army to instruct soldiers on how to handle their weapons with more dexterity.[29] Together with Alexander Church, Gilbreth also studied the possible benefits to production work of well-placed artificial lighting. They supposed that reflection of lighting caused tiredness among workers and analyzed the relationship between high-intensity electric lighting and production. In line with Gilbreth's approach, William Durgin, one of Edison's lighting specialists, carried out detailed tests in 1918 to highlight the link. His experiments in four different factories encouraged General Electric to increase its interest in the promotion of factory lighting, which incidentally was supposed to strengthen its dominant market position by selling more factory lighting equipment.

Scientific Management – or Taylorism as it became called – intended to create a clockwork world of tasks timed to the hundredth of a minute, of standardized factories with men and women working in formalized ways like machines (Kanigel 1997). It reinforced chrono-centrism, the belief that the manager's perception of time is the true and superior concept of abstract time and promoted an approach, a state of mind, applicable to every aspect of life. Its purpose was to take all control from the hands of the workmen and place it in those of management. It helped instill an obsession with time, a fixation on productivity and efficiency. Modern, large-scale, American industries might incorporate "not only the mechanization of tools and equipment but also the mechanization and systematization of the workers and their transformation into an interchangeable part of a working force" (Dan 1992: 38; Andrew 1981).

In fact, only a handful of American firms actually adopted the Draconian schemes in their entirety. In the 1920s, about 1 percent of all industrial workers in the United States labored under the full-fledged Taylor system. Most managers found its proposals for intensive productive energy studies and minutely detailed work instructions far too complicated to implement. Taylorism, with its standardization, routinization, and synchronization of labor processes had virtually nothing to say about what different machinery might do for efficiency. "Managers applying Taylor's principles mainly riveted on commandeering the craftsman's knowledge" (Montgomery 1987: 233).

After the death of her husband in 1924, Lilian Gilbreth continued their work. From 1925 to 1927, she conducted a series of motion study courses for participants from Johnson & Johnson, Macy, General Electric, and Cadbury (the British chocolate manufacturer). As a consultant, she sought work with Sears Roebuck, began lecturing about management issues and eventually became the "First Lady of Engineering" (Lancaster 2004). Taylorism provided the first management-oriented industrial ideology. It synthesized and promoted a coherent framework for changes that were already taking place in a piecemeal form in various industries. Fordism followed next.

Henry Ford

Henry Ford was born July 30, 1863, in Dearborn (Michigan). By the age of 16, he had left home and moved to Detroit, the place that was rapidly becoming the machine capital of America. In addition to automobiles, Ford was also fascinated with clocks and watches and developed an ability to repair a variety of timepieces.[30] One of his first paid jobs was repairing, maintaining, and using a neighbor's small steam engine. This led to a job as a local representative for Westinghouse, the manufacturer of steam engines. Along with this job, he continued evening and weekend work as watch repairman. He then took a job in Detroit with the Edison Illuminating Company as a mechanical engineer in a power station.

While repairing machinery, he realized that the "Silent Otto," an internal combustion engine, could be redesigned to drive a vehicle.[31] After some experiments, he took his first car onto the streets in 1896. Three years later, the Detroit Automobile Company was formed. In January 1900, the first vehicle, a delivery van, drove out of the factory. However, the company went bankrupt. Ford's second car company, the Henry Ford Company, opened business in November 1901. As Ford had won the famous Grosse Pointe Race, his new company was focusing mainly on producing racing cars. Henry M. Leland, one of the investors, did not like this development, fired Ford, and turned the company into the Cadillac Automobile Company which, in 1909, became the General Motors Corporation.

With the support of another investor, Tom Cooper, Ford started a new company that soon came out with the Model A. In this period, it was common practice to leave the manufacturing of the car outsourced to subcontractors with the factory concentrating on final assembly. By 1903, his company produced almost half of all US cars. The new company assembled 25 cars a day and, in 1904, three new models (B, C, and F) were launched. In 1906, Ford announced that they would produce 10,000 Model N vehicles at a price of 450 dollars. This model, together with its enhanced model R and S, generated spectacular growth. In the winter of 1907, Ford began to design a completely new car, his twentieth within five years, to make a stronger, lighter, and more reliable vehicle. For its mass production, an assembly line was designed by his technicians, Charles Sorenson and P. E. Martin. Ford's mass production consisted of a "just-in-case" system with huge stocks of everything.[32] Ford wanted a simple, standardized, and cheap vehicle for the public at large. In 1908, after several years of experimenting and improvising, the first Model T Fords were assembled with replaceable standard parts.

The modern Ford Motor Company – Highland Park Factory – opened in 1909. The assembly line, which made control of production easier, was introduced in 1913. The idea was to keep the workers in place and let the machinery pass by them. Myths exist regarding the greatest inventor of all times. It is usually said that Ford invented the modern age because of his invention of standardized cars and his assembly-line mass production. However, the assembly line predated Ford and was in operation in the Venetian Arsenal by the late fifteenth century, long before the first of nine different body styles and the initially green Model T appeared.[33]

One of Ford's engineers, William Klann, had also observed how, in the Chicago meat-packing business, whole carcasses hung from a moving line and were progressively disassembled and butchered as they were moved along past the workmen on a conveyor belt. Mass production was also already being utilized in the manufacturing of Colt revolvers and Singer sewing machines. Nevertheless, Ford would become an icon of the twentieth century and, at the same time, was a man with a mass of contradictions, "a leader whose organizational success rested not on an engineered organizational clarity but on invented organizational paradox" (Grint 2000: 189).

Taylor's policy of breaking a task down and timing each element of the task was pursued to the extreme by Ford's engineers. The conveyor belt and continuous operation of the assembly line eliminated challenges and surprises as the product moved along with every step worked out beforehand. The production process was further streamlined by observing every stage, determining the minimum movements necessary to complete all tasks, and instructing the workers how to execute their job. The process deprived workers of the opportunity to select the sequence of actions to complete an operation (Kern 2001).

Mass production of cars in a completely synchronized production process was born. The workers did not move themselves over to the material; the assembly line took the material to the workers. From then on, the pace of the conveyor belt and assembly line determined the work rhythm. Similar to the editing of pictures into a film, the process of the assembly line was synchronized through the eyes and minds of the civil engineers. Based on Scientific Management, the required choreography of the workers became programmed. The timing of the production process was built into the speed of the line and imposed itself "objectively" on the worker (Dan 1992: 43).[34] The *choreo synchrone* definer of time had become a fact and Ford tried hard to implement Wedgwood's principles of discipline and "make such machines of the Men as cannot err."

It was no longer the foreman or the engineer inspecting the workman's activities with a stopwatch that determined the speed of work; the conveyor belt regulated the speed of the workman. In 1909, a fully equipped model T Ford took almost 13 hours to assemble and cost 950 dollars. In 1911, some 78,000 Model T cars were produced by almost 7,000 workers. In 1912, the workforce had doubled, and they produced double the number of cars. By introducing interchangeability, Ford was able to further reduce the time of production and speed up the production process. He replaced the outsourcing of parts with in-house production and built a vertically integrated organization.

Ford's displacement of skilled workers by machines also opened up a sizeable source of labor for employment. "The interchangeable parts were accompanied by the interchangeable workers in the interchangeable organization" (Womack 1991: 28, 229). The new technology embodied the primary disciplinary mechanism. The price reductions were achieved through a permanent reduction of the costs of materials and brought-in components. It was insourcing as much as new production techniques and the ever-expanding size of the production plants that contributed considerably to the foundations of Ford's success. By further mechanizing the production process and introducing labor-saving techniques, Ford was able to reduce construction time. More than ever before, standardization and routinization became the key issues and were supported by time-and-motion studies.

Nonetheless, the company struggled with a tremendous labor turnover among workers with 22 different nationalities and, by early 1914, absenteeism was running at 10 percent. Whereas Taylorism attempted to improve the efficiency of workers, Ford tried to eliminate labor where possible by using machines, but the growth of the company forced him to continuously recruit more employees. Yet the work became so boring, monotonous, and alienating that Ford was forced to come up with a remarkable formula to get commitment from the workmen. He decided to not only pay decent daily wages but "sweetened the pot" by also introducing a shortened work day.

On January 5, 1914, he made an unprecedented gesture by raising wages to five dollars for an eight-hour working day and replacing two daily nine-hour shifts with a three-shift, 24 hours per day, nonstop rotation. The five-dollar-a-day wage created a myth about the Ford factories. The *Wall Street Journal* commented on it as the most foolish thing ever attempted in the industrial world, while Henry Ford described his decision as one of the finest cost-cutting moves he ever made. Taylor commented about Ford's results to a body of 600 Detroit auto factory superintendents and foremen by saying that the company had been "the first to install the principles of Scientific Management without the aid of experts" (quoted in Montgomery 1987, 233).

By late 1914, labor costs had been cut in half, and the price had been lowered to 490 dollars; in 1916, the total vehicle assembly time was cut down to 1½ hours. The same car would only cost 360 dollars, a price which was equivalent of a teacher's annual salary. Approximately ten years later, the price of the car had dropped to 295 dollars. The reduction in hours was mainly realized by replacing subcontracted work with internal labor. "It was organization, not technology, that was mainly responsible for Ford's leap in productivity" (Grint 2000: 204). By 1920, more than two million Model Ts were rolling off the assembly lines every year – half of the cars produced and sold in the United States. The next step was to facilitate car driving within the spatio-temporal order by creating "passages" through the building of highways and interstates with all kinds of facilities such as petrol stations, restaurants, and hotels, which developed into standardized networks that eased the flow of transportation and turned "making miles" into the standard expression amongst car drivers.

European responses

A fierce opponent to the Taylor system was the French physiologist and psychologist Jean Marie Lahy. He pointed out that Taylorism had an inadequate basis in science and stated that Taylor's method of chronometry was rudimentary and biased (Rabinbach 1990). Based on an analysis of the work of typists in 1908, he concluded that the commitment and productivity of a female employee was as high as a male; they, therefore, deserved the same remuneration.

Notwithstanding this criticism, Taylorism became applied in major car-manufacturing firms like Renault and Peugeot. Around 1910, the Ford Motor Company had expanded internationally by building subsidiaries in Europe. It began production in the United Kingdom in 1911 and by 1913, controlled as much as 60 percent of the less expensive car market in Britain (Brinkley 2003). Yet implementation of Taylorism in the European automotive industry continued to be complicated. During a famous strike in the Renault factory in March 1913, the unions contested the time-and-motion studies as ferocious and barbaric (Fridenson 1987; Fridenson and Reynaud 2004). Although the strikers were defeated, Taylorism became the object of a passionate discussion in the French technical press. Timekeeping was defended by factory owners as the most appropriate way to determine the best conditions of work corresponding to a minimum of fatigue.[35]

In Germany, electrical firms like AEG, Bosch, Borsig, and Siemens embraced Taylorism as well. The Verein Deutscher Ingenieure (VDI) urged the imitation of American methods of factory organization. Company owners and engineers showed a readiness to regulate the time and motions of the worker and supervise the details of production to a degree previously unthinkable. In 1914, Gilbreth led a joint German–American team assigned to advise the Berlin lighting firm Auergesellschaft in its plans to rationalize its operations, but he failed to satisfy the requirements of its management.

During the First World War, a rapid integration of new workers, especially women and adolescent males, into the munitions and armory industry increased the need for Scientific Management as a method to deploy untrained workers. This greatly accelerated the integration of studies concerning fatigue and psychophysics, undertaken by the Science of Work, into the vocabulary of Scientific Management. "Despite growing strikes against the detested time-motion study, the premium wage system, and long hours of work in both France and Germany after 1917, the war brought Taylorism and Taylor's ideas into the mainstream" (Rabinbach 1990: 261). The Science of Work became fragmented into a large array of academic subjects like psycho-technical aptitude testing, industrial physiology and psychology, plant sociology, and personnel management, "which in general reduced labor to an economic, technical and psychological dimension without paying too much respect to its generic character" (ibid.: 275).[36] The Science of Work in continental Europe and Britain failed to put up a wall against Taylorism. Yet, different researchers and authors voiced their discontent.

During the last few years before the outbreak of the First World War, there was a wave of industrial unrest. Wage cuts, poor working conditions, and rapid inflation had left workers deprived and disgruntled. In Liverpool, sailors organized a successful strike for an increase in wages and union recognition which soon snowballed. A huge demonstration organized by the Transport Workers Federation (TWF) took place in Liverpool on May 30, 1911.[37] Union leaders like Ben Tillett and Tom Mann, cofounder of TWF, inspired the dockworkers and transport workers to hold on to their demands. In 1886, Tom Mann had published a pamphlet, *What a Compulsory 8 Hour Working Day Means to the Workers*, which helped to persuade a large number of people to support this measure. Mann also formed the Eight Hour League, which was influential in convincing the trade union movement to adopt the statutory eight-hour day as one of its core policies.

In other countries like Belgium, the Netherlands, Australia, and the USA, seamen organized strikes out of solidarity and the protest movement obtained a truly international character. The combination of leaders inspired by socialism and the increasing influence of national unions organized in confederations increased according to British historian Hobsbawm (1974). The interest grew for the implementation of the eight-hour day which without the Owenite movement and Chartism would certainly not have become such an international movement. On the European continent, some unions successfully advanced the case. In Amsterdam, the union of the diamond workers, ANDB, as one of the first, obtained the eight-hour day on October 1, 1911 (Karsten 1990).

The political climate and economic distress inspired anarchists and communists to radicalize and challenge existing property rights and power relations as the foundation of a bourgeois society. Yet others believed that democracy could advance with general suffrage and as such save it from elitist interests. Under the pressure of growing social-democratic strength, the focus shifted to strong democratic nation-states to prevent revolutionary class conflict. Out of fear for radical changes like the one that took place in Russia in 1917 with Lenin seizing power, politicians and business communities were willing to give in.

During the war, European governments had suspended protective labor legislation. Food shortages, longer working hours, and increased regulation of mobility during the war had fueled working-class discontent, which translated into industrial militancy. Under pressure of war, promises to the labor class such as a general introduction of the eight-hour day, either by general labor contract or by state regulation, were made as compensation for their efforts during the war. French labor had already secured this long-standing objective in various government enterprises before the war when coal miners won eight-hour contracts in 1913. The French government of Clemenceau did not hesitate to implement a general legal eight-hour day to prevent radicalizing of the working class. German labor achieved the same in November 1918 and Italian workers in February 1919 (Maier 1975; Geary 1981). The First World War, therefore, is regarded as a watershed in European history (Mayr 1986). Not only was it a traumatic break with prewar politics but also with the role of the labor movement and socialist parties.

The Russian Revolution, and the revolutionary mood in Germany after its defeat, created an extremist atmosphere. Beginning with an eight-hour day proclaimed by the Bolshevik government in November 1917, eight-hour decrees spread early in 1918 to Finland and Norway. After Germany, the eight-hour day reached the new states of Poland, Czechoslovakia, and Austria by mid-December. From the revolutionary regimes of Eastern and Central Europe, agitation for shorter working hours spread to Switzerland, where up to 400,000 men struck in December for the eight-hour day. Also in that month came militant strikes for an eight-hour day in British mining and railways, culminating in a general strike in Glasgow for the 44-hour week in January 1919. This was followed by successful strikes in most major British industries.

The movement spread to Italy in the metal goods industries in February 1919 and led to a wave of shutdowns in March and April in textiles, chemicals, and even in agriculture. The French initiated only a few strikes for reductions in working hours in the first quarter of 1919. This was more the result of much weaker unions and the more repressive war economy than less interest in free time. Still, by March 1919, French metal, dock, mine, and textile workers demanded an eight-hour day (Cross 1988).

Paralleling this European-wide pressure from below were efforts to make the eight-hour day an international law. On January 25, 1919, French president Clemenceau persuaded the preliminary peace conference to create a commission for an International Labor Legislation. During the first International Labor Conference in Washington – where the International Labor Office (ILO) was officially constituted – the principle of an eight-hour work day was enshrined in the peace convention of 1919. Whereas many countries in the industrial world supported the idea of legally limiting working time to eight hours, the United States and Britain were the only major industrial nations that failed to pass such a law. Yet the eight-hour day can be seen as the fruit of a generation of agitation for an improved distribution of work and leisure. It was also the outcome of an increasing standardization of working time in standard working conditions.

Proposals for broadly based recreational facilities were investigated to counter upcoming mass-market leisure activities. The demand for an annual paid vacation also emerged during these years. The goal was to provide workers with extended time free from work to cultivate family life and travel and take some distance from the overwhelming influence of bureaucratically structured working patterns.

When the legal eight-hour day was generally introduced after the end of the First World War, the Industrial Fatigue Research Board quietly abandoned its emphasis on a further reduction in the hours of work. Yet during the interwar period, discussions about the impact of Taylorism in Europe continued, especially in the French technical press. Henry Le Chatelier, Inspector General of the Mines and professor at the Sorbonne, published a series of articles favoring Taylorism. He defended the interest of management and pointed out that the workers did not reject the presence of time-and-motion studies per se, they only resented the fact that they had not been involved in the decision to employ such methods. He stressed that critics failed to understand the purpose of timekeeping and determine the best conditions of work by reducing fatigue. The Frenchman and anticlerical freemason Jean Marie Lahy, who was sympathetic to socialism, contested some of the one-sided views of Taylorism and became the founder and leading proponent of the movement for the "humanization" of work. He believed that science and the laboratory might provide support in solving industrial problems. The writer Franz Kafka, living in Prague and flirting with anarchism, was appalled about the possible enslavement of mankind if Taylorism became generally accepted and said:

> Time, the noblest and most essential element in all creative work, is conscripted into the net of corrupt business interests. Thereby, not only creative work, but man himself, who is its essential part, is polluted and humiliated. A Taylorized life is a terrible curse which will give rise only to hunger and misery instead of the intended wealth and profit.
>
> (quoted in Janouch 1951/1971)[38]

In 1924, German psychologist Otto Lipmann analyzed the problems of labor time and pointed out that a further reduction of labor time and more rest pauses might still improve work performance. German scientist Hugo Münsterberg[39] drew attention to the fact that psychology could contribute a great deal of knowledge to testing, selecting, training, and motivating workers to prevent waste of energy by those who were not skilled for the job. He reproached Taylorism for its overreliance on self-interest and coercion, its single-minded focus on the task, and its disinterest for the subjective experience and skill of the worker. He propagated an industrial psychology which is concerned with adaptation, motivations, and satisfaction of the worker.

Other researchers suggested that the best solution would be to combine "psychotechnics" and the principles of American Scientific Management. Yet scientists and researchers within the Science of Work generally lost their influence more and more in the European business community.[40] They continued to believe that scientific writings could be used to emancipate the working class. In contrast to their academic focus and laboratory experiments, Taylorism focused on the plant-based engineer with his expertise on the shop floor and his observations of workers. Taylorism assumed that by offering higher wages as a premium for productivity, the unhappiness of workers could simply be compensated.

Yet both views were suspicious about the atavistic attitudes of workers and the traditional paternalism of factory owners and functional foremen. However, when rationalization of production was propagated during the second half of the inter-war years, it was mainly predicated on the rationalization of the body. In Italy, Mussolini paid homage to Taylor's achievements. When the war economy went into full swing, the Nazis appointed permanent posts for psychologists in German firms. In their hands, technology became glorified as the solution to all social problems. Pro-Taylorist technocratic approaches gained a dominant influence in the war industries across the European landscape with people like Bedaux and his outspoken approach.

Charles Eugene Bedaux

Charles E. Bedaux (1886–1944) was a French-born American who combined personal flamboyance with blockbusting business tactics. He was born in Charenton-le-Pont near Paris where he received technical training as an engineer before emigrating to Cleveland (Ohio) to set up a business as a consultant. In his approach, he placed emphasis on securing increased productivity both by stressing the importance of new equipment and better manpower control.

Large American companies, including Kodak, visited Bedaux' offices for advice. This and other similar successful experiences enabled Bedaux to take his consulting methods to Britain where he established an office in London in 1926. Bedaux's contribution to Scientific Management was his ability to market and sell its methods, becoming an entrepreneur in Scientific Management (Nelson 1992).

Bedaux introduced a new rating assessment which was called the B: an hour of work contained 60 units of effort combined with rest. He insisted that a rest allowance was essential for the recovery from fatigue and included that into the basic calculations of any B unit value.[41] This new label was called "the principle of human power measurement." As each element in the time study was clocked, the time study man assessed the rating value for the speed and effectiveness with which the element was carried out. Bedaux was soon able to demonstrate to industrialists that "major increases in productivity, increases in hourly earnings, and reductions in unit costs could be achieved without substantial additions to capital expenditure" (Tisdall 1982: 25). Bedaux' system produced remarkable results and created a kind of a cult among American companies that applied his approach. During the late 1920s, British managers in particular became a receptive audience to the Bedaux system (Nelson 1992). By 1937, 500 American, 225 British, 144 French, 49 Italian, and 39 Dutch firms had purchased Bedaux's industrial services.

However, applying piecework to inject control within the culture of the shop floor often faced worker hostility. In 1933, the British metal window manufacturer

Henry Hope & Sons was faced with a strike as a direct consequence of introducing a Bedaux-style piecework system. A preoccupation by management with the advantages of a new costing methodology associated with the Bedaux arrangement was repeatedly attacked by employee representatives (Boyns and Smith 2005). Nevertheless, the British Bedaux company flourished and counted among its clients ICI, BP, and Avon Rubber. The international business community became overwhelmed by the approach of Bedaux. "If the Bedaux archives reveal anything, it is that managers and employers were literally bombarded by statistical evidence that their factories could be run much more efficiently" (Nelson 1992: 163).

Bedaux obtained access to a wide circle of high-society friends and acquaintances that included the Duke of Windsor and his wife. He also knew Goering personally and became an admirer of early Nazi experiments in industrial reorganization. He collaborated with the French Vichy government, saying that the Germans were the only people left to do business within occupied France.

The management approach he had designed as an elaborate system of mass control in production had an enduring appeal to the European military mind. Bedaux was arrested by the American Army in North Africa while working on an elaborate scheme to lay pipes across 2,000 kilometers of the Sahara from central Africa. He was taken to America, investigated, and charged with treason. While in custody, he ended his life by taking an overdose of sleeping pills.[42]

Stakhanovism

In the 1920s, even the Soviet Union with its communist regime became sensitive to issues of time-and-motion studies by which Taylorism improved company performance in capitalism. Lenin (1870–1924) and Stalin (1878–1953) were both impressed by the advantages of Taylorism. In May 1929, the fifth Congress of the Soviets of the Union introduced a major reform of the existing work week by proposing the so-called "uninterrupted production week." The traditional interrupted work week with a free Sunday would be replaced by the continuous production week. The main purpose was the obliteration of the temporal symmetry, the traditional form of coordination that involves synchronizing the activities of different individuals so that they would take place together. The Soviet reform replaced temporal symmetry with temporal complementarity, "an alternative pattern of coordination which involved staggering the activities of Soviet social life" (Zerubavel 1985: 38). However, the fact that only 80 percent of the workers would be at work caused considerable problems in the management of work itself. Most disruptive were the problems of having to organize replacements for workers on the days they were free. After two years of experimenting, it was concluded on November 23, 1931, that the calendrical experiment had failed. Between December 1931 and June 1940, every sixth day was regarded as a common day of rest in the Soviet Union. Yet, while the Soviet workers' ideal threatened to evaporate, the Soviet Union launched a mass movement with the proliferation of "time leagues" which would teach young workers how to budget their daily activities according to rigidly prescribed schedules. To improve output in the mines, it developed a powerful movement for more efficient methods of organizing work in the coal industry. The movement was called Stakhanovism after the Ukrainian coal miner A. G. Stakhanov (1906–1977). Much like Taylor's experiments with Henry Noll, similar experiments took place with Stakhanov. In 1933, he became a jackhammer operator at the Central Irmino colliery in Kadievka in the Donetsk Basin. He was able to cut 102 tons of coal during a night shift in just five hours and 45 minutes (14 times his quota).

On September 19, 1935, Stakhanov reportedly set a new record by mining 227 tons of coal in a single shift. However, this had not been an ordinary shift but a deliberate party-organized attempt to gain plaudits from the Soviet leadership. Stakhanov was helped by two co-workers in removing the coal. Nevertheless, he became the example of a very efficiently working miner comparable with Henry Noll. The purpose of Stakhanovism was a mass increase in labor productivity, and those who joined the movement became the people who would possess a high degree of proficiency at work. His example was held up in Russian as well as European and American newspapers and posters as a model for others to follow.

> Stakhanovites were able to achieve their remarkable results only after overcoming the resistance – at times very obstinate – of those of the managers and engineers who would not part with the old ideas of what were possible production capacities and rates of output.
>
> (Stakhanov 1939: 19–20)[43]

Fordism

In 1918, Henry Ford built the largest industrial complex in the world, the Rouge Plant, which evolved into a "mass-producing warship" (Grint 2000). His company was almost wholly driven by the need for production. Finance, marketing, cost accounting, and managerial structure were of little or no interest to Ford, who still exercised a pre-modern form of management. The organizational structure of the company was mainly served by Ford's personal control. As someone with monarchic power, his management of the company was based on a discipline and punishment policy by which the unreliable human being was turned into a reliable machine. Quality control was synonymous with search and punish in a "rational" way. He behaved as a leader whose organizational success rested not on engineered organizational clarity but instead on an invented organizational paradox. Ford employed large numbers of quality inspectors, shortage chasers, company spies, and company informers to compensate for the weakness of the technology. Ford, however, could not cope with people who had a talent equal to his. If Ford could not dominate a member of his staff, he sacked the person.

Management scientist H. T. Johnson explained that Ford's River Rouge plant worked like clockwork to produce a standardized product:

> Ford spoke proudly of turning iron ore, silica, and latex into finished vehicles in less than three and a half days at the lowest cost in the world. And proud he could be, for the lead times and costs achieved in his River Rouge plant by 1915 had never been equaled before – nor have they been equaled since [...]. But Ford's system [...] was not designed to deal with variety [...]. [I]n his original system, there was virtually no variety [...], only one Model T. In addition, he produced one model of a farm tractor, the Fordson, a centerpiece in his River Rouge system. [...] The low costs [were] achieved by repetitively producing homogenous output, and Ford Motor Company led the way in designing systems to do just that.
>
> (Johnson 1992: 37)

Initially, the system Ford had introduced was called the "Springfield system" since people were familiar with that company as symbolizing the practice of the American system of manufacture (Robbins and Barnwell 2002). The general public truly believed a new era was dawning.

In the practice of the modern factory system, American workers saw their contributions more and more reduced to one identical movement repeated indefinitely and expropriated of their own knowhow. Between 1880 and 1915, no fewer than 15 million new immigrants were recorded in the USA, many of whom were willing to accept a job at Ford. Within the modern factory system, their presence shifted the balance between skilled and unskilled workers, and managerial dominance asserted itself. Managers and directors, themselves also wage earners of the company, took over direction of the business. Their sole objective was to make companies as large as possible in order to attain economies of scale. Planning departments were installed to secure a stable production and speed up production processes. Their forecasts provided certainty about supply and demand and security of jobs. Short-term decreases in demand were met by laying off part of the workforce. Uncertainty due to fluctuations in production and exploitation of employees strengthened the unionized struggle for a further reduction of working time in order to share work and unemployment as well as obtaining free leisure time.

The appearance of so many cars promoted a sense of increasing mobility among the general public. However, by 1926, market share had dropped to 35 percent primarily due to the competition from General Motors with its successful Chevrolet.

Sloan and General Motors

In 1920, one out of three American households owned a car. Although the Ford Motor Company dominated the market, General Motors Corporation soon became a serious competitor. The firm grew into a large company after a range of acquisitions. As a young man, the founder W. Durant (1861–1947) had become interested in automobiles and established a successful carriage and wagon producing firm. After taking over 25 independent companies that included Buick, Olds, and Cadillac, Durant created General Motors (GM) in 1910.

After a second wave of expansion during which Durant incorporated Chevrolet, he built a respectable industrial empire, but it lacked adequate management. General Motors mainly pioneered in decentralized management, but Durant could not possibly run this large corporation on his own. Marketing policies, for example, operated without appropriate information about the markets.

In 1920, the wealthy businessman P. S. Du Pont, owner of the chemical firm Du Pont, took over the reins at GM after having purchased 29 percent of the stocks, and asked A. P. Sloan (1875–1966), an engineering graduate from MIT, to design a new company structure.[44] In sharp contrast to the centralized, functionally departmentalized organization pioneered by General Electric in the 1890s, Sloan designed a multidivisional decentralized structure with a general office and autonomous but integrated divisions (Figure 8.1). Each division obtained a distinctive product line like Buick or Chevrolet and the multidivisional structure, the M-form. This quickly increased efficiency through decentralization. Sloan extended control of production from the factory through his distributors and dealers to the consumer himself. Continual comparison based on an efficient and speedy throughput of registered data, information from dealer reports, and market projections served to control inventories, to calibrate internal flows, and to refine forecasting techniques (Freeland 2001).

Next to the divisionalization, i.e., the decomposition of the firm into a number of distinct divisions which were dedicated to specific product markets, the second defining feature of the M-form was the creation of a corporate headquarters to oversee the corporation. Headquarters executives were responsible for monitoring, evaluating, and coordinating the activities of the various divisions, and carried out long-range planning for the corporation as a whole. They were freed from routine operational responsibility, which gave them the time to analyze the throughput of information and focus on long-term planning (Chandler 1977). The head office was assisted by a large staff composed of financial, planning, and legal experts who standardized financial and statistical control mechanisms in order to improve the transmission of information from divisions to headquarters, which resulted in an efficiently operating administrative hierarchy.

For marketing reasons, the divisions insisted on differentiating their cars from their competitors and informed headquarters about their plans. To be assured of strong cooperation and reciprocal understanding between headquarters and divisions, Sloan introduced a number of interdivisional committees with division managers in participation that created a sort of participative decentralization, opened ways to discuss strategic matters, and improved effective central control. The purpose of these committees was to effectively deal with the tension between delegation and decentralization. The divisions were granted some discretion by delegating decision power to them, but by keeping close contact with the committees, Sloan continued to coordinate the overall strategy of the corporation.

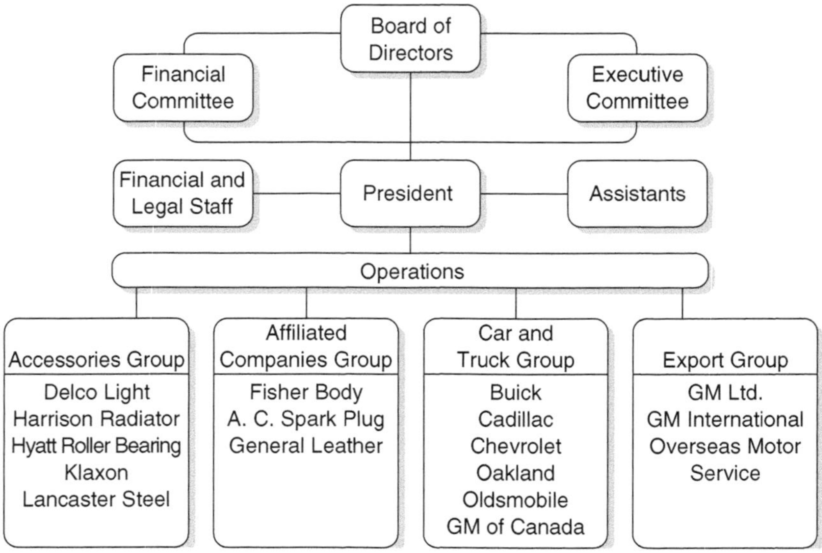

FIGURE 8.1 Decentralized management at General Motors, 1924. (Adapted from *Management and Administration*, 7 (May 1924): 525. Courtesy Houghton Mifflin.)

Monitoring was deemed necessary to ensure that goal congruence between the divisions was obtained and opportunistic behavior prevented (Freeland 2001).

The M-form installed a three-way division of labor among owners, general executives, and division managers. In the 1920s, owners pressed for the introduction of a standard M-form, making it difficult for divisions to dominate a strategic planning process. The shareholders disliked the maneuver whereby division managers became highly involved in strategic issues. They preferred a clear separation between strategic planning and daily operations and attempted to minimize divisional participation in the planning process as much as possible. Therefore, shareholders opposed the changes Sloan introduced with his committees that created participative decentralization and administrative centralization.

During the 1930s, the general office even intensified administrative centralization and became extensively involved in control of strategic planning in daily operations. The shareholders were not pleased with these developments.

General Motors continually struggled with a tradeoff between an organizational order of decision making required by the owners and a speedy throughput of information via the committees that divisional presidents and Sloan were propagating. This created a constant tension between the goal of economizing its activities in terms of efficiency and satisfying the owners. Freeland (2001) called this the tension between perfunctory compliance and consummate performance. If the primary imperative is to maintain order, executives may sometimes forgo efficiency for the sake of sustained cooperation and to satisfy the shareholders. In doing so, they may agree to decisions, policies, or even long-term organizational arrangements that they know to be suboptimal from an economic point of view. Hence, governance is not a simple matter of using the best information available to arrive at policy decisions that are then enforced by fiat and sanctions. Rather, it is an inherently different political process in which top executives must be willing to forge a compromise between actors in the firm in order to preserve cooperation and promote consummate performance.

Consummate performance is based on consent, i.e., voluntary acceptance, and rests on "non-rational commitments" to the wishes of the shareholders (Freeland 2001: 319). Consent encouraged divisional presidents to pursue organizational goals of shareholders which might be "antithetical to their own interest" and led to perfunctory compliance (ibid.: 24). However, rewarding compliance to organizational goals set by the shareholders sometimes led to serious internal struggles. Under the direction of Sloan, General Motors at times operated a "corrupted" version of the M-form, yet it led to continued success for a long period rather than economic decline.

By 1929, almost every household in America owned a car, and some car companies provided credit facilities to encourage Americans to buy cars. During the economic crisis of the 1930s, however, GM lost market share to Chrysler. Ford responded to the crisis by slimming the organization in order to reduce costs, but the organizational slimming almost turned the company into "organizational anorexia" (Grint 2000: 213). Sloan, on the other hand, diminished the influence of the shareholders by strengthening the position of the divisional presidents and, with the full support of the committees, was assured of an efficient production of materials. The Controlled Materials Plan (CMP) became the basic mechanism for integrating strategy and production of all materials.[45]

Between 1924 and 1958, GM did not stick to a textbook version of the M-form, which would have implied a strict division between responsibility for strategic and operational decision making. Instead, "GMs top executives sought to include division managers in planning and resource allocation" which repeatedly adapted the decision-making structure in the face of changing economic circumstances (Freeland 2001: 39). GM was not worried about the differences between its formal structure and the informal patterns within the organization operating with several different committees even though it sometimes caused difficulties.[46]

Nevertheless, Ford, GM, and Chrysler revolutionized the transport of the ordinary American citizen. The impact of the automobile and of all the accelerating technology that went with it was twofold: "it speeded up the tempo of current existence and transformed the memory of years past, the stuff of everybody's identity, into something slow" (Kern 2001: 129).[47] It not only changed the working life in the factory but also social life in the big cities. Around 1900, the German sociologist Georg Simmel (1900) began to write about the frantic life of citizens in the metropolis with all its distractions and noticed how the rising influence of the modern factory system also had an impact on the rhythm of social life in general.

Conclusion

The modern factory, it is true, still possesses strong rhythmic elements, but to the extent that they require regularly repeated motions, they possess an altogether different subjective significance than do the earlier work rhythms. Whereas this earlier rhythm corresponded to the inner demands of physiological/psychological energies, the present rhythm is related either directly to the indifferent objective movements of the machine or to the necessity for the individual worker who performs only a small part of the process to keep pace with the other members of the workgroup. Perhaps this brings about a deadening of the sense of rhythm as such.

Fordism created a new reality. In his *Brave New World* (1932), Aldous Huxley was so under the spell of Fordism that he coined a unique way of marking calendar time by replacing A.D. and B.C. by B.F. (Before Ford) and A.F. (After Ford). The Ford Motor Company became synonymous with the creation of a particular management style based on a bureaucratic organizational structure, hierarchical decision making, a strict functional specialization, and tightly defined jobs to speed up the process of throughput of information, an issue which will be elaborated on in Chapter 9.

The mass-production system promoted the professionalization of management in the automobile company. Ford needed a managerial hierarchy to design, install, and manage it, and to oversee the supply and marketing functions upstream and downstream from production facilities. It called for a managerial hierarchy that inserted itself between the owner, i.e., Ford, and the workers throughout the industrial corporation. Managers were the real beneficiaries of mass production in terms of monetary rewards and the power that they exercised over employees. Managers believed that if their production plans were good enough, they could be passed on to others who would implement them well without further contact with mangers. The clock metaphor once again symbolized a monochronic approach to planning: managers framed their general world view accordingly.

9

TIME AND EMOTION

After the end of the First World War, several international bodies tried to facilitate a resumption of free trade and liberalization of international markets. Yet the interwar period reflected a time of strict trade barriers and a process of further convergence to stimulate international economic progress stagnated, causing a serious backlash in the process of industrial globalization. However, in 1919, the international working class had obtained the eight-hour day through demonstrations and strikes. A clear distinction between time for the boss and time for personal development had either been marked by law or by general labor agreement.

In Europe, trade unions rapidly developed further and mobilized the workforce to prevent any intrusion on the new working time arrangement. In trades like iron- and steel-making, cotton textiles, coalmining, shipbuilding, and construction, the Great Britain craft unions were joined by so-called new unions which were less craft-based and more industry-based. This merging process led to the formation of broad industrial unions which, in the 1920s, organized about 30 percent of the industrial workforce.

In the automobile industry, British manufacturers rejected the American example of high throughput and capital intensive technology accompanied by the payment of relatively high wages. Fearing the resistance of workers to the introduction of labor-saving machinery, British automakers, for example, used more labor-intensive production methods that left workers considerable control of work processes. "Auto manufacturers in Britain balanced a relatively low productivity of their workers by paying lower wages than auto makers in America. British cotton textile manufacturers adopted similar strategies" (Blackford 1998: 153–154). During the interwar years, similar processes took place in other European countries.

On the other side of the Atlantic, American companies searched for ways to improve industrial output. The efficiency drive, which during World War I had been put in place to improve industrial output, however, had created resentment about too much control and lowered morale amid the workforce. There was a simultaneous growing interest in a more humanistic strand of management thinking (Crainer 2000). This became particularly clear in the high-intensity electric-lighting and information-processing sectors.[1]

Information-processing and electric appliances industries

During the period 1880–1930, a remarkable advancement in information processing occurred in the USA where an increasing percentage of the workforce became engaged in information handling. Technological developments like the wide implementation of the telegraph and long-distance telephone reduced the cost of spatial organization. On land, the American Magnetic Telegraph Company had expanded its infrastructure from New York to Philadelphia, Boston, and Buffalo. The Western Union Telegraph company achieved the same to the west and built an ever-extending network into a monopoly. It contracted Thomas Edison who improved the exploitation of the telegraph lines. At sea, the gigantic British ship *Great Eastern* set out to lay one single cable from Ireland to Newfoundland. After three failed attempts, the first successful trans-Atlantic cable connection was operating in July 1866. The great public celebrations that followed emphasized the public feeling that a new age of international relationships was coming about on both sides of the Atlantic.

As Alexander Graham Bell worked on the improvement of telegraphy, he invented a device to convey speech over distances by converting sound into electrical impulses sent through a wire. He demonstrated the telephone that he had developed for the first time in 1876 during the great International Exhibition held in Philadelphia celebrating the centennial of the American independence. More information about market opportunities could quickly be gathered, which stimulated the sales of telephones and increased the size of firms manufacturing all the required equipment. After the appearance of Bell's first telephone, Edison then switched his attention from lighting to acoustical instruments.

In 1877, the Bell Telephone Company was established. It produced new equipment that was an instant success, and the application of the telephone boomed as it began providing direct communications between offices and factories, and between stores and warehouses. In 1880, it provided telephone facilities abroad to Canada soon to be followed by the American-owned Edison Telephone Company, which installed the first European telephone exchange in London. Some large corporations introduced radio communication and, before the First World War, an affiliate of the British-owned Marconi's Wireless Telegraph Company dominated this practice in the United States.[2]

The American Navy however, became worried about British influences of distributing information and, in order to prevent the sector from falling into foreign hands, they found General Electric willing to be involved in the Radio Corporation of America (RCA) and take over Marconi's broadcasting facilities. Much of the infrastructure of the global economy – the telegraph, electricity, telephone, and radio – was soon put in place by international business enterprises (Jones 2007).

The dramatic growth in demand for telephones in the last decade of the nineteenth century and beginning of the early twentieth century placed tremendous pressures on the Bell Company. Since 1882, Bell had used Western Electric as its main manufacturer. As Western Electric had no competitor, Bell and Western Electric coped jointly with the crisis of capacity and built a new plant, the Hawthorne Works. The decision of the board to construct Hawthorne indicated a change in the strategy of the corporation: it departed from the electrical power business and put its energy into the telephone business.

In February 1905, the Hawthorne Cable Plant opened its doors and soon joined the ranks of giant works in American industry. In 1913, the plant employed 14,000 workers and was the single biggest employment site in the Chicago area. Western Electric had initially responded to the capacity crisis by renting additional space and operating around the clock – when workers were available. The company, however, was concerned about negative publicity concerning night working and overtime in their plants. A strike of the machinists in 1900 over the recognition of their union and the reduction of working time to nine hours had worried the board of management.[3]

The board decided to move the Hawthorne Works outside Chicago to a small town called Cicero where less-unionized employees could be found. It installed an organizational structure which complied with the new business challenges the company was facing. Processing information about its growing size of national and international markets forced the company, like so many other big firms, to increase the levels of formal bureaucratic structures. These larger firms invested in organization building and professional management (Chandler 1977). Leading firms in the capital intensive manufacturing industries made substantial investment in production, distribution, and management when markets were rapidly growing and emerging technologies presented opportunities to achieve economies of scale and scope in order to deal effectively with the speed required to serve potential markets. "Producers with the best control technologies could maintain the greatest speeds and produce at the lowest costs" (Beniger 1986: 238).

Around 1900, typewriters with modern Qwerty keyboards, punch-card tabulators, phonographs, and cash registers came into common use in American businesses to speed up the processing of data and information.[4] "The telephone and the typewriter together began the process of office mechanization which characterized the emergence of 'bureaucracy' in large firms" (Freeman and Louca 2005: 228). Next to the adoption of information technology, information-processing innovations concerned procedures like standardization, printed forms, consistent data collection, and record keeping (Chandler 1977). While managers and clerks constituted only 8.9 percent of the American labor force by 1900, their population had almost doubled by 1940 to 16.9 percent (US Bureau of the Census 1975: 139).

Big firms in information processing and electrical appliances internalized all steps involved in making and selling goods within their operations and became organized as corporations rather than partnerships where family or friendship ties prevailed. Vertical integration became a common response to the challenges of the market. Companies that were initially engaged in only one stage of the production and sale of a product integrated backwards to control their sources of materials and integrated forward to control the making and selling of its finished goods.

Horizontal integration where a number of companies joined together to control one step in the production and sale of goods in order to decrease competition and control the national market provided a second mechanism by which American industrialists structured their business. Many of these firms developed managerial hierarchies as time progressed. Manufacturing corporations were capital intensive and needed to raise vast sums of money to build their growing factories. Family capital was no longer sufficient and many of these growing corporations relied on big investors and investment banks for their expansion. Nevertheless, they did expand and, around the turn of the century, these big corporations "accounted for three quarters of the United States' industrial output" (Blackford 1998: 81).

With the growing volume of economic activities in America, large-scale modern businesses emerged and made administrative coordination more efficient than market coordination. Administrative coordination enhanced productivity and lowered costs but required a more intense involvement of management. The construction of an elaborate managerial hierarchy with clear lines of command and responsibility was the prerequisite for realizing the advantages of coordinating multiple units within a corporation. When more technologies were introduced, such as the assembly line controlling the pace of work, more intensive control was required.

The installation of an effective managerial hierarchy soon created its own source of permanence, power, and continued growth. These hierarchies became increasingly technical, professional, and independent of ownership. There was, however, constant struggle about the best way to improve the flow of information needed for effective coordination and to control the organization. Some kind of bureaucratic control would need to be introduced to smooth out the throughput of information.

> The defining feature of bureaucratic control is the institutionalization of hierarchical power. "Rule of law" – the firm's law – replaces "rule by supervisory command" in the direction of work, the procedures for evaluating workers' performance, and the exercise of the firm's sanctions and rewards; supervisors and workers alike became subject to the dictate of company policy. Work became highly stratified; each job was given its distinct title and description; and impersonal rules governed promotion. [...] The company promised the workers a career.
>
> (Edwards 1979: 21)

One question, however, became more and more urgent: How would the workers in these large factories respond to the hierarchical structures they were now supposed to respect?

Welfare work

Out of fear for the influence of unions, some American companies with paternalistic policies had begun to promote welfare work, which in 1916 was defined as "anything for the comfort and improvement, intellectual or social, of the employees, over and above wages paid, which is not a necessity of the industry nor required by law" (Jacoby 2004: 36). Welfare work was oriented towards visits from welfare directors at the homes of the workers to inspect how the workers spent their free time but also included insurances or housing programs tying workers to the company instead of the union. Henry Ford formed a sociological department to inspect living conditions of workers, provide sports events and dancing evenings, and organize social gatherings. Western Electric soon incorporated such programs as well.

Critics, however, did not believe all of these incentives paid off. Workers thought the welfare programs were a poor substitute for higher wages, and the union members were opposed to the anti-union connotation of the programs. Yet the welfare work movement grew in importance, and a professional subculture of welfare workers appeared. Even though production managers and foremen looked at welfare work negatively, welfare workers strove to gain a respected status in the managerial hierarchy.

The fact that some large firms had installed a welfare department did not have much influence in employment management, yet indicated a growing concern about the lack of control within and beyond the large companies. Western Electric, Bell Telephone Company, and General Electric appeared well suited for policies which could improve their overall control and involvement of the employees which would reach beyond a policy of employee welfare.

The General Electric Company was established in 1892 after a merger between Edison General Electric Company and Thomson-Houston Electric Company. The American Bell Telephone Company became fully integrated in 1899 when Western Electric was included in the merger. The newly formed entity was taken over by the American Telephone and Telegraph Company (AT&T), which the parent firm with full control over the federation of manufacturing facilities, operating companies, and research and development laboratories came to be known as the Bell System.[5]

In June 1919, AT&T stipulated that each operating company should appoint a manager to be responsible for personnel relations and the dissemination of modern industrial relations policies. It invited Walter Dill Scott, a psychology professor at North Western University who had introduced professional psychology to business. He gave a series of lectures to managers in the Hawthorne Club that had a positive impact. The company decided to apply psychological testing, which had been developed in the American army to recruit its own personnel. By 1921, a personnel department was installed to handle routine testing, test development, and job specifications. One year later, it administered psychological tests to nearly 6,000 employees and applied statistics to measure output results of the approximately

22,000 men and women working on assembly lines making telephone equipment.[6] Personnel managers were employed in each of the major branches at the Hawthorne Works.[7] The new focus on the "human element" echoed the highly popular Bell approach to encourage "the development of elaborate personnel policies" (Gillespie 1991: 24).

A few years earlier, in 1918, a lighting specialist at the General Electric Company, W. Durgin, had carried out detailed tests to highlight the link between lighting in factories and quality. His tests in four factories suggested that better lighting could increase production by 10–35 percent. The results were so encouraging that General Electric decided to increase its efforts to promote factory lighting (Donkin 2001). During the postwar downturn in business, however, warehouses of companies producing lighting facilities had stockpiled their lamps. To counter this stocking up of their lamps, some of the major companies had instigated the funding of research in order to prove that efficiency increased with electrical lighting. General Electric's Bridgeport Works and the Massachusetts Institute of Technology (MIT) began to conduct experiments to investigate the impact of lighting on human motivation.

Western Electric felt committed to AT&T's research policy and initiated a new series of studies to explore the possible causes of the unsatisfactory results of the lighting tests. Hawthorne researchers investigated different variables that could increase productivity, but the key issue was still the application of artificial lighting in industries. The explanation was simple: the company had a particular interest in stimulating the sales of electrical bulbs and if, by academic research, it could be proven that brighter electrical lighting would increase production, then certainly sales would go up.[8] Western Electric designed a policy to convince future consumers about the technical advantages of artificial lighting compared with natural lighting. The company initiated a campaign to demonstrate that using adequate lighting would save workers' sight, prevent accidents, and lead to an increase of productivity.

To come up with compelling evidence, the company was looking for results based on scientific evidence. For this purpose the Illumination Engineering Society asked for support from the National Research Council (NRC). "With the active involvement of government (NRC) and academia (MIT), this industry promotion effort bore the imprimatur of disinterested social science" (Adams and Butler 1999: 121). The NRC's Division of Engineering and Industrial Research, which was chaired by Frank B. Jewett, formed a committee to study the relationship between quality and quantity of illumination and efficiency. Western Electric's participation was no surprise as the company's president of research at the Bell Laboratories was the same Frank B. Jewett.

The Hawthorne experiments

In 1924, the first experiments conducted at Hawthorne focused on the usefulness of electrical lighting to increase the productivity of workers operating on assembly lines. These experiments were initiated by the NRC and fulfilled the role of coordinating scientific research in both universities and industry. The MIT researchers' goal, much the same as General Electric's Bridgeport Works, was to show that lighting would improve productivity, reduce accidents, and save workers' sight. The outcomes seemed promising. In both the group with improved lighting and in the control group, productivity improved. However, the problem was that productivity also went up among workers in the control group where lighting was not improved. One team managed to maintain output even when lighting was dimmed to the level of moonlight. Increase in production could not be credited to the increase of illumination. Some researchers alerted that supervision might have influenced the results (Gillespie 1991) suggesting that supervision had a bigger effect on productivity than lighting. As the results remained unclear, the final outcomes were never published.

In April 1927, a team of Hawthorne engineers initiated new research to determine whether the variables of rest periods and various hours of work might increase the efficiency of the operators.[9] "The relay assembly was chosen for study because it involved the kind of highly repetitive work with which fatigue was commonly associated" (ibid.: 49–50). A separate test room was installed with six women assembling, adjusting, and inspecting relays, a labor-intensive operation in which the speed of the individual operator determined productivity. Here, the engineers examined the effect of rest periods and working hours on the effectiveness of the worker.

The initial thought was that this experiment would last only a few months, but instead, the experiments continued until June of 1932. The research moved along the lines of wartime research into "industrial fatigue," but the researchers soon incorporated an interest in workers' attitudes. The women felt themselves a privileged elite among the relay assemblers. The atmosphere in the test room was very different from the one in a regular department. Supervision was very different as there was no foreman. Hence, detailed records were kept of the work the women performed in the test room. "When workers in other departments learned that tea was being served in the test room, they quickly dubbed it the T-room" (ibid.: 59).

During the process of investigation, a paradox in the position of the researchers became apparent. They were both objective researchers – objectively interested in the effect of changing different variables concerning working time – and supervisors of the working women from whom they wanted to squeeze as much productivity as possible.[10] Even after a year of studying the results, the researchers were unable to make sense of the outcome: They were unable to determine the cause of the increase in production. "They were industrial engineers and managers interested in developing more productive management techniques, not social scientists studying the social psychology of the factory" (ibid.: 65).

In 1928, two academics joined the group of company researchers: Clair Turner, professor of biology and public health at MIT, and Elton Mayo from the Harvard Business School who had developed an elaborate theory of the relationship between working conditions and attitudes to issues like fatigue. These academics began conducting research in a more scientific way, while the Hawthorne researchers were mainly interested in applying available scientific literature. Turner, however, was unable to develop a satisfactory alternative explanation for the increase in production.

Mayo emphasized that confusion was an intrinsic part of any scientific method. A new test room was established with mica splitters to enable the researchers to test the effects of rests, working hours, and test room conditions of workers who were paid on an individual piece rate. On several occasions, the Hawthorne engineers expressed their disappointment about the lack of concrete results of the research. Mayo and other academics, like Fritz Roethlisberger – Mayo's protégé – reassured them that, in the end, the outcomes would make sense.

Elton Mayo

Elton Mayo (1880–1949) originally came from Australia where he worked as a professor of philosophy. In his writings, he stressed that democracy failed to appreciate the social importance of knowledge. He arrived in America in the summer of 1922, where he soon caught the attention of the NRC. The University of Pennsylvania offered him a position in researching personnel issues. Mayo became interested in industrial psychiatry – or as he called it psychopathology – which led him into the field of mental illness, arguing that worker radicalism was irrational and an outcome of psychiatric disturbances. He was convinced that physiological states due to fatigue could transform into psychological states rendering the worker irrational and apt to reveries.

Sometime later, Mayo found a position at the Harvard Business School and obtained funds for research from J. D. Rockefeller, Jr. The dean of the Harvard Business School, Wallace Donham, came in contact with people from Hawthorne through the secretary of the Rockefeller Foundation, A. H. Young. Donham believed that Mayo's research might help transform business management into a science-based profession (Gillespie 1991). Mayo was very good in "selling" his experiments to the business community. He was able to keep a balance between the people who provided funds for research on the one side and finding time to test his own theories on the other. The funding institutions provided an important network to disseminate results of the Hawthorne experiments.

By 1929, Western Electric was so pleased with Mayo's contributions that the company offered him a job as director of a new research program to interview all workers at the plant and design a technique of social control that could deal with problems of industrial unrest.[11] At the Industrial Research Division of Hawthorne, Mayo initiated two programs: an interview program and a supervisory training program. The goal of the interview program was to identify existing problems between workers and supervisors. His theory was that all worker unrest originated in psychological disturbances and that experts could help workers adjust to the reality of modern industrial relations which had broken down the traditional social cohesion of groups of workers. The loss of this cohesion translated into a restriction of output by the worker which was understandable in terms of the solidarity and emotional security of the group facing threatening changes. As workers did not know what their real problems were, it would take educated personnel managers to assess what the workers really needed, thus the reason he also began a supervisory training program. This training program was intended for 2,500 supervisors to teach them how to avoid these problems and handle workers in a friendlier way.

In the summer of 1930, Mayo sent F. J. Roethlisberger to the Hawthorne Works to teach the researchers at the plant about his interview techniques designed to retrieve information about emotions and frustrations among workers. More than 10,300 interviews were conducted with workers who mainly complained about long working hours, poor pay, and bad supervision. In 1931, following the worker interviews, Roethlisberger then began conducting interviews among the supervisors to examine their attitudes. Mayo's supposition was that "supervisors had infantile and primitive attitudes" (Gillespie 1991: 147). His policy was to provide training based on the latest scientific knowledge of Human Relations.

However, his hopes of the new supervisor being a psychopathologist approaching the subordinate as a patient with a veneer of paternal concern remained a dream. Due to the economic depression of the 1930s, the company could no longer afford the expensive range of interviews, and Hawthorne was, therefore, forced to cut the budget for these major programs. Yet a more general interest in the outcomes of the Hawthorne research continued, and Mayo sailed nearly every summer to Great Britain to visit British social scientists and industrial researchers to inform them about the Hawthorne research.

Bank-wiring test room

Some of the Hawthorne researchers were less interested in Mayo's and Roethlisberger's focus on the emotional aspects of work behavior. Hawthorne researchers stressed that workers were making rational decisions about the way they executed their work. Economist William Dickson and anthropologist Lloyd Warner – both Hawthorne researchers – began to notice the significance of informal social structures in the test room. The workers apparently had a collective interest that conflicted with that of management.

Together with Art Moore, another researcher at Hawthorne, Dickson established constructive relationships with the workers and became aware of the background reasons for workers to restrict output. Warner delivered a report about the social structure within which the operators executed their job in which he noticed the conflict between formal industrial structures and informal shop-floor organization. Restriction of output was the outcome of group solidarity amongst the workers to deal with this conflict. Dickson saw the rationality of this behavior, whereas Mayo would have attributed it to emotional maladjustment of the workers.

In 1931, a new research project began: the bank-wiring test room. The supervisors in the bank-wiring test room did not want conditions to vary significantly from the normal department as they did not agree with the conclusions from the relay assembly tests that workers worked harder when there was less supervision.

On the contrary, they believed that workers could not work without a foreman. Twelve workers and two inspectors were put together in the test room and research began on November 13, 1931. Yet Mayo influenced the final reports about the bank-wiring test room by shifting the attention away from an analysis of informal organization to the role of the individual operator and his emotional maladjustment which caused a slowdown in working rhythm. The economic depression limited further elaboration of the results and, in May 1932, the bank-wiring test was terminated and the workers laid off.

Outcomes of the experiments

Despite Mayo's interventions, the experiments at Hawthorne generated the so-called "Hawthorne Effect": although the workers' productivity rose when, for example, the factory lights were brightened, their productivity also rose when the factory lights were dimmed. The explanation was that the social interaction between managers and workers by listening and discussing certain problems may possibly have as much a positive impact on efficiency as improvements in technology that engineers generally favor.[12]

One of the top executives of AT&T, Chester Barnard (1886–1961), became particularly interested in the Hawthorne results and the characteristics and tensions between the formal and informal organization. As president of New Jersey Bell, Barnard learned from the experiments at Hawthorne that people's commitment is determined by the sense of involvement supervisors can reinforce. Human beings do not only behave according to rationalist standards but also with an emotional involvement and cooperation once the goal of their contribution is clearly set. He seized upon the concept of organization as a social system and pondered about the differences between the formal and informal organization. The commitment of the individual, his everyday willingness of cooperation relies on a "zone of indifference" where orders are accepted without being questioned.

In a formal organization, authority is the character of order by which any member of the organization is governed.[13] But he noticed that there also exists an informal organization which is as essential to understand as the formal one. The informal organization is "the aggregate of the personal contacts and interactions and the associated groupings of people" which influence the throughput of information. Its function is "the maintenance of cohesiveness in formal organizations through regulating the willingness to serve and the stability of objective authority" (Barnard 1968: 122).

In 1937, Barnard was invited by the Lowell Institute in Boston to lecture about his views concerning the nature of cooperative systems and the way managers make decisions. He noticed that there was a lack of understanding of the human cooperation in formal organizations. A year later, his lectures were put together in *The Functions of the Executive* in which he articulated a practitioner's theory of management. His definition of a formal organization is "a system of consciously coordinated activities or forces of two or more persons" (ibid.: 81). This cooperation among men is conscious, deliberate and purposeful and takes place within the context of time and space. The dimensions of time and space are, therefore, of prime importance. "Temporal relationships and continuity" are primary aspects of organizations. "Any series of concrete acts will take place somewhere and since the acts of organization are those of persons (by definition), they will take place where the acting persons are" (ibid.: 133). People cooperate in order to do what they could not do alone and when the purpose is attained, their efforts are effective. However, individuals also have personal motives, and the degree to which they continue to contribute to formal efforts is a function of the satisfactions they personally derive. For Barnard, efficiency was mainly the outcome of the organization's ability to offer enough inducements to satisfy individual motives in the pursuit of group purposes (ibid.).

With his study, Chester Barnard definitely took a step away from the dominant machine metaphor and its mechanical view on men and organizations. He stressed that neglecting the informal side of the organization might slow down the effectiveness of a corporation's operations, the throughput of information, as well as the sharing of information. To prevent such a slowdown, executives should constantly facilitate the involvement of the employees in an organization by sharing information.

Barnard drew an analogy between the organization and the brain in relation to the rest of the body. The brain exists to maintain the bodily system more effectively by directing those actions which are necessary to adjust to the environment, but it can hardly be said to manage the body, a large part of whose functions are independent of it and upon which it in turn depends (ibid.).

Elton Mayo became impressed by Barnard's work and used the metaphor of an organization as a cooperative social system in his later writings (Wren 1987).

> The strength of Barnard's concept of cooperative systems and of its explanation of the essential conditions for rarely attained effectiveness and efficiency lies in the idea that purpose is central. He believes the definition of organization purposes to be particularly an executive function, made necessary to give meaning to the rest of the environment, to serve as unifying principle.
>
> (Andrews 2005, xxi)

But to reach that purpose successfully also depended on the morals of the leader.[14] Barnard had high expectations about managers leading the firm. The moving creative force in an organization is moral leadership. Leaders must hold a moral code, demonstrate a high capacity for responsibilities, and be able to create a moral faculty in others. He stressed professional moral leadership to enhance the effectiveness of an organization as well as the well-being of its people (Wren 1987). Although focusing less on moral leadership, *Human Relations* attempted to pursue a similar policy of well-being of the workforce while simultaneously advancing the effectiveness and efficiency of the organization.

Harvard and Human Relations

Since Mayo had obtained a grant from the Rockefeller Foundation, he offered Dickson the opportunity to work at Harvard and finalize his reports. The experiments were passed from the managers of Western Electric to academics/consultants at Harvard. Under the leadership of Mayo, the reports were drafted into a report of a remarkable scientific discovery. Between 1929 and 1939, several different publications about the Hawthorne experiments appeared. Twenty-five of those articles were written under the supervision of Mayo who "manufactured" the latest insights about the experiments in line with his preconceived views.[15] In 1933, Mayo himself published *The Human Problems of an Industrial Civilisation*, in which he mixed general diagnoses of personal and social maladjustment with Hawthorne material. He pointed out that the behavior of workers was dictated by the "logic of sentiment" and that emotions drove their behavior.

When, in 1939, Roethlisberger and Dickson published *Management and the Worker*, their analysis was much more thorough and reliable, and they dismissed the idea that workers' beliefs were irrational and based on emotional needs. The goals of the formal organization could never be reached if managers ignored informal social organization. Managers had left the organization of work mainly to engineers who were only driven by the logic of efficiency. They ignored the illogical consequences of their actions, and managers were asked to correct this biased way of organizing large-scale organizations. "Left to their own devices, engineers had become a source of interference in the organization" (Gillespie 1991: 203). Hence, the term "Human Relations" became the new umbrella label embraced by general managers.

The Harvard Business School became heavily involved in the dissemination of the Human Relations approach which advocated talking to workers about their grievances and taught foremen to treat their workers in a friendlier manner. Between 1935 and 1941, P. Cabot, professor of public utilities, organized weekends to share the ideas of Human Relations with business executives. Roethlisberger introduced training in the techniques of Human Relations in 1938 and conducted research to better understand the need of foremen to consider social contacts in the workplace. Talking with the workers instead of only keeping them in the drive system could improve productivity.

The impact of this new approach even reached boardroom level. As one of the influential top executives of AT&T, Chester Barnard was well informed about the outcomes. Although he had not been involved directly in the experiments, he praised Roethlisberger and Dickson for their work. In a letter to Roethlisberger dated October 10, 1939, he wrote:

> Both in detail and as a whole it seemed to me that the presentation was sound both because it checked with my personal adjustment based on experience, but more particularly because the important statements were based upon, and illustrated by, factual information as to the concrete

> behavior. At last we have something that really is the basis for Scientific Management.
>
> (Quoted in Gillespie 1991: 209)

The Hawthorne experiments and the subsequent Human Relations movement diffused the idea that an improvement of satisfaction of workers might increase production output. The Human Relations perspective legitimized the view that an increase in production was the outcome of good management. Human Relations advocates tried to convince general management of the important role of supervisors who needed good training in listening and explaining the production procedures and routines to workers. Yet the ideology behind the Human Relations approach was that only managers could see what would be best for the workers while the workers' understanding of the workplace was flawed due to their limited knowledge of the circumstances in which they had to work. American labor unions should not be allowed to penetrate the shop floor as they were viewed as only relevant for collective bargaining where grievances could be discussed. Further instrumentalization of Human Relations with working time being balanced out with leisure time so that an emotional misbalance would be prevented, however, was interrupted by the Great Depression.

The Great Depression

After the collapse of Wall Street in 1929, an economic crisis reversed the cycle of production, earnings, and consumption. The Great Depression caused a new wave of protection measures against international trade, slowed down the process of convergence among industrialized nations, and hampered further industrial globalization. Companies like General Motors, Standard Oil, and Sears Roebuck limited production by cutting back weekly hours to 40 in 1931 and then to 30 in 1932. Reduction of working hours was no longer seen as an opportunity for workers to obtain more free time but as a way to save employment. In August 1932, the Industrial Conference Board concluded that 50 percent of American industry had shortened hours to cut back on production and save jobs (Hunnicut 1988).

The American government developed a plan for a general 30-hour week to spread out employment and put people back to work. The policy stated that employers would hire back those workers they had previously laid off. However, as soon as the 30-hour legislation passed the Senate and the option of enactment became realistic, the business community presented fierce opposition (Schor 1992). Nonetheless, a movement for shorter hours continued to raise its voice and stress the necessity to reduce working hours by legislation to "share the work." The American Federation of Labor pressed President Hoover to call a national convention on shorter hours, but he only favored voluntary efforts. In 1933, America was deeply troubled as the unemployment rate rose to 25 percent, and farm prices fell by 50 percent. Banks in many states were closed and mortgages were foreclosed by the tens of thousands.

In order to restore employment, earnings, buying power, and demand, the new president, Franklin D. Roosevelt, initiated a program of large-scale public works. The New Deal was enacted in two phases and contained an extensive package of economic programs with the goal of providing Relief to the unemployed and farmers, installing Reform of business and finance, and promoting Recovery of the economy by bringing "industrial democracy" to the forefront of the business agenda (the three Rs).

The initial phase began in March 1933 and aimed at meeting the needs of impoverished groups in society by introducing the National Recovery Administration (NRA), which prepared a National Industrial Recovery Act. However, Roosevelt blocked the thirty-hour bill which the Senate was still discussing to "share the work." Yet the issue of working conditions was put on the agenda, and the National Labor Board (NLB) was created to settle a wave of strikes against existing working conditions. "Industrial Democracy" became the key label of the recovery program and offered a great opportunity for both social scientists and engineers to promote Human Relations.[16] William S. Knudsen, vice president of General Motors, declared his support for the New Deal "without reservations" and introduced Human Relations in his company.

As the broadening inequalities between rich and poor were hardly dealt with, the second phase of the New Deal during 1934–35 included relief programs, the introduction of a Social Security Act, and the Wagner Act to promote the role of labor unions. Roosevelt lost patience with business voluntarism and began to use public works to stimulate re-employment. In 1938, the Fair Labor Standards Act passed Congress and set maximum working hours to forty and fixed minimum wages for most categories of workers, but the bill was a watered-down version "with virtually all of its work sharing teeth pulled" (Hunnicutt 1988: 235). The proposal to introduce a legal thirty-hour work week launched by the shorter-hours movement never passed into law.

In Europe, the shrinking world markets in the 1930s brought a resurrection of the shorter-hours movement. Many faced joblessness or experienced overtime and more intense labor while employers sought to reduce unit costs. Labor unions pointed out that the recent economic rationalization reinforced economic distress. Increased output per worker was not balanced by an increase in income, job security, or more leisure.

The British Trades Union Congress (TUC) called for a forty-hour week with no reduction in pay. During the conference of the International Federation of Trade Unions in Stockholm in 1930, the forty-hour week was advocated as a more general measure for all of Europe.[17] The French Confédération Générale du Travail (CGT) launched a forty-hour campaign in 1931. Despite the pressure from national confederations, the International Labor Organization (ILO) in Geneva failed to establish an international context to organize a general support for an international reduction of work time. The TUC continued to press for a forty-hour standard through channels of collective bargaining but failed to win in any important sector. "Britain had clearly lost its leadership in advancing the labor standard; the traditional tactic of collective bargaining was no longer effective" (Cross 1988: 170).

In France, the situation was more favorable. Shortly after the election of a new French government based on the Front Populaire – a coalition of several leftist parties – a wave of strikes forced the enactment of a forty-hour law. However the collapse of the government by May 1937 was followed by measures which suspended the law. In Germany, the refusal of the unions to accept wage cuts due to the economic crisis "led to the collapse of parliamentary government and, eventually, to the nomination of Hitler as head of government in Germany" (Hobsbawm 1995: 137). Once in power, the Nazis began the destruction of the German labor movement, which severely weakened the prospects for labor law legislation with a forty-hour working week.[18]

In the Soviet Union, the situation was completely different. On October 5, 1927, the Central Executive Committee issued a manifesto for a seven-hour workday and a 42-hour working week, which would provide the Soviet workers with the shortest workday in the world. Although the elimination of many religious and tsarist holidays had meant that workers worked six more days per year, the numbers of hours worked per year declined by a quarter (Chase and Siegelbaum 1988). The purpose of the seven-hour day was to introduce an extra shift in sectors like textile manufacturing operating on a three-shift system.[19] To further improve production output, the fifth Congress of Soviets of the Union decided in May 1929 to introduce the continuous working week (*nepreryvka*) which meant a complete absence of common days off like Sundays and religious holidays. In practice, this meant that husbands and wives had different days off, causing considerable frustration among the workers. Management was asked to arrange work schedules in such a manner that family members in general would share the same days off so that their motivation to work would not be hampered.

Yet the biggest problem to improve productivity was not the continuous working week, per se, but the poor performance of the production facilities. Since the Soviet Union had introduced a Five Year Plan, Fordism was introduced in industrial sectors like the automotive industries in the Soviet Union which produced tractors, trucks, and cars. Fordist practices however, could not increase efficiency due to a systematic delay of deliveries of spare parts to keep machinery in continuous operation. "Faced with such a massive noncompliance, the central authorities eventually retreated, and Stalin gave his imprimatur for many enterprises to abandon the continuous workweek system" (ibid.: 205). Nevertheless, the Commissariat of heavy industry and other economic commissariats insisted that production output with a seven-hour working day could improve if Taylorist and Fordist practices would be implemented within a technically correct and economically viable infrastructure. By 1931, an increasing number of industrial workers in sectors like light and heavy industry were at a seven-hour day.

Despite vigorous efforts to improve labor discipline, however, the increase in labor productivity slowed down. The Stakhanovite movement promoted the introduction of new production techniques and provided a pretext for substantial moral commitment to production increases, but in the long run, many Stakhanovites could not sustain the level of output, which "led to accusations of sabotage and wrecking against enterprise management and technical personnel" (ibid.: 208). A lack of work discipline, poor production quality, mediocre supervision, and an irregular pace of production[20] forced the government to replace the seven-hour day in 1940 with an eight-hour day and a 48-hour working week.[21] Due to a lack of an efficiently functioning bureaucracy, central authorities remained unable to systematically control effective working time. A regime of oppression was put in place which created an atmosphere in which the Soviet workers felt that human dignity was becoming less respected.

The Marshall Plan

During World War II, a booming war industry increased the pressure for longer working hours. In the American war industry, in particular, the "manufacturing workweek rose more than seven hours between 1940 and 1944" (Schor 1992: 77). Alternatively, there was a labor shortage problem, and too many inexperienced and unskilled workmen became supervisors. The business community asked for an adequate supervisory training program to properly instruct all of the supervisors that were needed. In 1940, the American War Manpower Commission introduced the Training-Within-Industry program (TWI) to improve job instruction for supervisors and transfer this knowledge to the workers, teach them about job methods and the techniques of production supervision, and make supervisors familiar with job relations by instructing them in the basics of Human Relations.

The personnel relations manager of Western Electric, J. W. Dietz, became an assistant to the director of the TWI program and invited several researchers from Hawthorne, including Roethlisberger, to design the program and assist in its practical implementation. The TWI program trained nearly half-a-million supervisors on how to motivate their workers.[22] Human Relations was made attractive among American managers and also became one of the popular managerial approaches in Europe after World War II. The Marshall Plan and other American initiatives to rebuild Europe's industrial base provided an attractive basis to convey Human Relations across the ocean. "The Marshall Plan initiated the remaking of Western Europe in the American image and, directly and indirectly, made a decisive contribution to the *take-off* of the expansion of world trade and production of the 1950s and 1960s" (Arrighi 2010: 305).

The Marshall Plan and the subsequent rearmament due to the Cold War became the means to integrate separate domestic economies of national European nation-states into a single market. At the same time, Western Europe also became a fertile ground for the transnational expansion of American corporations. Together with the international position of the dollar and military supremacy, American multinational corporations became the cornerstones of American hegemony. General Motors became the symbol of *The American Century*, the era of American economic and military power.[23]

Concurrently in the United States, the immediate postwar decades saw remarkable changes in America's labor situation. The American business community attempted to restrict union activities which had been stimulated by the National Labor Relations Act of 1935. In 1947, Congress passed the Taft–Hartley Act which limited union activities. As long as the economy was experiencing long-term growth, management was generally willing to grant higher wages, but only if the unions gave up any aspirations to influence how executives ran their companies.

Due to this "conservatism of the post war years" (Schor 1992: 78), a rational bureaucratic organizational model was maintained as the best approach to coordinate large-scale organizations. Throughout the period of the 1940s–1960s, Human Relations policies and research remained an influential domain in these large-scale companies. Administrative coordination became a powerful concept once the size of the corporation required an hierarchy to effectively streamline the set of responsibilities managers would need to take to plan the activities and execute the work. A balanced coordination relied on standardization of planning practices, production processes, goods, and work activities. Standardization of the production led to standardization of working schedules and activities. Factory workers were kept in strict working schedules, and their output was rigorously supervised.

The modern factory system being practiced in large-scale corporations, however, continued to lead to conflicts about the pace and intensity of the work as well as about the length of the working day. Discussions about initiatives to limit working hours to 40 per week emphasized the physical limitations of a man's ability to work longer hours in a strict regime. Legislated limitations on working hours, however, became mainly justified on the basis of psychological and physiological necessity, not on the basis of workers' rights to leisure. Human Relations tried to humanize work but could not solve the problem of the intensification and speeding up of work in the modern factory system.

Peter Drucker

The consultant Peter F. Drucker (1909–2008) stood at the forefront of those who pleaded at boardroom level for improvements of working conditions in the highly bureaucratized large-scale corporations. Originally from Vienna, he traveled to the United States in 1937, where he became interested in the way business managers could reconcile capitalist with corporative values (Nelson 1992).

In 1942, Drucker was called by the public relations director of General Motors, who asked him to study GM's top management structure. There he became acquainted with a managerial tool that Donaldson Brown (1885–1965), a corporate director of GM, had introduced to assess standard financial ratios (return on investment and return on equity) and flexible budgeting, which allowed GM to effectively manage its decentralized corporation. President of GM, Alfred P. Sloan (1875–1966), commented that this approach "combined the initiative, responsibility, development of personnel decisions close to the facts, and flexibility of decentralization with the efficiencies and economies of centralization" (quoted in Nelson 1992: 216).

Such a type of bureaucratic operation seemed to fit well with Sloan's multidivisional form. Drucker was impressed by what he had seen at GM but was worried about the involvement of the employees in such a large organization. In 1955, for example, GM alone would employ 625,000 people. He subsequently became involved in operations at General Electric where he cooperated with Harold Smiddy, vice president of General Electric, who had been impressed by Drucker's book on GM, *Concept of the Corporation* (1946).[24] Smiddy's approach of participative management techniques and the things Drucker had learned from GM evolved into the concept of "Management by Objectives" (MBO), which indicated that clear goals could contribute not only to entrepreneurial effectiveness of managers but also to the employees' commitment.

Drucker compared the role of the managerial task to that of an orchestral conductor: the conductor selects the piece, i.e. the goal. Each division subsequently plays one instrument, whereas the conductor recombines all the parts so that the goal will be achieved. Drucker liked the metaphor because it assumed that, in bureaucratically structured organizations, harmony came from consensus rather than from control. In his view, authority was not a submission to force but the rule of right over might. In 1955, Drucker published his ideas and experiences in *The Practice of Management*, which popularized management as a learnable discipline and set off a management boom across postwar industry (Donkin 2001).

The quest for an alternative to both Taylorism and rigid bureaucracy led to a set of new cooperative principles of management. However, living in the United States, which valued liberal democracy, Drucker felt that he had difficulties legitimizing this rule of right. He was searching for ways to mitigate despotic bureaucracy in large-scale corporations of the 1950s where workers experienced strict working conditions under social contract which was "both an agreement of association and one of subjection" (Nelson 1992: 213). His therapy for the bureaucratic disease was for managers to integrate planning with doing and synthesizing tasks. It called on managers to stop treating people as factors of production and begin treating them as members of a community. He stressed that higher production levels would be an outcome of higher satisfaction levels amongst the factory workers.

However, his own MBO concept did not work quite the way Drucker had expected (Nelson 1992). His work formalized the approach of managing time in a strictly bureaucratic manner primarily focusing on the prioritization of activities and the deliberate concentration on the prioritized actions until they were finished. Prioritizations of schemes became the organizing structure (Slaven and Totterdell 1993).

Although Drucker's views were well received during the prosperous decades of the post war boom, critics stated that even MBO was too bureaucratic to achieve any corporatist virtue (Tosi and Carroll 1973). Nevertheless, MBO became one of the first fads in management literature to mitigate the negative side of bureaucratic organizations that alienated the common workmen from their work. Drucker warned that managers became ivory tower number crunchers and bureaucratic parasites instead of people who would participate in active collaboration with their subordinates.

In the 1960s, Drucker pointed out how the consensual decision making of Japanese corporations was the best way to set goals and realize them. He underlined the importance of empowerment practiced by Japanese firms:

> The assumption of managerial responsibility by the individual employee, the work team, and the employee group alike for the structure of the individual job, for the performance of major tasks, and for the management of such community affairs as shift schedules, vacation schedules, overtime assignments, and, above all, employee benefits.
>
> (Drucker 1979: 273)[25]

Yet, overall, bureaucratic structures in large-scale organizations were maintained well into the early 1970s. US-based firms which had initially expanded within the large domestic American market continued to rely on hierarchical structures with strict rules and standardized reporting procedures[26] even when they expanded abroad. "Between 1945 and the mid-1960s, the United States alone may have accounted for 85 percent of all new foreign direct investment outflows" (Jones 2007: 148). During the period of the so-called Thirty Glorious Years (1950–80), American multinationals became imitated in their bureaucratic approach by European companies which were also broadening their horizons (Whittington and Mayer 2000).

There was, however, one problem with this approach: headquarters mainly focused on the coordination of the external factors in terms of central planning. Bureaucratic approaches held the environment constant while searching for universalistic principles of structure, planning, and control to coordinate large-scale organizations. Yet these bureaucratic management practices were suddenly out of sync after the oil crisis of 1973 which generated uncontrollable dynamic changes in the environment. The corporations lacked flexibility to adequately respond to such changes. The consequences will be discussed in Chapter 13.

10

THE ACCELERATED TRANSFER OF PEOPLE AND CARGO

Railroads and steamships had drastically reduced travel time and transporting costs. New transportation technologies such as automobiles, buses, tramways, subways, and improved infrastructures strengthened the expansion of industrial globalization. The fascination for mobility was translated into words such as speed and velocity, which became the emblems of modern times. The advent of faster transport by air shipping and the introduction of containers on ocean vessels created a spirit of time saving in the aviation industry and of increasing transfer of cargo attained over sea. As will be shown in this chapter, aviation became the emblem of overall acceleration during the twentieth century.

With new transportation facilities in the air and at sea, the number of large multinational corporations grew tremendously and, by 1980, their number had grown to about 10,000 with about 90,000 affiliates. In the early 1990s, the number would rise to 35,000 and 170,000, respectively. By the late 1990s, the top 1,000 of these multinationals covered almost 80 percent of the world's industrial output. The economy of speed drove these corporations in their expansion abroad.

Initial steps to aviation

During the American Civil War, German Count von Zeppelin (1838–1917), who had achieved the first aerial crossing of the English Channel with his airship in 1875, had served as a volunteer in the American army and became interested in the use of military captive ballooning. He experimented with various forms of motive power which were attached to balloons to make them move in a particular direction independent of where the wind came from. Airships, or "dirigibles" as they were called, fascinated the public, but their speed was slow and the airship was primarily seen as a sport vehicle. As the dirigibles had very limited practical application, however, the attention shifted to airplanes with their coordinated mobility, and that new concept created the mood of a more modern time. Gliders were then equipped with engines to see if the flying machine could be kept in the air.

Pioneering pilots such as Alberto Santos-Dumont, a rich young Brazilian aerostatic sportsman living in Paris, experimented with flying airships. In October 1901, he flew such an aircraft on a circular trip from the Parc de Saint-Cloud near Paris, successfully rounding the Eiffel Tower and returning in almost 30 minutes, for which he won the Deutsch Prize. Then in 1906, he flew 220 meters in a heavier-than-air plane,[1] a new record in Europe, which created quite some excitement.

In 1904, while dining at the Restaurant Chez Maxim in Paris, Santos-Dumont complained to his friend Louis Cartier about being unable to effectively maneuver his plane and the difficulty of checking his pocket watch simultaneously to time his performance during flight.[2] Cartier was willing to design an alternative that would allow him to keep both hands on the controls and designed his first wristwatch for Santos-Dumont in 1904. Soon, the wristwatch gained wide popularity.

Wristwatches

Cartier further developed his wristwatches, called "tanks," for the US army. The word "wristwatch" did not become popular in English, however, until 1910. The growth in popularity of the wristwatch was stimulated by concepts concerning sports and modern rapid means of transportation. A German wholesaler offered wristwatches in a 1909 announcement and mentioned that it was appealing to horsemen, officers, and motorists. Women at that time were wearing wristwatches most often for travel or sport.

The wristwatch gained worldwide use during World War I. "Soldiers on both sides quickly realized that the conditions of modern warfare did not allow unbuttoning the overcoat and jacket when it was necessary to look at their watch" (Kahlert *et al.* 1999: 16). Since then, the wristwatch became a quasi-universal aspect of dress in the modern world and an intrinsic part of daily apparel. With the appearance and wider applicability of wristwatches, a practice was introduced "for the most detailed and definite arrangements and measurements that imparts an otherwise unattainable transparency and calculability to the contents of life, at least in regards to their practical management" (Simmel 1990: 445–446). Living on clock time by using the *handcuff of modern times* has become a defining characteristic of the modern lifestyle. The proper maintenance of one's own wristwatch was eventually seen as a sign of character.

Aviation

The Age of Aviation began in a bicycle repair shop of the brothers Wilbur and Orville Wright in Dayton (Ohio) who opened their shop in 1893.[3] They became increasingly interested in flying and, in 1894, Wilbur discovered a photographic reproduction of "The Flying Man" in an illustrated magazine, *McClure's*, showing a man dangling from a hand glider who had launched himself into the wind from a hilltop. Wilbur was so inspired by the picture that he went to the local library of Dayton to study *Animal Mechanisms* by Frenchman Etienne-Jules Marey, the same man who had inspired Gilbreth for his motion studies. From that time forward, the brothers began designing gliders with the conviction that only by becoming lighter than air was flying possible.

In 1896, upon hearing of the crash and death of German Otto Lilienthal, who had constructed aerodynamic gliders, Wilbur began to study Marey's book more intensely. With the support of their bike shop mechanic, Charles Taylor, an engine was designed to keep their glider in the air. On December 17, 1903, they succeeded in keeping their *Flyer* in the air for twelve seconds. Their construction being "heavier-than-air," but using the density of air as a lift mechanism finally made it happen. Their *Flyer* became the first powered airplane that could take off and land without crashing (Vance 1986). The *physical* experience of challenging time and space in a new form of flying was realized. "The Flyer was a triumph of human ingenuity and determination, but it was not yet a practical airplane" (Evans 2004: 199). Their biplane was not yet ready to transport others, but soon a feverish competition to build airplanes for mail, cargo, and passenger transportation grew on both sides of the Atlantic. Aircraft builders began to rival each other to manufacture the fastest, highest-flying aircraft.

Much like the Portuguese navigators in the fifteenth century, the first pilots were bold and courageous inventors who as "gypsies in the sky" risked their lives to fly their own airplanes. The Wright brothers offered their services both in Europe (France mainly) and America. American aviation pioneer and distinguished astronomer and physicist S. P. Langley received a War Department grant to develop a piloted airplane which he called an "Aerodrome" (coined from two Greek words meaning "air runner"), but his initial attempts at manned flight failed. As an academic, Langley focused more on universal principles and formulas than on practical application (Bereiter 2009: 239). He attempted to cooperate with the Wright brothers, but they politely turned his offer down as they preferred to advance progressively by realistic experiments instead of relying on purely scientific aeronautics.

The Wright's monopoly was soon challenged by Graham Bell, the inventor of the telephone who also had a feverish interest in flying like a bird and experimented with giant kites. In October 1907, Graham Bell formed the Aerial Experimental Association with four engineers. Undisturbed by this competition, the Wrights continued to improve their *Flyer*.[4] The Wright's performance at the Le Mans racetrack caught the attention of the aeronautics community. Frenchmen such as H. Farman, F. L. Delagrange, G. Voisin, and Louis Blériot, who crossed the English Channel by air in 1909, made successful flights with monoplanes which gained a lasting popularity in France and turned them into serious competitors of the Wrights.

The Wright brothers, in the meantime, had formed a French company in order to increase their own business opportunities. On the other side of the English Channel, British aircraft manufacturers such as A. V. Roe and T. O. M. Sopwith formed similar companies and introduced landmark designs manufactured at

Vickers Ltd and Hawker Aircraft. Pilots in other continental European countries pursued similar interests and filled the aerial space full of airplanes. In Germany, engineer Gustav Otto established his Gustav Otto Flugzeugwerke in 1910. On October 1, 1910, the Dutch pilot Henri Wijnmalen flew to a height of 2,780 meters in the Netherlands and set the first world record. In the same month, he also won the Paris–Brussels–Paris race.

Soon, the first schools for instruction and training were launched, such as the Guggenheim School of Aeronautics, and American pilots were required to obtain a certificate to be allowed to fly. The practice of the trinity of devices such as wrist-watches, instructions, and printed manuals, which had shaped the habitus of the Portuguese navigators, Maurits' commanders, British factory managers, and American railroad conductors, was also imposed onto the pilots and made air transport better controlled and safer.

The French government initiated an international conference to settle an arrangement about the state's sovereignty into airspace. The issue on the table was the right of a nation-state to adequately safeguard its territory against aerial intrusion; however, a regulatory framework was not initially embraced by the 17 European countries present.

World War I spurred the rapid development of the aviation industry. Greater distances could be covered by building more powerful engines, which increased thrust, and by designing wings that improved lift and stability. In March 1916, Eduoard Nieuport, who initially constructed a small airplane called the *Bébé*, designed a fighter plane with excellent maneuverability together with Franz Schneider. It was fitted with a suitable synchronizing machinegun which allowed for safe firing through the spinning blades of the propeller. In Germany, companies such as Rapp Motorenwerke (which became Bayerischen Motoren Werke BMW in 1917), Dutchman A. Fokker with his Fokker Aeroplanbau and Austro-Daimler provided the German military with new airplanes. In the USA, the Wright-Martin Aircraft Corporation obtained sufficient capital to begin producing airplanes on a larger scale. The first scheduled passenger air service was introduced in America in 1914 when the first three-seaters appeared for commercial aviation and flew at a cruise speed of 75 km/h.

After the war, the focus shifted towards designing more commercial aircraft. Leftover bombers from the First World War turned into a challenging source of equipment. The French Farman and the British Handley-Page firms created new designs to carry passengers. The Paris–London route offered a tempting opportunity, and the Farman Company soon began providing the service. Fokker, who had initially built high-wing monoplanes in Germany, moved to the United States to become a manufacturer of trimotors. Commercial aviation advanced with the introduction of the Lockheed Vega which could carry six passengers across a distance of about 805–1,449 kilometers at a speed of 177–217 km/h (Vance 1986; Hallion 2009).

Unlike the United States where mainly private companies were involved in passenger transport, most European countries witnessed the creation of subsidized national airlines. In October 1919, Albert Plesman established the Dutch airline company KLM – still the oldest airline in the world – to open an Amsterdam–London route with Fokker aircraft. In Belgium, Sabena began in 1923; in Germany, Lufthansa began operating an airline in 1926; and the Polish airline LOT was formed in 1929. France, Britain, and the Netherlands soon opened extensive networks of air service to transport passengers and serve their colonies such as Indochina, India, and the East Indies. It took a Fokker Pelican flying at a speed of 200 km/h four days to cover the distance from Amsterdam to Jakarta.

While in Europe some initiatives were taken to transport passengers by air, the American government initially focused primarily on airmail service. The mail service from city to city serving long-distance routes was soon considered a lucrative business for owners of airplanes. The US Post Office operating as an arm of government behaved like an entrepreneur and opened the first flying services between 1919 and 1925.[5] Even though the flights were only taking place during the daytime, two army lieutenants, D. Bruner and H. Harris, constructed a lighted airway between Dayton and Columbus: their flashing markers and rotating beacons facilitated the beginning of night flights (Heppenheimer 1995).

In 1925, the Kelly Act made it possible for private carriers to bid on the airmail business. It awarded airmail contracts to companies that connected cities in a north–south direction along with the already established transcontinental route (Mayo *et al.* 2009). Commercial aviation was handed over to private hands which, with their speedy airplanes, could compete with railroads. The Ford Motor Company increased its involvement in airmail and investors such as Rockefeller, Vanderbilt, Marshall Field of the department stores, the oil tycoon James Talbot, and newspaper publisher Harry Chandler became interested in airlines. The American Air Commerce Act of 1926 issued licenses to operate, and pilots and crews were obliged to pass examinations. General Motors became so extremely interested in aircraft business that it took a large stake in Fokker Aircraft Corporation in 1928. However, one of the most influential drivers of the aviation business was Trippe with Pan Am.

Pan Am's global expansion

Challenges soon reached beyond national borders. In March 1927, a major of the American military air force, H. H. Arnold, founded Pan American Airways to provide airmail services between the Caribbean and the USA.[6] Similar services were provided by Britain's Imperial Airways, France's Aéropostale, the German Lufthansa, and the Dutch KLM. Pilot Juan Trippe was able to establish the airline company Aviation Corporation of the Americas with the backing of Lehman Brothers and other influential friends from the Vanderbilt and Rockefeller families he had met while at Yale. This company introduced the international Key West to Havana airmail route. He pursued a policy of consolidation of the routes to Havana and the Caribbean and turned Arnold's Pan American Airways into a subsidiary of his own company. He was then also able to turn the airline Atlantic, Gulf, and Caribbean into a subsidiary and create a holding company. He became the president and general manger of the new Pan Am in 1927.[7] Due to Pan Am's abilities to comply with the latest technical requirements, the American government, i.e., the post office, granted the company every foreign airmail route for which bids had been set out to level the playing field with European competition (Nohria and Mayo 2007).

To further encourage the development of commercial airlines and airmail, the American government – in agreement with the post office – decided to offer contracts which combined carrying capacity for mail and passengers as well. Aircraft builder William E. Boeing, who had inherited a prosperous business in timber and iron ore, had begun building seaplanes as a hobby. He established the Pacific Aerial Products in 1916 which would later become the Boeing Airplane Company in Seattle. He offered such a low bid for airmail transportation in the USA that he had to improve the efficiency of his air-cooled engines to make a profit. Henry Ford also began to build planes and schedule mail operations in the northern part of North America along with providing passenger seats. Boeing built a Model 80 plane with a metal skin and skeleton which was designed for transport of passengers and mail. A clear distinction between commercial airlines and airmail contractors was indiscernible.

In 1929, Boeing became chairman of United Aircraft and Transportation Company which soon became known as United Airlines. As the US government, based on antitrust law, assumed illegal practices were involved, it investigated whether the company had become too dominant. It decided in 1934 that the company should be split between a manufacturing and an air transport business. Boeing became so disenchanted with this interference that he resigned.

In the meantime, stunt pilot C. A. Lindbergh had accomplished the first solo trans-Atlantic flight with his plane *The Spirit of St. Louis* from New York to Paris, a distance of almost 6,000 kilometers, in just 33.5 hours. He had never flown over a large body of water and brought two compasses and maps with him but actually navigated by mere reckoning without a watch.[8] His success fired up the public's imagination in 1927, and investors as well as passengers volunteered to the airline industry in increasing numbers. His success turned aviation into a respectable business (Vance 1986; Mayo *et al.* 2009). After his transatlantic flight, Lindbergh wrote a letter to the director of Longines, describing in detail a watch (Hour Angle Watch) which would make navigation easier for pilots by determining the exact longitude during long-distance flights.

As Pan Am was seeking further expansion of its services to South America, Trippe invited Charles Lindbergh to become the company's technical adviser and consultant to map out the most efficient routes and establish a network of landing strips. Trippe deployed twin-engine amphibian planes with stronger engines to cross larger distances and extend the routes. The Sikorsky Company was able to deliver an aircraft to Pan Am that was capable of intercontinental travel. The Sikorsky S 40 (the American Clipper, as Trippe called it) was able to transport 40 passengers 1,126 Kilometers miles. Together with Lindbergh, Trippe managed to build an almost monopolistic empire in airline routes to Latin America and began to impose rigorous training of pilots and crews for these long distance flights. With the Sikorsky S42, carrying 32 passengers on a 1,600-kilometer, nonstop journey, the company was able to fly from Miami to Buenos Aires and reduce travel time from eight days to five. With the introduction of these new aircraft, large ocean liners and other forms of transportation now faced very serious competition.

Planes at that time could travel just over 1,600 kilometers, but Trippe's interest also continued in routes across the Atlantic and Pacific.[9] He was looking for opportunities to set up flights to Europe and discussed options with the Norwegian airline DNL. Negotiations failed when Trippe broke off discussions with Aéropostale and Imperial Airways as they demanded reciprocal concessions which would allow them to set up similar Atlantic services. Eventually, on June 28, 1939, Pan Am succeeded with the first regular passenger service across the North Atlantic from New York to Southampton with the flying boat Boeing B314 – called the Yankee Clipper – transporting 22 passengers across the Atlantic in just 22 hours.

By island hopping across the Pacific, Pan Am established a route to mainland China in 1934 with the China Clipper (A Glen Martin M130 plane). It was able to do so following the acquisition of the American-run China National Aviation Corporation, a company serving large cities on China's coast, which allowed it to extend its services. Trippe received support of the American Navy, which was worried about Japan's policy of increasing its military presence in the Pacific. Pan Am served as a cover-up: infrastructures and landing facilities for his seaplanes at remote islands between the USA and the Philippines were established which could also be suitable for naval uses (Nohria and Mayo 2007). Navy personnel installed and operated long-distance communication for aircraft on this route and introduced advanced navigational and weather-related systems.

Radio stations were also implanted at different sites in the Caribbean and Latin America to improve the coordination of planes. This allowed Boeing to introduce automatic control technology for commercial aviation. With these new technologies, a comprehensive supervision of aerial space was realized. While Pan Am was pushing the aviation industry toward faster and more comfortable ways of intercontinental traveling, investors and businessmen saw new opportunities in the manufacturing of airplanes and the establishing of more airlines.

Airplane producers

In 1925, Pratt & Whitney established a separate Aircraft Company to build new and better engines. The company had originally been created in 1860 by machinists Francis Pratt and Amos Whitney who had both been working at Colt Armory, known for its accuracy of machines and weaponry and the pioneering in interchangeable parts and standardization. It built engines such as the Wasp and Cornet, both delivered in large quantities to the Navy as well as to the civil aviation market. It established cooperation with General Electric's plant for aircraft engines to improve the capacity of the engines and make them to fly faster. MIT engineer Donald W. Douglas, who initially worked for Glenn Martin Company, established the Davis Douglas Airplane company and, with an order from the army air services, began to construct very reliable planes that could make flights around the world (world cruisers). Transcontinental and Western Air (TWA) solicited Douglas to develop bimotor airplanes which led to the introduction of the Douglas commercial DC1 which was powered with two engines and a peak speed of 257 km/h. TWA began operations[10] and broke the speed record in 1930 from Newark to Chicago with a DC2 aircraft. Soon thereafter, TWA introduced a nonstop service with an 18-hour flight from the east coast to the west coast of America with stops in Kansas City and Albuquerque.

The introduction of the DC3 plane which accommodated 21 passengers had a cruising speed of 240 km/h and, in 1936, made practical daytime coast-to-coast flying possible in the United States. The DC3 became the icon of the 1930s aeronautics. It had much in common with the Model T of the Ford Motor company. They were both simple, rugged, and inexpensive and introduced Americans to their respective modes of transportation "[a]nd just as the Model T spurred a nationwide surge of road building, so the growth of aviation, paced by the DC3, meant that commercial travel would take on a professional cast" (Heppenheimer 1995: 108).

Next to Pan Am, three other trunk carriers – TAT, United Airlines, and American Airlines – established their own airlines.[11] Travelers demanded reliable schedules and greater speed. By the end of the decade, the DC3 began to dominate American aviation and was profitable by simply hauling passengers. "Schedules were 15 hours eastbound and 17 hours westbound with the time difference in the westbound flight more than compensated for by the gain in clock time due to the passage over three time zones boundaries" (Vance 1986: 559). Although traveling by plane was still more expensive compared with trains, business people were putting a money value on their time and increasingly favored aviation transport. After the Great Depression waned, passenger air traffic in the USA truly boomed. "It more than doubled between 1938 and 1940, and in that latter year over three million people took the air. The DC 3 posed no threat to the railroads, which carried 456 million passengers in 1940" (Heppenheimer 1995: 72).

Jet engines

The Second World War forced rapid improvements in aircraft and an enlargement of transoceanic flying. Douglas became America's largest producer of military aircraft, manufacturing more than 3,000 planes (Sobel 2000). Pan Am's clippers were deployed as military carriers which frequently flew 12 hours a day. A war of technologies was launched in which both sides invested vast resources in an attempt to build superior weapons and faster means of transportation. The aircraft industry switched from a small-batch craft industry into one of quantity production. Boeing, in Seattle, increased the scale and speed of the production of bombers. Although it delivered just five Flying Fortresses in 1941, by August 1944, the company increased its capacity to 360. It could do so by introducing an early version of an assembly-line production and a "just-in-time production strategy, which was later made so famous in Japan" (Freeman and Louca 2005: 287). Britain also put forth full effort into the building of fighters and bombers. The British Vickers-Wellington bomber was soon assembled in just less than 24 hours.

Germany and Britain became the only European countries with technical capabilities to develop and build new aircraft. German engineers Hans von Ohain and Max Hahn built the first turbojet and, together with plane builder Ernst Heinkel, the world's first jet plane was launched in 1938 at the time when Nazi Germany entered Austria.[12] The Nazis trusted capital to other competing firms as well and by 1942, the German Messerschmitt 262 was powered by a Junkers engine, which became the world's first operational jet airplane. "Had the Germans pushed the plane into quantity production, it could have severely hurt Britain's air defenses and even forestalled the Allied invasion of Europe in 1944" (Botticelli 1997: 121).

Although Rolls-Royce utilized quantity production to best serve the British army in wartime, it was determined to ensure maximum serviceability of Rolls-Royce engines.[13] Frank Whittle of the Royal Air Force, although a flight officer, spent significant time on designing a jet engine. With financial support from the British government, the Gloster Meteor was built in 1940, but the British government also trusted capital to the De Havilland Company to build a similar jet plane. Both in Germany and in Britain, military and naval engineers continued to press forward to build a jet engine designed for high speed to outperform the enemy. The Messerschmitt Company exploited thousands of war prisoners as slave labor to reach that goal. Its ME 262 was the first operational jet, flying faster than its nearest rivals. Yet in 1944, American B17 and B24 bombers, during raids, destroyed many of the facilities where the new planes were built.[14]

Following the war, many aircraft companies saw their contracts with the navy and army canceled and some of them faced bankruptcy. Douglas Aircraft had a lesser setback than its competitors and launched the DC6 in 1946 with a cruising speed of about 480 km/h. A growing number of civil passengers began to appreciate the advantages of this speedy air transport. The big air companies in the United States as well as national European carriers were looking for possibilities to extend air travel beyond their own borders and connect all adjacent capitals. To improve transcontinental flying, the power of the four engines on the big airplanes needed to be improved to make a reduction of the travel time feasible. Rapid evolution of international civil aviation became feasible with jet engines. "Successive developments of the jet engine have consistently improved its efficiency and propulsive power" (Doganis 2002: 9). (The British Overseas Airways Corporation (BOAC) demonstrated the effectiveness of the turboprops on fairly short-distance operations in terms of time, but these planes consumed great amounts of fuel.) In 1949, the British company De Havilland began marketing the first jet plane, the Comet, with a Rolls-Royce Avon engine, which BOAC introduced for aerial services. The Comet became an innovative aircraft which was flying with a cruise speed of 640 km/h.

With support from the Marshall Plan, Europe had initially been envisioned as a major export market for the American aviation industry (Heppenheimer 1995). However, a more balanced relationship in the aviation sectored was soon realized.

FIGURE 10.1 The Pan Am Stratocruiser passing the *Queen Elisabeth* in 1957 en route to England. (Courtesy Special Collections Division, University of Miami Libraries, Coral Cables, Florida.)

There was a remarkably quick "take-off" of the commercial jet aircraft practice in the 1950s, and, together with railroads and ocean liners, airlines took a sweeping role as long-distance carriers. The first eight Comets that BOAC bought clearly indicated a strong commitment to the future of jetliners. The Comet began service in 1952, connecting London to Johannesburg. Its performance astonished the general public. Air France used the Comet for a route between Paris and Beirut. UAT, an independent French carrier, used Comets between Paris and destinations in West Africa. After some serious accidents caused by construction failures with the De Havilland planes, however, the Comet was banned.

In the United States, Pan Am was one of the early airliners that decided to invest in jet engine aircraft. The main argument that jet engines consumed too much fuel to be economically viable was soon falsified. Trippe showed that the Boeing 707 with Pratt & Whitney jet engines flew almost twice as fast, at 972 km/h, as the propeller-driven Stratocruiser it replaced (Figure 10.1) and with twice as many passengers. In 1959, Trippe put the Boeing 707 in service for intercontinental lines and provided nonstop flights to Europe as well as a round-the-world service. He also set up a venture with IBM to introduce a fast reservation system called Semi-Automated Business Reservations Environment (SABRE). In 1964, Pan Am introduced a worldwide airline reservation system, Pan Amac, which was also developed by IBM.

Trippe's initiatives took other American legacy airlines into the jet age. Boeing built the 707 with jet engines that were based on aerodynamic insights learned from German constructors. Its large scientific and engineering department was using analog as well as digital computers for simulation and calculations for the design of this modern plane.[15] Soon, the new developments with jet engines spurred a boom for the commercial aircraft carriers which made it possible to fly at even higher speeds. Further improvements of the jet engines and streamlining of the body of the plane produced the highly successful B727, which remarkably reduced travel time. Boeing took over the De Havilland market and introduced the B737 for routes of shorter distances, while the B707 remained Boeing's long-range jetliner. In 1966, Trippe ordered the Boeing 747, a plane which initially had been built as a bomber for the American Army for civil services and began operations four years later when the Clipper Young America flew from New York to London at a cruising speed of 900 km/h (Vance 1986). At the same time, Douglas lost market share and could no longer keep up competition with its DC8. It was eventually taken over in 1967 by McDonnell Company which was still successfully manufacturing military aircraft.

> The arrival of the turbo jet engine had a twofold impact. In the 1960s, the turbo-jets led to a dramatic increase in speed whereas the size of the aircraft did not increase appreciably. In the later 1960s and early 1970s, there was no appreciable increase in speeds because existing speeds were approaching the sound barrier, but there was significant increase in the size of aircraft, particularly with the introduction of wide-body fuselages.
>
> (Doganis 2002: 9)

Like American aircraft producers, European companies also continued to realize considerable technical contributions with the jet engine. Rolls-Royce, a British industrial champion, won an exclusive contract with Lockheed in 1968 to deliver 540 engines in three years' time. Yet the company needed heavy governmental support to survive amidst American and Russian competitors developing similar large jet engines.[16]

Although Vance (1986: 545) concluded that the European aircraft companies "were very seldom able to shape an outstanding commercial aircraft," the building of the Caravelle, a small jet airliner with Rolls-Royce turbo jets, as well as the building of the Concorde, flying at Mach 2.2 (2,695 km/h), showed otherwise. This Anglo-French cooperation featured the fascination for ever-greater speed and laid the path to the creation of the European Airbus Consortium, This consortium was based on a multinational cooperative and featured governmental financing. Airbus and Boeing, which relied greatly on military contracts, began to build larger planes, known as jumbos, by cutting costs and carrying more passengers. In 1974, Airbus introduced its A300 as the first short-haul, wide-body aircraft.

Rolls-Royce, together with American Pratt & Whitney and General Electric, dominated the western market for jet engines with more propulsive power and delivered to both Airbus and Boeing. The companies were striving in different ways to improve quality and on-time production schedules based on *lean* principles (Womack and Jones 1996). With the introduction of lighter composite materials in airframe construction and larger passenger capacity – the Airbus A320 could carry 180 passengers – the hourly productivity of the planes improved even further. Manufactures noticed that the larger an aircraft, the lower in general direct operating costs per unit of output. Next to size, aircraft speed also had an effect on unit costs "since hourly productivity is the product of the payload and the speed, the greater an aircraft's cruising speeds the greater will be its output per hour" (Doganis 2002: 122).[17]

The trend towards medium-sized aircraft for long-haul services led to the introduction of the Airbus 340 in 1993 and Boeing 777 in 1995, planes which were entirely designed by computer and were able to transport as many passengers as the bigger B747. The increase of passengers was not only due to the technological improvements in cutting flying time, but was also stimulated by the impact of deregulation. More passengers began to fly more frequently which fed the overall impression that social life was accelerating at an ever-increasing level.

Deregulation

During the period 1919–1949, a framework of international regulation of air transport had been enacted. The American Civil Aeronautics Act of 1938 had clearly stated that regulation of both international and domestic air services was required to control unregulated competition. The International Air Transport Association (IATA), which was an organization of the airlines, not of the governments, improved as a cartel the coordination of international aviation services and facilitated the increase in size of international flights. Until the 1970s, however, an international regulatory framework remained in place. There was a set of wide-ranging controls as well as a whole range of technical standards concerning safety in airline operations which had to be respected.

Although some European countries remained reluctant to embrace free-market principles in the aviation industry, in the 1970s the issue of "open skies" was raised. Political and consumer pressures for liberalization increased as it was felt that this might favor lower tariffs and more efficiency. The other general argument was that inefficient carriers were kept alive, and new airline companies were excluded from access to long-haul trunk line markets. In 1977, the Carter administration favored deregulation and, in 1978, the Airline Deregulation Act was signed. Gone were the days of airlines under regulation "that contributed to safe, accessible, and reliable air travel for a broader array of consumers" (Mayo *et al.* 2009: 189). At that time, six carriers still dominated the American air: United Airlines, American Airlines, Delta, Eastern, TWA, and Pan Am. It was expected that the new Act might increase US airlines' share within the international air transport and restore its dominant position which had eroded. Yet it also opened the door for the main players to be taken over, traded, sold, and taken apart. In the United Kingdom, a similar policy became embraced as the Civil Aviation Authority allowed for a more liberal approach in its licensing policy. The Dutch and US governments signed the first "open skies" agreement in 1992 which inaugurated a new phase in overall deregulation of the aviation sector.

Airliners increasingly replaced surface vessels as the principle medium for overseas travel. The number of passengers traveling by plane increased more than those traveling by boat: flying internationally became a common practice for an increasing part of the mainly Western community. Yet the new policy did not turn out to be favorable for every airliner. Pan Am went bankrupt due to a combination of several different factors. The company struggled with: the effect of the oil crisis of 1973; excessive orders for new Boeing 747 jumbo jets while an increase in air passenger lagged behind; an increase in competition of cheap airlines due to the Act of 1978; and finally, an expensive purchase of National Airlines in 1980. For similar reasons, Eastern had to sell its Latin America routes to American Airlines. TWA also declined in importance.

Beginning in the 1970s, however, labor unrest in the aviation industry increased remarkably with strikes for better working conditions including shorter hours as compared to labor conditions at the railroads and the shipping – docking industry (Silver 2003). In the early 1990s, United, American, and Delta dominated three-fifths of the American domestic market (Heppenheimer 1995). In the meantime, the member states of the European Community also favored liberalization and introduced directives for deregulation. Some of the European airline companies such as British Airways and Lufthansa were able to strengthen their position, whereas others such as Swiss Air and Sabena lost market share during the process of consolidation; others again such as Air France and KLM merged to survive.

Air cargo

Improvements in speed and service delivery also took place in the area of air cargo. Transport of cargo by air first began around 1910 and obtained an official classification after World War I. In the early years, air freight was a method to fill spare capacity on passenger aircraft, but the high growth rate in the 1960s of air freight introduced new competitors. The combination carriers were soon challenged by all-cargo carriers which operated more flexibly and preferably flew at night. In order to reduce freight transit times, all-cargo carrier Federal Express (FedEx) introduced the concept of hubbing which relies very much on excellent punctuality. "Effective hubbing requires that flights from different airports, which are at the spokes of a network, arrive at the hub at approximately the same time" (Donagis 2002: 254).

The number of all-cargo carriers, such as Korean Air, increased continuously, especially when the electronics industry with their light semiconductors and chips began to boom (Putzger 2009). The center of gravity shifted towards East Asia and the Pacific. Next to Korean Air, Cathay Pacific and SIA increased their freight transport capacity, extending the international air freight business. Undoubtedly, the introduction of new information technologies increased an effective scheduling of planes and the reliability of air freight flying schedules. FedEx, CargoLux, and Lufthansa Cargo remained dominant players in this field and developed transport for supply chain cargo, perishable goods, live animals, sensitive and high-technology equipment, as well as oversized freight shipments.

Together with FedEx, which is by far the largest, UPS (United Parcel Services) and DHL developed a door-to-door service from shipper to customer with goods for which shipping time is important. As integrators, these companies have been able to remarkably reduce rates for air freight. The North American routes have especially manifested serious declines in prices. "In 1980, the real costs of airfreight had fallen to about a quarter of its level on the eve of World War I" (Crafts and Venables 2003: 329). The result has been "a more than tenfold increase in the ratio of air to ocean shipments in the years since 1962" (Findlay and O'Rourke 2003: 50).

Although the integrators cover only a small part of the overall international air cargo market, their policy of offering speed, on time delivery, and reliability has been rather influential. It is estimated that air cargo currently covers some 35 percent of world trade movements (Kasarda *et al.* 2006). The quality of services segments the air freight business and prices are generally calculated on the basis of speed of delivery. Doganis (2002) expects that integrators and combination carriers will establish closer cooperation and advance into global alliances, providing a speedier expedition of air cargo.

Sea transport

Whereas the aviation industry facilitated a technological acceleration on an unprecedented scale, the sea transportation business would need to wait for a fundamental transition until the innovation of the container, the cheapest way to move products around the globe even today. The introduction of the container ship for intercontinental trade was another key factor for the acceleration of industrial globalization. Nevertheless, it took some time before shippers and large internationally operating manufacturing firms would discover this advantage.

In the seventeenth century, the British Post Office introduced an ocean-shipping service for mail and postal packages in ports such as Dover, Harwich, and Falmouth. With two vessels, a transatlantic postal service to New York was set in motion in the second half of eighteenth century. After American independence in 1776, Benjamin Franklin became America's National Postmaster General and was highly involved in the regular transportation of packets of collected mail, from which packet boats earned their name, as well as newspapers and magazines over land and sea.

Yet "the invention of regular transatlantic cargo shipping came from a company headquartered in America" (Winchester 2010: 293). By 1812, cotton broker J. Thompson and textile manufacturer B. Marshall established a cotton transportation system with small vessels from the southeast ports to New York from where cotton was sent in a regular manner to Liverpool. With the support of ship owner I. Wright, who secured enough cargo space across the ocean, their four ships were called "transatlantic liners" as they traveled across the ocean on a fixed and regular timetable. During the late nineteenth century, the capacity and frequency of these liners increased with the introduction of steam-powered ships and the opening of the Panama Canal in 1914, connecting the Atlantic with the Pacific Ocean, which cut the travel time in half for ships.

The international seaborne shipment of cargo in bulk began in the mid-nineteenth century. Bulk cargo was traditionally packaged in bags or, in case of rain, stowed with separations in the holds of a cargo ship. "The first types of bulk carriers for iron ore and coal were developed in the beginning of the twentieth century, but the modern bulk carrier as we know it today was developed in 1955" (Wijnolst and Wegeland 2009: 205). The throughput time at the harbor was significantly reduced as labor-intensive loading and unloading of a ship with heterogeneous cargoes was replaced by homogenous-mass cargoes which could be handled by cranes.

In the oil sector, a remarkable development also took place in transporting. Before the first oil tankers were introduced around the 1860s, Standard Oil Company of Rockefeller stored and transported oil in wooden barrels of 159 liters. Although it invented the pipelines as a reliable way of transporting oil, for the transatlantic shipment of the liquid, it continued to cling to transporting oil in barrels.

The Swedish brothers Nobel who, by the early 1870s, became involved in the exploration and exploitation of the Baku oilfields on the Caspian Sea, built their first tanker steamer by 1878. By 1882, the Nobels used 12 oil carriers on the Caspian Sea "well ahead of the shipping developments on the Atlantic Ocean" (ibid.: 179). Yet when vertically integrated oil companies put intermediate storage in place near harbors, the use of bulk ships became a more common practice. *Glückauf* became the first oil tanker in 1886 with a hull uniquely used as a containment system and was built in England for a German subsidiary of Standard Oil. The introduction of the diesel engine in 1897, whose application became diffused quite quickly in the maritime industry, enabled further innovation in shipping and certainly propelled oil tankers much faster.

By 1907, Marcus Samuel, founder of Shell, introduced tankers which passed through the Suez Canal with kerosene for the European market. More oil was transported from newly found pits in Mexico (1901), the Middle East (1909), and Venezuela (1914) to the industrializing nations. During the period 1950–1970, oil began to replace coal as the major source of energy and seaborne oil transport increased remarkably.

Two successive closures of the Suez Canal in 1956–1957 and 1967–1975, however, forced oil tankers to sail from the Persian Gulf along the African coast, which caused longer voyages and pressed for an increase in the size of the tankers. The oil industry was forced to move into the construction of super tankers. The first one was built by the American tanker pioneer Daniel Ludwig who built the *Universe Apollo* in Japan. "The tanker sector thus became the largest segment in shipping; it accounted for half the world's deadweight capacity of the merchant fleet" (ibid.: 188).

The biggest improvement, however, for cargo transport in large quantities was introduced in the mid-1950s. Efficient storing and abiding to strict schedules was the brainchild of the American trucking executive M. McLean from North Carolina with his concept of applying large steel boxes, i.e., containers, with a standard size which could be put onto trucks or railroad flatcars once they were discharged from a ship. Containerization not only became a different way of loading cargo onto ships, but the whole chain of (sea) transport over land changed and facilitated rapid, predictable, and cheap shipping.

In April 1956, Malcolm McLean's company, Pan-Atlantic, launched the first container ship – a converted American navy tanker – to move goods in 58 containers loaded on truck trailers on the spardeck from Newark (New Jersey) to Houston (Texas) (Dicken 2007). Based on this experience with a roll-on, roll-off ship for coastal transport of his trucks, Pan-Atlantic moved into full container operations. The company renamed itself in 1961 to Sea-Land Service Inc., and its successful innovation was soon imitated by others.

For a smooth manner of transport, harbors needed to adapt their infrastructure to handle transport by containers. The harbor authorities of Rotterdam, for example, introduced their first container terminal in 1965 with specially made shore cranes. A year later, the European Container Terminus (ECT) was established, which was taken over by Lloyds three years later. The company, like a few others, subsequently invested heavily in container ship innovations to reach economies of scale and reduce the round-trip time of container service in comparison with conventional service (ECT was taken over in 2002 by the Chinese Hutchinson Whampoa Limited).

During the Vietnam War, the application of containers by the American army intensified and became an example for other shippers and harbor authorities to improve their transport facilities. New ports were built in places such as Singapore, Hong Kong, and Shanghai, which established its first container ship dock in 1972 (Gipouloux 2011). Using containers became an entirely new way of handling freight through an extensive transport system. Every part of the system – ports, ships, cranes, storage facilities, trucks, trains, and the operations of the shippers themselves – was adapted.

As shippers were still operating with their own container sizes, quick transfer was delayed and efficiency was hampered. Therefore, in 1970, a full set of international standards was adopted. Remarkable, however, is the fact that after the introduction of containers, shipping prices rose during the 1970s, were peaking in 1985, but fell sharply thereafter (Findlay and Rourke 2003). Nevertheless, containerization transformed the shipping business from a romantic atmosphere with tides and waves and the smell of tar into an intermodal system whereby the ship is just one part of a long chain of different types of transport (rail, road, sea, air). In 2004, a new generation of container ships, each capable of carrying 8,000 containers, was introduced (Dicken 2007).

Traditionally, large multinational companies which were utilizing vertical integration would obtain raw materials from their own mines or oil wells and move them to their factories by using their own trucks, ships, or rail cargo equipment. The raw materials would then be put through a series of processes to refine them and turn them into finished products which would subsequently be exported again. As freight costs began to plummet and as the rapid exchange of cargo from one transportation carrier to another became routine, manufacturers discovered that they no longer needed to do everything themselves. They could contract with other companies for raw materials and components, locking in supplies, and then signing transportation contracts to assure that their inputs would arrive when needed. So, in the end "integrated production yielded to disintegrated production" (Levinson 2006: 265). Since then, manufacturing firms have learned through global sourcing that components can be made on one continent and assembled in another, a transformation which would not have been possible without the systematic reduction of transport time and costs. Global sourcing is a procurement strategy to find the best source at the best price for a product. "The cost of a minute of time for a seamstress thus becomes the object of extremely precise comparisons" (Gipouloux 2011: 189).

Container shipping generated significant economic advantages. The difference in container shipping compared with traditional types of shipping is its high degree of mechanization. It eliminates time-consuming storage and has doubled the speed of loading and unloading. A container can be lifted directly from the ship to a railroad flatcar or truck chassis. This new approach eliminates large inventories and raises quality and efficiency. Containers have speeded up the process of unloading and further transporting commodities to the local destination.

Outlook

For less voluminous and lighter goods, transport by air is still favored despite the fact that transport by air is about seven times more expensive than transport by ship. However, international businesses began to favor air transport in many circumstances as speed and timing became the main target for trading and transporting companies (Deardorff 2003). Shipments from one continent to another are expected to arrive on schedule. Just-in-time (see Chapter 12) has become a common policy even in the transportation sector. Companies began to arrange relationships with a small number of carriers which are able to meet stringent requirements for on time delivery. Railroads, ship lines, truck lines with large route networks, and sophisticated cargo-tracking systems together with warehouses have the edge.[18] The organization of interim transport has developed into a chain of global management. By delegating the logistics services to an outside firm, the logistics are outsourced.

In 2009, the five biggest container companies, shown in Table 10.1, transported almost half of all the containers in the world.

TABLE 10.1 The five largest container companies of 2009

Company	*Country of origin*	*Market share (%)*	*Containers transported (millions)*
Maersk	Denmark	17	13.38
MSC	Switzerland	10	10
CMA-CGM	France	7	7.68
Evergreen	Taiwan	5	6.20
Hapag-Lloyd	Germany	4.4	5.45

Source: Wijnolst and Wergeland (2009: 245).

The British author Alain de Botton has eloquently described how, for example, the center of England's logistics hub functions:

> The single mindedness of the goings-on in the logistics park are most transparent at night, when the appearance of the moon questions the significance of efficient courier services from an interplanetary viewpoint, as does – from the perspective of eternity – a slender church spire built in the late fourteenth century, visible as a pitch black arrow on the far side of the motorway. [...] Nightfall used to be the time when members of our species would acknowledge their physical limitations and huddle together to mitigate their fear of ghosts and witches. The logistics hub, however, makes few concessions to human frailty, the spirit world or the primacy of natural rhythms. Floodlights come on to compensate for the sun's retreat, bathing the area in the nocturnal orange glow familiar from airports and military installations. Workers are dropped off by bus at a central reception area and clock in before seven [...]. The largest warehouse in the logistics park belongs to a supermarket chain, which throughout the night receives dispatches from food suppliers and recombines them for onward delivery to stores across the country. The aisles of an average supermarkets contain 20,000 items, four thousand of which are chilled and need to be replaced every three days, while the other 16,000 require restocking within two weeks [...]. Components of the national diet race around the building on conveyor belts high above the ground: thirty cartons of crisps for North fleet, 1,200 chicken drumsticks for Hams Hall, sixty crates of lemonades for Elstree. [...] Time is of the essence. At any given moment, half the contents of the warehouse is 72 hours away from being inedible, a prospect which prompts continuous struggle against the challenges of mould and geography. Clusters of tomatoes still attached to their vine, having ripened to maturity in fields near Palermo at the weekend, are exchanging the destiny seemingly assigned to them by nature to try to find buyers for themselves on the northern fringes of Scotland before Thursday.
>
> (Botton 2009: 40–42)

Over the years, large organizations have been outsourcing transportation which made logistics management more important in the delivery of the right product at the right time, in the right quantity, for the right costs, without damage or waste, to the right customer. The standardization of cartons, boxes, and crates, as well as the inter-changeability of containers, has enhanced efficiency, but these advantages would not have been attained without the necessary steps of improved coordination. While fewer workers were needed, their skills for information handling have become more important. The information related to physical movement of materials and goods required adequate transmission and manipulation. Coordinating the various sources of information between ships, trains, trucks, and warehouses at national and international levels to deliver goods has required new skills in the domain of exchanging information and knowledge sharing.

Computer software has improved the efficiency of the processes of handling goods by introducing barcoding technology which IBM, for example, has developed for Universal Product Scanning (UPS). These new technological advantages forced large companies to handle information in a more structured and efficient way not only within their own internal setting but also between suppliers and transporters, as well as logistic companies at national and international level.

Since the late 1980s, logistics management – the process of planning, implementing, and controlling the efficient, cost-effective flow and storage of raw materials in process inventory and finished goods – has become a routine business function for manufacturers.[19] Retailers discovered that they could manage their own supply chain, cutting out the wholesalers who stood between manufacturers and consumers.

Transport organizations have grown into logistics service providers striving to improve inventory management, warehouse, and material handling as well as transporting activities. Whereas the traditional physical distribution had been labor intensive and provided standard services, the "new physical distribution" approach has become knowledge intensive, depends on the latest information technology, and operates within a partnership of demanders and suppliers based on client-specific services (Stock and Lambert 2001).

The nature and efficiency of distribution systems based on strict schedules accelerated the principle of "Time is Money." Today, computer-based navigation courses for ships carrying 15,000 containers moving at a speed of almost 30 knots (almost 56 km/h), being unloaded by cranes timed to the second, has led to a practice which almost "doubled the speed of ocean shipping" (Crafts and Venables 2003: 351).[20] By the late 1990s, roughly 90 percent of total world trade was moved in containers (Dicken 2007). Automation of the onshore and the onboard cargo-handling processes have contributed tremendously to a further reduction of time and cost (Wijnolst and Wergeland 2009). With modern information technologies and container shipping, any manufacturer can design its own products, order production elsewhere in the world with all the details required, and have the goods delivered to its distribution center closer to the time that they need them. "This precision performance would have been unattainable without containerization" (Levinson 2006: 267).

The dramatic decline in transportation costs through air cargo, containerization, and the streamlining of an (inter)nationally operating railroad network extended and intensified the impact of industrial globalization. Without the technological advances created by modern Information and Communication Technology (ICT), however, many multinational corporations would not have realized the extent of expansion that they experienced because of it. ICT introduced facilities such as telecommunications, the Internet, e-mail, voice mail, and the cell phone, which significantly reduced time limits and increased speed-up of transmission and transfer processes.

PART III
Electronic globalization

The increasing speed and volume of industrial output made it almost impossible to adequately control production and distribution processes. American companies pursued a policy of increasing the levels of formal bureaucratic structures to better process information about growing sizes of national and international markets. Large and leading firms in the capital-intensive manufacturing industries invested substantially in organization building and professional management (Chandler 1977). They could do so as emerging technologies presented opportunities to achieve economies of scale and scope in order to better serve potential markets. By introducing new technologies of transmission of information, standardizing procedures and routines, as well as introducing specific management concepts to keep employees committed, conglomerates attempted to prevent a crisis of control and accelerate their activities by designing an administration as an efficient timesaving instrument. Technological improvements intensified the creation of an *empire of speed*. Although initially "increases in the speed and reliability of throughput outpaced the development of information-processing and communication technologies adequate enough to control the larger systems" (Beniger 1986: 248), the revolution in information technologies eventually provided an adequate solution.

In the 1970s–1980s, Japanese corporations contributed to this development by introducing *timing* as a key new concept. Japan became a new global economic player with remarkable innovative production methods and restructuring of work organizations based on "just-in-time." With information and communications technology (ICT), Western and Japanese multinational corporations have been able to institute worldwide transportation and communications networks which have evolved into a global telecommunications infrastructure. They have created an electronic globalization which hints at a process of integration across frontiers of market economies at a time of rapidly falling costs of transport and communication, suggesting the world is becoming smaller, faster, and more connected (Wolf 2005).

Computers have become capable of processing in *real time* by shrinking existing process cycle time as small as possible, and worldwide infrastructures have facilitated the transport of information on a global scale.

Part III sketches the contours of the intensifying technological and social acceleration in the Western World. It first focuses on the introduction and impact of the revolution in ICT. While the story is a complicated one, its development will be illustrated by highlighting IBM's role in the organization of the speed-up of transmission of information. After a description of the formidable social and economic changes that took place in Japan, the attention will shift towards the cultural consequences of the technological and social acceleration and the intensifying social interactions faced by internationally operating managers who are expected to be adequately capable and possess a substantial cultural intelligence to deal adequately with a large variety of cultural time patterns.

11

INFORMATION AND COMMUNICATION REVOLUTION

Usually literature distinguishes three stages in the extraordinary development of technologies which have improved the transmission of information. The first stage began during the French Revolution when a system of mechanical semaphore signals, *le télégraphe*, was constructed by installing a vast network of relay stations able to transmit messages across the country in a few hours. Yet the functioning of the system, which served only state interests, was hampered by bad weather, darkness of the night, and human errors. The first electric telegraph lines built next to railroad networks became operational in the 1840s. By 1851, the first submarine cables were laid that connected Britain with continental Europe and, by 1866, a transatlantic cable linked Britain with America. British imperial power subsequently pushed the extension of the electric telegraph network even further. Although the telephone was invented in the 1870s and the radio around 1900, their application only came into widespread use from the 1920s onwards. While telegraphy and telephony made the transmission of words nearly instantaneous, the radio did the same for sounds. Basic new communication technologies that improved international distribution of information included "a Pacific cable (1903), transatlantic wireless telegraph (1907), transcontinental telephone (1915), radio facsimile (early 1920s), transatlantic telephone (1927), high speed telegraph (1929) and teletype (1931)" (Beniger 1986: 324–326).

Along with the German company Telefunken, The Radio Corporation of America (RCA) was one of the prime movers and was established as a joint venture of three leading American producers of electrical and telecommunications equipment. O. D. Young, vice president of GE, was successful in pulling the Marconi Wireless Company of America[1] into the joint venture, turning RCA into a dominant player in consumer electronics. RCA marketed the radio receivers that GE and Westinghouse produced under its own label. European companies such as Marconi, Philips, and Telefunken signed patent agreements with RCA to join the "unexpected explosive growth of radio receiving and broadcasting equipment" (Chandler and Mazlish 2005: 18). The invention of the transistor by Bardeen, Brattain, and Shockley at the Bell Telephone Laboratories after World War II is an often-told story that does not need to be repeated here[2] but definitely influenced the fast transmission of signals in telephone devices, transistor radios (1954), and circuits. The new technology that replaced vacuum tubes by transistors improved the quick transfer of information. In 1956, the first transatlantic telephone cable became operational. The market for semiconductors initially began with the US military, and the Cold War strongly nurtured this industry (Langlois and Steinmueller 1999).[3]

A second wave of transmission of information began in the 1940s and 1950s when television became popular and affordable due to rising mass incomes. The experiences learned in commercializing radio laid the foundations for consumer electronics and the commercialization of television.[4] With the development of the technology of electronics, it became possible to transmit pictures in a more general way. RCA invested massively in the design of a complete television system. In 1939, it had demonstrated its first commercial system at the New York World's Fair by broadcasting President Roosevelt's opening speech. After World War II, black-and-white television reached maturity and became an instrument in mass persuasion, transforming daily life (McNeill and McNeill 2003). The next challenge was the development of color television and, by 1959, RCA made its first profits from color TV. "Recording technologies were introduced early in the United States as the difficulty of relying on live telecasts in a nation that spread over four time zones gave studios extra incentives to push nascent videotape capabilities" (Friedel 2007: 522). By the end of the 1980s, satellites broadcasted directly to private homes, thus distributing multichannel television. The installation of private communications satellites speeded up the transmission possibilities tremendously. The introduction of digital transmission allowed for more channels to be provided to the public. By 1996, more than 840 million households worldwide had a television set, which was two-thirds of the world's households (Cairncross 1999).

A third wave of improvements in the transmission of information took off after the invention of the computer.

> Between 1942 and 1946, researchers at the University of Pennsylvania created the Electronic Numerical Integrator and Computer (ENIAC), the predecessor of the computer which, after a series of improvements, laid the ground for a very successful computer industry in the United States.
>
> (Freeman and Louca 2005)

The first modern, high speed/stored program, all purpose digital/reckoning devices computers had been invented in order to meet the demands of the Second World War and were utilized by British and American intelligence services to crack secret codes being used by the German enemy.

> Wartime innovations in high speed calculators, cryptanalytic devices (for code breaking), and in servomechanisms (as used in controlling gunnery

> fire) set the stage for the invention of instruments to carry out complex mathematical calculations at unprecedented speed and accuracy for specific military needs.
>
> (Chandler and Mazlish 2005: 83)

The development of radar during World War II had provided the impetus for a major reorientation of the relationship between the American state and the academic world. It was strongly believed by policy makers that a major investment in scientific research was necessary for the future security and financial prosperity of the American nation. From 1947, the annual federal expenditure for research and development multiplied by a factor of ten over the next 50 years (Lewontin 1997). Due to the space and arms race with the Soviet Union during the period of the Cold War, the American government promoted research on computer technology. The Defense Advanced Research Projects Agency (DARPA) financed the development of a computer network. RCA quickly became interested in data-processing computers and established a separate division for these new computers along with its consumer electronics' division. This transition took place at a time when large Japanese firms such as Matsushita and Sony, and the Dutch company Philips began to concentrate on improving their electronics products and processes, "many of whose technologies they had initially received from RCA" (Chandler and Mazlish 2005: 37).[5] RCA shipped its first digital analytical computer to the US Army in 1956 to provide speedy and accurate information about its inventories. The magnetic disk provided computers with random access memories which were reliable and fast. After the Russians launched the Sputnik in 1958, the US Army asked DARPA to prevent any future similar technological surprises, so, with the availability of a large budget, DARPA invited scientists from all over the country to participate in the project. It backed an experiment at the company Bolt, Beranek, and Newman in Cambridge (Massachusetts) to connect computers across the country (Cairncross 1999).

The building of the network was fueled by the coming of the router, the World Wide Web, and the browser. Electronic links were created to make it possible for researchers in different universities to share resources and results. In 1963, Joseph Licklider described a more practical way for staff members working at different locations to use the then still minimally available calculating power of computers. In 1969, the first network, called Arpanet, was introduced, initially connecting only the four largest of the military-funded university computer centers by cables. Still, the Arpanet was restricted to non-commercial use. Soon, however, it was expanded into a nationwide network linking university computers. In the meantime, Telenet was launched in 1975, and provided commercial computer services.

The further advancement of computer application was enhanced once J. Kilby of the Texas Instruments Corporation and R. Noyce of Fairchild Semiconductor Company devised the means to apply silicon for integrated circuits. The application of silicon in digital switching applications proved superior especially in terms of speed. Silicon provided the means for miniaturization by building all the components of a circuit on a single piece of semiconductor. Improvements by Intel Corporation led to the introduction of the 4004 chip in 1971 and generated a series of remarkable improvements. Once the microprocessor revolution got underway, providing the opportunity for the manufacturing of the small personal computer (PC), the public Internet became accessible to the larger public.

In 1983, the Defense Department required all host computers on the Arpanet to use the same protocol and establish a common language, i.e., a format to package data for transmission over the Internet.[6] Arpanet became part of the Internet once the Defense Department had withdrawn its own access to the network, which made the Internet publicly accessible. The networking of computers then shifted from a largely military and scientific project to the modern Internet.

The computer industry, with companies such as Microsoft (founded 1975) and Apple (founded 1976), soon saw PCs become a booming business.[7] The quickly spreading information and communications technology (ICT) facilitated the combination of two distinct technologies: communications technology, which is concerned with the transmission of information, and computer technology, which deals with the processing of information. With the power of the chip multiplying and the price of computing power dropping, computer capacity increased tremendously. "The constellation of new, very fast-growing industries which had initially developed quite independently[8] of each other in the nineteenth and twentieth centuries had now crystallized into a new technology system which became generally known and recognized as *ICT*" (Freeman and Louca 2005: 319).

In the early 1990s, Tim Berners-Lee and Robert Cailliau, while working at CERN (the European Organization for Nuclear Research) in Switzerland, developed the World Wide Web (WWW) and introduced the first web browser. The WWW, or global hypertext system, provided a common way to label and address any information on the Internet and was made public for free in 1994. The evolution of browser software initially designed by M. Andreesen and E. Bina at the University of Illinois in 1993 facilitated access to the web. The web sprang to life with the introduction of a graphical web browser, which accommodated online graphics, sound, and moving pictures. In 1995, the Internet became fully privatized when Netscape Navigator went public with the first commercial search engine.

By the end of the twentieth century, "the electronic digital computer became as much a communications device as an information processor" (Friedel 2007: 515). Further improvements of the computer propelled the third wave of information sharing, and the intensity and velocity of the traffic of information facilitated the appearance of the stage of electronic globalization, wiring the developed world together. In 2000, approximately 360 million people had access to the Internet; by March 2011, this number reached two billion (US Bureau of the Census 2011). Giant new firms emerged such as Microsoft, AOL, Yahoo, Amazon, Google, and Apple, which created new facilities for a wider application of Internet and facilitated the birth of a vigorous new industry, which extended its application and services quickly worldwide. Yet, it is noteworthy that just as railroads did not invent the automobile, telephone companies did not invent the Internet (Cairncross

1999). Nevertheless, by the mid-1990s, the Internet as a product of the public sector seized public imagination. Communication companies grew exponentially and evolved into giants such as IBM, a corporation which nicely illustrates how the recent development toward electronic globalization relied not only on technological innovations but also on the size of large-scale corporations and their particular strategic activities in a booming industry.

IBM

The origins of what became IBM are remarkable. In the 1880s, Willard L. Bundy (1846–1907), a jeweler in Auburn (New York), became interested in timepieces and invented a time-recording clock in 1888 that would register employees' time. He established a firm with relatives in order to produce his invention. His brother, Harlow Bundy, organized the Bundy Manufacturing Company and began mass-producing clocks. Willard, who was more of an inventor, had obtained several patents for mechanical devices that he had designed but was forced to split his partnership with his brother after discussions arose regarding the ownership of these patents.

Willard claimed the patents and kept them for himself and his family. He moved to Syracuse where he purchased the Dey Time Recording Company from a Scottish physician and mathematician named Alexander Dey, who also had developed time-recording systems that made it possible to check in and out of work with precision by punching a clock. This procedure determined the temporality of work in the factory. Bundy renamed the company the International Time Recording Company (ITRC). Yet, Willard was not the only one operating in this business.

Herman Hollerith (1860–1929), a German-American statistician, developed a mechanical tabulator based on punched cards – which had a long history beginning at the time Jacquard had introduced them in looms – in order to rapidly tabulate statistics from millions of pieces of data. Hollerith built these machines under contract for the Census Office, which used them to tabulate the 1890 census. Whereas the 1880 census had taken eight years, with his machine it only took one year.[9]

In 1896, Hollerith began his own business, the *Tabulating Machine Company*. Improving on his own technology, he invented the first automatic card-feed mechanism and the first keypunch which allowed a skilled operator to quickly punch 200–300 cards per hour. His Type I Tabulator, introduced in 1906, could execute different jobs without having to be rebuilt: it was the first step toward programming. Most of the major census bureaus, as well as major insurance companies, leased his equipment and purchased his cards. His inventions laid the groundwork for the modern information processing industry.

By 1907, nearly all of the leading time-card stamping-system manufacturers were bought by ITCR (Levine 1997). After a merger with several other firms producing business clocks, including Hollerith's firm, the company was renamed in

1911 as the Computing-Tabulating-Recording Company (CTRC).[10] The deal had been put together by financier Charles R. Flint, who ran the new company from his offices in New York. In 1914, Flint hired a 40-year-old unemployed executive named Thomas J. Watson to run CTRC. Ten years later, Watson demonstrated a clear vision about the future of the company and renamed CTRC to the International Business Machines Corporation (IBM), even though, at that time, the firm was actually not very international.

IBM marketed its products by appealing to the virtues of punctuality and stressed that time clocks would save money, enforce discipline, and add productive time. In 1924, the company set up a large manufacturing plant in Europe. It opened several different laboratories overseas to improve record-keeping machinery. Throughout the 1920s and 1930s, Watson gradually moved IBM's focus away from low-tech machines and into state-of-the-art tabulating instruments: the punch-card machines.

IBM and its competitor, Remington Rand, dominated the American tabulating equipment market as duopolists. In 1928, Remington Rand was still three times the size of IBM. Though the stock market crash of 1929 and the subsequent Great Depression put intense pressure on the profits of most American companies, IBM was barely affected. This was mainly due to the National Industrial Recovery Act which necessitated extensive data gathering and tabulation by the government, for which IBM received extensive contracts. By 1935, IBM had 85 percent of the world's market for sophisticated office machinery and would become a prime mover in the process of electronic globalization (Chandler and Mazlish 2005). Remington covered the other 15 percent, though by 1939, IBM and Remington Rand were roughly equal in size. By 1938, IBM had begun to develop a programmable calculating machine (Friedel 2007). In 1939, it facilitated the Harvard professor H. I. Aiken to collaborate with the corporation to design the Harvard Mark I calculator.

During World War II, the entire military establishment began to use IBM cards.[11] For the US Navy, IBM provided an improved Mark in 1943, which was able to solve computing problems, but its electromechanical system demonstrated inherent limits in terms of speed and reliability (ibid.). After the war, the demand for American manufactured goods boosted supporting industries such as banking, retail, and insurance which were big customers of IBM. Watson implemented a leasing strategy which proved quite popular among its customers as they could not afford to buy the machines. The leasing system strengthened the ties between IBM and its customers, as well as assuring that the company continually received a solid source of cash flow.

In the late 1940s, Watson, Sr was succeeded by his son, Tom, Jr. After reviewing all of the projects the company was heading, he came to the conclusion that IBM was still in the electronic Dark Ages (McCraw 1997). He initiated an in-house evaluation to assess the enthusiasm of IBM's scientists and engineers for using a new technology of electronic vacuum tubes in a variety of circuits. However, he was also focused on the increasing customer demand for faster and less cumbersome tabulating machines, and his staff stressed an intensifying competition.

Based on this information, Watson, Jr. finally decided to begin the production of computers in June 1950.[12] He chose an aggressive strategy for the computer market, and by 1955, IBM became the world's leading producer of computers. By 1960, foreign markets accounted for 20 percent of IBM's total sales and continued to be very important contributors to IBM's further expansion. (It held nearly 40 percent of the Japanese market.) The 1960s became a period of rapid growth for the American computer industry, and IBM designed the world's first automated transistor production line (Langlois and Steinmueller 1999). IBM's strong commitment to research and development (R&D) left competitors behind as it placed formidable constraints on the technological options for its rivals. IBM's data-processing revenues were significantly higher than those of its rivals, and it marketed several different classes of computers.

Even so, everything did not remain well. Although the corporation offered eight computers, each with a different internal architecture, there was a lack of compatibility which prevented IBM from achieving optimum scale efficiencies in engineering, manufacturing, and sales. Watson, Jr. became aware that it would have to move faster or risk being overtaken by rivals. In response, IBM used its scale and scope to develop a quicker mainframe computer called System/360, which was introduced in 1964. This machine became one of the most important products of the Electronic Revolution or, in other words, the Third Industrial Revolution, and for many people obtained the same status as Model T Ford did for the Second Industrial Revolution.

With the System/360, IBM set the standard and stimulated unprecedented growth in the use of computers throughout the world.[13] It established itself as the leading-edge supplier and servicer of commercial mainframe computers. "At the time, the 360 was the largest privately financed, commercial project ever undertaken" (Collins and Porras 1997: 101). When Tom Watson retired in 1971, the demand for computers had penetrated most of American businesses, and IBM had a more than 60 percent share of the world computer market. The corporation did not rely on outside low-cost suppliers, but favored its own manufacturing capabilities. Computer users worked at terminals connected to giant mainframes. The terminals displayed information and transmitted instructions to the mainframe.

Further technological developments such as the invention of the microprocessor made it possible to put the calculating power in smaller computers. Digital Equipment Corporation (DEC) – which was established in 1957 – challenged IBM by manufacturing a masterpiece from 1965 onwards: the minicomputer with a low price and high performance. "By the early 1970s, DEC was becoming the most dynamic enterprise in the computer world" (Chandler and Mazlish 2005: 106).

Hewlett Packard (HP) – a producer of electronic analytical instruments since 1938 – was another successful challenger to IBM with its minicomputer and soon became second to IBM in maintaining worldwide competitiveness in new product sectors that evolved after the coming of the microprocessors.

In the development of this new industry, Europe was lagging behind. Some European governments pursued policies to protect their own infant computer industries and encouraged them to consolidate to successfully resist the intrusion of the large American corporations. Yet "the European computer industry was unable to achieve the output scale of American firms, notably IBM, who often produced in Europe in order to circumvent a 17 percent ad valorem tariff" (Langlois and Steinmueller 1999: 37). At the same time, boosting dynamics in the new industrial sector caused by the appearance of rivals at home as well as some management mistakes put IBM under tremendous pressure. The biggest threat to IBM was the standardization of the semiconductor technology, which allowed independent suppliers to provide parts to a host of new computer makers. Barriers to entry fell and competition proliferated. Product cycles shortened, new entrepreneurs took the opportunities that the new growth segments offered, and brought their products to market with a speed IBM could hardly match.

In the 1980s, IBM encountered serious competition from many sides, including Japanese firms such as NEC, Fujitsu, and Toshiba. IBM was forced to reduce its labor force worldwide by more than 50 percent. The corporation understood that its computers would be in serious jeopardy in the marketplace and that continued success for IBM would depend on its ability to remain at the leading edge of technology. Its product and marketing methods were redesigned and old ones made obsolete (Pugh 1995). By doing so, IBM remained the world leader in computer manufacturing.

Mainframes continued to be a giant market, but with the arrival of low cost memory and low cost processing, the entire computer industry fundamentally changed. The advent of PCs drastically changed the customer–supplier relationship. PCs were easier to understand than were their "black box" predecessors. Although IBM introduced its own PC in 1981, it was confronted with harsh competition from Intel, the company that sold the world's first microprocessor.

The Intel Corporation was founded in 1968 by Robert Noyce and Gordon Moore, who became leaders in California's electronics industry long before Silicon Valley became a well-known area (Jackson 1997). After Intel launched their second microprocessor, the business slowly, but continually, expanded.[14] In 1983, one million PCs were sold in the United States. In 1985, Intel launched a new 386 microprocessor, which offered much more speed than the IBM chip.[15] So-called IBM PC clones such as Compaq, Dell, HP, and Acer adopted the IBM standard but then proceeded to attack IBM by introducing computers that applied that same standard better, cheaper, and with more technological back-up. Unix, which had been developed by AT&T employees working at Bell Labs, was an "open" operating system supported by Sun, a vendor of software and hardware, and Hewlett-Packard. It offered customers the first attractive alternative to IBM's mainframe computers.

Initially, IBM had failed to see that PCs would be widely used by business and enterprises. Then, in the interest of speeding up the process of launching a PC to the mass consumer market, IBM decided to cooperate with outside suppliers and produce software independently from the hardware. IBM then bought about 20 percent of Intel and decided to put the Intel processor into its own PC (Jackson 1997). However, the strategic alliance between Compaq and Intel also placing the new technology of the Intel chip in a Compaq laptop became a serious challenge to IBM. Compaq's focus was centered around the increasing speed of its computer and took a large chunk from IBM's PC business (Friedman 1999).

IBM also handed control of the operating system over to Gates' Microsoft, which had moved from Albuquerque to Seattle in 1978. Initially, this cooperation was successful (Jackson 1997).[16] Soon, however, IBM's leadership position began to disintegrate. Intel and Microsoft were allowed to license to others which made their chips and software the industry's standard, and both companies received "the most lucrative franchises in the history of American industry" (Chandler and Mazlish 2005: 137). Fujitsu, Digital Equipment, and Compaq became the competitors for hardware components and were catching up fast by approaching the customer from a direct service angle.

> The years 1980 to 1985 were critical in the evolution of the U.S. data processing computer industry and, indeed, of the larger U.S. electronics industry. In that half decade, when the Japanese consumer electronics industry had all but completed its conquest of world markets, the Japanese computer industry had all but destroyed the U.S. memory chip business, shutting down the main facilities of Intel, Advanced Micro Devices, and National Semiconductor, the core enterprises of the Silicon Valley Nexus. However, the introduction of the microprocessor at the same historical moment swiftly recast the industry worldwide. It permitted the U.S. industry to meet the Japanese challenge and to retain its position of global dominance.
>
> (Chandler and Mazlish 2005: 117)

In 1986, sales of PCs had increased to ten million in the USA and reached a level of 44 million six years later (McNeill and McNeill 2003). Similar developments also took place in other parts of the industrialized world. Compaq masterfully introduced the procreation cycle, i.e., "moving a product from research to design to development to manufacturing to sales and ultimately to profits and then going through the whole cycle again" (Friedman 1999: 173). NEC, the Japanese computer giant, based its manufacturing on the orders received and designed a pull strategy based on demand instead of clinging to the traditional push strategy based on product forecasts.[17] The appearance of PCs on a multitude of desks within corporations required the creation of internal enterprise networks to connect individuals within and between departments. The increasing demand for PCs kept the US computer industry, including IBM and its clones, Microsoft and Intel, in a dominant position (Chandler and Mazlish 2005).

In the early 1990s, however, IBM lost its comfortable market position at a very rapid rate. In April 1993, Louis Gerstner was appointed as CEO and had the task to transform the "elephant" IBM in such a way that it could "dance" and regain its competitive position (Dittrich *et al.* 2004). It mastered its survival, but new challenges followed one after another with an increasing speed. Competition increased as PC manufacturers were forced to include the latest-generation microchips in their PCs in order to keep up with the competition. Each product generation introduced a new era of computing due to the large increase in processing power. Some began to suppose that the large-scale corporations would be too ossified to respond adequately to the continuous stream of new challenges.

Yet IBM, with its large size, remained stunningly flexible. Two American psychologists, James Collins and Jerry Porras (1997) looked at IBM and found that it continued to survive due to specific organizational capabilities. Through an intensive management development process, they compared a number of big American corporations, some of whom had a strong focus on historical continuity and a particular sense for the development of particular capabilities. They asserted two perspectives to identify the main characteristics of some of these American corporations: they were either "clock-builders" or "time-tellers."

They determined that IBM was a clock-builder. The concept of a clock-builder describes an organization as an entity "with a strong, cult-like as culture that transcends dependence on the original visionary founders" (Collins and Porras 1997: xix). Clock-builders are organized on the principle of a simultaneous drive for progress and change. Firms that have been able to succeed were driven by renewal but continued "to preserve the core and stimulate progress" simultaneously (ibid.: 82).

Clock-building companies excel in "developing and promoting highly competent managerial talent from inside the company, and thereby attain greater *continuity* of excellence at the top through multiple generations" (ibid.: 34). Setting such goals as progress and change did not imply that these goals had to be fully reached, but they certainly reinforced the level of commitment to realizing them, creating tangible mechanisms aligned to preserve the core and stimulate progress. Clock-builders concentrate primarily on building an organization that will prosper "far beyond the presence of any single leader" (ibid.: 23). The perseverance of the organization is "the company itself" which is truly achieved by clock-building leaders. In 2004, IBM announced the selling of its whole PC division to the Chinese computer company Lenovo, the largest Chinese PC manufacturer in 2005, to create a new worldwide PC company. Yet IBM took 20 percent equity stake in Lenovo, creating a strategic alliance, which made Lenovo the preferred supplier of PCs to IBM.

On the contrary, a corporation which relies on a charismatic visionary leader alone is, according to the same authors, a "time-telling" corporation. Collins and Porras describe time-tellers as companies which primarily concentrate on having a great idea and a great leader – usually an outsider – at a certain time, which unfortunately might become obsolete at some other point. A time-teller might hit a market just right with a visionary product idea, exploit the growth curve of an attractive product's life cycle, and capitalize on the market opportunity. They suggested that it is better to build a visionary organization than to rely on a leader with a mere vision.

Compared with IBM, Burroughs, founded in 1892, initially had the technological lead, and was such a time-telling corporation. Yet, eventually it decided to focus primarily on older lines of accounting machines without massively investing in computers. Neither did the company invest in the provision of corporate values like IBM did. The company president, R.W Macdonald, stifled individual initiative which caused talented employees to leave the firm (Collins and Porras 1997). Although Burroughs was seen as a "mainstay of the office machine industry," the company can be seen as a "time-teller" organization as it relied too much on the inventor William Burroughs himself.[18]

Outlook

The information-processing technologies launched a digital revolution, which embarked on the highway of electronic globalization and fundamentally reshaped the functioning of large-scale corporations operating in ever-expanding markets. A global information superhighway was born. By 1997, an estimated 75 million people around the world were on the Internet (Cairncross 1999). Search engines such as Google further introduced remarkable innovations which stimulated the applications of computers and created an increasing demand for Internet connection. Submarine cable systems utilizing satellite and optical fiber infrastructures transformed communication technology, increasing the rapidity of establishing relationships on a global scale.

This new era of globalization, i.e. electronic globalization, which was dawning is usually described as an acceleration of the transmission of information across borders which has subsequently reinforced an intensification of the movement of capital, goods, and people. Castells (1996) has articulated a vision about this process of electronic globalization in terms of leading to the network society. Today's era of globalization is built around falling telecommunication costs – thanks to microchips, satellites, fiber optics, and the Internet.

With these new technologies, the global world seems to be woven together even tighter and operates in real time, or timeless time, according to Castells. Information networks describe the full potential of what information technologies can attain to speed up information flows but seem to primarily serve the interest of financial institutions, ICT companies, and other multinational corporations. Information networks shape a new artificial space in which a constant stream of data flows freely. ICT creates a non-territorial region in the world economy, a "decentered yet integrated space of flows, operating in real-time, which exists alongside the spaces-of-places which are called national economies" (Ruggie 1993: 172). The primary language of the Internet has become global English. The tremendous development of ICT has led to a historical shift from industry to services and from goods-handling to information-handling activities.

Japan has been highly involved in this development as well. Japanese computer companies such as Fujitsu evolved into a serious competitor and, in the early 1990s, became the world's largest computer maker behind IBM (Chandler and Mazlish 2005). The Japanese government initiated programs for improving manufacturing capabilities and the quality of products to challenge American dominance in the computer industry as much as in the car-manufacturing business. Japanese firms in the manufacturing of computer memory products, for example, produced chips of a markedly higher quality than those of their American competitors (Langlois and Steinmueller 1999).

However, the reshaping of the organization and a different style of management in Japan did not remain unnoticed. With the introduction of the computer, large scale American corporations mainly targeted on downsizing without changing their management approach. Top managers primarily realized that, with using computers, any corporation could accomplish the same amount of work with significantly fewer workers. To understand what the consequences were of such a traditional management approach in many American firms, we must shift our attention to the Japanese miracle which reshaped common perspectives about the coordination and control of management in internationally operating corporations.

12

THE JAPANESE MIRACLE AND JUST-IN-TIME

During the glorious 30-year period of the supremacy of the American economy, a Japanese miracle was dawning. Although the USA was able to maintain its world economic leadership, in the 1990s, Japan became the second largest economy in the world. Unlike the European miracle of the sixteenth century, the Japanese miracle was not based on international trade and explorations of unknown territories, although they did occupy Korea and Taiwan.

Instead, the Japanese miracle was based mainly on internal transformations facilitating a surprising economic progress which turned the country into a dominant global player following the Second World War. The role Japan began to play in the global economic arena was primarily achieved by the successful introduction and development of a new "just-in-time" production system. To properly apply these methods, firms needed to be assured of the quality of the spare parts and semi-finished products that suppliers were providing so as to guarantee the manufacturing of high-quality final products. To attain these successes, the Japanese society went through a few remarkable transitions before it could reach the state of a global economic player.

This chapter, therefore, addresses this Japanese miracle and the manner in which the car manufacturer Toyota introduced the concept of *timing* as a novel version of the application of clock time, which then paved the way for flexible production systems and a flexible workforce throughout the Japanese economy. Their advancement showed that speed alone was no longer a guarantee for success. Timing became the key to further advance the globalization of manufacturing, commerce, and trade.

Until 1853, Japan had maintained very little contact with the Western world. It had long cherished a self-imposed isolation from the West. When American ships approached the Edo Bay to force Japan to open its markets for trade, it felt too weak to resist and surrendered. The aggressive rush of Americans and Europeans into the Japanese market left deep marks and changed the power structure of the society.

In 1868, the Meiji government launched fundamental institutional changes and facilitated an industrial revolution (Bernstein 1997b). The country demonstrated a willingness to learn from abroad and adapt the most successful organizational structures and methods into Japanese practices. It began to reform the country's social and economic system to catch up with the West under the banner of "rich country, strong army" (ibid.). The government had at its disposal a "development fund," i.e., money taken from samurai aristocrats in exchange for government pension. This facilitated the government's ability to spark investments in shipyards, munitions, and railroads.

The principal Meji railroads, for example, resembled their Western counterparts. Full-time salaried managers were recruited initially from the government bureaucracy, but they soon came directly from universities. Most Japanese railroad companies patterned their administrative structures on the British model of centralized departmental organization. However, the average speed of trains lagged behind the Western counterparts. As a British traveler remarked in 1898: "all notion of speed, haste, or flurry are utterly foreign to the Japanese nature" (quoted in Ericson 1989: 62). By 1906–1907, the railroads were nationalized and efficiency improved.

The roots of contemporary Japanese management were not to be found in the railroads but instead in the nascent *zaibatsu*. These were groups of businesses owned by capitalist families and could count on large support from the government.[1] Japan's rapid industrialization was determined by favorable economic factors such as the buoyant world demand for silk. "By introducing quite small-scale technological improvement in silk weaving and processing, Japan was able to maintain her external balance of payment in reasonable shape" (Bayly 2004: 181–182). The mechanization of cotton textile production also expanded rapidly in the 1880s.

Time concepts

With the westernization of Japanese institutions, new values concerning time were introduced. During the Edo period, Japan had been using a seasonal or non-standard time system in which day and night were divided into six equal hours. Clock time varied in summer and winter. This agrarian sense of time was merged with Confucian moral teaching intended to keep farmers adhering to their proper calling. Under the influence of the British Empire, indigenous temporal orders began to collapse. Shortly after Japan had been thrown open to Western influences, clocks of a modern type were imported to the country and equal hours were introduced.

During the Meiji restoration, a series of systems involving time was reorganized according to the Western model. In the 1880s, Japan adopted two Western systems of temporal measurement: the system of international meridians and the use of the Gregorian calendar. The Meiji government adopted the Gregorian calendar to underline its international political orientation and to westernize its time system.[2] Until that time, the Japanese had no proper word for the hour or arriving at a specific time. The very term that signifies time, *jikan*, first appearing in an elementary school reader, was invented to replace *oki*, which indicated time in the temporal framework of the lunar calendar and the non-standard time system.

While the idea of grouping seven days into one week was alien to the Japanese, the concept of a week and its Japanese translation *shu* were also introduced. In order to train the Japanese to the new regime, new practices were initially introduced at school, and children at school were trained to respect the new temporal order. Compulsory elementary education entertained the youngsters to respect time regimes and punctuality. "The word at that time to describe this adoption of Western systems and values was *civilization*" (Ikuko 1997: 245). The indoctrination of punctuality, regular attendance, and diligence was critical for the future development of large-scale industry.

In 1872, the government built the first government-run factory to industrialize textile production. Textiles were primarily produced for export and, with the foreign currency these exports generated, the government was able to buy the necessary equipment for the development of its heavy industry. The workers were forced to labor long hours for low wages. Ninety percent of the Japanese labor force in the textile industry was made up of single females from farm villages. The atmosphere in these factories was initially undisciplined, and Japanese managers designed tricks similar to those of their British counterparts to cheat time. Stick-and-carrot tactics were put in place to extract maximum labor output from the female workers. The temporal consciousness associated with the early stage of Japanese industrial production required an abolishment of the traditional rural sense of time.

In 1884, a collection of national development plans *Kogyo Iken* (A Proposal for Economic Development) was published to further imitate the West (Bernstein 1997b). In order to foster a modern textile industry and offset the British textile import, the Japanese government invested in modern factory building and sold machinery to entrepreneurs at low prices. In 1894, the government ordered its functionaries to come to work in Western dress (Bayly 2004). Morosely, the westernization did not develop without interruptions, conflicts, and turmoil.

In the aftermath of the Russo-Japanese War of 1904–1905, many companies attempted to rationalize their factory operations, which led to industrial instability. Japan was confronted with serious labor unrest, which spread through many industries. To dampen labor resistance and avoid strikes, some industrialists resorted to strategies very similar to those used in Great Britain and the United States. Scientific Management became popular in the Japanese business community, but industrialists achieved only limited success in their efforts as Japanese workers resisted. They viewed it as a technique for speed-up and saw it as an infringement on workers' prerogatives. Control of the shop floor remained mainly in the hands of skilled workers (Blackford 1998).

Though the Japanese government recognized the right of craft unions, it managed at the same time to become a powerful imperialist nation by acquiring Korea and Taiwan as colonies and defeating Russia in 1905. During the Russo-Japanese war, the demand for looms increased and Sakichi Toyoda (1867–1930), an active inventor in the textile machinery business, became obsessed with improving handlooms. He gained his first patent in 1891 and founded the Toyoda Company where he continued inventing (Bernstein 1997a). After a visit in 1910 to the USA, he decided to establish a cloth mill, Toyoda Automatic Weaving. World War I increased the demands for textiles, and the number of orders at Toyoda Automatic Weaving called for expansion. In 1918, the new Toyoda Spinning and Weaving Company invested in China to avoid tariff payments on the produced cotton goods (ibid.).

After even more improvements, Sakichi Toyoda introduced the G-loom in 1924, which was the first automated loom with a non-stop automated shuttle change. Sakichi, together with his son Kiichiro, installed this loom in their Toyoda Automatic Loom Works in 1926. The new machinery was such a success that Platt Brothers, the dominant foreign textile machinery manufacturer in Japan, signed a licensing agreement with Toyoda in 1929 (ibid.). Japanese weaving firms were eager to install the new invention in their own factories.

With the development of *zaibatsu*, big business arrived in Japan, and bureaucratic management systems were adopted. Large and small firms increasingly began occupying different spheres of manufacturing with small companies either acting as subcontractors or producing as independents for niche markets. The Japanese government supported the development of the textile machinery, as it did in so many other industries, by promoting the establishment of *zaibatsu*. They became important factors in Japan's business life as they expanded the scope of businesses but remained decentralized in the management structure. *Zaibatsu* established new firms for which they retained the coordination with other firms, but the firms remained independent, small, and undiversified (Bernstein 1997b). Basic scientific and engineering technical knowledge to build an industrial infrastructure had to come from the West.[3]

Entrepreneurs were also beginning to show a growing interest in car manufacturing. The mechanical engineer Kiishiro Toyoda went to the USA and Europe several times to visit textile mills, during which time he became fascinated by automobiles. The Japanese automobile market was small and dominated by Ford and General Motors (GM). In 1929, there were 81,000 vehicles on Japanese roads, an amount similar to a weekly production output of the US car manufacturers together (Bernstein 1997a). Yet Kiishiro Toyoda believed there were opportunities for a locally made car. His brother-in-law Risaburo, the president of Toyoda Automatic Loom, supported this belief and, by 1933, a separate automobile department within the company was created (ibid.).

Relying on local mechanics with no experience in car manufacturing, they created cars by disassembling Western automobiles. The first Model A1 passenger car, launched in 1935, borrowed its body design from Chrysler but also contained parts from Ford and GM's Chevrolet. To minimize any reliance on foreign truck manufacturers, the Japanese government issued the Automobile Industry Law in 1936 (ibid.). With one stroke of a pen, GM and Ford became deprived of further operations in the Japanese market. Although the two companies attempted to form joint ventures with Japanese car manufacturers such as Nissan, the Japanese government refused these initiatives.

In 1937, the Toyota Motor Corporation was established. The first mass-produced vehicle, however, was not the A1 but instead were trucks ordered by the Japanese military (ibid.). During the Sino-Japanese war, the government became the biggest buyer of Toyota products, and when Japan entered World War II, the demand for trucks by the Japanese military boomed.[4]

Since the military had taken control over the whole Japanese economy, smaller firms and large companies were forced into cooperation. The large firms had to pay close attention to their suppliers and vice versa, as both parties depended on the military, which led to close subcontracting relations (Bernstein 1997b). Toyota formed the Joint Prosperity Association (*Kyoryskukae*) in 1943 in which all its suppliers participated. But due to the fact that Toyota was not assured of a proper quality of the required parts, it decided to form its own supplier network by turning previous Toyota workplaces for spare parts into independent new companies. Together these new firms would create the Toyota Group in the 1950s.

Reconstruction of Japanese society

After World War II, Japan became occupied by the Supreme Commander of the Allied Powers (SCAP) which took control of the Japanese economy and intended to defeudalize, demilitarize, and democratize the state (Bernstein 1997b). As General MacArthur, who was in charge of SCAP, wished to be able to speak directly with the Japanese people, Western Electric was asked to help re-establish the telephone, telegraph, and radio broadcast systems throughout Japan. Western Electric, AT&T's subsidiary that developed and produced its equipment, had established the Nippon Electric Company (NEC) in 1899. It had produced radio tubes, receiving sets, and broadcasting equipment but needed support to re-establish its production facilities in order to meet postwar challenges.

A small Japanese company formed by A. Morita and M. Ibuka in 1946, called Tokyo Telecommunications Engineering Co., became involved in re-establishing the radio industry. In 1953, they produced their first audio magnetic tape recorder and were licensed by Western Electric to produce the transistor. This company would later become Sony, a prime mover in electronic tape-based products and world pioneer in electronic miniaturization. Another important player in consumer electronics was Toshiba, whose foundations go back to 1885 when Shibaura Electric was established to build electrical telegraphic equipment. Since the 1920s, General Electric had transferred corporate knowledge and technical knowhow to this Japanese protégé. In the 1950s, Toshiba moved into electrical machinery and light office machines, finally moving into computers. By the late 1960s, NEC would become "the largest exporter of radio and microwave communications equipment and a world leader in digital switching and mobile communications" (Chandler and Mazlish 2005: 200).

With the outbreak of the Cold War, communists taking control of China in 1949, and the Korean War in 1950, the USA no longer had any interest in keeping Japan in a weak position. The American government supported the initiative to revitalize Japan as a bastion of democracy and capitalism. American military expenditures in Japan, especially for the Korean War, allowed the Japanese to finance a major increase in their imports.[5] The new American policy of support and revitalization was helpful in developing Japan into a partner state.

The Liberal Democratic Party was formed in 1955 and, while staying in power for the next 40 years, it created political stability, which provided the appropriate circumstances for rapid economic growth (Bernstein 1997b). The government enacted the Economic Planning Agency (EPA) in order to construct long-term economic plans. It also became responsible for the administration and planning of policies concerning price stability (Ikeo 2000). The Ministry of International Trade and Industry (MITI) that had been established in 1948 was entrusted with improving rationalization processes in the basic industries. MITI's mission was to encourage production in key industries without too much consideration of their profitability (ibid.). Nevertheless, the country quickly became included in what Dowrick and DeLong (2003) have called the "club of convergence."[6]

Although profitability might not have been an overarching concern, an increase in productivity was thought essential but could only be established when there were balanced relationships with labor. In the first post-war years, there was abundance of labor when production for the army had stopped and many skilled workers were available on the labor market. Japanese union membership exploded, and an industry-wide car workers' union was formed in 1947.

In the early 1950s, Japan experienced an upsurge of labor militancy which forced the automobile companies to depart from a Fordist style of mass production and introduce a multilayered subcontracting system that "simultaneously allowed them to guarantee employment to a core labor force while obtaining low cost inputs and flexibility from the lower rungs of the supply network" (Silver 2003: 42). Labor management relations evolved into a system based on two principles introduced by the Labor Management Consultation System: the consultation system sets the rights and responsibilities for both parties and collective bargaining determines the final outcome. Japan's labor unions became enterprise based and not industry based, which implied that both blue-collar and white-collar employees are represented in the same union which created tighter collaboration between management and union (Minoguchi 2000).

Next to the enterprise unions, which were important as a platform of dialog between labor and management, Japanese companies introduced permanent employment for core workers and seniority-based wages as a basis for stability. Business leaders became eager to obtain this stability as they believed it to be essential for high-speed economic growth, and labor received increased security and wage benefits. However, these practices were usually not extended to workers in small firms. The new policy tried to stimulate the productivity movement, which was established and supported by the government. Managers were regarded as the people who should treat enterprises as public institutions, while the owners were supposed to suppress their self-interest and direct their firms for the benefit of public welfare (Nishizawa 2000). Within this context, an increasing interest for quality became apparent.

The quality issue

In the late 1940s, a few members of the Japanese Union of Scientists and Engineers (JUSE) took the initiative to study technological development in the United States. This group called themselves the "Quality Control Research Group." After their return to Japan, they mobilized JUSE to improve manufacturing processes and invited a number of American quality experts to Japan. The main interest was to improve output and reduce overall costs by managing all available means based on a process-driven focus. Senior managers of the Japanese consumer electronics industry were invited to meet representatives of Western Electric and discuss some of the main issues of management including organizational control and quality control.

W. Edwards Deming (1900–1994), who had been working for Western Electric with Walter Shewhart,[7] left the company to teach statistics at the graduate School of Business at NYU, and then was invited to go to Japan in 1951 to lecture about quality control issues and improve the capabilities of any process in organizations.[8] Deming had designed Statistical Process Control charts as a key method for identifying special causes for a lack of quality. The charts assisted in the diagnosis of quality problems and enabled the production process to be brought under control. By using statistics, Deming could reveal "how managers mistakenly blamed systemic problems on workers who have no ability to fix them" (Hoopes 2003: 202).

While in Japan, Deming stressed that good quality and uniform quality had no meaning except with reference to the customers' needs. Deming's policy was to eliminate irrational tyranny in the daily practices of production and stressed that production processes involve a continuous change in time, a progressive motion from one point to another in a continuous development toward a specific end. To achieve a systematic focus on these processes, training in understanding these processes is required to stress the procedure through which results are obtained, not the results themselves (Locke 1996). Deming emphasized process as a fundamental concept which is opposed to a policy of individualized performance measurement. He advocated making improvement in the process in which the individual's work is immersed, not trying to eliminate individual "mistakes."

After his return to America, Deming sent a copy of Joseph Juran's recently published *Quality Control Handbook* to the JUSE (Butman 1997).[9] Juran had been recruited by Western Electric in 1921 where he had developed a career in quality issues. He worked at the Hawthorne Works in Chicago at the inspection branch and, after obtaining a law degree, moved up the career ladder and obtained a position as manager responsible for job evaluations, time studies, and wage and salary determinations at the headquarters of Western Electric. His slogan about quality was *fitness for use*, a label which stressed the respect for the needs of customers and a provision of product satisfaction (ibid.).[10]

Juran journeyed to Japan in 1954 to help spread the quality control idea in Japanese industry. He was invited for a seminar, toured factories in Tokyo, and listened to the issues raised by senior executives. He noticed that the working pace was greater than he had seen in plants in America. Juran's principal message was that quality was as much about management as it was about statistics and control charts, cycles, and surveys. He told his Japanese audience that they personally had to take responsibility for quality and that they must become involved in setting their own quality policy in developing products, in organizing for quality, and in making sure that quality was actually being achieved. Juran's inspiration was like a "welcome rain" to JUSE's crusade for the improvement of quality (Tsutsui 1998). Whereas Deming stressed statistical quality control, Juran taught the audience about total quality control as a responsibility of all employees of the organization. Although initiated in the consumer electronics industry, the ensuing movement for the improvement of the quality of products, however, found its most receptive ground in the car-manufacturing industry.

Quality and just-in-time at the Toyota company

In the 1950s and 1960s, the Japanese economy showed high-speed growth. Investment in plants and equipment rose 22 percent per year, and Japan's GNP grew by an average of 10 percent annually. Economic expansion opened new opportunities for business. Smaller firms obtained the possibility to grow in size, especially companies in heavy industry such as car manufacturing, chemicals, and steel, which expanded more rapidly than those in light industries. Legislation limited imports into Japan and restricted foreign investment. Large Japanese manufacturers were able to expand rapidly due to the increased use of subcontracting. By 1971, almost 60 percent of all the small industrial companies in Japan were acting as subcontractors for larger companies. The linkages between industrial firms and their subcontractors gave Japanese firms advantages over many of their foreign competitors.[11]

The Japanese government launched a campaign for the reinforcement of the domestic automobile industry. MITI's policy of protectionism favored the growth of Toyota and Nissan. Compared with Nissan, Toyota had a few advantages. Its plants were located in rural areas where land prices were cheaper. The plants and suppliers were concentrated in specific areas, and its factory union was more obedient than Nissan's. Moreover, Toyota put more effort into organizing the cooperation with its suppliers (Hong *et al.* 2009).

MITI also enacted a policy for Japanese car producers to learn from foreign firms and set up joint ventures. Toyota began negotiations with Ford, and Kiichiro's cousin, Eiji Toyoda, traveled to the USA to examine the car production process at Ford. After Kiichiro Toyoda's death in 1952, Eiji took over. The Japanese economy was growing, and the demand for passenger cars increased quickly. In 1955, Toyota introduced the Crown, and the Corona came two years later. Once it had taken the lead in Japan, Toyota entered the automobile market of the USA by exporting the Crown, but the results were continually disappointing.[12] To supply the booming national market, it opened a new plant with a monthly capacity of 5,000 vehicles in 1959 (Bernstein 1997b).[13]

To increase exports, Toyota needed to ascertain how to couple variety with low-cost volume production. Toyota solved the problem differently from the American car manufacturers. Taiichi Ohno (1912–1990),[14] who began his career at Toyoda Spinning and Weaving in 1933, had been working for Toyota since 1943. He shared Kiichiro's aversion to waste and had a sense of selecting what is necessary and what is not in the production process. Since accumulating inventory increased costs, Ohno – who was impressed by what he had seen at the Ford Motor Company – decided to design a no-inventory, continuous-production system which was geared to customer demand. He stressed the importance of just-in-time (JIT) for the identification and avoidance of waste in the shape of idle labor rather than of raw materials.

In 1947, Ohno introduced the practice of workers having responsibility for more than one machine, each working in a so-called L-layout at the Toyota machine shop (Driel and Dolfsma 2009). The company called this multi-machine handling "autonomation" (automation with a human face). [15] "Japanese management, engineers, and workers grew accustomed to thinking of the entire production process as a system and of thinking in an integrated way about product design and process design" (Freeman and Louca 2005: 280). Toyota increasingly used multifunctional workers, restructuring jobs so that their operations became more versatile but also demanded less skill than before. Toyota's policy was geared more toward considerations of reducing any reliance on skilled labor and less to considerations of scarcity of material resources.

Actual demand determined the pace of production. To introduce variety, the production line had to be stopped to allow for sudden changeovers needed for different products. The success of this strategy, therefore, depended on minimizing waste in terms of quality and in the time required to make the necessary changeovers since downtime reduced volume and raised costs. By 1954, quality had become an integral part of management, and quality-control circles were introduced by 1962 to fine-tune the sequences in the production process from planning to execution with final delivery to the customer in the most detailed way.

> The central production planning office drafts quarterly and monthly production plans for each factory based on its market demand forecast and presents corresponding procurement plans to outside suppliers. These prior plans provide only a general guideline, however. The integrated production-delivery plan for a ten day period is prepared by the commodity flow office on the basis of orders from regional and overseas dealers. In response to this plan, the engineering office of each factory prepares a sequence of daily production schedules. The daily schedule is then adjusted two days

> prior to actual production in response to actual customers' orders transmitted from dealers to the factory through the online network system (the "daily adjustment").
>
> (Aoki 1990: 4–5)

Toyota's main focus was to reduce total costs by consuming no more resources (material, labor, supplies, power, capital, and the like) than necessary to produce what is needed to serve customers (Johnson 1992). Toyota introduced a flexible production system, which could convert the line from one type of component to another. Quick conversion presses – so-called Quick Die Change (QDC) – reduced changeover times to a fraction of those that were customary in Western car manufacturers. Japanese companies focused on single-minute exchange of dies when setting up or changing machines.

In 1982, Toyota Motor Manufacturing and Toyota Motor Sales merged to become the Toyota Motor Corporation (TMC). The main reason for this merger was that manufacturing could produce a car in two days, whereas sales needed from 15 to 26 days to close the sale, transmit the order to the factory, get the order scheduled to the factory, and deliver the car to the Japanese customer. The new corporation wanted to become more driven toward marketing (Stalk 1988). By organizing teams combining product engineering, manufacturing, marketing, and purchasing throughout the development process of a new car, which allowed them to make decisions themselves, enormous reductions in time and expenses were attained.

Unlike the American system in which production schedules reflecting expert market forecasts *pushed* cars through the factory in long-run, big-volume spurts, the Japanese system was an actual customer demand system that *pulled* the product through the factory. Therefore, instead of long standardized production runs on the final assembly line, cars with a variety of shapes, colors, transmissions, and engines rolled one after the other along the line "seemingly at random." In response to the customer orders flowing on computer printout into the factory from dealers, a JIT delivery system from suppliers, which avoided inventories of parts, and a total quality control which eliminated the losses involved in the withdrawal of defective cars from the line, kept production costs to a minimum (Locke 1996). JIT production systems increased labor's workplace bargaining power because it increased the vulnerability of the production system to disruptions in the flow of manufacturing. The Toyota model offered employment security to a core labor force in exchange for cooperation, but it also created a larger buffer of less privileged workers without the same rights. In that sense, the model had a dual labor policy of core and periphery workers and became lean in the sense that, with a core of highly committed employees, transitions could be implemented quickly: lean and dual (Silver 2003).

The new system required the closest cooperation and coordination within and between groups involved – and a highly motivated workforce. This was true in preproduction (making the production system) and in production (constantly making improvements on the floor to the existing system) where customer orders, because of their variety, required a highly flexible response. Highly skilled and knowledgeable engineers were involved in the chain of production activities and continually improved the process. Their presence on the factory floor resulted in a remarkable superiority in reducing "average setup time" which was necessary to achieve manufacturing variety. These Japanese engineers worked closely together with production workers as well as preproduction workers and people in sales. Because Japanese engineers not only developed and designed products but also went to the shop floor and worked with regular employees to solve problems, they gained first-hand knowledge of the manufacturing process and obtained the quick feedback necessary for close and continuous study of the relationship between machine design and quick change processes (Locke 1996) This allowed the Japanese production people to achieve their great success in manufacturing. Keeping costly downtimes to a minimum permitted continuous improvement, fine-tuning, and product variety.

When Juran revisited Japan in 1966, he was astonished at the industrial progress. He was introduced to the *quality control circle* with which he was not familiar. Nothing quite like it existed in America: It was a Japanese invention. The quality control circles, small groups composed of workers, had achieved substantial cost savings and quality improvement in companies throughout Japan. Its membership was "voluntary," but there was an implicit pressure for workers to participate, and some of the work preparations and discussions were conducted outside of regular working hours.

To Juran, the quality control circles demonstrated that troubles with quality control had nothing to do with indifference among the operators. He decided to introduce these Japanese ideas to his colleagues in Europe and, during a conference of the European Organization for Quality Control held in Stockholm, he referred to his experiences. Juran saw himself as a "renaissance man" in the subject of propagating quality and promoted the necessity of "quality mindedness" among all employees and managers of a company (Butman 1997: 186). The Japanese government supported this view of quality mindedness and introduced a new policy, Creating Human Economic Society. The overarching goal was to improve Japan's international status in the world (Ikeo 2000).[16]

Due to its quality focus, Toyota achieved a tremendous advantage over Western car producers. In 1985, its export of cars had reached 20 million units. While in the USA the errors and defects were usually detected and caught at the end of the assembly line, at Toyota workers could stop the assembly line immediately if an error occurred that they could not repair. Employees could immediately appeal to colleagues instead of a supervisor and prevent any breakdown. This meant a big advantage in cost saving (Adler *et al.* 1997) and promoted a flexible production system. This form of production gained its power from the fact that inventory holding costs were cut down by reducing stock to a minimum through sophisticated systems of demand-oriented supply. The result was the development of lean production with teams of highly qualified members, flexible production facilities, and a well-coordinated flow of information (Womack 1991). This teamwork model pushed decision-making authority as far down the managerial ladder as possible (Rifkin 1995). Workers were even asked to reprogram the production line when needed. Whereas Western manufacturing firms usually needed a changeover night shift to implement a change, a Japanese car plant was able to reduce – "to the astonished eyes of Western observers" – the change to within 15 minutes (Jürgens *et al.* 1993: 65).

Toyotism attained its objectives mainly in a team-based "lean and dual" manner. The overall success of the Japanese way was not only related to technology but was as much based on an improved work organization. Changeover time flexibility was the result of an efficient reorganization of manpower deployment based on process flexibility.

> While change-over flexibility characterizes the capability to quickly change dies or transferring and clamping devices which are principally model or component specific, process flexibility refers to running tools or work pieces independent from the mode. Components no longer have to be combined into homogeneous lots. They can be processed in chaotic sequence.
>
> (Jürgens *et al.* 1993: 66)

With processing flexibility and flexibility of retooling, a higher level of flexibility was reached than with changeover flexibility alone. In American companies, flexible transfer lines could only handle differing components in a fixed sequence, an activity which was sometimes executed by industrial robots.[17]

Japanese firms developed computer systems which tied their sales people directly to the factory scheduling operations, which cut delays in sales and distribution and improved customer services (Stalk and Hout 1990). They developed the same practices for the supply of spare parts which were provided by their outsourced suppliers (Iyer *et al.* 2009). The overall responsiveness of these companies, therefore, improved nationally as well as internationally. During the 1980–90s, an emerging network production system was created between suppliers, manufacturers, and sales departments, referred to as the Toyota Production System (TPS). This system turned into a learning model and led to the movement called lean production. It was the second massive transformation, displacing mass production techniques, which had previously, in the first transformation, replaced craft-based production during the early twentieth century. The aim of TPS was to take waste out of value streams (Lander and Liker 2007).

The pinnacle of TPS was built on four main pillars that have been developed and perfected through the years:

- JIT production, which requires close collaboration within a network of core suppliers;
- a team concept amongst the employees to work at different workstations;
- the "zero defect" quality focus requiring an active real-time problem-solving ability from the employees;

- standardized work and *kaizen* (continuous improvement) which stimulates involvement of the workers, who were encouraged to share their ideas and suggestions for changes of the working process with others.

TPS had a serious impact on the use of human resources. Among the core workers these principles increased employee involvement and empowerment (Adler *et al.* 1997). Employees obtained control over issues such as the layout of production, personnel transfers, and the execution of tasks. In this way, the speed of model change became a competitive advantage of the Japanese car manufacturing industry (Womack 1991).

Working in teams also made it possible for employees working at different workstations to acquire knowledge about the production process in general. TPS became the standard in many Japanese companies. A Japanese manager at Honda once remarked that the Hawthorne experiments and Management by Objectives (MBO) had looked at human behavior from the wrong perspective. They should have been based on the idea of empowering workers, placing responsibility closest to where the knowledge resides. Instead, the MBO approach turned out to be an "inhibitor of learning" (quoted in Pascale 1990: 251).[18]

Ohno had discovered that most waste in an organization was related to the waste of human intellect (Watson 2009). The reduction of that waste was essential for improvement of the products the company delivered, which generally were already of high quality.

> Toyota's decision to follow Deming's philosophy was one of the most influential of the twentieth century. While Western companies used costly, large, and unhappy workforces to produce gas guzzling cars in the 1970s, Toyota was forging ahead with implementation of Deming's ideas.
>
> (Crainer 2000: 200–201)

Nonetheless, in 1982, Eiji Toyoda said to the head of the Ford Motor Company, Philip Caldwell, that Toyota had learned its famous lean production system from observing Ford's plant at River Rouge.[19] With JIT, the Western concept of time in terms of speed and acceleration had found a new translation and had undergone a serious process of Japanization. Although JIT initially suited the auto manufacturer, its application was soon widened to cover electronics and most other industrial sectors (Taylor 1999).

American response

It took a crisis to propel quality out of the factory and into the minds of America's senior managers. That crisis was the rise of Japan as an industrial competitor. The basic method of manufacturing automobiles had changed little between 1913 and the early 1970s. Car manufacturing had long been considered "the industry of industries" just like the textile industry in the old days, and the management approach in this industrial sector had long been an implicit point of reference for many. Until the 1970s, American management in general was not much interested in the quality issue. Whereas Deming's ideas were rejected by American management, he rose to prominence in Japan where he became closely involved in the development of quality issues. His methods were widely taken up there, and the "Deming prize" for contribution to quality was launched.[20]

In his later work, Deming seriously criticized the unwillingness among Western managers to seriously learn from the Japanese approach and codified the necessary shift in management mentality in seven fundamentally mistaken beliefs (the Seven Deadly Sins) about bad management practices such as a focus on short-term profit, job hopping, and performance appraisals which should be eliminated before Western styles of management could successfully implement any quality initiatives. His insistence on the desirability of continuous improvement, a long-term orientation, and the prime role of the service sector were hardly listened to. He stressed that if freight carriers would improve their services and decrease costs through containerization, these cost savings would then be passed on to the manufacturers and to other industries which would help American businesses to improve their market performance. But Deming was still seen as an obscure statistician. Despite the fact that he continually pointed at the human side of the quality issue and stressed again and again that most of the quality problems belonged to management, his ideas were not grounded in the Human Relations approach.

With the oil crises of 1973 and 1979, things began to change and worries about the decline of American business increased. During those years, the intensification of international competition and an increased volatility of international markets forced American multinationals – in particular the car manufacturing industry – to internally change their Fordist strategies. International operations were restructured to drive down costs and improve supply responsiveness. The Big three American car producers – Ford, Chrysler, and GM – were suffering from losses, and the rise of the dollar made it difficult for them to compete with Japanese competitors.

They initially had turned to the American government asking for help to cope with the "Japanese threat," a common label in the auto industry at that time. In 1981, the first voluntarily agreed upon Export Restraint Program was introduced to limit Japanese car import to about a 20 percent share of the market (Bernstein 1997b). Yet this was seen by many as only a temporary solution. While the National Broadcasting Company (NBC) was searching for subjects for a documentary about this decline and the "Japanese threat," it came across Deming, who told them that he was much honored in Japan. The NBC program evolved into a television documentary called *If Japan Can, Why Can't We*, which was broadcasted on June 26, 1980. The program focused on the issue of productivity both in Japan and in the United States. It was stressing that Japan was beating America and that something was fundamentally wrong with the United States. Deming featured prominently in the documentary and pointed out that it was essential to work smarter instead of longer.

The Nashua Corporation was the first American company to adopt Deming's quality management principles. Deming's prioritization of management before technology represented a reversal of the attitudes of many managers. The NBC documentary marked the turning point between Deming's academic obscurity in the western world and public fame. The documentary turned into a wake-up call. A few days after the broadcast, R. Hayes and B. Abernathy, two Harvard Business School academics, published "Managing Our Way to Economic Decline," a controversial article in *Harvard Business Review*. The message was clear: American managers had forgotten to direct their attention to the production of superior products. The authors insisted that customer orientation was a vital way to revert the negative trend of decline. The advice was to look to the East. Deming became a frequent speaker and a sought-after adviser to some of the nation's largest manufacturers including GM and Ford. But Deming was not alone in setting the quality movement in motion.

American quality management

Deming and Juran became well-known figures and stars of the quality movement in the Western world. Philip B. Crosby, a corporate vice president of ITT who later became a consultant, followed in their footsteps. He was one of the first authors in America who developed "quality management" as a particular new approach in business. He clearly indicated how defects could be reduced by eliminating deficiencies and, with his *Quality Is Free* (1980), stressed the need of a quantitative approach of quality topics. Like Deming, he also underscored a leadership role of management to prevent rather than cure deficiencies and focused attention on management as the key driver of quality. Zero defects as a policy was an achievable objective for which he introduced a 14-step program with items such as raising quality awareness and establishing quality councils composed of multidisciplinary teams, which would force managers and other staff to break out of their "comfort zones" and relinquish some of their expert and hierarchical position power. Crosby's influence was far reaching. Lee Iacocca, CEO of Chrysler, emphasized that Chrysler's Quality Institute in Michigan "was modelled after Crosby's ideas" (Bank 1992: 75).[21]

A succession of other books soon appeared which all examined the secrets behind Japan's success. Ouchi's (1981) *Theory Z* pointed at Japan's employment and managerial practices. Pascale and Athos (1981), two business school professors, published a bestseller called *The Art of Japanese Management*, which stressed that Japanese managers had visions, something which was notably lacking in the West. The study by the Japanese management author K. Ohmae, *The Mind of the Strategist,* originally published in 1975, was discovered in 1982 and showed the American business community that Japanese companies did not have pointless hierarchies or endless analyses but an art of strategy which was highly creative and intuitive. The comparison with Japan made American business leaders aware of the fact that their employees should be more involved in the company's goals to increase overall performance.

The 1980s also became a period where an interest for joint management–labor committees focusing on quality issues reemerged. Ford and GM introduced worker participation programs called Quality of Work Life (QWL). Similar programs occurred in Xerox, Bell Labs, Bethlehem Steel, and a number of rail and transportation companies. QWL programs focused on improvements in work environment and productivity through participation in decisions about job design and problem solving. However, in corporations such as Xerox and Bethlehem Steel, which were highly unionized, committees were prohibited from discussing economic or business problems or collective bargaining issues.

As American import restrictions had initially shielded the American market against large accesses of Japanese manufacturers, Japanese car manufacturers began establishing their factories in the USA and became involved in joint ventures with American companies.[22] In the early 1980s, the first Japanese production facilities appeared in the USA. With the presence of Japanese managers on American soil running production plants, things began to change as Japanese transplants built in the US automobile industry adapted to local conditions. Although they transferred parts of their production system, e.g., work teams, other issues such as consensus decision making and the seniority practice were not transferred. These new firms were, therefore, called "hybrids" (Abo 1994). During the 1980s and 1990s, Japanese companies ranging from Honda Motors to Nissan Steel to Bridgestone Tires built Japanese transplants in America's Midwest. They introduced business methods which were markedly different compared to their American competitors. Production teams, rotation of jobs, adoption of workers' ideas about production improvements, and reliance on just-in-time production systems were unfamiliar practices.[23]

Cooperation between American and Japanese manufacturers increased with the establishment of joint ventures. One successful example of such a cooperation was the establishment in 1984 of the New United Motor Manufacturing Inc. (NUMMI), a joint venture between Toyota and GM in Fremont (California). It transformed one of the most unproductive, strike-ridden plants of GM into a model of productivity.[24] The plant was originally run by GM but encountered a vast range of problems leading to strikes and low productivity. By GM's standards, the facility at Fremont became an enigma. It introduced a simple device much like a household clock to record downtime on the line, in contrast to the hundred-billion-dollar information system that GM had been unsuccessfully experimenting with for the same purpose. "Managers and workers moved around the large plant on bicycles rather than on automated self-guided vehicles" (Pascale 1990:73). The work organization at NUMMI stood as a model for the successful application of the team concept in GM. Under Japanese management, grievances decreased and productivity improved. The secret was that "Japanese do it primarily with *people*" (ibid.: 73).[25]

NUMMI also introduced the Japanese Problem Solving Circles where volunteers were working on a certain problem for weeks and then proposed a solution for it. Individual workers or teams made 16,000 suggestions in 1992 alone. The work at

NUMMI demonstrated how factory workers could use their own heads to exchange knowledge and opinions, which were then secured by internally solidifying new policies (Adler *et al.* 1997). TPS achieved positive learning and improvements within NUMMI and became widely imitated (Lander and Liker 2007). Team-working increased the organization's adaptability to dynamic environments while teams could handle more complex and variable production processes, have more transparent control structures, show an improved performance, and had fewer coordination costs (Molleman and Slomp 2003).[26]

American car manufacturers began to learn the secret of Japanese production policies by forming joint ventures with Japanese companies. Yet US automobile manufacturers selectively emulated Japanese organizational practices. Flexible work rules, JIT delivery systems, teamwork, quality circles, and subcontracting were widely adopted, but these firms did not promise job security to their labor force and geared their companies more toward a "lean and mean" policy (Silver 2003: 67). The focus was primarily on quality, safety, improving work methods, and savings in inventory and rework costs.

However, even with the limits of union involvement and no job security, the quality issue stimulated the appearance of a quality movement which had a broad perspective and encompassed many new approaches which were wrapped in attractive management concepts that increasingly caught the interest of business leaders and the mass media. It strengthened the fact that quality as a continuous improvement proved to be a rich and durable idea that could be a fundamental source of competitive advantage. The catchword phrase of the day became Total Quality Management (TQM) with a host of TQM institutes featuring an integral approach of quality issues.[27] TQM became the most influential slogan of the 1980s. In 1987, the American government created an equivalent of Japan's Deming prize, the Baldridge prize, awarded for long-term improvement in efficiency and effectiveness, so named after Malcolm Baldridge who served as Secretary of Commerce from 1981 to 1987.

TQM protagonists stated that the sharing of information among workers led to high efficiency and fewer product defects. TQM not only became popular in the car manufacturing industry, but also in businesses such as the consumer electronic industry where the first initiatives about quality control had been taken. Similar success stories were to be found in corporations such as Hitashi, Toshiba, JVC, Matsushita, Sanyo, Sharp, and Sony (Kenney 1999).

> One of the factors driving the success of Japanese firms in memory products in the early 1980s was the higher quality of the chips they produced. For Japanese chips, defects – the fraction of chips that prove to be defective – were probably half to one tenth the rates for American products. By the second half of that decade, however, American firms had dramatically increased expenditures for quality control, imitating Japanese practices such as total quality management (TQM).
>
> (Langlois and Steinmuelller 1999: 49)

By the early 1990s, TQM had become so popular that it was considered managerial heresy to argue against quality (Micklethwaut and Wooldridge 1990). No one dared not to talk about quality, which turned the concept into a buzzword.

Buzzwords

In 1982, two of McKinsey's consultants, Tom Peters and Robert Waterman, published *In Search of Excellence*. The book was an attractive response against the view that all the good news came from Japan. The authors stressed that American managers bore much of the blame for declining competitiveness due to their obsession with short-term interests and indifference to quality.[28] American managers talked about "empowerment" and similar popular management concepts but were actually hoarding power and proclaimed to re-engineer the organization while they were actually firing workers by the thousands. The persistence of bureaucratic organizations, while debating the relevance of quality and continuous improvement, actually created frustration and anger. American corporations had become disengaged from their customers and from their employees. The authors intended to inject humanity back into management with their book and tried to reassure the American business community that even amidst an American decline, not everything looked as grim as it might appear.

While generally ignoring Japan, Peters and Waterman demonstrated that some American corporations had become excellent companies while focusing on quality and on customer orientation. The two consultants pointed out that a different approach of management was needed, less in terms of engineering or cost reductions but more in terms of a shift of mentality among the staff of American companies. The authors introduced a framework called the 7S which related key company features such as strategy, structure, and systems with the four topics of staff, skills, style, and shared values. The applicability of the model focusing on the interrelatedness between the 7S was a task any top manager could perform as long as one had a vision about the company and the actual setting that the company was operating in. The book contained 14 case descriptions which illustrated the effectiveness of the model in a convincing and cunning way.

The oil shocks of 1973 and 1979 had clearly shown that strategic planning had become too much of a cerebral activity which took little interest in the messages it received from the company's eyes and ears in the marketplace.[29] Peters and Waterman opened a discussion about the relevance of "management by speech," actions to motivate the staff, get them involved, and embrace challenges, especially from Japanese competitors. They stressed that managers needed to: learn to persuade employees to behave differently by staying closer to the customer, "stick to the knitting," and facilitate autonomy and entrepreneurship.

Their book became extremely popular, was sold by the millions, and soon other management authors joined in a humanistic counter to corporate dominance. The success of the book produced a proliferation of management concepts with an increasing transience into fashions. In the field of management science, publications materialized on the emergence of different management concepts, packaged ideas with elements of a new management approach and a new value basis, equipped with a new language, procedures, and tools. Management concepts became the words and labels which were used to dynamize traditional constructs such as corporations and bureaucracies, which themselves had survived thanks to a particular linguistic discourse built on rational planning, statistically sound forecasts, and elaborated strategies "Management concepts made it possible to formalize knowledge, which rationalized a specific business practice and enhanced certain techniques" (Hatchuck and Weil 1995: 15).

The new concepts made organizational changes necessary by introducing new conventions and habits. Many of them, however, turned into fads as consultants developed a habit of telling their clients what the competitors had been implementing yesterday (Abrahamson 1991). Management fads follow a regular life cycle usually beginning with a particular discovery by a manager in a particular corporation, a consultant, or a management academic. The new idea is then formulated into a technique and published. Consultants generally grasp the idea and treat it as a universal panacea. It becomes widely promoted as a means of increasing productivity, reducing costs, or whatever else might be disturbing the managerial mindset.

In 1990, the former McKinsey consultant, R. Pascale, who had initially collaborated with Peters and Waterman, expressed his surprise about the tremendous popularity of certain management concepts. Reviewing the prevailing management literature, he noticed the ebbs and flows of many business fads. The sudden rise and fall of many, even conflicting, fads was alarming in his eyes, but he noticed that some of the management concepts that initially looked like fads such as TQM had eventually been adopted as an enduring way of improving business. Apparently the Japanese approach was able to add something essential to business practices.

Two years later, Grint (1997) noticed in *Fuzzy Management* that the situation described by Pascale had been inflated. He noticed that the relevance of these concepts was not determined by scientific criteria but by whether they could secure business results that were deemed legitimate. Even if their academic value could not be proven, management concepts were believed to result in an increase in productivity, efficiency, or profitability.

Throughout the 1990s, managers, consultants, gurus, academics, or management authors jumped on the band wagon of inventing and popularizing particular management concepts. Two Harvard Business School professors, Nohria and Eccles (1998: 279), even concluded that:

> if asked, most people would tell an interesting story about the variety of sources that have contributed to the ways they act and think as managers. Indeed, management knowledge comes from everywhere: it comes from a manager's own experience, from books and articles on a variety of topics [...] and increasingly from consulting firms.

The most remarkable fact, however, is that the popularity of management concepts has much more to do with the quality of the source providing the concept than with its truth. Managers are interested in ideas that are established by the reputation of a particular country (such as Japan), company (such as General Electric), manager (such as Jack Welch), consulting firm (such as McKinsey), educational institution (such as Stanford), or professor/consultant (such as Peter Drucker). That is the source of a particular concept.[30]

While many of the fads failed to be practically implemented and did not deliver the promised results, the business community became aware of the fact of how difficult it is to convert any bright idea into a sustainable practice. The adoption process turned out to be more difficult than expected (Fincham and Roslender 2004).[31] Management concepts were used in the Western business context as instruments and techniques to reinforce managerial interests instead of focusing on coordination practices in a broader perspective which are based on *interactions* between managers and other stakeholders in organizations.

Critical literature about management fashion began to focus on the roles and discursive strategies of fashion setters (e.g., consultants, gurus, and business schools) and clarify the development of persuasive management concepts (Clark 1995; Abrahamson 1991; Kieser 1997; Grint and Case 1998; Bos 2000; Heusinkveld 2004; Benders and Veen 2001). The fashion analysis further explored the cyclical development of management concepts by studying the networks through which new concepts are communicated and diffused.

The management academic M. Guillen (1994) remarked that there still was a lack of understanding about the social and relational processes that underpin and shape the concepts diffused through various networks. Some authors began to explore the socially constructed nature of the concepts among groups of managers and other staff who were interested in their possible application (Huczynzski 1993; Lillrank 1995; Child and Rodrigues 1996; Czarniawska and Sevon 1996; Botti 1997).

In Search of Excellence has been one of the most influential management books over the last 30 years, stressing the social over the rational and challenging the one-sided definition of corporations in terms of bureaucracies with powerful planning departments coordinating bundles of liquid assets and promoting excessive diversification. The book points out that excellent corporations have strong cultures and coherently composed shared values, which lead to an entrepreneurial set of capabilities and competencies.[32] The attractiveness of the 7S model was based on its label and its general statements about the way the 14 studied companies had tackled market challenges and become excellent firms. The case descriptions appealed to one's imagination and offered solutions to those who were striving to become excellent as well. All this turned the 7S model into a management concept that became extremely popular in the industrialized world and opened the doors for a proliferation of other management concepts.

Peters and Waterman (1982: 261) stressed the importance of adding new terms to the management vocabulary like metaphors as lenses to focus on excellent performance. They noticed, for example, that "most impressive of all the language characteristics in the excellent companies are the phrases that upgrade the status of the individual employee." The excellent firm was apparently able to translate a series of material and linguistic techniques into practical operational activities. British management author Holden (2002: 237) drew the conclusion that Peters and Waterman "can be credited with a major attempt to refashion the grammar and lexicon of management." Yet the main contribution to the improvement of the management of large-scale corporations did not come from these authors but from the Japanese contribution to the application of the clock time principle in a new version: timing. Impressed by the Japanese successes, two consultants from the Boston Consulting Group, Stalk and Hout, made a compelling case for time as the key topic for international business in 1990. In their study *Competing Against Time,* they stressed the importance of the link between time and international business.

Their experience with Japanese firms had led them to conclude that in international business, a timely response to the desires of customers leads to a competitive advantage. They were convinced that the best way for Western businesses to improve their practice of managing time should be to imitate Japanese multinationals, which pioneered the use of *timing* as a competitive weapon in the globalizing economy. For that matter, Ford of Europe introduced the company-wide Concept to Customer (CTC) in the 1990s as an ongoing reduction of the time required to conceive a new product, design it, and place it in dealer showrooms (Starkey and McKinley 1993: 135).

In the waves of Western management concepts which flooded the western business community and where many passed quickly through their life cycle and disappeared again, the Japanese JIT concept maintained itself as a solid rock withstanding the waves of fashions in management. Yet there was one other concept which was concomitant to JIT which also survived far into the twenty-first century: flexibility. Again, its roots were Japanese.

13

FLEXITIME[1]

Authors such as the futurist Zuboff (1984) foresaw that work would become more reliant upon information technologies (IT) in which the use of educated human capital would become much more central to organizational performance. Between the 1970s and the early 1990s, new information technologies spread throughout the globe with lightning speed. Competition between corporations would no longer be determined merely by the application of IT, since they were all using more or less the same hardware and software. He recognized that efficient organization and use of human capital were essential for the advancement of a corporation and successfully withstanding competition with rivals. He also noticed that the digital revolution marked a clear break with the manufacturing economy, shifting the focus onto the service economy. With the pervasive use of IT by banks, insurance companies, hospitals, warehouses, clinics, and retail stores, the era of domination by industrial mass production was approaching its end.

One other remarkable outcome in the service sector was the transition from private capital, still mainly operating nationally, to the internationalization of finance, which made private capital far less "national" and far more willing to search for the most profitable opportunities in the globalizing economy. The changes in technology, increasing world trade, and internationalization of financial markets appeared simultaneously in many parts of the industrializing world at an unprecedented pace.

Yet in spite of a substantial increase in technology inputs and acceleration in the pace of technological change due to the electronic revolution in data processing and communications, American businesses were still struggling with the slowing down of productivity growth rates. To shift the downward trend of productivity growth, western economic, business, and political literature launched new views on the need for the flexible firm as well as on the flexible labor market. However, institutional differences between Japan, the USA, and the European Union produced a remarkable divergence in approach to the establishment of a flexible labor market and a flexible firm.

Japanese challenges

Toyota, through the success of its Toyota Production System (TPS), became the icon of a corporation that introduced a new approach centered on flexibility (Watson 2009). As mentioned previously, in the US automobile industry, errors and defects were usually detected and caught at the end of the assembly line. At Toyota, workers could stop the assembly line immediately if an error occurred that they could not repair. This meant a big advantage in cost savings (Adler *et al.* 1997) and promoted a flexible production system.

The drive for flexibility was not only an outcome of internal company policies but was also based on the way the institutional setting supported such a development. Japanese companies could require flexible attitudes from their workforce as these emerged from the history and culture of the Japanese system of labor relations of high commitment and involvement of the workforce. Japanese labor relations were unfamiliar with adversial interests between employees and employers. Their institutions favored representation of worker interests to secure conformity and obedience (Wolveren 1989; Locke 1996). This institutional setting facilitated business policies to require Japanese workers to accept long working hours. As they were committed to their workers, major Japanese firms, on the other hand, provided extensive and ongoing training and education for their core employees from shop floor to top-level staff. As a result, Japanese businesses possessed self-educated employees able to adjust to changing situations. However, this system was limited to male employees working on a permanent basis for big business, representing only one-third of Japan's industrial workforce (Blackford 1998). Lifelong employment security was, therefore, only a privilege for a minority of workers and certainly not a main characteristic of Japanese labor relations.

Beyond the core of lifelong employees, Japanese car manufacturers hired temporary or part-time employees on the periphery of their own operations. They claimed access to their employees' "own time" by requiring many overtime hours, including the so-called "service hours," which were unpaid. The employees were willing to provide their own free time to solve small group problems or to participate in further training. This willingness to be flexible did not, however, lead to more leisure time. Unlike in Western Europe, Japanese workers continued to experience long working weeks. This form of temporal flexibility (about which more will be discussed later in the chapter) favored the smooth functioning of JIT production and made managerial requirements for improvements in the production process generally accepted even beyond the regular working hours.

Toyota developed a time-based strategy on a cycle of flexible manufacturing, rapid response, expanding variety, and increasing innovation in the attempt to shorten their production runs. The great transformations that Toyota plants in Japan experienced challenged the Fordist production system which was predominantly managed on low wages and economies of scale by maximizing production runs. This led to a struggle with long lead times and a reliance on forecasts with delays in the flow of information and products.

Toyotism, however, was offering its core workers functional flexibility in performing tasks while implementing just-in-time (JIT) production practices. The advent of these JIT production practices moved Toyota in the direction of creating a flexible organization which realized more variety in cars with lower total costs. The objective was to obtain the assent of the worker, humanize austere Taylorist practices, and involve the worker fully in the organization of work (Tsutsui 1998). The TPS improved quality, cost, timing, and delivery (Spear and Bowen 1999; Lander and Liker 2007).[2] Due to its overall flexibility, it could adapt many technological innovations quickly, a practice which caught the attention of other Japanese firms and was soon imitated. The ultimate principle of the TPS was to minimize "buffers" of people and material which was achieved by an integrative use of JIT. The flow of production was driving the speed and rhythm of work in order to deliver at the right time.

However, in Japan, people were also working 400 hours a year longer than most Europeans. On the other side of the globe, the Japanese society began to struggle with a new phenomenon: *karoshi*, or death from overwork (Donkin 2001: 282).[3] Forced, prolonged, and overloaded work could cause the collapse of mentality and body, resulting in a potentially fatal stroke or heart attack, or in suicide. The first case of *karoshi* was reported in 1969 with the death from a stroke of a 29-year-old married male worker in the shipping department of Japan's largest newspaper company. With the increasing public concern about this new phenomenon, the Ministry of Labor in Japan began to publish statistics in 1987 on *karoshi*. A major life insurance company investigated 500 male white-collar workers in top-ranking corporations in Tokyo, and found that 46 percent of their respondents were anxious about their own risk of *karoshi*, and 25 percent experienced complaints from their families being anxious about *karoshi*. In 1994, the Economic Planning Agency of the Institute of Economics of Japan estimated the number of *karoshi* deaths at around 5 percent of all deaths at work related to cerebrovascular and cardiovascular disease in the 25–59 age group.

Due to their tremendous success in terms of sales, the four biggest Japanese car manufacturers (Toyota, Nissan, Honda, and Mazda) needed to extensively increase their capacity. The employment volume, however, grew at a much smaller extent than the production volume. The growth in production was mainly due to an enormous increase in labor productivity (Jürgens *et al.* 1993). From 1994 to 2001, the labor productivity in Japanese car manufacturing increased by 20 percent on average, whereas it remained generally the same in American manufacturing, while the British car plants noticed a decrease of 13 percent (Oliver *et al.* 2002). Japan's lean management practices made its manufacturing system ideally suited to take advantage of the new computer-aided design and manufacturing technologies, which were quickly applied (Biesebroeck 2007).

Toyota's export opportunities also increased remarkably as its cars, in addition to their quality, were relatively cheap. The price advantages of Asian products stemmed from a number of factors including a weak position of trade unions, the introduction of lean manufacturing systems (Womack 1991), as well as strong government expenditure in infrastructural areas such as roads, ports, airports, and the application of telecommunications, which reduced the costs of intermediate inputs such as transport. Investment in large container terminals cut the cost of shipping Japanese products to Western markets. In that sense, Japan pioneered innovations in production as well as in transportation and marketing. It turned timing in trade and commerce to its advantage and, due to favorable exchange rates between the dollar and the yen, Toyota's as well as other manufacturers' exports boomed (Deardorff 2003).

Although there was a common belief that the TPS, or lean production, could only be successful in companies which were producing highly standardized products for mass consumption, it soon turned out that this was a limited view. Many more companies imitated the practices the car manufacturers had developed.[4] The concept of timing in relation to flexibility especially drew notable attention.[5]

Timing

Timing became the cutting edge and turned out to be essential in situations where companies were to operate in an ambiguous, uncertain, fast-changing, and complex marketplace. On top of adopting speed in their processing of information and in improving their production processes, Japanese car manufacturers had introduced timing as a new concept. Toyota's management discovered that it took more time and money keeping the cars in the distribution line than in the production line itself. JIT ensured that the right products of the right quality and amount were delivered to the right place at the right time.

In contrast to the American way of rigid future analysis and planning, Japanese management took into account the elements of surprise and unpredictability. Changes in the consumer market had to be easily catered for. In order to be prepared for this "surprise time," managers and employees were expected to be flexible in acquiring new skills, handling technological changes, new production methods, and outlet possibilities. To deliver the next generation cars to potential customers, timing was translated into strategies to encourage flexible production, alertness, increasing variety, and innovativeness. Such a translation could only become successful if workers and managers turned their joint working attitudes into a team setting and primarily shifted their focus to the customer.

Davis (1987: 14–15) pointed out that, in order to adhere to the principle of timing, flexible responses were essential and, therefore, any corporation embracing such a view had to move its industrial perspective from one of being a producer to one serving the customer: "When the corporation focuses externally on the customer, it transforms its sense of time. [...] Its transformative quality lies in the elimination of any waiting at all: *zero-based time*." Management literature displayed a growing interest in the combination of timing and a flexible approach and launched new concepts such as Time Based Competition (TBC) and Time to

Market (TTM), which were in great demand to reduce the time needed to provide the required product or service.

TBC was first introduced by Stalk (1988) in the *Harvard Business Review* and reflected the extension of JIT into every facet of the product delivery cycle. It stressed that flexible firms were not only fast to produce and fast to market but did so in a concerted way. These firms were quick in responding to customer demands and reduced the throughput time in manufacturing and sales by reducing lead time throughout the system. Time-based firms limited all unnecessary paper work, speeded up the order entry process, and compressed the manufacturing process (Stalk and Hout 1990; Stalk 1998). Stalk and Hout (1990) looked at time as the main competitive asset for many companies and noticed that time based companies gained distinct advantages over their competitors.

Management methods designed to improve company performance abounded: JIT, flextime, the shortening of design and production cycles, speed of work, and deadlines became common jargon. Firms using this terminology were capable of evaluating all managerial decision processes assessing alternatives in terms of time (Karagozoghu and Brown 1993; Vickery *et al.* 1995; Yeh *et al.* 2000). Time-based firms were eventually capable of increasing the overall value of their manufacturing and delivery systems.

In 1995, TTM was introduced for the first time during a conference in London as another version of the same principles of TBC. It stimulated the reduction of scheduling time, increased speed, minimized resources, and created flexibility (Smith 2005). TTM's focus was not to blindly pursue a speeding up of manufacturing but to align the mastery of time with a clearly formulated business strategy (Kahn 2004). Yeh *et al.* (2000) introduced the zero-time organization as having the ability to act instantly and adapt quickly to rapid, continuous change. These and similar management concepts demonstrated how timing became a key concept for strategic success. With Toyota's example on display, the international business community as a mass turned its attention toward flexibility.

American responses

Spurred on by the "Japanese threat," American car manufacturers took the initiative to introduce flexible automation systems to produce different models and vehicle variants. Flexible technology pushed mechanization forward in areas such as small-lot production, complicated assembly, and joining operations in mass production. Automation was meant to replace labor and to reduce direct production wage costs. General Motors introduced robots to replace inefficient working activities (Jürgens *et al.* 1993). Yet in spite of a substantial increase in technology inputs including robotization and acceleration in the pace of technological change, growth rates in American productivity during the 1980s–1990s continued to slow down (Madrick 2009). American corporations lost comparative advantage as they neglected to promote entrepreneurship and innovative organizational forms. While many larger firms – multinational corporations as well as medium-size companies – felt a pressure to cut labor costs and trim the size of the labor force, they began to restructure their bureaucratic organizations to increase productivity and competitiveness. Vertical integration of corporations was supposed to discourage flexibility. Headquarters' bureaucracies came under attack as their inappropriate business culture was favoring bureaucratic management instead of entrepreneurship.

In order to adequately respond to rapid technological advancements and increased global competition, American companies began to downsize, reallocating manufacturing to countries with cheap labor and eliminating many jobs in their country of origin. The slogan for more flexibility became used as a *leitmotif* to reallocate resources quickly and smoothly in response to necessary changes. Downsizing, delayering, and deregulation challenged corporations to re-center their activities in their core business.

Countries such as Indonesia, Vietnam, and South Korea, where the workforce was docile and their governments were authoritarian, were yearning for foreign direct investment, and American multinationals grasped the opportunities offered. The fact that the workers in the subsidiaries toiled in unsafe and inhumane environments was not regarded seriously. It was considered strategically sound to position subsidiaries in regions with nation-states which embraced "flexible economic policies." This form of flexibility was seen as key to improving American international competitiveness (Buckley and Casson 1998a; ibid. 1998b).

Next to technological solutions, however, the 1970s had also laid the ground for a growing interest for new approaches to organize work practices. Along with automation of production processes, an increase in staff mobility, as well as a strengthening of services to customers, required the promotion of innovative, managerial practices. The modern Fordist factory system with its clearly defined personnel policies still enacted a strict discipline based on a rigid working-time regime, and this became seriously challenged. American sociologist Richard Sennett pointed out that western corporations continued to operate like armies "in which everyone had a place and each place a defined function" (Sennett 2006: 20). The organization of work was still based on an elaborated pattern pinpointing the workforce when to be on or off duty. Rationalized use of clock time was the key characteristic of this almost military form of organizing work which had been further institutionalized in the corporate world since the Human Relations approach. In order to shift the attention toward a new way of managing the employment relationship and create a flexible firm, the focus on the employee approach in companies shifted toward Human Resource Management (HRM), which comprised four tenets (Legge 1995; Clark and Mallory 1996):

- integration of managing employees with corporate business strategies;
- development and enforcement of a strong corporate culture through HR policies and practices;
- increasing focus on the individual as opposed to collective employee relations issues;
- devolution of the employee management to line managers.

While striving for *efficiency* had dominated management science and practice until the 1970s, and *quality* had taken over in the 1980s, in the 1990s, practitioners as well as researchers adopted *flexibility* as the dominant topic in order to cope with increasing uncertainties. Yet the new concept obtained a plethora of different meanings.[6] At the most abstract level, one could say that flexibility refers to the capacity to adapt to change.[7] American companies which maintained a large part of their manufacturing in their country of origin were challenged to reengineer their activities before they might introduce the latest technological improvements and new personnel policies. Business Process Re-engineering (BPR) advocates stressed re-engineering business processes from functional departments to process teams to promote flexibility (Hammer and Champy 1993).

Downsizing companies signaled the end of the Fordist vertically integrated production structures. The discussion about its consequences for flexiblization, however, moved into different directions. Boyer (1987) and Blyton and Morris (1991) noticed that flexibility began to turn into a multiform and ambiguous generic term; it became employed as a summary concept for a wide range of developments, although it was sometimes presented as a unitary concept embodying no essential conflicts of interest. In some cases, it opened the way for the introduction of HRM policies with a broad spectrum of flexible work organizations with a higher level of self-regulation not uncommon in the Japanese context.[8] In the majority of cases, however, the search for flexibility mainly addressed changes in the organization of work due to technological and decentralized forms of production forcing the American workforce to accept an extension of working time, as can be seen in Box 13.1.

BOX 13.1 WORKING TIME 1[9,10]

Beginning in the late 1950s, it was generally believed that with an increase in the range and depth of automation in American corporations, "the worker will be released from work permitting development of his inherent human capacities" (Diebold 1952:162; Zuboff 1984). In the 1960s and 1970s, many commentators predicted a substantial further reduction in working time and an increase in leisure time. It was expected that working hours in America could decrease to a four-day work week (Pollock 1955; 1962). Yet the American institutional setting, especially the labor market, moved into a different concern about working time. Employers in capital-intensive, continuous-process industries mainly responded to increased demand of their products by introducing shift work. Employers in machine-tending and mass-producing industries primarily introduced overtime. In response to the need to maximize the use of factories or offer round-the-clock service, employers increasingly moved to more and more variable working arrangements. Overtime became one of the most common practices to increase output (Casey *et al.* 1997) but frequently operated on an informal basis.[1] Overtime had traditionally attracted premium rates of pay, and this policy increased as the base working time of forty hours per week decreased. Many American employees on fixed hours worked overtime every week, and overtime (paid or unpaid) turned into a very common form of flexibility.

During the 1980s, the American economy combined rising unemployment with longer working hours for those in jobs. In a climate of economic conservatism, a reversal of the shorter working hours tradition appeared. Until the 1940s, the average hours in both Europe and United States had been declining in tandem for nearly a century. Juliet Schor (1992), an economist at Harvard, calculated in *The Overworked American* that the American workforce spent the equivalent of four weeks more a year on the job in 1989 compared to 1969. The average American worked 52 hours a week – 46 hours in the office and a further six at home. Compared with Germany and France, American manufacturing employees were working about eight weeks a year more. This difference was due to the fact that similar international pressures for flexibility set in motion fundamentally different struggles in different contexts. Flexibility in the American context was mainly translated as a renegotiation of work rules and more freedom for American managers to hire, fire, and lay-off workers (Locke and Thelen 1995).[2]

The longer working hours in the United States, however, did little to increase productivity. In October 1982, American unemployment hit 10 percent, the first time it crossed the double-digit barrier since the Great Depression. During the Reagan administration, the country feverishly looked for solutions and began looking for approaches that could reverse the downturn (Bernstein and Adler 1994). Some were convinced that minimizing the costs of paying workers could be one of the best options. Some of the large corporations, however, had been enjoying the ability to compete on grounds other than price alone. They followed a model of cooperative management stressing that "productivity depends more on the effective interaction of many people than on the motivation and effort of single individuals." Yet the competitive model of management grew in popularity and favored easily firing many workers, subcontracting work to low-cost firms, and demanding a faster pace of work from the employees who remained.

[1] In such a case their extra effort was likely to be unpaid (particularly for managerial or professional staff).

[2] The United Kingdom soon followed extensive overwork and longer working hours (Donkin 2001).

The flexible firm

Internationally operating companies began to focus on flexibility as an integral part of HRM which could provide an appropriate way to manage time, space, and people more effectively by aligning outputs with demand and respecting sufficient adaptability to cope with change. Organizations needed to deliver services and products with maximum efficiency, just in time, and at the lowest cost while realizing appropriate labor utilization. Confronted with the upturns and downturns of an increasingly globalized economy, organizations were required to implement flexibility which would rely on the way the workforce was willing to deal with the issue.

In the modern factory system, the traditional approach had been stereotyped as *hire* when additional labor or new skills were required and *fire* by shedding workers when there was no longer demand for their labor or particular skills. Short-term increases in demand were usually met by using overtime, as Box 13.1 indicates. For a significant period, the typical full-time, permanent, nine-to-five, Monday-to-Friday job with one employer had prevailed as the dominant work arrangement. Yet in the 1980s, inspired by the Japanese examples, the western management literature introduced a new model introduced of an ideal organizational design with a set of core workers with solid labor contracts and a periphery composed of workers with mainly temporary contracts. It was supposed that such a design could effectively cope with all the contingencies of an uncertain globalizing economy.

At the British Institute of Employment Studies, researcher Atkinson (1984; Atkinson and Meager 1986) introduced the flexible firm as the best way to manage appropriate labor utilization. It gained broad attention but soon also received some serious criticisms. Other researchers found little evidence of this type of organizational form in western business practices (Marginson 1989). It was found that not every British employer, for example, was in favor of the new approach (McGregor and Sproull 1991). The central concept of core and periphery was ambiguous in practice (O'Reilly 1992) and was seen as deeply flawed while being too prescriptive and static (Pollert 1991).[11] Yet the concept of core and periphery created an increasing interest in the distinction between functional and numerical flexibility.

Functional flexibility covers topics such as multiskilling, job rotation, teamwork, and similar forms. It refers to the facility with which an employee can be moved from one task to another. Sometimes the focus was on jobs-based flexibility, which is related to issues connected with the design of work, jobs, and even organizations at large, such as with BPR. However, Sparrow (1998) found that functional flexibility translated in terms of downsizing to avoid duplication of jobs by making one person do what otherwise would have been done by two or three (Cohen 2003).[12] Functional flexibility promised a panacea to the interests of employer and employee alike but was frequently difficult to attain (Beatson 1995; Dyson 1992).

It was supposed that a tripod of functional flexibility, quality, and teamwork would lead to more success (Wickens 1987) but Legge (1995) warned for an alternative "tripod of subjugation" through work intensification, control, and peer surveillance. Echteld (2007) questioned whether current evidence on functional flexibility really justified the belief that new forms of work organizations such as a post-Fordist version would indeed improve working conditions.

Along with the focus on the internal restructuring of large organizations by introducing functional flexibility, the impact of IT advancements was seen as another driving force to strengthen flexibility of firms.[13] Bridges (1994) forecasted that IT would change the workplace dramatically: more flexibility would lead to more part-time work, more freelance work, work on commission, and, consequently, less certainty and predictability. He focused primarily on what is called numerical flexibility by temporarily recruiting staff for certain activities without being offered multiskilling opportunities, training, and empowerment or working within teams. Castells (1996–1998) also stressed the impact of IT and suggested that the upcoming flexitimer would become an inevitable outcome. Brewster *et al.* (1997) found evidence that larger organizations were increasingly using this numerical type of flexible work forms and advancing a high use of part-time/temporary workers on a nonstandard basis.

The impact of IT raised the central concern as to whether the division between traditional permanent employment and increasing nonstandard employment might introduce a two-tier system in the labor market with good and bad jobs.[14] By varying the size and the structure of the workforce with primarily nonstandard employment arrangements, numerical flexibility turned into an almost accepted phenomenon introducing an increasing variety of working-time arrangements. Numerical flexibility introduced a policy of varying labor contracts with, as an extreme, employees spending their entire career on fixed-term contracts (FTCs) for the same employer (Meager 1985: 11). FTCs[15] became a type of temporary employment mainly in the form of *external* flexibility. They are used as a strategy to deal with uncertainty, but their application also reflects a polarization of a permanent use of the most highly skilled, whereas the low skilled are often employed on a casual basis. The apparent advantages may be outweighed by the disadvantages. Although this type of flexibility appears to provide a buffer against uncertainty, it relies heavily on the availability of a ready supply of appropriately skilled workers. Some organizations might become dependent on temporary workers for even core activities due to drastic pruning of the permanent workforce.

With reference to working hours, as described in Box 13.1, there has been considerable pressure from both employers and employees to vary the length of the workweek in the form of *temporal* flexibility. The term is frequently used to refer specifically to employment arrangements where the working hours and pattern may be varied (Deery and Mahoney 1994). This form may occur within both external and internal flexibility arrangements.[16] Employees seek a reduction of hours and/or a pattern of hours to suit their personal needs such as balancing work and family life, which made flexitime and part-time patterns become more common. Employers might seek a reduction of hours whenever the company has to adapt to sudden external changes which is illustrated in Box 13.2.

BOX 13.2 WORKING TIME 2[17,18,19]

In 1993, stiff competition and new work technologies forced Volkswagen, at that time already one of Europe's largest car manufacturers, to adopt a four-day workweek to save 31,000 jobs that might otherwise have been lost. The employees supported the initiative of management. German unions were successful in participating in work reorganizations, and saw shorter work hours and flexibility as a means to combat high structural unemployment by redistributing jobs. They allowed employers to schedule longer regular hours during periods of higher demand as long as they were balanced by shorter hours at other times (Locke and Thelen 1995).Volkswagen became the first multinational corporation to move to a 30-hour workweek. The shorter workweek was an equitable alternative to mass permanent lay-offs. In November 2003, Opel in Germany took the same initiative for a period of a year in order to save the jobs of 21,000 employees. Part-time work as an example of temporal flexibility required some care in categorization because some arrangements were highly secure (located in internal flexibility) and others highly temporary and precarious (located in external flexibility).[1]

Sometimes new arrangements – along with part-time – were introduced to differentiate between minimum and maximum labor time per day or week.[2] However, formal rostering of these flexible working schedules was frequently a stumbling block.

Another interesting case appeared in Finland where a few firms in the vicinity of Helsinki, for example, introduced a six-hour dayshift model that

[1] Some authors have put forward further categorizations of flexibility. Sparrow and Marchington (1997) proposed no fewer than seven categories. But these additional forms do not provide conceptual clarity as more categorization leads to more overlap and more ambiguity. Some part time work may not be subject to variation at all (just simply a contract for fewer weekly or annual hours) and should, therefore, not be considered to be flexible employment! Functional and numerical flexibility are prominently used as the description of flexible employment.

[2] Despite all the recent evidence that non-standard work forms are increasing – with populist predictions towards a jobless future – we do observe considerable continuity with traditional approaches. Casey et al. (1997) found that overtime was the most common form of flexibility, but Dex and McCulloch (1997) claimed that the tide has turned and that even in Britain full-time standard employment was increasing and temporary work declining. Although this might be true for the British private sector with its decreasing use of non-standard forms, in the public sector they are still on the increase. Cost-cutting policies and uncertainty remain very high in the British public sector. In other countries, however, such as the Netherlands, the job machine is based on an expansion of part time employment. One of the key struggles is how people organize their entry into and exit from various roles over their life course. On the family level, timing involves the synchronization of individual life transitions with collective family ones.

comprised two six-hour shifts, one starting at 6 a.m., the other at noon. The daily working time dropped by two hours. Overtime was no longer necessary. The basic wages remained the same, but total income decreased while extra overtime bonuses disappeared. The new model facilitated the proper combination of work and free-time activities (Kandolin and Kauppinen 1996).

In 2002, The Dutch University Medical Centre of Utrecht began an experiment with working hours in order to anticipate the ever more acute shortage of qualified staff. Staff members who were also parents at the same time were recruited on a so-called "mother contract," which included that working times became synchronized with childcare and school hours. The wishes of parents with respect to working hours were taken as the point of departure. Inspired by Swedish experiments with the software program Time Care, which facilitates employees' self-rostering, the Utrecht hospital introduced a similar practice in 2007.

These examples illustrate an increasing variety of part-time or variable contract arrangements across Europe. Part-time offers the advantages of flexibility to parties but traditional values, institutional arrangements, and cultures in many companies have hampered the ability to provide part-time work on an equal status. A leading example of a major increase in flexible forms of work is the Netherlands with 42.9 percent of the working population in part-time jobs (Sels and Van Hootegem 2001). Temporary employment agencies also began to play an important intermediary role in the labor market in the Netherlands. In 1997, this sector with around 18,000 agencies constituted the second largest employer in the Dutch private sector. This rise in temporary labor must be seen as part of an increase in flexibility in the labor market in general (Data from European Foundation for the Improvement of Living and Working Conditions 2002; http://www.eurofound.europa.eu).

Based on an agreement with the social partners in April 1996, the government placed pressure on employers to adhere to a package of basic requirements to protect the interests of flexible workers, which is similar to those of standard workers (Visser and Hemerijck 1997). At the same time, an exchange was agreed upon of a shorter working week (36 hours) for increased working-time flexibility in sectors such as department stores, banking, insurance, railways, the dairy and food industry, and chemicals and printing. Increasingly, working time became annualized with a corridor of 32–40 hours per week. Industrial unions introduced "vari-time" or "working hours corridors," and in the service sector, unions negotiated *à la carte* collective agreements. In 1999, further legislation (the Flexicurity Act) was enacted to adequately match employers' demand for flexibility and employees' demand for working-life security and income. Currently, almost 70 percent of working women in Holland have part-time jobs (SCP and CBS 2006). Although it is gradually

(Continued)

becoming more widespread, many employers and employees still show reluctance to embrace such forms of temporal flexibility.[3]

[3] The findings of Casey et al. (1997) have supported the conclusion by Hunter et al. (1993) that employers in general prefer to use full-time permanent employees because they believe that they are easiest to manage, the most committed and the least likely to leave. Geary (1992) examined firms which used a dual strategy with permanent and temporary workers and noted that the managers were averse to this policy because of all the problems that arose. One of these was the tension between temporary and permanent workers, which has been noted in other studies (Filipzcak 1997). Some businessmen (usually supported by consultants and politicians) go so far as to argue that organizations should not offer any sort of career but, instead, the commodity of employability. It is dubious whether such kinds of deals are what most employees seek, and there may be serious resultant damage to the employment relationship. This could create problems and costs for employers (Bryson and Barnes 2000).

America continued to work long hours. The International Labor Organization (ILO) in Geneva calculated that, in 2002, the average American still worked 1,815 hours per year, whereas in France average working hours reached 1,545 and in Germany 1,444. Labor economists pointed out that these differences were due to different institutional patterns. Europe with its strict work regulations kept Europeans from longer hours and facilitated preferences for more vacation time. The economist E. C. Prescott (2004) suggested that the main reason for the difference might have to do with the American inclination for consumption now over consumption later – "keeping up with the Joneses" – which forces them to work more hours and accept overwork. According to Prescott, the American tax system favors this inclination and as long as the European Union is not doing the same, the imbalance will continue.

Others, such as Whitely, and Hall and Soskice, however, have pointed out that an Anglo-Saxon institutional setting with a weak role for unions and a traditional managerial inclination for long working hours has created a labor market situation where pressure for extending working time has become common practice. The British researcher M. Young (1988: 217), for example, noticed in a study among senior executives that British employees in general were dealing with extreme time pressure.

> Executives were likely to overstretch themselves by committing themselves to an 8 day or a 40 hour week, always trying to do too much. When their elaborately planned futures arrived, they were rushing from one event to another without time to savor anything or make a well-considered decision; they did not allow themselves time to think about the future of their companies, their colleagues, or themselves; they were continually harassed, even though they were never supposed to look as though they were; and they had no holidays from the pressure of time.

In spite of the consequences, Anglo-Saxon countries showed an increasing acceptance of these working practices. Outsourcing became seen as an appropriate way of flexibilizing organizations. For example, during the presidency of J. Welch, General Electric fired around 30,000 employees in one year, and job security of many other American factory workers was put in jeopardy. This form of flexibility was perceived as a proper way "to attract and retain good employees in a labor market that is steadily becoming more competitive" (Olmsted and Smith 1989: vii). Flexibilization became advocated as a means of utilizing labor more efficiently.

Yet the discourse surrounding the relevance of a flexible work organization in American companies became primarily driven by a dominantly managerial language about the flexitimer. "The grueling schedules that used to be typical only for top corporate management and self-employed people became more common in one occupation after another" (Perlow 1998: 331). These practices instigated a long hour culture, which operates at the top of these organizations, predicated on the assumption that management staff has to work long, full-time hours to be effective in doing their jobs.

> Big businessmen, who have never in their lives had to worry about putting food on the table, lecture as authorities on the proper conduct of working life. They preach about the necessity of hard work and putting in long hours as though these were the solution to the economy's woes.
>
> (Sennett 2009: 164)

This culture is hostile to family and private life and has caused a counter-response to the boardroom attitude, especially among highly trained American women. The American journalist Lisa Belkin (2003) signaled that well-educated and successful American women who had jobs as bankers, medical doctors, lawyers, and business women decided to follow the "mommy track" or the Opt-Out Revolution. The female CEO of Pepsi-Cola left her job to spend more time with her children; the female advisor of President Bush, Jr. quit her job to reunite herself with her family in Austin (Texas).[20]

Rousseau (2005) has pointed at another new phenomenon: I-Deals. She noticed that, in America, highly skilled employees bargain idiosyncratic deals and negotiate individualized conditions of employment for themselves. These deals are arranged between worker and employer in such a manner that work arrangements become flexible enough to meet each other's needs. She is convinced that the conventional labor management relationship where the company determines the work conditions and the worker, due to an hierarchical, asymmetrical relationship with only one party in charge, cannot but accept or refuse, is disappearing and will be replaced by a setting where the employee has power to bargain and work in a setting which fits both interests.

This new habitus becomes especially apparent among knowledge workers. The American telecommunication industry, for example, calls for a much more radical change. Some of the entrepreneurs in this sector such as Ginn, Sarin, Lee Cox, and others have cultivated a partnership with employees. They eschewed the adversarial history of organized labor and have sought instead to motivate employees by giving them a stake in the business. They engendered loyalty at all levels of the organization and created an environment that encourages innovative attitudes (Galambos and Abrahamson 2002: 260). Under such circumstances, I-Deals might be more adequately arranged and include flexibility "on the part of an already highly supportive employer" (Rousseau 2005: 14). Knowledge workers might be on the forefront of "achieving both individual flexibility and collective fairness, and fulfilling unique personal needs while promoting mutuality and cooperation at work" (Rousseau 2005: 21).[21] Yet these options seem valid only for the well-trained and highly skilled. For many others in the American labor market, nonstandard working arrangements increased remarkably. By 2005, one out of five employed people worked mostly at nonstandard times, i.e., during the evening, at night, or on rotating shifts (Presser 2003; Golden 2005; Presser *et al.* 2008).[22]

The debate around the flexible firm centered mainly on whether corporations were willing to adopt the model as a strategic choice. Although few would now support Atkinson's narrow use of core and periphery, the forms of flexibility he suggested have become more accepted as common parlance or discourse about the subject and have left a strong legacy of terminology (Blyton and Morris 1991).[23] The discourse did not focus only on the internal development toward a flexible firm, but also on the issue of creating a more flexible labor market which could reinforce the establishment of the flexible firm. This has fired the interest in flexibilization within the European Union.

The European Union approach

During the golden age of sustained economic growth (1950–1970), identical working patterns in the Western economies seemed to abound. For a significant period of time, as stated previously, the typical full-time, permanent, nine-to-five, Monday-to-Friday job with one employer prevailed as the dominant work arrangement. After the oil crises of 1973 and 1979 and increasing labor unrest, however, Western European short-hour movements re-emerged to tackle increasing unemployment.

> In an era of increasing technological unemployment, reduced work time became a means of sharing scarce employment; and in an age of feminist inspired sex role changes, shorter hours became a way of reducing the domestic burdens of two income families. Yet [...] efforts to lower the hours of working time have largely been failures. [...] Instead of working shorter days, most people preferred longer vacations and other extended blocks of time free from employment. As a result, there was little unity over a call for, say, a 35 hour week.
>
> (Cross 1988: 173)

During the 1980s with Ronald Reagan's administration in the USA and the government of Prime Minister Margaret Thatcher in Great Britain, a new Anglo-American production regime was introduced which focused primarily on the organization of production through markets and market related institutions. This led to what were eventually called Liberal Market Economies (LME) (Soskice 1999;

2009). These economies are characterized by competition and formal contracting with flexible labor markets and dispersed shareholder corporate governance. Australia, New Zealand, Canada, and Ireland followed this pattern in similar ways. Post-Fordist workplaces in these Anglo-Saxon contexts began to offer flexible work schedules with more autonomy and challenges for employees but, at the same time, continued to force workers to spend more unpaid hours at work and thus increase the profit practices of these firms (Perlow 1998; Echteld 2007).

Proponents of economic globalization presupposed that convergence toward a minimalist, individualistic, US-style model of industrial relations would be the best option for further global economic progress. They also shared the belief that "the international forces of economic globalization and capital mobility have severely constrained the room in which national governments can maneuver" (Weishaupt 2011: 60). Western nations would be forced to deregulate labor markets and sail toward the same shores by recalibrating their national labor market policy regimes toward the US-style model.

The Organization for Economic Cooperation and Development (OECD) became an influential spokesperson for such a flexible, active labor market policy. Its *Jobs Study* (1994) propagated the expansion of nonstandard forms of employment, such as part-time, weekend, and home work as well as more self-employment. The third article of the *Jobs Study* recommended an "increase in the flexibility of working time." The *Jobs Study* has been interpreted as an American or neo-liberal prescription which heavily focused on an improvement of labor market flexibility by removing rigidities caused by labor unions and governmental regulations. With regard to recommendations concerning social and labor market policies, the OECD enjoyed an almost monopoly-like position in the discourse about flexibility.[24]

The appearance of this new regime, however, also intensified a debate over the question whether Fordist production practices everywhere were gearing toward a convergence of work organizations irrespective of differences in institutional context where the manufacturing took place. It was stressed that this development indeed entailed that national institutional frameworks of industrial relations were challenged by putting the role of the state at arm's length, reducing the role of the unions by deregulating the labor market, and promoting market coordination between companies.

Since the 1990s, the European Union gradually emerged as an important new actor in the dissemination of novel ideas and approaches of a flexible labor market, sharing a common policy of reaching high levels of employment and social protection by raising the standard of living and quality of life.[25] In line with these perspectives, it focused less on deregulation of labor markets and decentralization of wage negotiations but put more emphasis on capacity-building state interventions and social partnerships. It also stressed the relevance of a more flexible organization of work in terms of functional flexibility by training employees and increasing career perspectives through internal labor market policies within firms.

Warnings for the negative outcome of the Anglo-Saxon market approach (LME) were clearly expressed:

> A society without institutions, i.e. a totally flexible society, is anomic, and anomy frustrates successful adaptive behavior of individuals or organizations as much as total ritualistic regulation. An obsession with immediate responsiveness to markets is a recipe for chaos and confusion but unfortunately reflects the patterns and policies we observe.
>
> (Streeck 1988: 417)

Continental Europe began to opt for a different development with businesses operating in a variety of what Hall (1999) called Coordinated Market Economies (CME). In these economies, businesses depend more on nonmarket relationships to coordinate their activities. They rely much more on collaborative rather than competitive relationships. "These economies emphasize vocational training, regulated labor markets, and bloc shareholder corporate governance" (Soskice 2009: 134).

Within CMEs, the role of the institutional perspective – even the one of the European Union – was clearly stressed and insisted upon the relevance of a more varied functioning of the market. Divergent production regimes within CMEs deal with hegemonic workers' parties, widespread collectivist and egalitarian principles, and strong welfare states which strongly conditioned how national economies were responding to pressures of globalization and socio-political changes the European Union began to promote.

Although it was recognized that Europe contained a variety of institutional patterns of political and economic governance, the European Union took up the topic of flexibility in the 1990s to provide a European-wide institutional development, which would favor positive forms of flexibilization while at the same time respecting the targets of a social Europe. In 1993, the European Union introduced the European Working Time Directive (WTD) which set targets to limit the working hours for men and women to adjust work and flexibility to the wellbeing of mankind (Supiot 1999, 2001).[26] Its main provisions intended to limit maximum working hours and establish minimum entitlements to rest periods and paid annual leave for most workers in the EU.

Most member states had national working-hour legislation in place that was more restrictive than the WTD. The spirit of the WTD, however, forced employers to take account of the general principle of adapting work to the worker with a view in particular to alleviating monotonous work and work at a predetermined work rate depending on the type of activity and of health and safety requirements especially in regards to breaks during working time (Article 13 of the WTD).[27]

Over the period 1985–95, the European Union showed a 15 percent increase in flexible employment. Brewster *et al.* (1992) found evidence of an increasing use of flexible employment in nine out of ten European countries (the exception was Sweden) but with a considerable diversity in the forms of flexibility.

Flexibility was propagated as a way to facilitate an appropriate combination of work and life by institutional measures offering employees the possibility to adjust their work schedules to the rhythmicity of leisure/personal training and caring tasks at home. As the dual-income family gained adherence, it put the topic of a work–life balance at the forefront (Fagan 2001). Critics ventilated a serious concern: the relationship contained an antithetical dimension in the sense that the balance between the sphere of work and the sphere of family life presupposes an ideal quantitative allocation of time to each sphere. They believed that it would be impossible to arrange this properly at an institutional level and emphasized that individuals should solely perceive their life as a career, i.e., a career-centered perspective (Hakim 2004). Others, however, insisted that there is a societal dimension to the topic of flexibility which the European Union should regulate to prevent home/work boundaries from becoming too permeable.

In the late 1990s, the European Foundation financed a study on employment options and labor market participation which indicated that over half (54 percent) of the European workforce (both self-employed and employed) would prefer to work fewer hours if they had the choice and to balance their work with care responsibilities, thus reducing the average working week to 34.5 hours (Oeij and Wiezer 2006; Doorne-Huiskes *et al.* 2005). A study on the labor market in the European Union showed that one-third of the employees considered that they had no influence in the workplace over their working hours or even when they took their annual leave, while another third felt they had no influence over both issues. Similarly, on the specific issue of start and finishing time, just over 40 percent of employees felt that they had some influence over their start and finishing times, while the remainder had the start and finish time set by the employers. The degree of working-time autonomy varied markedly according to occupational status and was generally highest among managers of both sexes and male professionals and lowest among nonskilled manual (blue-collar) jobs (Fagan 2003).

Between 1994 and 1999, almost two-thirds of the new 6.8 million jobs across the European Union went to women (Webster 2001), but women accounted for 80 percent of part-time workers, and 13 percent of them were employed on short-term contracts (European Commission 1999; Webster 2001). In the UK, a quarter of the workforce occupied part-time jobs. Most of them were accounted for by a large number of women working part-time (45 percent of female employees) compared with just 10 percent of men. Yet the growth of part-time work in the 1990s among men was also quite remarkable. The pattern of part-time working between the sexes showed that men tended to be part-timers at the beginning and end of their careers and women in the middle or throughout. However, as long as "deregulation is occurring in response to high levels of unemployment and the political quest for more flexible labor markets, part time employment is likely to be marginalized, even if full time standards are also reduced" (Fagan and O'Reilly 1998: 24).

Much of the growth of jobs in Europe has indeed been in part-time work and many such jobs have also been temporary. In the Netherlands, for example, many part-time workers are on permanent employment contracts and, as a rule, the number of hours worked by part-time workers is fixed. Hence, part-time workers are not faced with the uncertainty of continued or reduced earnings of temporary workers or workers with variable hours contracts (Visser 2000). Although ideologically the nine-to-five view was still present in the Netherlands, the percentage of people working irregular hours increased from 53.9 percent in 1992 to 64.2 percent in 2000 (Delsen 2000).

Despite the policy of the European Union and the national policies that the European members have put in place, flexibility did not go hand-in-hand with a reduction in working time that facilitates a proper balance. Bielenski (1999) reported that the preferences for working hours were not primarily dictated by the need to combine paid work and family duties, specifically the care of young children. The majority, i.e., 77 percent, preferred more time for themselves and for their own activities. A substantial percentage, 53 percent, had a preference for part-time jobs to overcome the strain of working full-time. Although an increase in leisure time or free time might in principle slow down the pace of life, paradoxically that did not seem to be the case. Instead of creating an abundance of time, somehow the techno-economic development of speed-up processes and acceleration of technological and social change generated a feeling that one was continuously exhausted. A concomitant new pattern was the pressure of global competition to practice long working hours and extended working schedules as was the case in Japan and the USA.

Contrary to what was forecasted about more flexibility and less security in the information society, the standard employment relationship in Europe was still the predominant employment pattern. The proportion of employees with fixed-term contracts actually rose between 1992 and 2003 from 11.1 to 12.8 percent. The proportion of fixed part-time employees rose over the same period from 14.2 to 18.2 percent. Employees with fixed contracts favored flexibility which altered their starting and ending times of work on a daily basis to such an extent that it helped them to balance their work and family responsibilities. It became clear that not all forms of flexible policies employers had enacted were to be considered "family friendly" (Peper *et al.* 2005; Tietze *et al.* 2006). For part-time jobs, national legislative changes were introduced to end abuses and exploitation, and the progress of equal opportunities did much to enhance the rights of part-time workers.

Lisbon

The 1997 *European Employment Strategy* (EES) institutionalized a policy at the level of the European Union of inviting the social partners in all member-states to negotiate flexible working arrangements, examine the possibilities of more adaptable types of work contracts, and facilitate equal opportunities for all members of the workforce by taking the appropriate employability measures. The EES document became a "real alternative to the prescription of the OECD *Jobs Study*" (Weishaupt 2011: 164). The ESS was an essential building block of the subsequent Lisbon Strategy which was launched in March 2000 by the European Council in Lisbon where it declared that its main objective was to become the world's most dynamic and competitive knowledge-based economy with greater social cohesion in 2010 by making more investments in people and activate social policies which prevent all forms of social exclusion (Giddens 2006).

The Lisbon agreement put forth that an appropriate work–life balance for European citizens is basic to the EU policy guidelines. National policy makers were invited to promote the balance issue among the younger generations by eliminating gender inequalities in the professional sphere of life, promoting an equal distribution of time over the life course which enhances the quality of the life of its citizens. Although Europe reflects a large variety of approaches concerning standard working-time arrangements and flexibility, the overarching image was that its employees prefer to live under conditions that prevent the blurring of the boundaries between work and life. The Lisbon guidelines have been reiterated in subsequent summits in Barcelona, Stockholm, and Laeken (Belgium) to stress the relevance of care issues as well as the relevance of collective responsibility (Boulin *et al.* 2006).

In 2003, the relevance of the Lisbon Strategy was again emphasized by the study *Jobs, Jobs, Jobs* which urged for a policy that would enhance adaptability and flexibility of labor markets to increase flexible working-time arrangements *for* the workers rather than *of* the workers. The European Commission (2004) clearly underlined that a Working Time policy should improve the balance between working and family life.

The Council continued to pursue a proactive labor market policy based on lifelong learning and a "destandardization" of the life course. The Lisbon agreement stressed that work–life balance is basic to EU policy guidelines, and it targeted to increase the participation of women: by 2010, 60 percent of European women would be participating in the labor market. Yet even after introducing European citizenship policies which might include that men and women will adhere to gender equality and a fair sexual division of labor, the European Commission had to observe that the greater commitment of women to paid work has not been accompanied by any significant redistribution of household labor, with women performing more than 80 percent of household tasks in all but the Nordic members states and the United Kingdom (European Commission 2000).

To further improve a gender balance and social cohesion, the European Council also continued to address the issue of the balance between work and family life for men *and* women.[28] Some of the work–life policies that the European Union propagated have led to new organizational practices ranging from formal to informal arrangements which should allow employees to manage their own roles, responsibilities, and interests in their life as complete persons who are potentially engaged in work and non-work spheres such as family, friendships, personal development, and lifelong training projects, political, associative, spiritual and sports activities, and leisure (Thevenet 2001). However, the overall power structure within the business community still allowed managers to control the working time of employees.

The European Union, however, is a supranational institution that can only pursue its goal if it receives effective support from its European members. To attain that support, the member states of the European Union act in principle in concert with one another through the Open Method of Coordination (OMC) (Giddens 2007).

Yet it turns out that there still is a large diversity among the European member-states that have been the result of historical developments reflecting specific divergent patterns. The European Union is challenged to properly deal with this diversity.

Convergence versus divergence

European policies concerning working-time arrangements – as laid down in directives – have been elaborated by its member-states in a variety of ways depending on the their own national welfare state regimes. Esping-Andersen (2000) has designed a detailed typology by distinguishing between liberal, conservative corporatist, and social democratic regimes to better understand the difference in European welfare states. Post-war Western European welfare states have provided generous benefits to workers and the elderly.

The research also showed that the European welfare state has worked best in small, homogeneous countries. On the other hand, Germany's job market became the archetype of approaches replicated across Western Europe with shorter working hours and seven weeks of vacation time. In some countries, social democratic policies were grounded in programs of social legislation; in others, such as in Britain, it amounted to a series of pragmatic policies aimed at alleviating disadvantage and reducing extremes of wealth and indigence.

In order to focus more specifically on flexibility within Europe, Anxo and O'Reilly (2000) have, in line with the approach of Esping-Andersen, differentiated three forms which are determined by differences in the national welfare regimes. Sweden, Germany, and the Netherlands are examples of *negotiated flexibility* countries as they respect an extensive involvement of the social partners through negotiations and a moderate level of statutory working-time regulation. France, with its statutory working-time norm and the state's ubiquitous regulating role, is an example of *statist flexibility*. Britain has an industrial relations system where statutory working-time regulation is still absent. Unions have a right to free collective bargaining, but employees may work either very long or very short hours. This regime is an example of *externally constrained voluntarism*.[29] These countries show a large variety in temporary work and part-time jobs as forms of flexible work arrangements and have generally been very unevenly distributed. These arrangements are much higher in Latin countries such as Spain, France, and Italy, compared with Scandinavia and the UK. However, in the Netherlands, the part-time policy has favored an increase in productivity (Figure 13.1).[30]

While the percentage of women in the active working population in Western Europe increased rapidly with more women working in more flexible work arrangements, flexibility arrangements within Europe still showed remarkable differences and struggled to provide a balance between work activities and domestic responsibilities. There is still the fear that flexible work forms like telework will bring the work organization's time-reckoning system into their home, allowing billable hours to crowd out other ways to reckon time while at home. A recent study about time pressure showed that the outcome of more flexibility and the

FIGURE 13.1 Link between increase in prosperity, productivity and number of hours worked in the Netherlands. (Source: Central Bureau of Statistics.)

quality of life still reflects a large variety of patterns within the European Union (Lippe 2009).

One of the latest examples of the Lisbon Strategy to promote flexibility and social cohesion was the introduction of Flexicurity (European Commission 2007) as an avenue for securing work for the individual employee who is willing to take a less secure, flexible job but who is still secured for unemployment provisions in case they are needed.[31] On December 5, 2007, the Employment and Social Affairs Council adopted some common principles about Flexicurity which originated from the awareness that globalization and technological progress are rapidly changing the needs of workers and enterprises. Flexicurity intends to promote a combination of flexible labor markets and a high level of employment and income security. The concept moved away from a job security mentality to an employment or employability security mentality (Business Europe 2007).

A recent report of the European Commission, *Equality Between Men and Women* (2008) concluded that although Flexicurity principles have now been embraced and the female employment rate had reached 57.2 percent in 2006 – which should bring the objective of 60 percent by 2010 within reach – there are still large gaps between men and women concerning the reconciliation of professional life and private life.

Outlook

In order to stimulate the establishment of a European civil society, the notion of European citizenship has lately taken on greater prominence as a way to reconcile the balance between work and family and gender equality. This may facilitate the development of a work–life culture, i.e., a set of shared assumptions, beliefs, and values regarding the extent to which an organization supports and values the integration of employees' work and family lives (cf. Thompson *et al.* 1999: 394).

The desire for a stronger social Europe stimulates the social dialog about the necessary balance between work and family as a precondition to further promote flexibility. Although Europeans might show a keen interest in more flexible working-time regimes (Oeij and Wiezer 2006), it is still a challenge to see how a European citizenship policy might encourage a shift in the domestic division of labor (Lister 2003) and put the spotlight on care issues while pulling the private domain into the discourse on the importance of care to society (Kremer 2005). In order to reach that goal, Cass proposed that "men should not be accorded full citizenship if they do not fulfil their responsibilities for caregiving work" (Cass 1995: 60). Although this might sound a bit too radical, she wants to move beyond the constant struggle between the sexes and European citizenship policies that might become the building block for a discourse, which reaches beyond this conflict.

Although The European Commission continues to propagate a roadmap for "equality between women and men" which intends to "encourage men to take more part in family life" (European Commission 2008: 5) while stressing that equality between women and men is an essential quality component of work, this essentially requires innovative and flexible arrangements for work and leave at company level. A major problem, however, for macro-level determination of flexibility is that it is based on aggregate measures which may have been collected on different bases and using dissimilar definitions which do not provide a coherent framework to analyze the flexibility situation at the organizational level. Yet, for the European Union of 27 member states, the internal Flexicurity instruments such as time sharing, work-in-training, and wage moderation could help maintain an updated Lisbon agenda. Also, adopting a decent work agenda may "strike a new balance between competition, capacity building and sustainability" (Rodrigues 2010: 239).

In this sense Giddens (2006: 85) has proposed a radical shift by saying that the EU can only establish the objective of economic development and social cohesion if it seriously promotes a further reduction of working hours:

> No country has yet come close to what we might take as a guiding ideal, both of gender equality and lifetime flexibility – a situation in which, as measured over the lifespan, both men's and women's jobs would converge around, say, a thirty hour week (with periods of longer and shorter working hours for both).

His view is in line with what Thomas More stated in his *Utopia* of 1516, where he stressed that six hours would be enough to enjoy a civil life which is "the virtuous object of all human efforts" (More 1516/1978: 92).

The European Union will replace its Lisbon Strategy of 2010 by EU 2020. It has recently been admitted that not all of the goals of the previous document have been attained. Investments in more flexible labor markets have been hampered by the recent financial crisis of 2008 and forces of globalization in terms of moving companies to other parts of the world have frustrated an internal European social cohesion.

The issue of flexibility highlights an urgently felt need for transformation. The industrialized world has apparently been moving away from a once generally embraced combined notion of regularity, standardization, and central coordination, which during the industrialization of Western society had uniformly arranged working-time patterns. Over the centuries, the clock as a symbol of abstract time had dramatically influenced the organization of social life "by shifting the emphasis of everyday living and working patterns from variable rhythms to invariant ones" (Adam 1995: 47). Admittedly, the experience of full-time, permanent, secure employment by the mass of the workforce had mainly characterized the appearance of large-scale, highly capitalized economies with long-range processes of economic growth (Felstead and Jewson 1999).

The establishment of the standard workweek and standard workday has been central to the evolution of our modern work organizations. Due to technological and social innovations, which were rooted in a frantic drive for greater organizational efficiency and flexibility through quality improvement and customer satisfaction, the industrialized world has introduced a greater diversity in flexible working-time arrangements. Industrialized countries seem to move away from standard employment arrangements and promote less regularity and create a setting for "portfolio people offering their skills to many different clients and customers, moving between different working places, working from home using modern technology" (Felstead and Jewson 1999: 3). Notions of regularity, standardization, and central coordination, which have been embedded culturally and economically in different institutional settings, now become liable to shift under new specifically socio-cultural changes and acceleration of technological change. The permanent, full-time, weekday, daytime hour arrangements are in decline – albeit more slowly than politicians and management gurus would have us believe.[32]

The industrialized world is currently undergoing a process of de-synchronization, which means that the so-called flexible production regimes "involve much more intense coordination, but sharply diminished requirements for standardization and regularity" (Glennie and Thrift 1996: 287). Glennie and Thrift (1996) have shown that time awareness and time orientation which, in everyday life, structure living patterns in a regular way did not have to await the introduction of the factory to become common practice. Neither did the need for central coordination only arise once standardized factory production had become a general feature. National institutional frameworks have embedded this three-pronged intensity of time discipline with standardization regularity and coordination in a variety of forms. Over the centuries, a varying pattern of reduction in the time spent at work has become visible, a decline as much as an increase (Blair *et al.* 2001). Although from country to country the details may have differed, there has been a time that, at least in Europe, the general direction was the same: working time diminished.

The flexibilization of working time, however, is beginning to change this general development. Part-time and shift work, as well as weekend and evening work, represent new forms of working patterns. The increase of nonstandard, deregulated, and irregular forms of employment will not fit the existing temporal schemes of standard employment and central coordination mechanisms. The drive for more flexibility will create new organizing practices matching new working-time patterns such as those being caused by part-time work. Ingrained habits and attitudes of traditional patterns may change, and companies will be challenged by the new situation.

In the heydays of Taylorism and Fordism, punctuality was a *sine qua non* for the survival of these production systems, but IT currently facilitates an unraveling of its dominance. The triptych of collectivism/homogeneity/regularity could be replaced by the triptych individualism/heterogeneity/irregularity (Boulin 1992). It has become less important to be "just-on-time" but more important to be "just-in-time" as the Japanese miracle has shown. Flexibility distances itself not only from regularity, standardization, and fixed routines but also from punctuality. Taylorist and Fordist production systems are being replaced by Toyotist, flexibly specialized production systems (Lash and Urry 1994). Others fear that in post-industrial and service-oriented countries, low-paid employees might become the victims of the dissolutions of regularity and standardization and will no longer enjoy adequate labor relations.

According to Sennett (1998: 57), flexible organizations experiment with varying schedules of time (flexitime): "Instead of fixed shifts unchanging from month to month, the working day is a mosaic of people working on different, more individualized schedules." In such a setting, employees may be perceived as free actors who relate to organizations through the market, leaving responsibility for arranging and managing time with them.

The actual development of flexibilization is a challenging one for management and employees. The creation of the flexible firm (a version rather different from Atkinson's) with its automated production processes which allow for small-scale production almost reaching near personalization of products can certainly have more positive effects: the opportunity to apply new organizational principles to labor and to innovate technology. Literature has begun to show that the use of IT can be associated with centralization and/or decentralization in firms with more flexibility and/or more rigidities leading to electronic scientific management. The more malleable IT becomes, the less technical management is required, and the more people management is required. All will depend on how the firm will turn human resources into capabilities and competencies (Hagström 2002). Sennett noticed that corporations are shifting from being dense, often rigid, pyramidical bureaucracies to more flexible networks in a constant state of inner revision. The command lines are changing and the deregulation of the workplace shifts responsibilities downward. At IBM, for example, this practice "marks a stark contrast to the paternalistic tightly organized chain of command which orchestrated the corporation for most of its history" (Sennett 2000: 187).

HRM is faced with the challenging task of adequately catering for a new awareness of time with flexibilization *of* the employee put in parallel with flexibilization *for* the employee. Working atypical hours and weekends will change the temporal structure of family life, constraining the time that family members spend with one another, and threatening the quality and stability of relationships. Working nonstandard hours is associated with greater health risks due to changes to an individual's circadian rhythms (Presser *et al.* 2008). Management will need to broaden its scope and recognize that it operates with a complexity of different times of individuals and in multiple timeframes where the context specific timing and tempo of interactions and transactions matter. An adequate weaving in and out of different forms of time is an accomplishment that does not yet tend to form part of the explicit repertoire of management tools. Metaphorically speaking, in organizations the warp of specific stabilities such as flexibility *for* employees needs to be shot through the weft of flexibility *of* the employees. Then, securely organized companies may establish *sustainable flexibility* which prevents rigidities by (Mayrhofer 1997):

- not only designing organizations as totally flexible learning systems but also by focusing on prevailing stable company structures too;
- effectively combining company-wide standardization (coherence) and local differentiation (diversity);
- making organizations lean in terms of a fit between resources and tasks, while not endangering their capability to adapt to changing environments.

The search for greater flexibility may lead to a range of hybrid, mixed, and intermediate forms of productive organizations with more fluid forms of contract codification facilitating workplace changes which reflect significant continuities and well-organized patterns of authority. The main challenge will be how management deals with uncertainty while coordinating flexibility.

The internal management of flexible human resourcing cannot deny the challenges of different temporalities that multinational enterprises encounter in the global arena. . Usually, however, the majority of literature on organizational culture implicitly addresses the clock time perspective (see Rollins and Roberts 1998). The fact is that flexiblization is only possible on the basis of some stable and unchanging cultural orientations and institutions, which impose certain limits on flexiblization. Yet some of these institutions and cultural orientations are under pressure of erosion in late modernity, as will be shown in the next chapter.

14

TIME AND CULTURES

In 2004, management researchers Earley and Mosakowski raised in the *Harvard Business Review* (2004a) the question why some people can act appropriately and effectively in new and unfamiliar cultural contexts while others flounder. To analyze management behavior in unfamiliar cultures, they introduced the concept of *cultural intelligence* (CI) as an ability that individuals may have to easily adopt the modes of someone they are unfamiliar with, change their behavior when a cross-cultural encounter seems to require it, and adapt to the lifestyle of a different culture. These authors presuppose that it depends on individuals themselves whether they are able to adapt to specific lifestyles which are ingrained in cultures.

There are also, however, structural components determined by time patterns in cultures which need to be properly understood. This issue has become highly relevant in the twenty-first century where an increasing number of people are acting and living in multicultural contexts which are structured on different temporalities. Anthropologist Edward T. Hall (1959) has elaborated how the rules of social time constitute a "silent language" within a culture. These rules might not always be explicit but "exist in the air." They are either familiar and comfortable or, vice versa, unfamiliar and a source of misunderstanding. As a culture builds on shared norms and values, differences in these norms and values distinguish the members of one group from those of other groups. Other social scientists such as Barbara Adam (1994) have also emphasized that people rely on social norms, shared cultural values, and common understandings to value social time. They view social time as the hub of culture and have drawn attention to the way social time is perceived and organized. They have stressed that the Western clock time's prevalence does not preclude other forms of social time.

Within the globalizing economy, a growing number of multinationals require their staff to understand and respect diversity and to know how to act within a context where many different cultural backgrounds with different time patterns are present. Although there are infrastructural as well as technological developments which have facilitated an easy access to other cultures in the global economy, there are also differences in cultural patterns dealing with social time which will need to be understood and respected appropriately to make it possible for people to cope with the challenges and boundaries of increasing cross cultural interactions. That will be the topic of this chapter.

Multinational enterprises

The number of multinational enterprises (MNEs) has continuously increased since the late nineteenth century, and they continue to prosper. Before World War I, MNEs flourished and became essential agents in the internationalization of production (Lang 2006). Stopford and Dunning (1983) have calculated that in 1980, their number exceeded 10,000 with approximately 90,000 foreign affiliates. *The Economist* (March 27, 1993) estimated that in 1993, this number had increased to 35,000 and 170,000, respectively. In the following ten years, the amount almost doubled to 63,000 with 821,000 affiliates (Gabel and Bruner 2003). In 2009, about 82,053 MNEs had been registered with 807,363 affiliates (UNCTAD 2009).[1] According to estimates in 2008 by GMAC Global Relocation Services, in between 30 and 40 million expatriates are living and working globally (GMAC 2008; White *et al.* 2011). Some protagonists of globalization believe that the major MNEs are the most important and "most powerful globalizing institutions in the world today" (Sklair 2002: 7). Others like Lang (2006: 909) remarked that international trade today is not noticeably greater than during the era prior to World War I. Yet "the world has moved towards greater economic integration since World War II." In his view, globalization returned to historic peaks in the late twentieth century. Trade and new transportation technology once again generated global convergence. MNEs are seen as the most visible entities in the vast process of economic transformation that increases the speed and scope of technology diffusion and accelerates their activities to join the worldwide market expansion. They have been able to put in place worldwide transportation and communications networks which have evolved into a global telecommunications infrastructure (Geyer and Bright 1995).

Without the technological advances and facilities created by the modern IT, many MNEs would not have been able to spread as far and as wide as they have actually done. The economy of speed, acceleration, and timing drove these corporations in their expansion abroad. Due to the information and communication revolutions, time in the electronic age is now even organized at speeds beyond the feasible realm of human consciousness. Computers and networks are operating in nanoseconds and have become capable of processing in real time by making existing process cycle times as short as possible (Yeh *et al.* 2000).

MNEs have improved the speed of transfer of information, capital freight, and people. Worldwide infrastructures currently facilitate the transport of information, capital, goods, and services as well as human beings on a global scale. Today, they can operate worldwide while satellites circle the globe and provide almost instantaneous communication, which improves the timing of their activities. Globalization, in other words, hints at a process of integration across frontiers of market economies at a time of rapidly falling costs of transport and communication (Wolf 2005). The cell phone, the beeper, e-mail, and voicemail keep people connected 24 hours a day and seven days a week. These latest communication tools, together with Internet and an elaborate telecommunications system, have extended the geographic reach of an increasing part of the population of the globe and have increased the intensity of connection between people located at so many different points in the world.

Electronic globalization also promotes an increasing interconnectedness and interdependence. It organizes the world in new ways and creates new opportunities for large-scale organizations. Many of the MNEs transcend existing national boundaries to operate and communicate at a transnational level (Bartlett and Ghoshal 1998). They establish coordinated networks in which the parent company, as well as the subsidiaries and partners, transfer knowledge and expertise to cope with foreign markets in a timely manner. MNEs, therefore, continue to play an essential role in the dissemination of the beats of the metronomic society across the globe. Organizational and technological changes in the twenty-first century increasingly break down distinctions of night and day, the working week and the weekend, home and work, and leisure and work.

These incessant MNEs are operating more and more on a 24/7 pattern. In 1867, Marx had already called night work a new method of exploiting labor, but the practice was at that time still in its infancy. Initially in the mines and the textile industry and later in the steel-making and petro-chemical industry, continuous production with shift work became a common practice. When more capital-intensive industries appeared which needed to operate 24 hours a day, the empire of the day extended into the evening and increasingly colonized the night as well. MNEs operating 24/7 throughout the year became detached from societies which operate within institutional frameworks with other time patterns and arrangements.

These international corporations are not only colonizing the night but family time, too.[2]

> Today, more people than ever are active outside their homes at all hours, engaged in all sorts of activities. There are all-night supermarkets, bowling alleys, department stores, restaurants, cinemas, car repair shops, taxi services, bus and airline terminals, radio and television broadcasting, Rent-A-Car agencies, gasoline stations. There are continuous-process refining plants, three shift factories, post offices, newspaper offices, hotels, and hospitals. There is unremitting provision of some utilities – electric supply, staffed turnpike toll booths, police patrolling, and telephone service. There are many emergency and repair services on call: firefighters, auto towing,

> locksmiths, and suppliers of clean diapers, ambulances, bail bondsmen, insect exterminators, television repairers, plate-glass installers, and funeral homes.
> (Melbin 1987: 5)

Protagonists of globalization insist on the benefits of economic development and technological innovation. Globalization stimulates economic development and does not stop when political systems change.[3] But globalization's influence is supposed to reach further. Globalists such as Japanese management author K. Ohmae (1990) have even predicted that the proliferation of MNEs will also have a positive impact on societies and that global forces such as those of the MNEs are becoming so powerful that even nation-states may disappear. This point of view, however, has been contested. Global business activities may have altered the traditional relationships between economic organizations and nation-states. Yet these last ones, together with the civil society, are still essential in determining the rules of the free-market system, and institutionalize the rights and obligations of the economically powerful (Korten 1995). The emergence of these globally operating corporations has required transnational institutions to regulate the global market development. Multilateralism has been institutionalized by, for instance, the General Agreement on Tariffs and Trade, GATT (1947) and the World Trade Organization (WTO). To what extent this development will press toward a convergence is still debatable. Economists Dowrick and DeLong (2003) noticed that during the period 1950–2000, the process of globalization within what they called the "convergence club" of industrialized nations took a remarkable turn.

> The convergence club both expanded and contracted massively, as for the first time many economies joined and, also for the first time ever, many economies dropped out. In Latin America, countries like Venezuela, Peru, Argentina, Chile, and Uruguay exhibited stunning relative declines over the first half century. Coastal West Africa, French North Africa as well as the former Soviet Union fell out of the convergence club.
> (Dowrick and DeLong 2003: 200)

Whereas globalization is presumed "to reinforce the convergence trend" (ibid.: 208), it failed to do so. Their explanation is that open economies have grown substantially faster than closed economies which are locked in "poverty-trap conditions of subsistence level income, low savings and investment, low levels of education and high fertility." Although a large number of the poorer countries have opened their economies since 1980 and have created access to the globalization process, they do not share the global "convergence club" (ibid.: 217). India, Brazil, and China, on the other hand, are currently countries with impressive growth rates, but whether they might become members of the circle of convergence is yet to be seen.

The increasing number of MNEs may stimulate the process of convergence and may eventually downplay cultural diversity as a less influential role in managing these large corporations. MNEs certainly contribute in making the world "shrink" and become increasingly interconnected and interdependent. The functioning of electronic globalization is based more and more on the application of universal time. A clock-driven, chrono-centricity suggests that it makes the globe "frictionless" (Leamer 2007: 123). Yet even though globalization facilitates the transfer of ideas and shifts of technology, it also generates increasing uncertainty about and re-evaluation of relevant cultural values and norms which maintain a continuing variety of temporalities (James 2009). The further shaping and formation of a global economy might also grow in complexity while it generates an increasing diversity and divergence. Does the worldwide communality of the watch lead to a global convergence toward one global lifestyle?

The handcuff of time

In the 1960s, the Japanese watch-making industry introduced the quartz watch: mechanical regulation was eliminated and replaced by micro-electronics. They developed precision watches using quartz crystals that vibrate almost thirty-three thousand times per second. "The vibration of a crystal (like quartz) became used to calibrate microcontroller circuitry" (Langlois 2007: 58). The mass production of microelectronic chips enabled the large-scale manufacturing of digital watches.

Although Greenwich Mean Time (GMT) had been accepted since 1884 as *world time*, astronomers continued to insist on improving the precise measurement of time. This led to the further development of the quartz crystal clock and the cesium atomic clock. The most accurate timekeepers of today, the atomic clocks of the national observatories, use quartz crystal oscillators (Landes 1983).[4] However, the rotation of the earth around the sun measured with the atom clock slows down at an irregular pace. In 1972, GMT was replaced by Coordinated Universal Time (UTC) as the international time reference. UTC is coordinated by the International Earth Rotation Service in Paris[5] and maintained by an ensemble of atomic clocks around the world. UTC is no longer measured by the motion of the earth but by the oscillations at the atomic level of cesium providing an atomic time scale which approximates GMT with a tolerance of 0.9 second. Since its introduction, the Earth Rotation Service has added an extra second 24 times to remain accurate.[6] Pilots of commercial aircraft and ground controllers, as well as space missions and military aerial and naval activities operate on UTC which is called "Zulu Time." The corrections to be made, however, worry some organizations. The International Telecommunication Union (ITU) has been asked to decide whether these corrections are still needed because they require costly interventions in the computer and network systems.[7]

The concept of the mechanical clock time has ended up on the electronic highway where the processing of information takes place in the order of nanoseconds. The mobility of information (i.e., facts, ideas, and beliefs) today is tremendous, and it seems as if the speed at which information is exchanged determines the pace of progress. Digital technologies have created a computerized electronic network where speed is determined by the standard of instantaneity. Clock time, i.e., the homogeneous "empty" time of hours, minutes and seconds, has become an almost universal phenomenon in the contemporary age of electronic globalization. According to Castells (1996), organizations are now able to operate in real time on a planetary scale.[8] The modular timeframe facilitates manufacturing businesses operating in the real economy to assemble products wherever it turns out to be most cost-efficient. The network economy is further emptying out time in its specific forms by pushing the natural borders of day and night to the point of disappearance. The development toward a global economy suggests an increasing homogenization of time patterns. Exchange of electronic information generates a "timeless time," which is disconnected from the natural rhythms of day and night. "The dominant tendency is toward a horizon of networked, an a-historical space of flows, aiming at imposing its logic over scattered, segmented places, increasingly unrelated to each other, less and less able to share cultural codes" (ibid.: 428).

This process of emptying out of time, however, is not colonizing the whole globe. We still cling to certain natural rhythms. One is the Circadian rhythm.

Humans are regulated by the circadian clock, a biological rhythm that is detached from external time indicators. The circadian clock regulates processes at all levels, from genes to behavior, in a pattern which is closer to a rhythm of 25 hours a day. The discrepancy between biological and social time is called "social jetlag." The circadian rhythm can be exemplified by the characteristics of certain chronotypes. There are individuals who rise early in the morning, whereas others prefer to stay up late and work in the evening. The early risers work more efficiently during the day, whereas the late risers seem to be more efficient later in the day. These social jetlags have consequences for the way different chronotypes see themselves scheduled in work organizations.

In the Western world, the clock has generally taken precedence over biological patterns such as the circadian rhythm (Levine 1997). The Western *metronomic society* is primarily focusing on abstract, cosmological time as overarching all kinds of natural as well as culturally determined and socially subjective times (Young 1988). But biological events are still present and have to do with the way humankind has embedded the impact of natural phenomena such as circadian patterns in their social lives.[9] Factories, schools, and offices structure their coordination to coincide with the mechanical clock. Their functioning is not achieved so much by the circadian clock, but "it is perilous to override our biological rhythms which are tied in to astronomical rhythms" (Young 1988: 254).[10] With the increasing expansion of the 24/7 economy, chronobiology is focusing on the temporal program driving certain behavioral patterns in a cyclic environment. It studies how phasing the biological clock asynchronously with the social clock results in misalignment and causes problems of health, productivity, and quality of life. It also explores how the organization of society has to take the existence of these chronotypes into consideration (Wittmann *et al.* 2006).

Another example of the way clock time is perceived in a cultural way is the way the modern watch has become a fancy *handcuff of time*. Whereas traditionally a watch was an expensive object which was only bought once in a lifetime, today it has become a gadget. The traditional watch-making industry in Switzerland lost its tremendous market share after the Japanese innovation of the quartz watch. Yet in 1981, the consultant Nicolas Hayek who received an invitation by the Swiss banks to intervene and save the Swiss watch industry was able to introduce a new market strategy. He took the initiative to merge several different Swiss watch-making companies into a large one, Société Suisse de Microélectronique et d'Horlogerie (SMH), and introduced a new marketing strategy of low-end Swiss brands such as Swatch (Langlois 2007). Watches evolved into more than just the casual cachets adapted to the dress code of the day. The versatility of the handcuff of time is pictured in Figure 14.1, where the watch has become a consumer good with really

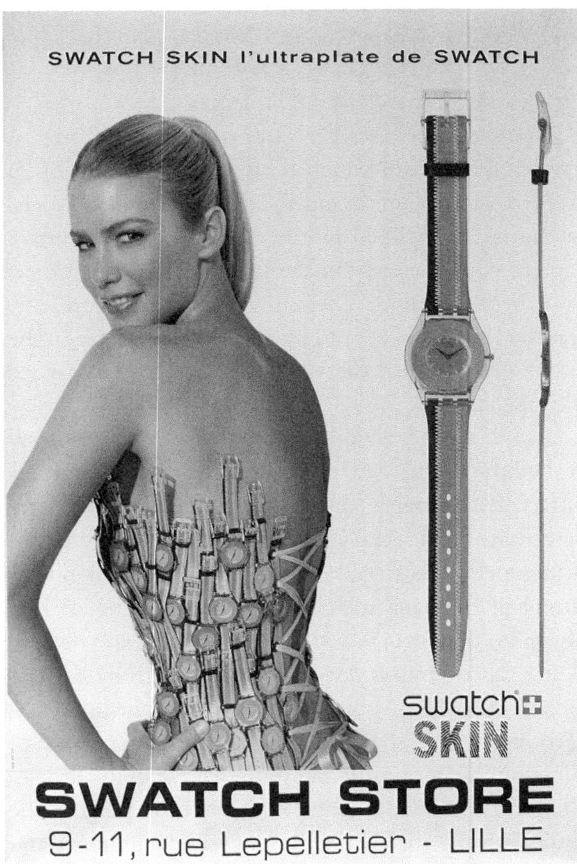

FIGURE 14.1 Advertisement for Swatch wristwatches.

no specific features related to respect for time.[11] Swatch no longer "sells" time but a particular image and style instead.

And finally, the practice of Daylight Savings Time (DST) challenges the prevalence of a fully synchronized universal time, or summertime. While MNEs were trying to extend uniformity, both in time keeping and the application of the Greenwich meridian to coordinate their international communication and transportation facilities, an unexpected approach to the use of social time was launched (Bartky 2007).

Daylight Saving Time practices

During the period of industrial globalization, regulating life in accordance with time zones not only became common practice but was also challenged by the introduction of Daylight Saving Time practices. DST challenged the standard time of each time zone and is usually memorized as "spring forward, fall back." It became a social convention of advancing clocks so that afternoons would have more daylight and mornings less.

Around the same time, two men in particular, both living in the British Empire, took initiatives for the use of DST. The New Zealander entomologist, astronomer and Fellow of the Royal Society of New Zealand, George Vernon Hudson (1867–1946), had a shift-work job which allowed him leisure time to collect insects. He began to notice the advantages of shifting daytime in accordance with the seasons. Yet plans to introduce this practice nationwide did not materialize. It is the British builder and outdoorsman William Willett (1857–1915) who has been credited as the father of today's practice of advancing the clock periodically. Willett observed with dismay how many Londoners slept through a large part of a summer day and advocated that during the spring and summer months, the clocks should be put forward so that people might enjoy the early morning light. In 1907, he sent a pamphlet entitled *The Waste of Daylight* to the British Parliament in which he proposed to improve health and happiness by advancing the clock 20 minutes on each of the four Sundays in April and retarding them by the same on the four Sundays in September. Willett saw economic advantages in his proposal while diminution of "wasted daylight" might save illumination costs. His proposition, however, encountered considerable opposition in parliament. Winston Churchill, president of the British Board of Trade, was surprised about the way members ridiculed the plan, pointing out that cows would not adapt and give milk one hour earlier.

When a bill was drafted in 1909, Churchill, now as Home Secretary, did everything in his power to convince members of parliament to vote in favor, yet the bill failed.

However, things changed quickly during World War I when Germany and Austria-Hungary advanced all clocks by one hour on April 30, 1916, in order to reduce the use of petroleum, gas, and electricity expenses for lighting. Other belligerents and neutral countries such as Holland soon followed the example.

Belgium, Holland, Denmark, and Sweden all advanced their clocks one hour from May until the end of September. Great Britain and Ireland immediately followed, and France adapted in June (Bartky 2007). Unlike most European nations, the United States did not have one national time. Still, American retailers and manufacturers led by the industrialist Robert Garland lobbied for DST but were opposed by the railroads. Although DST was seen as a potential way to save energy, reports about its effects were contradictory. The use of electricity turned out to be greatly affected by geography, climate, and economics, which made it difficult to generalize from single studies.

Yet, several different American Chambers of Commerce were convinced that implementation of federal time zones and advancing the nation's clocks by one hour during the summer would certainly lead to savings. The president of the American Federation of Labor, Samuel Gompers, also supported the initiative as he was convinced that it would improve working conditions. An act was passed, making the new approach effective from January 1, 1918, beginning on the last Sunday in March and ending the last Sunday in October. Yet the war's end swung the pendulum back. Agitation against the act increased, and it was soon repealed. With the end of the war, many other countries also abandoned the system. During the interwar years, however, some primarily northern American states continued to apply DST.

During World War II, the expected saving of coal and electricity through the advancement of time was generally regarded as a desirable collective goal which could support the war effort (Zerubavel 1982).

American Congress reinstated DST on February 9, 1942, and clock time was advanced one hour to save energy. However, after the war, the act was repealed again. A DST coalition composed of American companies such as fast food chains, greenhouses, and sportswear manufacturers believed that shifting the hour of daylight would successfully enhance their sales and continued lobbying in Congress to shift the onset of DST to the first Sunday in April. In 1966, the USA adopted a uniform time act but it was not fully implemented. American states requiring an exemption from DST were allowed to pass a local ordinance. In Europe, some countries adopted DST to conserve energy. A British Summer Time (BST) – later renamed British Standard Time – was enacted in 1968. But the persistence of darkness in the winter mornings forced a reversion in 1971.

On both sides of the Atlantic, studies found no evidence of net energy savings resulting from DST, but there were indications of an increased number of accidents (Bluedorn 2002). A focus on the advantages of DST in terms of public safety did not provide conclusive results. Research focusing on advantages in terms of health only showed that changing shifts disrupted sleep and sometimes reduced efficiency due to adaptation of the circadian rhythm. During the period 1968–1971, the United Kingdom and Ireland continued with experiments with DST but abandoned the project due to increasing unpopularity. Although the complexity of the application of DST continued to hamper a favorable decision, the European Union finally decided to implement a European summertime in 1969 from the first

Sunday in March to the last Sunday in October. In 1996, the European Union standardized the European summertime, and it is now being observed from the last Sunday in March and lasts until the last Sunday in October. European citizens now live according to a "spring forward, fall back" practice.

After the Arab oil embargo of 1973, America put most of the nation on extended DST for two years to save additional energy. Although it was decided to end DST after two years, again retailers, sporting goods manufacturers, and other businesses continued to favor DST, stating that the public might benefit from extra afternoon sunlight, shop longer, and participate more intensively in afternoon outdoor sports. From 1987 onwards, most of the United States and Canada respected DST with starting and ending dates from the first Sunday in April to the last Sunday in October. Since 2007, DST has been observed from the second Sunday in March to the first Sunday in November.[12] Although DST suggests saving a good percentage of energy consumed by lighting and appliances, proof is still inconclusive.[13] The impact of DST is currently being extensively studied by researchers in chronobiology to improve the balance between the biological, i.e., circadian and social time.

Beyond the influence of natural phenomena such as the circadian and seasonal rhythms, there are also sociocultural patterns that demarcate our ways of dealing with social time. Humankind is living in a world where social time is everywhere, undetectable, taken for granted, and finally difficult to describe (Saunders *et al.* 2004). It is also living in a more-than-capitalist world and, therefore, clock time's prevalence does not preclude other forms of social time or temporalities. Social time has its roots deep in culture and people's daily behavior is influenced by social time.

Timeframes in cultures

American anthropologists Edward T. Hall (1914–2009) and his wife Mildred E. Reed Hall (1925–1994) have conducted several studies about intercultural communication in which they stressed the role of nonverbal communication and the ignorance of the "silent language" of behavior. They labeled the rules of social time as "silent language" which is more than body language (Hall 1959). It is the complex product of key conceptions of time, space, and culture. Serious misunderstandings of the silent language result from the fact that one starts to feel and reason based on premises which are determined by one's own culture. This may lead to a false perception of the intentions of a person with another cultural background. Edward and Mildred Hall have stressed that organizations are cultures which reside in someone's individual interpretations. "Silent language" has some common characteristics often hidden in indiscernible aspects of social life. Yet it also generates consensual interpretations among people of how things are and sustains their views of the world and images in the world.

Hall has drawn several different distinctions to better understand the impact of a diversity of views on social time and the way cultures deal with it. These distinctions will be introduced here as they have inspired many other researchers to apply them in order to improve our understanding of the meaning and impact of different temporalities in cultures.

High- and low-context cultures

Hall (1976, 1983) was the first to introduce a distinction between high-context and low-context cultures. The author referred to the cross-cultural differences observed in various fields including social interactions, communication, and problem solving, and identified, in this respect, these two main kinds of cultures. Social orientation is one of the most fundamental differentiating aspects of high-context societies. People in such societies are deeply involved with each other and create strong bonds between family members, friends, and colleagues. These bonds determine members' behavior. People in these societies have a generally higher commitment level which is mainly due to their stronger social orientation. Complex interpersonal bonds influence what one is allowed to do, leaving limited space to follow individual interests.

These people are more committed to common goals and more willing to cooperate with fellow members, but they are also more characterized by conformity and group orientation. They tend to use high-context messages in their daily communication. These messages contain in-group references, which may be only understandable for group members who share similar experiences and expectations. High-context societies, therefore, are strongly group oriented, and people have a close connection over a long period of time which enables them to communicate quickly and efficiently using a large set of insider references through common knowledge and beliefs. Social interaction within the high-context cultures requires more time.

High-context communications or messages are carried out with most of the information either contained in the physical context or internalized in the person and very little in the coded, explicit or transmitted form. In high-context cultures, things unsaid may be more important than things said (cf. Kim *et al.* 1998). The context in which the message is delivered provides just as much information as the actual content of the message. One of the pitfalls of the high-context version of communication, however, is that it presumes common tacit knowledge.

Low-context societies, on the other hand, are typically more individual based. Their bonds are less likely to add up to a large and interconnected network which would influence their actions. As a result, people in low-context societies tend to make their decisions based on personal judgment and according to their own interest. Relationships are easy to enter into, but they are less likely to become deeper, and they bond the participants only moderately People in low-context cultures are likely to use low-context messages, referring mainly to explicit knowledge and facts with the mass of the information vested in explicit codes. These communicative interactions consume much less time. People tend to engage themselves in shorter interpersonal relations through information which is public and easily accessible. Knowledge appears in a more transferable form and is usually independent of context. Low-context cultures such as the USA pay more attention to the meaning of words as carriers of information. Actions are settled within the context of written rules and regulations. Personal knowhow is more important than networks. Investing time in relationships is less important than earning money.[14]

Kim *et al.* (1998) have fine-tuned Hall's initial distinction and identified certain areas such as social orientation, commitment, responsibility, confrontation, communication, and how to deal with new situations to describe where and when high-context and low-context cultures differ significantly. People in high-context societies, for example are more comfortable within the boundaries of their own culture but are less effective when exposed to new effects. Low-context cultures emphasize more on individual rather that personal relationships, a direct meaning carried by a word rather than ambiguous word is preferred. Behavior is ruled by written rules and regulations, and one gets straight to the point without wasting time (Trompenaars and Hampden-Turner 1998). In all the areas mentioned, social interaction within the high-context cultures requires more time. Deals are not only made on the quality of the product or the best price but also on whether the person is deemed trustworthy (cf. Saee 2005). To arrange a deal is much more time-consuming.

Figure 14.2 shows how the high–low-context continuum is closely related to the degree to which one is aware of the selective screen that one places between oneself and the external world. It indicates the relationship between high and

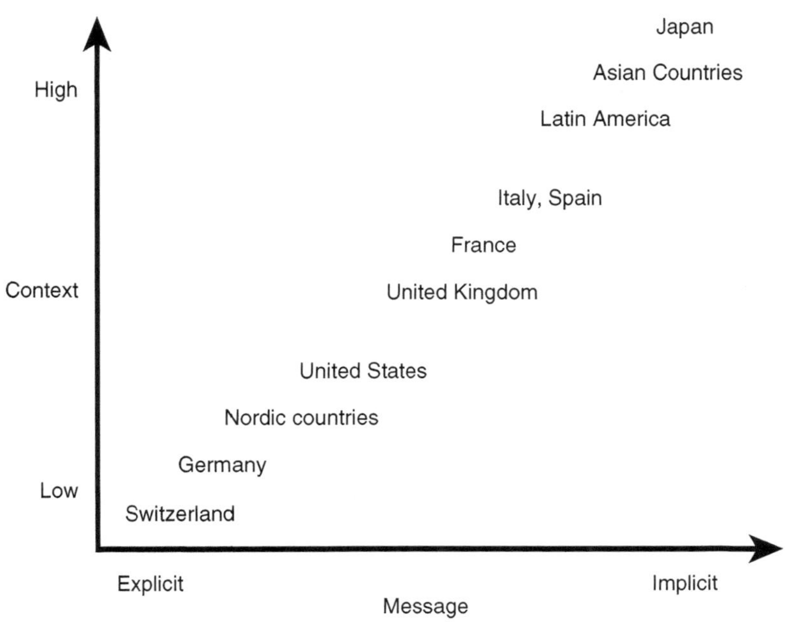

FIGURE 14.2 Communication patterns (Demorgon and Molz 1996).

low-context cultures and the implicit and explicit status of messages in different countries. The higher the context, the more implicit the message will be. As one moves from the low to the high side of the scale, the awareness of selective processes increases.

Hall has pointed out that, although no culture exists exclusively at one end of the scale, some are more high-context while others are more low-context. According to him, American culture is more toward the lower end of the scale and China, as owner of a great and complex culture, is more on the high-context end of the scale. However, Hall did not consider any of these cultural forms to be superior. Both have their benefits and pitfalls, but once differences are recognized and acknowledged, the cooperation among persons from high- and low-context cultures may become easier.

Clock time and event time

Hall has also drawn a remarkable distinction between clock time and event time. As social time is deeply rooted in culture, clock time is related to a certain cultural background. Most commonly, clock time prevails in low-context cultures, is associated with a higher level of industrialization and development and seen as a scarce and limited resource (cf. Ballard and Seibold 2004). "Time is Money" has taught people in low-context societies that time can be spent, saved, used up, possessed, wasted, planned, and invested (cf. Lee and Liebenau 1999).[15] Clock time in industrialized countries has been integrated in infrastructural provisions by subsequent revolutions in transportation, communication, and productions which presently underpin a worldwide infrastructure. It has introduced concepts such as punctuality, deadlines, efficient-time-use, and a faster pace of life (cf. Brislin and Kim 2003; Whipp *et al.* 2002).[16] Schedules symbolized as a *clock* represent official, formal, and task-oriented temporal concerns. In that sense Hall once noticed that:

> People of the Western world, particularly Americans, tend to think of time as something fixed in nature, something around us and from which we cannot escape, an ever present part of the environment just such as the air we breathe. That it might be experienced in any other way seems unnatural and strange, a feeling which is rarely modified even when we begin to discover how really differently it is handled by other people. Within the West itself, certain cultures rank time much lower in overall importance than we do [...]. As a rule, Americans think of time as a road or a ribbon stretching into the future along which one progresses. The road has segments or compartments which are to be kept discrete ("one thing at a time"). People who cannot schedule time are looked down upon as impractical.
>
> (Hall 1959: 28)

People behaving according to clock time tend to rush from one place to the next in order to be punctual. They get irritated when others are late or indicate carelessness of schedules and appointments (cf. Schriber and Gutek 1987; Macduff 2006). Punctuality is always linked to schedules, deadlines, and efficient use of time (cf. Ballard and Seibold 2004). Great importance is attached to time saving, and high efficiency of time use is well appreciated and pushes people to accept a life with speed, acceleration, and timing.

Modern industrialization with its fierce competition situated in a modern society with a mainly urbanized life has imposed pressure on people to deal with more than they are able to deal with effectively. The anthropologist R. Levine (1997) has ranked 31 countries on their overall life pace, and concluded that Japan and Western European countries scored fastest overall (see also Levine and Norenzayan 1999). He also examined the walking speed of pedestrians and noticed that seven of the top ten countries from Western Europe still rank in the top ten.

The picture is different in high-context cultures where an event-based view of time is prevalent. As clock time is not so much integrated in high-context cultures, it is quite natural that people have no sense that time is limited and precious and should be saved. In these cultures, the pace of life is slow and people focus less on punctuality, scheduling, deadlines, and efficient time use. Events start when people arrive and end when people leave. Hall (1959) has called this a diffused point pattern. If an appointment is set at 8:30 a.m., the attendants might show up anywhere between 8:25 a.m. and 8:45 a.m. While the clock dictates the beginning and end of activities, the natural course of events determines the occurrences in event time. These high-context cultures are still predominantly based on the inheritance of agricultural traditions in which activities occur by cycles, such as seasons (cf. Palmer and Schoorman 1999).

The French sociologist P. Bourdieu (1963) once studied a farming society – Kabyle – in Algeria where any semblance of haste in their social affairs was despised, regarding it as a lack of decorum combined with diabolical ambition. In such societies, peasants and their families were living their lives at a rhythm which was determined by the division of the ritual calendar. That rhythm was based on a cluster of contrasts between complementary principles such as day and night, light and shadow, the rising and the setting of the sun, and spring being opposed to autumn.

> All of these pairs of opposing notions, which define the calendar of agricultural labor and crafts, also underlie the principle of the division of labor between the sexes and the distinction between the moist foods of autumn and the winter and the dry foods of the spring and summer.
>
> (Bourdieu 1963: 56)

Peasants lived the rhythm of social life, the season of feasts, and the cycle of pastimes without haste and without punctuality. Social time was associated with the experiences of events and the space within which these events took place. Clear distinctions between past, present, and future were rarely made. Farmers and peasants did not dissociate their work (the present) from its economic results (the forthcoming).

The notion of the future was formulated as "foresight," an anticipatory vision, "born of the logic of the situation" (ibid.: 66).[17]

The clock was referred to as "the Devil's mill" and was mistrusted. Rushing from place to place was not significant, neither was time efficiency in completing tasks. In Algeria, social meetings are not seen as social obligations as they are in the Western vernacular. Similar resistance against a clock time regime in factories can be found in Tunesia (Hibou 2011). Levine (1997: 94) noticed that many Mediterranean Arab cultures define only three sets of time: "no time at all," "now," and "forever." As a result, "American businessmen have often encountered frustrating communication breakdowns when trying to get Arabs to distinguish between different waiting periods, for example, between a long time and a very long time."[18]

Whereas clock time is connected with a low-context culture, event time is linked to high-context cultures. In contrast with clock time, which focuses on formal concerns and on tasks, event-time patterns focus more on the interpersonal relationships among people (cf. Brislin and Kim 2003). But even for some individuals living in the modern post-industrial world, event time can still prevail. They are not reckoning their life by abstract clock time but by breathtaking events. When event time dominates, scheduling is determined by activities. People who act according to event time organize their days around different events and participate in one event until it reaches its natural end. "Event time is based on social, economic and environmental cues, and, of course, of culture values" (Levine 1997: 91; cf. Usunier 2000).[19]

Event times depend more on nature-related patterns, such as the seasons, the weather, and even on the position of the sun in the sky, and can be found in South America, South East Asia, Africa, and other developing countries. These regions still extol event time as a philosophy of life. In Mexico, for example, there is a popular saying: "Give time to time" (*Darle tiempo al tiempo*). Across Africa, you still can hear people using popular slogans such as "Africa invented time, Europe the watch" and "Even the time takes its time" (Levine 1997). Africans apprehend time as a looser concept, more open, elastic, and subjective. The Polish journalist R. Kapuscinski (2001: 16–17), who had widely traveled in Africa, remarked: "Time is something that man can create outright, for time is made manifest through events, and whether the event takes place or not depends, after all, on man alone."

Taking time, therefore, is very important in African social relationships (Box 14.1). People in Africa live in sequence, implying that something has to happen before the next step can be taken (Mbiti 1969). During workdays, Africans take the time to socialize, listen, and communicate. In Burkina Faso, for example, it is imperative to make time for the people in one's life. If one does not, it is seen as an extremely wasteful use of time. Any conversation starts with asking about the family situation and how someone's family is doing. Taking time to talk with and listen to one another includes that one respects the fact that more time for social encounters increases the quality of the interaction and creates a sound basis for trust-based relationships.

Calendars and daily schedules may cause misunderstandings for an outsider. In Ethiopia, for example, it is one o'clock at 7 a.m., two o'clock at 8 a.m. and so on.

Visitors and expats need to master cultural intelligence to be aware of this when making an appointment at one o'clock since then one has to get up really early. [20]

Kapuscinski concluded that for Africans:

> Time appears as a result of our actions and vanishes when we neglect or ignore it. It is something that springs to life under our influence but falls into a state of hibernation, even nonexistence, if we do not direct our energy toward it. It is a subservient, passive essence, and, most importantly, only dependent on man.
>
> (Kapuscinski 2001: 16–17)

BOX 14.1 SOCIAL TIME IN AFRICA

The African researcher M. Dia (1996) observed that the tensions of labor-management relations in sub-Saharan African firms for a long time were largely the outcome of the adversarial attitudes of colonial employment relations that excluded any consideration of local cultural patterns. Jackson (2004) organized a large-scale cross-cultural African management analysis and remarks that post-colonial management approaches in African firms may develop more hybrid forms removing bureaucratic control mechanisms and emphasizing a more humanistic view of organizations serving the ends of multiple stakeholders and respecting the heterogeneity in African firms. In Africa, life can hardly be compartmentalized into professional life, family life, and afterlife as has become common in the Western world. Though, bureaucracy and hierarchy may still slow down decision-making in African firms. Jackson, however, also notes that by depending on the hybrid forms of African management perspectives, short-term decision-making may be facilitated in order to obtain relatively quick results (Jackson 2004, 89). As long as firms in Sub-Saharan Africa require low skills and little training time, the workers might have little autonomy and intensive supervisor control. With more modern production operation and improved Human Resources Management policies, jobs with more variety and less supervision may facilitate hybrid management forms which facilitate the integration of African and Western connotations of time patterns (Kamoche 1999). Developing effective and appropriate management in Sub-Saharan Africa is key to designing responsive and responsible organizations that can deal with a variety in time arrangements. Eritrea, for example, functions according to the Geez calendar which means that November 6, 2009, is actually October 27 of the year 2002. Eritreans celebrate New Year on September 11. The orientation toward the past, present, or future is also a complicated issue (Gibson *et al.* 2007). In a Tanzanian language Schambala, no words exist for future or for past. It is either "today" or

(Continued)

"not today" (Du Plessis and Venter 2007). The empress of African song, Miriam Makeba (1932–2008) – who had been expelled from her homeland of South Africa during the apartheid regime and returned when Nelson Mandela became president – illustrated in her autobiography (Makeba 1988, 2) how western and African societies differ in temporal focus with regard to past, present and future. In comparing the West's temporal orientation with that of the South African people, she wrote:

> In the West, the past is like a dead animal. It is a carcass picked at by the flies that call themselves historians and biographers. But in my culture the past lives, my people feel this way in part because death does not separate us from our ancestors. The spirits of our ancestors are ever present.

The African emphasis on the past, present, and future creates a different temporal focus. Brislin and Kim (2003) point out that if someone wants to understand his own cultural time, it is essential to analyze the orientation to the past, present, and future. This is usually called an orientation toward the time horizon (Macduff 2006).

Similar visions on event time can be found in South America and South East Asia where the focus on clock time is less integrated in work life (Levine and Norenzayan 1999). Levine (1997) explored the pace of life within 31 countries and came to the conclusion that Syria, El Salvador, Brazil, Indonesia, and Mexico are the countries with the lowest pace of life.[21] In Indonesia, for example, people do not name a time when setting an appointment, but they use a day or a part of the day, such as "Pagi" (between 5 a.m. and 11 a.m.), "Siang" (between 11 a.m. and 3 p.m.), "Sore" (between 3 p.m. and 6 p.m.), or "Malam" (after dark). They also use "rubber time": "Jam Karet." This kind of social time is not fixed but, instead, very flexible and can be stretched for whatever reason.

Levine (2004) surveyed 200 colleges in California and Brazil to understand what it means to be "early" or "late." American students believed they were "late" for a lunch date when they made a friend wait for 19 minutes; for Brazilians it was 34 minutes. In a similar vein, Levine asked pedestrians in both countries who were not wearing watches to estimate the time. The errors Brazilians made were twice as large as those by Americans. The results of these two surveys show that the elasticity of social time varies across low-context and high-context cultures.

As for the pace of life, it is usually stated that Western Europeans combine a fast pace with ample leisure, while Latin Americans take their time both on and off the job. Eight of the nine fastest countries were located in Western Europe, "with Japan the lone intruder on this monopoly" (Levine 1997: 132). People pursuing a lifestyle with a fast pace constantly struggle with overloaded tasks as there is the sense of a continuous shortage of time.[22] The last eight ranks were occupied by unindustrialized countries from Africa, Asia, the Middle East, and Latin America. "The slowest of all were the great cradles of *amanha*, rubber time, and *a manana*: Brazil followed by Indonesia, and in last place, Mexico" (Levine 1997: 136). Many people in these countries were completely out of touch with clock time. Levine (1997) also noticed that Mexican-Americans differentiate between *hora inglesa* – which refers to the actual time on the clock – and *hora mexicana* – which treats the time on the clock considerably more casually. He discovered in one of the neighborhoods in Los Angeles that the unemployed Mexican-American residents had almost no perspectives toward the future. He also noticed that some African-Americans have the attitude of casualness as expressed in "Colored People's Time" (CPT) which means that coming late to meetings is the norm and "contrasts sharply with an Anglo-European penchant for punctuality and timeliness" (Levine 1997: 86).

In a similar study regarding the application and accuracy of clocks in banks, Levine noticed that in several different countries the banks did not respect the same synchronized standard. Switzerland ranked first followed by Ireland. The only "trailer" from Western Europe was France where clocks of banks were less accurate. The American bank clocks were significantly more accurate than their counterparts in other parts of the world such as Brazil. The differences between clock time and event time affect people's daily behavior, and these differences in social time perspectives are to be recognized and acknowledged to improve one's cultural intelligence (Jones 1988; Earley and Mosakowski 2004b; Macduff 2006). Yet in industrialized countries, there is a prevailing attitude to determine cycles of work in line with the clock time regime, that is, in allocating time to specific tasks and weeding out event time such as rubber time as much as possible (McGrath and Rotchford 1983). Levine and Norenzayan (1999) noticed that punctuality was greater in countries with colder climates, with more productive economies, and with more individualistic or low-context cultures. They noticed high correlations between these three variables and punctuality. Yet even without taking into consideration climatological circumstances, one may notice a gradual transition from event time to clock time.

Polychronic and monochronic time

Hall (1959) was also the first to develop the distinctive concepts of monochronic and polychronic time, which provides a fundamental description of temporal differences focusing on the complexity of temporally infused behavior.[23] Hall termed the people who prefer to complete one task before becoming involved in another task as having a monochronic orientation. Activities are scheduled as separate items: doing one thing at a time.

> Years of exposure to other cultures demonstrated that complex societies organize time in at least two different ways: events scheduled as separate items – one thing at a time – as in Northern Europe, or following the

> Mediterranean model of involvement of several things at once. The two systems are logically and empirically quite distinct. Like oil and water, they don't mix. Each has its strengths as well as its weaknesses. I have termed doing many things at once: Polychronic, P-time. The North European system – doing one thing at a time – is Monochronic, M-time. [...] For polychronic people, time is seldom experienced as *wasted*.
>
> (Hall 1983: 45–46)

Hall considered both concepts as another relevant distinction indicating the impact of the "silent language" within a given culture that determines different forms of interaction and communication which affect important temporal concepts such as appointments, schedules, and acquaintance time, which may cause serious misunderstandings. Hall (1983: 113) reported, for example, how a French polychronic manager, who saw himself as monochronic – which is "consistent with a systematized linear culture in the Cartesian sense" – could not deal with a monochronic German colleague.[24]

In monochronic time (M-time), all significant activities are strictly scheduled. M-time focuses on segmentation and promptness and implies doing one thing at a time. Hall (1976) asserted that making schedules strengthens monochronism which is characterized as a short-term and linear view of time and an attitude of people to compartmentalize time which denies context. For monochronic patterns, schedules are binding and punctuality prevails. There is a clear preference to perform tasks sequentially. He noticed that Americans have a drive for scheduling and alleged that, if it had not been for schedules, industrial societies might not have come into existence. Monochronic cultures are renowned for their drive toward punctuality and are obsessed with a practice of separate and sequence activities (cf. Macduff 2006).

Cultures around the Mediterranean, on the other hand, are more polychronic. People living there prefer an involvement of several things at once. Polychronicity means "the extent to which people prefer to be engaged in two or more tasks [...] simultaneously and believe that their preference is the best way to do things" (Bluedorn 2002: 51). Polychronic people pay less attention to schedules and are more focused on interpersonal relationships (cf. Zakour 2004). Polychronic cultures in Latin America are also less focused on punctuality and place more significance on the goals, not so much on the means or stages to achieve the goals (Trompenaars and Hampden-Turner 1998). People from polychronic cultures live their lives more quietly and much more "nature-paced." It is as if *temperance,* which was once so highly esteemed in medieval Western Europe, is still well respected there. Polychronic patterns of behavior respect the fact that every activity takes its own time. In countries such as China, India, and the Arabian countries, there is a tendency to view social time more flexibly and to attach less importance to time consciousness or the urgency to respect appointments. In such polychronic cultures, employees must to be constantly persuaded and pressed to meet deadlines.

Hall initially concluded that the two orientations of polychronic and monochronic patterns were logically and empirically quite distinct and acted like oil and water, which do not mix. Yet in later years, Hall admitted that things were changing and that the direction of the change is difficult to gauge because the forces promoting the change are pulling in opposite directions (Bluedorn 1998). He added that, historically, the monochronic culture in America was related to the fact that the European settlement of North America was initially dominated by people from Northern European cultures with traditionally monochronic cultures such as England, Germany, and Scandinavia. Yet the prominence of Native Americans and African Americans during the European colonization created a variety of polychronic subcultures and situations where different cultures became harbored within one country at the same time. People with a Hispanic polychronic cultural background who arrived much later in large numbers are now becoming a pulling force toward more polychronic American attitudes. As for monochronic forces, he has stressed that certain technologies such as televisions and computers are still pulling toward monochronicity.[25]

Zerubavel (1981) likewise has remarked that the focus on polychronic behavior in Western society has also increased since the distinction between work time and family time has changed. He noticed in America a sharp difference between work and home for low-status workers, i.e., clerks who go home on time, and professionals such as managers who work late. Technological changes such as the use of computers, however, have resulted in a less sharp distinction between the home and the workplace as people today can often work from home. The growing interest for flexible work time arrangements such as telework might also shift the balance toward more polychronic patterns. Yet domestic labor, with its locally constructed event times such as "dinnertime" and "bedtime" for children, has a significantly different time pattern than that of paid work, which causes tensions within the home.

Hall suggested links between high and low-context cultures, clock time, and event time as well as monochronic and polychronic time.[26] Monochronic time emphasizes schedules, segmentation, and promptness and is usually considered to prevail in low-context cultures. In the Western world, where low-context cultures prevail, man finds little in life that is exempt from the iron hand of M-time. In contrast, P-time is more connected with high-context cultures. Yet Kaufman *et al.* (1991) found Americans more polychronic when looking at their attitudes toward multitasking work habits. Brislin and Kim (2003) also noticed that the strategy of "killing two birds with one stone" is used by many Americans. So, there is no overall answer to the question of which orientation – M-time or P-time – is better. Yet monochronic time does not seem to pay sufficient attention to the polychronic aspects of society, while, on the other hand, polychronic time could cause inefficiencies (Levine 1997).

Hall, however, admitted that in today's post-industrial societies, a balanced combination of M-time and P-time as well as respecting the coexistence of clock time next to event time may be the best accommodation to the way we presently live (Bluedorn 1998). Japan is, in that sense, a nice illustration. The country currently demonstrates an extremely complicated temporal pattern, which is shaped by the coexistence of rigidity and casualness (Tsuji 2006). Japanese live in a high-context culture which includes that, in meetings and other social encounters, they do not get to the point quickly but presume that intelligent people will discover the point of a discourse from the context (cf. Hall 1983). Yet since its industrialization, Japanese society has become familiar with the tyranny of the clock, which regiments people's activities at workplaces and schools as well as in public transportation systems by demanding speed, punctuality, and timing. Tsuji (2006: 182) noticed that in terms of monochronic time, "Japanese people endeavor to be on time, do things swiftly, and avoid wasting time."

At the same time, however, the Japanese do not always honor these values because polychronic time also has a significant impact on their lives by giving top priority to social relationships rather than punctuality, speed, and productivity.

> Japanese people respect punctuality in the morning, but while they do not draw a clear line between work and social life, they apply great flexibility in the evening and do not return home at set schedules. During the day, clock time is respected and Japanese commuters perceive a 90 second delay of trains behind any set schedule a serious delay. Event times, however, which are shaped by social interactions, require flexibility. Japanese people may not plan an appointment with a medical doctor or hairdresser ahead of time and are willing to wait for quite a while at the doctor's office or at the hairdresser. Japanese discourse of time, therefore, is not monolithic but contains inconsistencies and contradictions.
>
> (Tsuji 2006: 191)

There are apparently tensions between clock time and event times, yet the Japanese seem able to live with the fact that event times are always embedded in specific – historically determined – sociocultural settings.

Brislin and Kim (2003) have looked at the possible combination of M-time and P-time respectively with both clock time orientation and event time orientation, as shown in Figure 14.3. The four boxes illustrate possible combinations, but as Bluedorn *et al.* (1999) have indicated, the two dimensions are not fully distinct concepts, but reflect opposite poles; several different combinations can be imagined. The USA, for example, has traditionally been a combination of M-time with clock time (Square A), however, it is now moving toward the combination of P-time and clock time (Square C). On the other hand, modern and industrialized Asian countries, such as the Japanese and Korean societies, are moving to the combination of clock time and M-time (Square A) from a traditional event time and P-time (Square D) as they have begun to realize the importance of speed, timing, and deadlines when doing business in the global market. Brislin and Kim (2003) expect that Latin American and Mediterranean countries will eventually converge toward clock time/M-time because they cannot escape from the hegemony of watches and clocks without isolating themselves from the shift to the stage of electronic globalization. In high-context cultures such as China, event time still seems

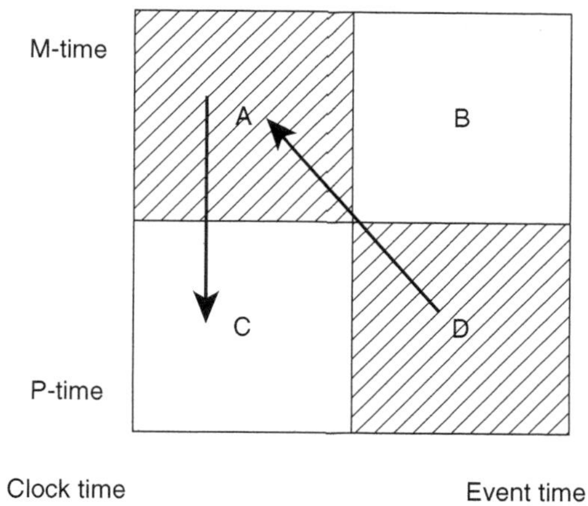

FIGURE 14.3 M-time and P-time versus clock time and event time. (Designed by Li Yanjie.)

to prevail, and waiting time is referred to as valuable since it is an opportunity for exchanging information with others in an informal atmosphere. Yet China might, just like Japan, move to a hybrid form in between Squares A and C.

Corporate culture and time

Hall's studies have inspired many authors to further explore the rules of social time which constitute the "silent language" for a given culture. Many different definitions of the complex concept of culture have been provided over the years. The one which still stands out is the widely accepted definition of culture provided by Kluckhohn (1961: 73):

> Culture consists of patterned ways of thinking, feeling and reacting, acquired and transmitted mainly by symbols, constituting the distinctive achievement of human groups, including their embodiment in artifacts; the essential core of culture consists of traditional (i.e. historically derived and selected) ideas and especially their attached values.

Although anthropologists approach organizations as cultures, while sociologists emphasize that organizations have cultures which can be studied in a functional way, they both stress that people rely on social norms, shared cultural values, and common understandings to value social time (Elias 1985). They view social time as the hub or the core aspect of culture (Schein 1992) and have drawn attention to the way social time is perceived, managed, and organized (Adam 1994). In this context "fruit time events" have become quite famous.

Ethnographic studies such as the one by Roy (1960) have provided accounts of the social construction of social time in an American machine shop. Roy examined social interactions in the workplace and explored daily workgroup rituals. The work was both long (12 hours, 6-day week) and tedious (simple machine operations) and required monotonous work tasks. He discovered that the daily grind was interrupted by informal breaks that were determined by the consumption of a particular fruit. There existed within the shop a "game of work." The workers had established their own event-based time reckoning system for structuring the day. As the working day stretched, the group punctuated it with several event times, each of which was the signal of a particular form of social interaction. The regularity of peach time, banana time, window time, pickup time, fish time together with the specific themes which accompanied each time meant that instead of the day being an endless *sequence,* it was transformed into a series of regular symbolic activities. The "beast of monotony" (Roy 1960: 215) was, in the eyes of the workforce, mastered. A large part of the "game of work" was based on tacit knowledge. Although formal lunchtime is, in general, more scheduled to the presumed internal rhythms of individuals (the onset of hunger), external social rhythms can bring about that even the definition of lunchtime becomes a group decision. The events of fruit time defined the overall experience of social time. Time was *in* the event. The fruit breaks became qualitatively different from the hours that preceded and followed it.[27]

The practice of event-based time reckoning gives meaning to the rhythm of the workday. Similar studies have been conducted by Ditton (1979), who noticed a distinction between the linear time orientation of management and the cyclic time orientation of employees. Cavendish (1982) rendered in her account of women assembly workers the difference between clock time and event time that the women experienced in line with the situation the work group faced. They viewed Monday as a good day, time-wise. Everyone was "fresh" and willing to work hard, whereas Friday was a slow day.[28]

Organizational cultures embed not only rigid clock time practices but also specific social temporal norms such as deadlines, punctuality, the pace of work, speed of actions, and a work/life balance (Blount and Janick 2001). The distinction between M-time and P-time, for example, inspired Perlow (1998) to study a team of software engineers in an American firm who were struggling with the structuring by management of social time. Their work required individual efforts for specific clients as much as collaboration within the firm. The reward system, however, was primarily based on their contribution to solving more general problems during interaction time which managers required the engineers to attend. They were expected to bring in solutions during the moments of collaboration when a particular managerial issue was put on the table. Many of these interactions were spontaneous and initiated by management but disrupted the efforts of the engineers for their clients. Perlow discovered that the social context such as work relations among engineers and supervisors were responsible for disorganized workdays. She noticed that engineers were framing their work in terms of "real engineering" and "everything else." The existence of "everything else" caused a lot of stress that management wished to be resolved. Perlow solved the problem by suggesting to management to discontinue pursuing their cultural practice of constant interruptions and instead introduce a schedule of "quiet time," i.e., to solve more general problems at specific moments in the weekly time schedules during which the engineers were not to be disrupted. They knew ahead when these quiet times were scheduled. Quiet time became an uninterrupted interval where they could focus on one specific engineering problem for one of their customers.

Prior to the introduction of quiet time, the engineers' working day was quite polychronic with constant interruptions which forced them to weave back and forth among multiple tasks. Instituting quiet time created moments of monochronic work within a flow of polychronic interactions. Overall productivity improved when a culture of both types of social time alternating during a week was imposed.[29]

The distinction between the polychronic and monochronic concepts is not absolute and, due to societal as well as technological developments, can become more fluid. Bluedorn (2002: 61) found "that polychronicity and speed values correlate." Yet he also noticed that monochronic individuals are happier and more productive in monochronic organizations, whereas polychronic people do better in polychronic work organizations. And in a polychronic working environment, an individual may be required to change a monochronic working habit toward a more polychronic one.[30]

Lindquist and Kaufman-Scarborough (2007) conclude that an individual currently has a general polychronic/monochronic tendency that is very versatile and which makes it possible to detach from the traditional 8-hour, 5-day work time and be able in the 24/7 global economy to attend virtual meetings wherever one is located geographically (cf. Teagarden *et al.* 2008). These outcomes indicate that there is a growing interest in examining if, for example, employees' preference for M-time or P-time might correspond to the time-use preferences of employers and to what extent the ratio between work and leisure time is culturally determined.[31] These findings are also relevant for a better understanding of the consequences of an imposition of M-time practices originating from headquarters with a low-context organizational culture onto subsidiaries operating within a in high-context culture where P-time prevails (Bluedorn *et al.* 1992).[32]

A remarkable experience has been illustrated by Brazilian entrepreneur and consultant Ricardo Semler (2003) who also has challenged traditional management practices of scheduling time. In 1982, he took over the company Semler & Co., a marine-pump-manufacturing firm founded by his father, removed management layers and corporate staff (firing almost two-thirds of most senior managers), moved into banking and environmental services, and renamed the company Semco. The company disposed of quality control departments and transferred this responsibility to the workers themselves. The economic success of the firm was based on "replacing control and structure with democracy in the workplace" (Semler 2003: 12). Semler considered "babysitting people just to make sure they clocked in on time" uninspiring (Semler 2007: 14). He encouraged personnel

> to rearrange their week, drop the traditional notion of a weekend, and divide the seven days among company time, personal time, and idleness. Anyone who can eliminate the stress of an overtaxed schedule, arrange her work week so she can sleep according to her biorhythms rather than a clock time, and enjoy a sunny Monday on the beach after working through a chilly Sunday will be a much more productive worker.
>
> (Semler 2003: 23)

Semler supposes that with such flexitime, *temperance* as a virtue can be restored and offer space to think and reflect. He encouraged employees to determine when they would come to work and select their own bosses. He presumes that his staff will form new habits and avoid routine. Elastic work schedules would allow his personnel to take into consideration their own circadian rhythms. They are given "some latitude when it comes to designing their own work schedules" (Semler 2007: 15). Semco employees began to customize their workdays and came to their work earlier or later than traditional schedules allowed. Their working hours became determined by joined interest instead of by company dictates. The company achieved greater productivity, effectiveness, and flexibility by creating more employee-centric workplaces.[33] The punctuality of the employees became *a new mindset*, not one which was uniquely related to watches (Semler 2003: 38).[34]

This approach is just one example among many more which challenge the attitude among businessmen and managers "who lecture as authorities on the proper conduct of working life. They preach about the necessity of hard work and putting in long hours as though these were the solution to the economy's woes" (Sennett 2009: 164).

International corporate cultures and time

In the age of electronic globalization, corporate culture has become an essential asset of MNEs which intend to become multicultural entities and, therefore, need to build and expound multicultural attitudes. The current internationalization of business requires MNEs to also focus on cultural differences of host countries, coordinate their efforts with local staff, and prevent that only the domestic culture of the MNE prevails (Selmer 2001). The corporate culture of an MNE provides a code of practice above the simple production or service function it performs (Wilson 2001).

Yet globalization also questions taken-for-granted traditions and opens ways for multiple options of working habits, values, and lifestyles. The distance to a host culture is seen as a vital criterion for organizations operating abroad. Corporations may favor foreign markets in countries that are more similar to their home country in terms of low-context or high-context cultures (Johanson and Vahlne 1990).

This view has, however, been contested when some international firms from a low-context culture prefer to operate – and do so successfully – in high-context cultures (Hotho 2009). The initiatives of the first MNEs such as the Dutch East India Company (VOC) would otherwise not have been conceivable. A multinational corporation with a strong corporate culture might demonstrate less striking cultural differences. Yet Laurent (1983) alleged that managers working for a single MNE strongly maintained their national cultural distinctiveness, whereas managers working for different firms throughout their career were more acquainted with a broader distinctiveness between cultures.

The growing body of research on corporate cultures has stimulated a growing interest in the importance of cultural distance between the corporation and its subsidiaries (Hofstede 2001). Cultural distance creates a need for more knowledge about other cultures but, on the other hand, also creates barriers to knowledge flows between the country of origin of the corporation and targeted markets in other countries (Barkema *et al.* 1997). Anderson and Gatignon (1986) asserted that a larger cultural distance increases information costs and impedes the transfer of management knowhow.[35]

Hofstede's (2001) work also drew a clear distinction between the constructs of corporate and national culture.[36] Within the globalizing economy, MNEs have expanded and intensified their business beyond the border by setting up subsidiaries and joint ventures – becoming manufacturers without factories – and sending out expatriates to represent the interests of the MNE. In the context of globalization, MNEs seem to facilitate the emergence of a post-national order where homogenization, convergence, and synchronization prevail as differences and variety seem to disappear.[37] That, at least, was the perspective when MNEs began operating internationally and stimulated the north–north as much as north–south trade (Davis 1997). They generally shared an ethnocentric perspective and hardly respected the cultural context within which their subsidiaries were located. Box 14.2 gives a nice illustration of the ethnocentric culture prevailing in the Ford Motor Company of the 1920s when they established their own rubber plantations in Brazil.

BOX 14.2 FORDLANDIA

In 1917, Ford Motor Company decided to set up production in Latin America. When it turned to local rubber production for its tires, it soon became confronted with the consequences of its ethnocentric perspectives on coordinating working time patterns.

In 1925, Ford decided after a meeting with his friend Harvey Firestone, the tire company founder, during which they discussed British initiative to create a rubber cartel among British plantations, to set up his own rubber plantations in Brazil. In 1927, an industrial-agricultural utopia called Fordlandia – similar to the ideas of Robert Owen – was set up, based on Midwestern American values, with clapboard houses around village squares.

Ford managers started to build a plantation between Brazilian Belém, the Amazon's major Atlantic port, and Manaus, more inland at the Amazon River

(Continued)

in the jungle. A large area was cleared, at least a thousand acres planted, and more than 5,000 residents installed. The work of the rubber tappers was extremely harsh. Rubber had to be tapped in the morning starting before sunrise when the latex flows freest through the thin vessels that run up the trees. Local residents preferred seasonal labor, tapping latex for a specific period (Grandin 2009). They still lived off a river economy that was based on payment in kind, and credit. Neither cash wages nor working according to a fixed schedule was familiar to them. The transition from agricultural time to industrial plantation time caused tensions: the local residents lived time, they certainly did not measure it. Yet early in the morning, a whistle blasted to summon the workers to their work. Complaints about food and a lack of service in the eating hall led to protests, ending in a serious Labor conflict in 1930. The workers destroyed everything breakable including tractors and trucks and "turned to the time clocks smashing them to bits" (Grandin 2009: 230). The workers hated the time clocks that were impractically placed too far from their workstation, making it difficult to punch-in as required under penalty of losing wages. Resistance against the clock time regime continued and, in November 1945, Fordlandia, as well as another Ford rubber plantation, Belterra, were sold to the Brazilian government and nationalized. The government promoted the creation of tapper cooperatives and helped them sell the latex on the international market accepting a more locally common social time practice.

Even today, Western corporations coming from a low-context culture prefer clock time and M-time practices even when operating intensively in countries with high-context cultures favoring event time and P-time practices. More MNEs produce less and less at home and have set up production facilities by developing alliances with locally owned factories and distribution centers which serve as affiliates and subcontractors that have their own organizational cultures operating as partners of MNEs (Friedman 1999). The ethnocentric attitudes within Western corporations, however, cause misunderstanding and conflicts in manufacturing units. The Indonesian anthropologist Warouw (2004, 2006) who studied the position of Indonesian migrant workers in subsidiaries of western corporations in the leather and rubber industry noticed how much overtime these workers were made to endure which caused a feeling of being chased (*keteter*). Lavatories and store rooms were used to take a nap as protest against "their suppressed aspirations on the shop floor."

The German car maker, Volkswagen (VW), neatly illustrates how subsidiaries in high-context cultures have to adopt post-Fordist characteristics of outsourcing, numerical flexibility, and an accelerating obsolescence of products. VW invested in Brazil and, by 1980, realized an annual production of over a million vehicles per year. Yet it struggled with the habitus of Brazilian car workers.[38] When the VW

Group intended to improve performances of its subsidiaries in Mexico and Brazil by increasing productivity, quality, and flexibility, it instructed its expatriates sent to the plants to cling to the monochronic, clock-time-based operational policy and ban high-context, event time issues which characterized the Brazilian way of doing things and where the spheres of work and private life were hardly differentiated. The German expats concluded that it was difficult to keep deadlines, while the Brazilian workforce complained that the German expats had "an inflexible, plan-orientated working style," whereas their own style was flexible and fit to react to changing situations (Mense-Petermann 2006: 312–314). The researcher found out that VW instructed its expatriates to remain a "stranger" and to not adopt local patterns of perception, interpretation, and legitimization but instead to "irritate and question" specific work time habits. They were typically employed to handle the latest technologies and install complex administrative systems. Yet the indigenous Brazilian business practice proved less than willing to absorb such foreign requirements. By sending expatriates with a broad sense of cultural intelligence, such clashes might have been prevented and might have increased the intertwining of cultures. Therefore, she pleaded for recognition of other temporalities within reorganizational measures and for sensitizing Western management for the consequences of subsidiaries operating in high-context cultures. In order to do so, their chronomanagement attitude would need to change.

Chronomanagement

In Western managerial practices, clock time has been utilized to create standardized, regular, centrally co-coordinated work organizations, typically called Fordist work organizations. Chronomanagement has been seen as the most appropriate approach to make these work organizations operate smoothly, and this vision has taken the upper hand with regard to other visions of social time within management, such as event time (Reinmoeller and Chong 2002). The superiority of the monochronic temporal orientation prevailed broadly over any polychronic orientation (Bluedorn and Denhardt 1988).

Chronomanagement is based on a monochronic machine-paced orientation:

> Managers use explicit schedules, sequencing patterns, and deadlines to coordinate people's activities within an organization [...]. Managers make schedules that allocate temporal resources within and across subunits such that their interdependencies among functions, groups, and individuals can be directed in a rational manner, thus minimizing conflicts over time. These organizational schedules serve many functions for the individual; they reduce uncertainty regarding when interdependent activities will be accomplished, when shared resources will be accessed or consumed, and when changes or transitions between activities will occur.
>
> The second element of the temporal structure involves implicit cycles and rhythms of behavior that exist within organizations. These cycles and

> rhythms are often associated with concepts of entrainment. Borrowing from biology and physics, entrainment theorists describe how actors adjust the phase and/or periodicity of their behavior to align with other entities. Familiar examples include the synchronized chirping of crickets and flashing fireflies. In organizational settings, many authors have observed how groups that interact frequently tend to entrain their cycles of activity [...].
>
> The third element of an organizational temporal structure includes the temporal norms embedded within its culture, for example, the degree to which (1) deadlines are strictly adhered to, (2) punctuality in arriving to work is valued, (3) a fast versus slow pace of work is preferred, (4) speed versus quality is valued in decision making, and (5) people are expected to allow work demands to interfere with personal time. Organizational cultural norms about time express implicit values regarding how time is to be perceived and evaluated within a particular work environment.
>
> (Blount and Janicik 2001: 569–570)

Chronomanagement reflects a one-sided perspective on the coordination task of management.[39] Managing on clock time is primarily focusing on active planning, monitoring the execution of planned actions, and on manipulation. The Western clock time concept has been translated in conventions such as appointment time, meeting times, and arrival times which operate as markers to convey temporal meaning to people. Managers use schedules, sequencing patterns, and deadlines to coordinate people's activities within and between organizations. Specialized software programs like "Time and Control" keep records of the time employees have spent on a given task and allow management to be informed instantly. Next to explicit schedules, managers also operate implicit cycles and rhythms of behavior as the above quote has stressed. In May 2011, the British Chartered Management Institute (CMI) found after a poll that, due to smartphone apps, British managers themselves show an increasing readiness to switch between work-related and leisure activities which have added an extra three weeks to their working year.

Yet the clock time prevalence has obscured many other qualitatively and culturally different aspects of social time such as rhythms and cycles based on event times which are essential to understand cultural heterogeneity and the presence of diversity when it comes to cooperation between many different partners within and between MNEs (Crossan *et al.* 2005). Managing on event time, however, implies a stronger focus on flexibility and allows for improvization, which guides managerial action in a more spontaneous way (ibid.). This raises the following question: Will high-context cultures with event time patterns where subsidiaries of MNEs are located for manufacturing purposes still be able to maintain local time conceptions or will they transit gradually toward clock time?

The mere existence of so many different social time perceptions will affect cross-cultural organizational interactions which, if not acknowledged, can lead to misunderstanding and eventually to clashes between organizational cultures (Baird *et al.* 1990; Palmer and Schoorman 1999). Intensive, time-consuming communication is required to prevent such confrontations. The language to be used is not one which can be seen as a system for transferring thoughts and meanings from one brain to another, "but a system for organizing information and for releasing thoughts and responses in other organisms" (Hall 1976: 57). International managers, and in particular expatriates, with an elaborated cultural intelligence and strong intercultural competence might make themselves familiar with the distinctions Hall has drawn to be able to adequately respond to these tensions and even accept the possibility of reverse innovation in the sense that high-context cultural practices be balanced out with low-context cultural ones.[40] When multicultural MNEs shift from hierarchical structures to webs of diverse inter- and intra-firm relationships, they may leverage resources and specific competitive advantages in their coordination with subsidiaries in order to create, disseminate, and share knowledge effectively in a global context (Ghoshal and Bartlett 1990; Gupta and Govindarajan 2000; Kogut and Zander 1993; Kostova and Roth 2003; Nohria and Ghoshal 1997; Tsai 2002). Yet researchers have found that expatriates reported a decline in their knowledge of the local environment during their stay abroad (Pedersen and Petersen 2004).

A nascent literature on heterogeneous forms of social time in organizations, however, stresses the increasing presence of "pluri-temporalism" (Yakura 2002). The ability of international managers to improve their cultural intelligence will determine how their internationally operating corporations may adequately deal with the social time differences Hall has introduced. Social scientist Barbara Adam (1995) has stressed that, although clock time may still be dominant, it does not eliminate the fact that other social time perceptions, which are highly context dependent, are also present in high-context societies. She has noticed that a holistic view of time, integrating multiple social time views, has disappeared in the drains of industrial, clock-time-based domination. The anthropologist I. Sabelis (2004) has equally emphasized that patterns based on event time have been gradually replaced by clock time, while this transition has been mainly associated with a higher level of industrialization and progress. But the question is whether the implicit association between a western ethnocentric perspective on coordinating social time and economic progress is justified.

Weiss (1994a, 1994b), as well as Glennie and Thrift (1996), have argued that it is not correct to believe that the ethnocentric Western views are the only possible approach of social time and that global Western management is the only reliable approach for international management practices. Macduff (2006) likewise stresses that the beliefs and practices of the Western industrialized world can no longer be seen as the only norm. The Western pattern of dealing with social time is not culturally universal nor constantly present (McGrath and Rotchford 1983).

Time infrastructures as developed by the railroads, telegraphy systems, telephones, computers, and wireless networks, as well as social time systems such as the Gregorian calendar and GMT and UTC, do have a tremendous impact on local experiences of social time, but they do not fully determine social time patterns. Globalization as a complex interactive system offers space for a broader focus on other kinds of arrangements of social time, despite the fact that in industrialized

Western cultures, clock time has become the dominant conception of time (Levine and Norenzayan 1999; Whipp *et al.* 2002).

Globalization currently confronts many more individuals with an expanding world which challenges them to look at social time through various temporal lenses, that is, taking into consideration different perceptions of social time (Mosakowski and Earley 2000). Yet the development toward the global economy has long been associated with a ceaseless drive for detached mobility as the ultimate symbol of modernity. Yet this kinetic mobilization only favors progress as a "movement for the sake of movement" without taking into consideration the intensifying presence of cultural differences (Sloterdijk 2005). As long as this continues unimpaired, clock time increases its hegemony and maintains a *chronolatreuic* sphere in the Western world where individuals behave like slaves of social time (Sloterdijk 1999). To put a halt to this slaving of social time, management should challenge the one-sided dominance of clock time. Traditional chronomanagement with its command and control practice may not be able to ensure a broader approach which encompasses a connection with and coordination of a wide variety of social time arrangements. Deregulation of collective norms, diversification of the length and pattern of working time, a blurring of the distinction between working time and leisure, the increase of polychronic work patterns, the relevance of *quiet time*, event times in multicultural teams, and slow time abound. National governments introduce measures to promote part-time work and provide equal treatment for part-time work and other atypical workers. Attempts to combine business time with increased time sovereignty for employees operating in a multicultural context will inevitably raise questions about the traditional structuring of work organizations acting in the global arena.

Timescape management

Indian author H. K. Bhabha (1994) has stressed that the epoch of industrial globalization has generated a growing interest in multiculturalism. Yet for a significant period of time, imperial and colonizing empires, internationally operating commercial companies such as the VOC as well as more recently the MNEs have, based on an ethnocentric perspective, imposed their Western time frames onto local non-Western settings. During the times of imperialism, Western colonial traditions reinforced attitudes toward the *other* as being backward local people. Native languages and local time frames were replaced with the language and time frames of the colonizers. It was hardly taken into consideration that the Western way of thinking could be disruptive and partial (Bhabha 1994). The non-Western world simply had to adapt to the Western way of timekeeping. Locally embedded concepts of social time were excluded or destroyed but certainly not integrated.

However, in the age of today's post-colonialism new cultural forms emerge which he describes with the label "hybridization," which transforms our understanding of cross-cultural relations. During the colonial times, terms such as Western and Eastern became common parlance and have long been seen as concepts which properly described the nascent global economic and political order equating capitalism with universalism. Yet the Bhabhaian perspective (Frenkel 2008) stresses the need to re-evaluate the geopolitical power relations between dominating and dominated forces which, for example, in MNEs negatively influence the transfer of knowledge from headquarters to subsidiaries, while the predominant culture is still one of wanting to operate in a Euro-American way which is based on the same ground of temporality. Developing countries where MNEs became active were still seen as places with a *premodern* culture with other temporalities. Cultural distance could only be overcome by rejecting indigenous heritage and embracing modernity.

Yet a concept such as the *late developer* is an example of distancing, interchanging a quantitative marker (time in this case) into an attribute. Cultures in less developed countries were inserted into geographical locations categorized and ranked according to spatial distance from centers of industrial or technological supremacy.

Anderson (1991) raised a similar concern and set the stage for a broader discussion about the relevance of comparative studies in today's stage of the globalization and the links that can be drawn between space and time. He argued that with the establishment of European nation-states, nationalities became cultural artifacts which during the ages of colonialism and imperialism were exported to the less developed world as a model of national modernity. For him, a nation is an imagined political community, for which people were willing to fight and die. Appadurai (1996), in a similar vein, designed a terminology which would reach beyond concepts of nationalities and divisions between West and East to overcome the limitations of area studies and cross-cultural approaches. He stresses that modernity strives after universal applicability for itself, without the national context. Appadurai sees the global as the unraveling of the monopoly of autonomous nation-states over the project of modernization. Individual as well as social groups' interests and aspirations nowadays transcend those of the nation-state in transnational encounters. The global cultural economy is developing into a complex overlapping and disjunctive order moved by mass media and migrants. Centers of power dictating the world system of images are appearing to unravel. Imagination itself has become an organized field of social global practices. In today's world of mass media circulation and mass migration, irregularities are created because both people and information are in simultaneous circulation and neither fit into audiences that are easily bound with local and national spaces. This interaction makes people want to appropriate the global into their own practice of modernity by means of their imagination.

For that reason, Appadurai proposes a theory of rupture that uses media and migration to explain their joint effect in the work of imagination that ultimately defines modern subjectivity as opposed to nationality. He sees modernity as an idea that is unevenly experienced. Media and migration currently have a joint effect on imagination (desires for different truths) which ultimately defines modern subjectivity, i.e., *what is modern to whom*.

To show that modernity is experienced by uneven perspectives of different actors, Appadurai (1998) has introduced the suffix "scape" to characterize new conditions within which sociocultural developments are taking place. He replaces the symbolization of the social world in terms of maps which illustrates stable social aggregates with fluid flickering screens representing cultural flows that crystallize into specific scapes such as technoscapes, ideoscapes, and mediascapes. A scape points at spaces of action with social practices and social groups and refers to fluid, irregular shapes.[41]

The notion of scape is important as it indicates that time is still inseparable from space and that context matters. Scapes are not objectively given relations that look the same from every angle of vision but, rather, are deeply perceptual constructs, inflicted by the historical, linguistic, and political situatedness of different sorts of actors (Appadurai 1996). This framework represents the global irregularities created by the existence of mass media and mass migration. As ethnicities, media, technology, finances, and ideas know no boundary and are not static in the globalized world, they are explored via ethnoscapes, mediascapes, technoscapes, financescapes, and ideoscapes, which are perspectives of different actors which create *imagined worlds* that coexist unevenly and that define modern subjectivity. An ethnoscape, for example, is a landscape of persons who, as migrant workers, deal with the reality of feeling forced to move (*human motion*) and as such constitute a shifting world. Appadurai underlines the notion that homogenization can no longer be regarded as Americanization or Westernization because as these forces come into contact with new societies they become indigenized, so the new global cultural economy is a complex, overlapping, and disjunctive order. He proposes a framework in which those disjunctures are explored via these so-called scapes, which are perspectives of different actors, creating imagined worlds similar to what Anderson (1991) called "imagined communities." This is intrinsically connected to the appearance of deterritorialized groups that are living around the world and seek through imagined worlds a connection with their homeland. Appadurai concedes that the idea of nation-state is in decline because it can no longer contain the ways in which minority populations in every country link themselves to wider movements of religious or ethnic affiliation, therefore public spheres are no longer typical, exclusive, or necessarily national.

People and information are in simultaneous circulation and neither fit into audiences that are bound with national spaces. This interaction makes people want to annex the global into their own practice of modernity by means of imagination. He further explains that culture should not be thought of as a substance, but as a dimension of phenomena that belongs to an embodied and situated difference, Appadurai (1996), therefore, alleges

> that the noun *culture* as used by Western authors appears to privilege the sort of sharing, agreeing, and bonding that fly in the face of the facts of unequal knowledge and the differential prestige of lifestyles and to discourage attention to the worldviews and agency of those who are marginalized or dominated.
>
> (Appadurai 1996: 12)

He stresses that cultural dimensions of globalization point at diversity. He, therefore, prefers to use the adjective "cultural," as this takes into consideration "the realm of differences and contrasts" (ibid.). "Cultural" points at situated and embodied differences.

Inspired by the ideas of Appadurai (1996), Barbara Adam (2004) calls for a "timescape" reflecting the increasing heterogeneity of social time patterns. She notices that the traditional public industrial timescape has comprised features such as rigid time frames, tempo, timing, strict time patterns, prescribed time sequences, and time extensions. The key issue is that this industrial timescape is dominated by "the control over the objectified resource, i.e. abstract time, which leaves no temporal feature untouched" (ibid.: 145). All of the features of the industrial timescape are subject to control that is the speeding up and slowing down, the rearrangement of sequence and order, the evening out and accentuating of peaks and troughs.

> The variable, rhythmic times of life are regulated and disciplined to conform to uniform invariable temporal patterns. Through socioeconomic technologies, transport, transmission and transplantation, time is compressed and processed. Clock time is used to regulate and rationalize the pace and seasonality of organisms and beings, social activities and institutions. With its aid, rhythmicity is transformed into a rationalized beat.
>
> (Adam 2004: 145)

Globalized production affects the time, tempo of work, and timing. Start times and end times will be aligned with different time zones. The current tempo of work will be aligned to the beat of the 24/7 drum of global production, and the synchronization of work allows for collaboration across multiple time zones (Ladner 2009).

By combining infrastructures which have stimulated the mobility of information, capital, goods and people, and familiar elements of Western cultures with unfamiliar temporal elements from other parts of the world with different cultures may, however, create a new hybrid timescape (Bluedorn 2002). Social time practices from American, European, Japanese, Asian, Latin American, and African cultural settings may, in combination, develop into a global timescape which challenges the prevailing, standard industrial timescape.

The European Union, for example, attempts to facilitate the appearance of a post-industrial timescape and proposes the development of patterns of social time which harmonize natural, technological, and sociocultural rhythmicities of individuals with time arrangements that still prevail in companies. Hybridization is seen as a "sociocultural process in which discrete structures or practices, previously existing in separate form, are combined to generate new structures, objects, and practices" (Canclini 2005: xxv). This hybridization challenges the traditional conceptual pairings of local versus global, as well as traditional versus modern. Such a hybrid timescape increases the ability of MNEs from low-context cultures when operating in high-context cultures to absorb their influences and enrich their own corporate culture. Indisputably, such a new multi-temporal timescape will be a challenge on "how to be productive enough to be comfortable, to minimize the temporal stress on which this achievement is built and to simultaneously make time for caring relationships and a civilized society" (Levine 1997: 220).

In the context of MNEs, flexible multicultural organizations can focus on the creation of attitudes that recognize the relevance and advantage of a broader timescape without the intention of governing the body and the soul of the workforce in a traditional manner (McKinlay and Taylor 1998). In the age of electronic globalization, timescape management may become materialized by bringing people together from many different cultural backgrounds in flexibly operating corporations and developing multicultural policies which are appropriate for de-standardized, irregular, rhythmically co-coordinated social time (cf. Ettighoffer and Blanc 1998; Alis *et al.* 2006). The need to coordinate across a variety of temporal practices is an organizational challenge not free of difficulties (Ancona *et al.* 2001a, 2001b). Fast-paced practices which are dependent on slower-paced practices require coordination skills which take into account the existence of cultural differences. Yet MNEs operate incessantly in countries and communities of the world, which irrespective of bloc or region, appear to be more and more bound up with one another through the extension of transnational practices of these corporations (Sklair 2002).

MNEs eager to keep their presence in local markets will continually update their knowledge about these markets and their institutional, as well as cultural, differences. Multicultural multinationals, that is, cosmopolitan conglomerates, comprise a high percentage of different nationalities in their staff that are diverse enough to respond adequately to local tastes but united enough to amount to more than the sum of its parts. They have introduced diversity management and multicultural teams to improve transfer and implementation of knowledge adequately (Mosąkowski and Earley 2000). Multicultural organizations pay particular attention to the cultural sensitivity of their staff and its customers and create a cadre for global managers to excel.[42] For expatriates such a multicultural context is a challenge and requires intellectual flexibility (Crossan *et al.* 2005).[43]

Apparently, MNEs not yet turned into multicultural organizations still do not look at country differences in terms of diversities in cultures, religions, and political systems, as well as in different concepts of social time which have a challenging impact on the way the flow of information and communication is coordinated.

15
EPILOG

Throughout this book, a narrative has been told about the increasing interconnections between globalization processes and time issues. During the late medieval period, space-bound Europe became familiar with the mechanical clock which, beginning in the Italian city-states, swiftly diffused across Europe. Unequal hours turned into equal ones, while the time of the clock varied no longer by season. The proper manner of keeping time spread out among the citizens of commercial cities and to rural areas where monasteries were located. The partitioning and structuring of the day into segments and stretches of time became perfected in the numerous monasteries of medieval Europe (Whipp *et al.* 2002). Monks executed their daily obligations less by dependence on private impulses or nature but by a fully regulated regime with strict clock time patterns.

The Italian city-states also became the breeding ground for the rising of commercial capitalism. International trade and local manufacturing made these cities wealthy and autonomous. Naval power had been a decisive factor in the remarkable shifts in economic fortunes. However, as Arrighi (2010) has sketched, it was characteristic of the first great commercial powers that they foreshadowed economic decline in their late stages. High finance in the hands of Venetian and Genoese bankers withdrew from the area of production and shifted their interest toward the Iberian Peninsula which offered more promising commercial and financial opportunities. Inspired by adventurers, monarchies on the Iberian Peninsula expanded their horizons and began to conquer new territories.

The first part of this book highlighted the impact of the Treaty of Tordesillas in 1494 being the first meridian which – with papal approval – facilitated the rulers of the Iberian Peninsula to divide the unknown world into two segments ready for further terrestrial occupation. It marked the beginning of a European perspective on the division of the world into different spheres of influence. The improvements of sailing ships provided the Iberians increasing opportunities to maintain that division of unexplored places. After 1500, a spirit of exploration and expansion became ingrained in other Western maritime societies. Trade contacts with large-scale territorial empires in Eurasia such as the Mongols and their descendants, the Mughals and Ottomans, shifted toward the Atlantic Ocean. The blows the Dutch Republic struck on the Spanish monarch at sea as much as on land devastated his empire. The region of the Low Countries became the new driving center of the second stage of terrestrial globalization. The first joint-stock chartered trading company, the Dutch East India Company (VOC), under the protection of the Dutch nation-state began to sail the oceans. Its success pulled other European countries into a heavier involvement in commercial capitalism. Seaports became the key nodes for international commerce. Yet as in the case of Venice and Genoa, the Dutch Republic lost its hegemony at sea to the English nation.

During the seventeenth and eighteenth centuries, European trading companies played a key role in expanding their territorial scope and reinforced the exclusiveness of the European system of sovereign nation-states. Occupying new territories and creating new trading places consolidated the epoch of terrestrial globalization. It also became the period during which living by the clock developed into a common Western pattern. Clock time depicted a continuum of linear, infinitely divisible moments which turned into objective, quantifiable units. These units of time were homogeneous, uniform, regular, precise, deterministic, and measurable. Years were divided into equal days, days into equal hours, hours into equal minutes, and minutes into equal seconds. Over the centuries, older concepts of social time based on the position of the sun, the seasons, and the cockerel gave way to the tyrannical two-handed clock which structured social life in commercial Western societies. Authors such as Jonathan Swift (1726: 38), however, mocked the fixation on the mechanical clock; in *Gulliver's Travels*, Swift expressed severe criticisms of his native English country. During his first voyage, Captain Gulliver was washed ashore after a shipwreck and found himself prisoner of a race of minute people – about six inches tall – who were inhabitants of Lilliput. Their size signified that their motives, acts, and humanity were of the same, dwarfish kind. After the shipwreck, Gulliver was chained and his pockets emptied. One of the Lilliputians described how they found on this tall man a pocket watch.

> He put this engine to our ears, which made an incessant noise like that of a water-mill: and we conjecture it is either some unknown animal, or the god he worships: but we are more inclined to the latter opinion, because he assured us [...] that he seldom did anything without consulting it. He called it his oracle, and said it pointed out the time for every action of his life.

The British hegemony became established on a policy of joined commercial and industrial expansion. The wars with the Dutch, among others, created opportunities to expand British market opportunities. During the epoch of industrial globalization, the British factory system habituated the factory workers to become subordinate individuals. Rural adults and children were introduced to new mechanisms of power. Disciplining "docile bodies" (Foucault 1975, 1977) sustained the upcoming regime of industrial production (Braun 1991). Factory workers saw their agrarian bodies transformed into capitalist instruments and, as human machines, were forced to respect the precision of task execution measured by the clock. Their movements became synchronized with the rhythm of the shop floor comparable to the synchronized rhythms the Dutch ruler Maurits had imposed onto the soldiers of his army.

The factory system of the nineteenth century was built on a thorough application and exploitation of clock time. In 1867, Marx pointed out that even night work became a new method of exploiting labor. Although the practice was still in its infancy with the development of gas lighting in factories, extended working hours and overtime increased in practice. Gas lighting facilitated a double-shift practice. The empire of clock time, which reigned the day, was extended into the night. Although the order of the clock was questioned by the workers who toiled in the factory system, the general public believed in its usefulness.

The transition from terrestrial to industrial globalization was paralleled – though not synchronically – with speed-up processes which initially took place mainly in the Western world but spread globally during the expansion of the industrial globalization. The second part of the book emphasized how technological improvements in Britain sustained the realization of infrastructures which facilitated faster processes in transport (train), communication (telegraph, radio, and telephone), and production. The historian Tony Judt (2011: 60) remarked how tremendous the impact of the train has been on the shaping of the modern world and the mobility of the populace: "More than any other technical design or social institution, the railway stands for modernity." Whereas the pre-modern world had been space bound, the modern world became time bound. The railway timetable which appeared in the 1840s represented advancements in safety, commerce, and convenience. Although "we no longer see the modern world through the image of the train, [...] we continue to live in the world the trains made" (ibid.).

The progress that British society experienced during the epoch of industrial globalization of the world system through commercial, industrial, and transportation improvements was eventually taken over by the United States where railroads and communication networks began to require further standardization of time and introduced a common civil time across geographical time zones. The American watch industry with companies such as the Waltham Watch Company merely took over the market but lacking the pool of skilled workers and a web of existing manufacturers which were present in countries such as Britain and Switzerland, the American industry developed new capabilities in terms of substituting human capital with machinery that was able to manufacture interchangeable and standardized parts in large numbers (Hounshell 1984). The revolutions in transportation and communication further facilitated the extension of standardization and have underpinned a worldwide infrastructure dominated by clock time dictating the beats of the modern metronomic society.

The modern factory system that F.W. Taylor and H. Ford designed institutionalized a geography of work that was based on monitoring, observation, and surveillance. Presence and visibility became attributes of managerial control which were effective as long as the workforce obeyed strict working-time schedules. The landscape of factories and offices symbolized the regulation of workforces through control over space (Felstead *et al.* 2005). The decisions of managers within organizations became dominant in shaping patterns of working time.

> In the corporate workplace, the worker steps into a setting where the machines, spaces, and job instructions have all been managerially studied and configured for maximal profit. Because of this, it is not just the contractual power of the boss, but the various constraints and enticements placed in the workspace by the industrial engineer that lead employees to comply with the corporate behavioral agenda.
>
> (Dawson 2003: 142–143)

Management developed into a profession, and formal training of future managers was institutionalized in business schools. Managers' vocabulary became packed with time references such as time-and-motion studies, speed, mobility, precise time schedules, and acceleration.

By the end of the nineteenth century, the formation of communications-based systems of control began to envelop the world in global circuits of power (Geyer and Bright 1995). Around the turn of the same century, the German sociologist G. Simmel noticed how much the adage of "Time is Money" began to prevail and how the "Zeitgeist" had turned into a "Geldgeist," an eagerness for money.

> The mathematical character of money imbues the relationship of the elements of life with a precision, a reliability in the determination of parity and disparity, an unambiguousness in agreements and arrangements in the same way as the general use of pocket watches has brought about a similar effect in daily life. Like the determination of abstract value by money, the determination of abstract time by clocks provides a system for the most detailed and definite arrangements and measurements that imparts an otherwise unattainable transparency and calculability to the contents of life, at least as regards their practical management.
>
> (Simmel 1990: 445–446)

A wide range of speed-up processes, not only in the technological field but also in the economic and socio-cultural fields, became noticeable. Citizens in the metropolis began to feel that the process of interaction indeed accelerated and the *homo accelerandus* became a new species. The power structure in organizations allowed managers to require the workforce to comply with fine-tuned schedules, faster routines, and intensified working practices. Western management itself became mesmerized by a cult of long working hours.

> Such people may have lost leisure rather than gained it. Whatever the other advantages of their lives, they are the victims of time; they are harassed at work and away from it, even though they are able to spend a great deal of money on the fruitless task of trying to *save* time.
>
> (Young 1988: 219)

Since the 1970s, the sudden rise of Japan as a global player has reinforced this pressure for acceleration. Management vocabulary was further enriched with concepts such as Total Quality Management (TQM), Just-In-Time (JIT), time-to-market, market-based competition, timing, and flexibility. These labels encapsulated the significance of Japanese production practices and represent the internationalization of Japanese management.

In the age of electronic globalization, as discussed in the third part of this book, information networks began to shape a new artificial space in which a constant stream of data freely flows across the globe. The multinational enterprises (MNEs) operating in networks drive globalization, which is based upon the homogenizing of world rhythms that are set to the Western time concept and its concomitant chrono-technology. International managers of global corporations are seeking to coordinate their work organizations in such a manner that the nation-state system will have a less profound impact. Clock time practices are without serious problems and transferred to other parts of the world without spatial distinctions mattering much. Institutions which hinder synchronization from a corporate interest perspective are put under pressure and deregulated wherever possible.

The blurring of spatial distinctions seduces some authors to a discourse of the "timeless time," where traditional arrangements of different time patterns do not matter anymore. This seems rather exaggerated, while space itself is as much a product as a modifier of numerous social relations (Castree 2009). Natural definers of time as well as calendar systems shall still determine the regulation of social time. It cannot, however, be denied that the nature of those patterns of social time may change.[1]

Without any doubt, social time has ended up on the electronic highway where the processing of information takes place in matters of nanoseconds. Digital technologies allow for 24-hour trading and instant communication. The blurring of spatial distinctions and the appearance of timeless time make the traditional arrangement of different time patterns matter less and less. The economic historian J. Mokyr expects that the modern factory and office system is in retreat "not only as a physical central location of activity, but also as a time-organizing institution in which work begins and ends at given times and the lines between leisure and labor are firmly drawn" (2002: 153). Due to the intensifying transmission of information via computer and satellite communication, the relationship between time and movement across space has changed. Duration has been compressed to zero. The intensive electronic present is no longer part of chronological time. Real-time has to be conceptualized as chronoscopic time (Adam 2004: 132).

The proponents of the 24/7 economy hail the positive effects of the Third Industrial Revolution and are of the opinion that day and night rhythms, as much as the weekly and seasonal patterns, will transform into one homogeneous global pattern. Some hyper-globalists such as Ohmae (1990) have even predicted that global forces such as the MNEs are becoming so powerful that even nation-states may disappear. Although global business activities may have altered the traditional relationships between economic organizations and nation-states, these last ones, combined with the civil society, are still essential to determine the rules of the free market system and institutionalize the rights and obligations of the economically powerful (Korten 1995). MNEs as global players do create an economic world which becomes increasingly interconnected and interdependent but also grows in complexity. These corporations have become main actors in the implementation of abstract time within a regime that has become internalized as a hegemonic metric.

American MNEs in particular have to a very large extent contributed to the expansion of electronic globalization. They have embodied the latest stage of capital accumulation. Arrighi (2010) has emphasized that during each of the successive stages from the Italian city-states, via the Dutch Republic and the British empire, the swings have become larger and more complex than the previous ones.

> The speed of each swing, as measured by the period of time that it has taken each regime to form, to become dominant, and attain its limits has increased steadily with the scale and scope of the leading agencies of systemic processes of capital accumulation.
>
> (Arrighi 2010: 341)

Arrighi has described how only a fraction of the liquidity created by the American monetary authorities found its way into new trade and production facilities. During the 1970s, there was a resurgence of private high finance with capital turning into petrodollars and Eurodollars. With the rise of computerized electronic networks not only has the Eurodollar market become part of a vast stateless banking system, but there actually appeared a globalization of finance capital. "The timing of information and money has fused together into a digitally driven instantaneity of finance capital" (Hope 2006: 277). Today, finance flows far more freely across national borders than trade. Yet the recent financial crisis demonstrates that the capital market operating in real time is not able to function without market oversight.

The end of the fourth wave

Arrighi (2010) expects that the most dynamic center of the process of capital accumulation on a world scale will move from North America to East Asia. Japan's catching up has clearly been the most sustained and spectacular one so far, but he believes that the East Asian region will become the epicenter of the following wave of capital accumulation. The geographer P. Dicken (2007) also believes that the significant global shift in the geography of the world economy must be located in Asia. The sinologist M. Jacques shares the same opinion: while Japan, after the crisis of the 1990s, has been unable to regard itself as "a model for others [...], China might displace Japan as the leader of East Asian economic expansion. China is actually the largest exporter to America and has invested its surplus capital in various forms of US debt" (Jacques 2009: 69; 189). Indeed, China's impressive growth, its investments in technology, its foreign exchange reserves, and its remarkable infrastructural developments might demonstrate that progress relies not only on the Western model, and this is not only valid for China. The miracle of the Asian Tigers may be tarnished in view of the financial and political events since the late 1990s, nevertheless, the economic growth in these countries is often referred to as an "Asian economic miracle." Arrighi (2010: 381) is convinced that "East Asia might become the center of the world market society buttressed, not by superior military power as in the past but by the mutual respect of the world's cultures and civilizations." Whereas Jacques and Arrighi expect East Asia to become the new epicenter in the twenty-first century, the political scientist Francis Fukuyama, who in 1992 became famous with his statement about the universalization of Western liberal democracy, which would come about after the collapse of communism, presupposes that next to the USA and the European Union, with China, India, and Brazil, a polycentric, multi-polar development will occur. China has a GDP which is now reaching number two in the world, and China holds 1.17 trillion dollars in US debt. Yet, China's growth still remains based on the manufacturing of low-end products produced by cheap labor. Together with China, other developing countries account for about 80 percent of the world's population, representing a substantial market and workforce. China and India together have a population of over two billion inhabitants. The developing world provides an array of products and services to the Western world and is a very large market for this developed world as well (UNDP 2000, 2005).

In terms of globalization and time, however, the East Asian countries do not yet show remarkable new characteristics. China is actually constructing a new Silk Road based on the world's most ambitious high-speed rail line of 8,000 kilometers, linking 17 countries. A first part is now in operation, running the *Harmony Express* from Wuhan to the harbor city Guangzhou, covering a distance of 1,068 kilometers in just three hours and ten minutes with a speed of 357 kilometers per hour. The planned high-speed rail network will improve China's ability to transport crucial energy resources and provide reliable transport facilities, and may restore some of the grandeur of the traditional Silk Road, once one of the first grand avenues of globalization. Eventually, the high-speed railway could be capable of transporting passengers and cargo all the way to London.

Yet the high-tech marvel seems to be married to a traditional view of the world where science is not fully embraced. The actual railway system is still applying central Beijing time with one time zone across its vast area. Western regions such as Xinjiang and Tibet remain very dark while their clocks, which are set to Beijing time, indicate eight o'clock in the morning (Hongladarom 2002). Recently, three semi-official time zones have been introduced which, however, conflict with the five zones China should be using in order to comply with the generally accepted Greenwich time zone pattern.[2]

In a similar vein, it can be remarked that China has, together with other East Asian countries, the longest working hours: "Thanks to the shorter workdays and longer holidays Europeans work shorter hours. Between 2000 and 2009 the average American in employment worked just under 1,711 hours a year. The average German worked just 1,437 hours – fully 16 percent less" (Ferguson 2011: 265). Yet in China, like in other East Asian countries, the situation is quite different. Although, in February 1994, China imposed its Regulations on Working Hours of Workers and has enforced, since January 1, 1995, a Labour Law which protects the health and safety of workers, lower-level workers still work extremely long hours to secure a decent income (Zeng *et al.* 2005). East Asian countries may continue to struggle with an unbalanced and complex social timescape.

Even the introduction of more flexibility seems to be of a limited kind. In the 1990s, China, for example, became second, only preceded by the USA, as location for Japan's overseas foreign direct investment. Yet the overseas Japanese subsidiaries have only moderately transferred Japanese production and work organization such as TQM and JIT practices. The subsidiaries mainly customized Japanese practices to local Chinese conditions which meant application of numerical flexibility and recruitment of temporary staff during peak times mainly (Taylor 1999).

These and similar cases indicate how the global and the local can still be temporally opposed to each other and may cause inter-temporal conflicts (Hope 2006). If Beck's (2002: 17) definition of globalization as "a non-linear, dialectic process in which the global and the local do not exist as cultural polarities but as combined and mutually implicating principles" is correct, it is then to be seen how the East Asian region may provide new dimensions to the timescape by translating this dialectic into a new synthesis.

Cosmopolitanization

The expansion of the global market place is reshaping the arena for international business. Management approaches and practices are challenged as they exhibit a significant *Western* flavor. These organizational views may become less applicable to the further integration of the global economic world.

The networks formed by MNEs transcend national boundaries, identities, and interests. These corporations have the power to coordinate and control operations in more than one country, even if they do not own the subsidiaries. Yet one of their main characteristics is their potential geographical flexibility to switch operations between locations at a global scale for their own interest. In that sense, they string networks together which primarily favor economic globalization. Former prime minister of France Lionel Jospin (1999), however, warned against the outcome and once said that there still are huge divides among nations which engender large inequalities: "We have to accept a networked world, but not a world dominated by networks, because then it would be dominated by private interests."

Despite their power, MNEs operating in international networks will have to respect the institutionally bounded realities of the world economy (Castells 1996).

The Network Society that Castells sees rising should not be determined by the spatial organization of dominant managerial elites alone. "Articulation of the elites, segmentation and disorganization of the masses seem to be the twin mechanisms of social domination in our societies" (ibid.: 415). Castells presumes that these managerial elites will develop cosmopolitan profiles. However, we have seen that their actual behavior is more of a cosmocratic kind. Cosmocratic managers primarily focus on the further homogenization of clock time. They will only promote innovation to improve productivity and stimulate flexibility which increases competitiveness (Castells 1998). As such, they will certainly exhibit less respect for the larger variety of temporalities which actually prevail in local cultures. Still, those internationally operating managers will continue to experience a business environment in the developing world which is very different from what they know from their industrialized home country.

For MNEs, one of the greatest challenges to face in the increasingly international business is how to manage human resources in networks operating on a global scale while understanding the "silent language" of cultures as Hall has described. The presence of the mechanical clock and calendar time are indeed deeply embedded in the fabric of Western industrial societies. The existence of clock time, however, no matter how dominant, does not obliterate "the rich sources of local idiosyncratic and context-dependent time awareness which are rooted in the social and organic rhythms of everyday life" (Adam 1995: 20–21). Cosmopolitan managers might be able to understand this silent language and use their cultural intelligence to properly understand multiculturalism. Cultural intelligence pertains to the ability to systematically combine generally shared values with cultural idiosyncrasies. It merges an understanding of global interconnections across boundaries with a large variety of temporalities in high- and low-context societies. Cosmopolitan managers may look at the increasing political-economic expansion at global scale as a project which calls for a legal, institutional, and socio-cultural framework based on generally shared values and norms to embed and regulate globalization.

Eventually, more individuals and social groups may henceforward enjoy the opportunities globalization offers in terms of improving world economics and the general quality of life without undermining human security, social equity, freedom of the individual, and democracy. Technology can make life and work easier but will finally be determined by the way social time in organizations is being coordinated. Properly coordinating social time requires the recasting of the human machine into a sensible human being as La Mettrie pleaded for. This can be realized if we remind ourselves of the words of the late medieval adviser Christine de Pisan who stressed that a sensible human machine needs to be guided by the most noble virtue of *Temperance*, the moderator, to know and rule oneself. Only then may globalization become beneficial for all and can cultural idiosyncrasies be valued.

With the invention of the mechanical clock at the end of medieval period in Europe, time ceased to be personal and local. The commodification of time developed over the centuries into a common fact that began to rule the lives of more and more people. Globalization, which developed through different stages, has furthered the hegemony of clock time and homogenized more and more forms of social life. Temporal innovations in transport, production, and transmission in the western world have compressed time to its limits. Instantaneity and increasing connectivity reveal temporal limits which we are now barely able to control. The intensifying cross-cultural exchanges between peoples and communities of the world, which are also a consequence of globalization, are, however, tempering the effects of clock time, while they reveal other rhythmicities of social life. Globalization and the locally bounded life of the majority of the world population ask for harmonization. Globalization, therefore, invites a reappraisal of a range of other temporalities. Internationally operating managers encounter forms of social time which will require a border discourse to better understand the complexities of our modern timescape (Adam 2004). This book intended to contribute to this discourse by sketching the historical roots and the universal impact of the mechanical clock.

NOTES

Introduction

1 Only after the introduction of electric technology was the intricate link between transportation and communications systems dissolved.
2 The German philosopher Peter Sloterdijk has stressed that the phenomenon of globalization has an even longer tradition. Classical Greek philosophers had already discussed the phenomenon of the globe as a sphere representing the geometrically perfect circle called Cosmos or Uranus. Absolute beauty (*sphairee*) became perceived in terms of the geometrization of space, lines being drawn across an imagined globe. For Aristotle, imaging a measurable cosmos was the first move toward thinking in terms of globalization (Sloterdijk 2005: 199). Observing and analyzing the globe within space and time led to the development of what he calls cosmic globalization which, however, remained a primarily philosophical exercise.

1 The coordination of time

1 www.businessdictionary.com, retrieved May 5, 2010.
2 This not only happened with the agrarian population who suddenly had to work in factories under conditions they were not familiar with. It also occurred when the Atlantic slave trade plunged Africans into a dis-embedding process, away from their local cultures with their own rhythms of social life, being sold to American plantation owners who drove them to long working days throughout many years of captivity. They became re-embedded in new social interactions and institutions where concepts of time and space were applied in a different geographical, cultural and societal context.
3 Technology is not just a way of organizing knowledge for practical purposes in a neutral way , but is promulgated by individuals and groups (firms , corporations but also military and political leaders) with special power interests. Rosa does not explicitly elaborate the issue in how far technology is a system of rationalized control over large groups of people and machines by a small group with special interests who operate through organizational hierarchy (cf. Boersema 2009).

2 Exploring the globe

1 The cities embodied justice for all its citizens and became well suited to the new market economy. An intellectual revolution among the urban elite weakened the power of religious traditions on the upper crust of Greek society.
2 The *vilicus* system does not imply the slave mode of production.
3 Rome's Christian residue in Byzantium was a Christian Greek empire that had managed to hold off a Turkish advance into Europe for several centuries. With the shift of Arab operations to Baghdad, the balance of power altered: the Arabs occupied the Zoroastrian Persian Empire that reached through Central Asia as far as India.
4 Gibraltar is named after him: the mountain of Tariq or Jebel Tariq.
5 It would take centuries, almost 1,250 years, of nation-state formation, wars, and nationalist sentiments before the contours of the European Union with a single market and common financial and political policy became established in 1994.
6 Yet the Muslim world did not create universities as learned centers; their madrasahs kept scientific enquiry outside their walls.
7 Some have asked why the Americas did not explore and conquer Europe. J. Diamond (1997: 177) stressed ecological reasons. The key axes in Africa and the Americas run from north to south with tremendous varieties in climate preventing the establishment of extensive empires. This axis orientation affects the rate of the spread of crops and livestock and possibly the diffusion of inventions. The north–south orientation of the Americas did not support the transfer of technology because innovative crops and livestock that would flourish in specific latitudes would not survive elsewhere.
8 The arsenal lost its influence once ships were constructed on a skeleton of keel with rib-and-plank constructions.
9 Portolan charts are navigational maps based on realistic descriptions of coastal features, ports and harbors which were frequently drawn on sheepskin.
10 It is thought that the Mongols probably brought the Black Death to Europe in the fourteenth century.
11 The Hansa was expelled from England in 1598 and prohibited Baltic merchants from maintaining the northern trading frontier on English shores.
12 Other famous narratives were those of John of Plano Carpini who reached the capital of the Mongols and wrote about his experiences in *History of the Mongols* around 1248; in 1253 the Franciscan friar Rubruck had been sent to the Mongol emperor for support to fight the Muslim aggression in the Mediterranean; and the Chinese monk Rabban Sauma was sent to Europe in the late thirteenth century to become familiar with European religious and cultural life and report this information back to the Chinese emperor.
13 The commercial routes Polo had followed eventually became known as the Silk Road, the label a German geographer, Ferdinand von Richtofen, introduced in 1877.
14 Whereas the Arab nomads conquered in the name of the Islam, and European crusaders fought in the name of Christendom, the Mongols had no specific ideology.
15 During the Tang dynasty, Chinese navigation had become familiar with the invention of a handy ruler and a star-measuring ruler, special devices for astronomical observation-based navigation. Based on the principles of trigonometry, the devices estimated the latitude of ships' positions by measuring the elevation and bearing between the horizon and the Polaris or other stars at any given time. The farther north the observer was, the higher the pole star; the farther South, the lower. The star-measuring ruler became a standard tool during Ming times (Deng 1999). Later, Chinese nautical astronomers focused more on Polaris, the new device, the polar Star aiming plate that was later adopted by the Danish

astronomer Tycho Brahe (see footnote 22). This device was first referred to by Marco Polo as consisting of 12 square plates in diminishing sizes with a range of 24 square centimeters to two square centimeters. European astronomers of Polo's time employed a different configuration based on the horizon and the sun's motion through the heavens (the ecliptic).

16 Although the printing press and gun powder are without exception Chinese inventions, the Chinese did not develop the mechanical clock.

17 It was 37 feet high (over 11 meters), and the time was indicated both aurally and by "jacks" that appeared in apertures in the tower. There was an armillary on the top and a celestial globe inside; both were rotated by the clock mechanism once in a sidereal day.

18 Another famous observatory stood in the capital Maragha of the Mongol ruler Húlegú (Rossabi 1992: 63).

19 Since the earth's orbit is elliptical and the axis of rotation is inclined, the length of the natural day, measured against the 24 equinoctial hours of the clock, varies by half an hour over the course of the year. The difference between an apparent and mean solar time is known as the equation of time (Holford-Strevens 2005: 9). Only many years later did the mechanical clock bring about the displacement of the apparent solar time shown on the sundial by mean solar time shown by the arms on a clock.

20 By 1400, with the disintegration of the Mongol Empire, the spread into Central Asia of Islam, the isolationist policies of the Ming dynasties, and the development of sea routes between Europe and China, the use of the Silk Road fell into decline.

21 Menzies indicted that Professor Wei of the University of Nanjing had written a book about the same topic entitled *The Chinese Discovery of America*. Unfortunately, he died, and the work still needs to be published and translated to English.

22 The Chinese used the pole star, which was regarded as the heavenly equivalent of the position of the emperor on earth. Polaris sits directly above the pole star. The Chinese, however, could not determine their latitude south of the equator (Menzies 2003: 161) where Polaris was invisible. They needed a star in the southern hemisphere and selected two: Canopus, the second brightest star in the sky, for a latitude and the Southern Cross, which points to the South Pole but is not directly above the pole, for navigation. The Chinese had not yet mastered the use of the sun to obtain the longitude. To solve this problem, they measured the distance traveled from home using hour glasses of sand. The day was divided in ten parts with each hour glass equaling 2.4 hours, the length of one watch for the seamen on duty.

23 Its remnants have only recently been excavated in Shanghai.

24 Whereas the scientific revolution of the West that started in the fifteenth century had its focus on mathematical reductionism and was much more mechanical, Chinese scientists shared an organic materialism.

25 The Pope dispatched Jesuit Matteo Ricci to Macau, and he eventually settled in Peking in 1601. Ricci acquainted the Chinese for the first time with the mechanical clock. The Chinese court, however, appreciated the mechanical clock only as a dispensable curiosity and treated it as a toy. In Chinese society, hours and minutes did not really matter. The self-ringing bells that Ricci demonstrated were seen as mere amusement. "Although some Chinese were ready to pay for European time pieces, they were still relegated to the status of oddities and trivia" (Landes 1983: 48).

26 He became an "important pioneer in the history of the transatlantic slave trade" according to Thomas (1997: 52).

27 For a long period of time, it has been contested whether Sagres actually established a school to study geography and navigational challenges as there was still a long and subtle way to go to acquire relevant knowledge about ocean currents and winds as well as the geographical perception of the world.

28 This map of the world contained a description of all known parts and was brought to Venice by Nicollo da Conti who had returned from his own travels to the East. Menzies (2003: 396) supposes that Dom Pedro, Henry's elder brother, had come across this chart during his twelve-year odyssey hunting after a dream of uniting Christians in Africa, India, and China by voyages of discovery. When he returned to Portugal in 1428, he brought with him this map disclosing new land in the western Atlantic.

29 They probably rediscovered these islands which the Phoenicians might have explored. During the Roman Empire, Roman sailors had reached these islands too. During the early fourteenth century the Portuguese discovered the Canaries.

30 The Waldseemuller world map of 1507 became the first to chart latitude and longitude (Menzies 2003) and for the first time contained a reference to the continent America.

31 The sun's maximum height varies day by day throughout the year, and the difference between this daily height and height at its lowest point in the midwinter is called the declination of the sun. This, however, was a very unreliable practice which the Portuguese first solved in 1474 and enabled them to measure longitude in the northern hemisphere. The Arabs had discovered that the sun's declination gave the latitude of a place in the northern Hemisphere. Regiomontanus, the Viennese astronomer who had attended Henry's court, had produced a set of tables giving day by day declinations of the sun. A captain in the distant oceans merely had to measure with a quadrant – a rudimentary type of sextant – the altitude of the sun at its meridian passage and then consult the tables to find the sun's declinations for that day. By subtracting declination from altitude, the mariner knew how many degrees south (and hence how many miles) he was from home.

32 In order to measure the latitude of important coastal objects, competent observers had been sent out who explored the field and reported back. It provided a range of astronomical and geographical data that were essential to convert unreliable measurements of the altitude of the sun into data about earthly altitude. This information was translated into tables with the sun's declination on different days of the year.

33 Yet linear distances still bore little relation to distances as measured by time. Ships taking the route from West Africa to Europe were obliged to sail near the equator, catch the south-east trade winds and westerly current and reach Europe via the Azores.

34 The Ottoman Empire was defeated in the battle of Lepanto in 1571 due to the superior firepower of the Christian fleet. In the years after 1580, Islam and Christendom disengaged in the Mediterranean, one gradually to introvert, the other to explore.

35 Paolo Toscanelli, who had met Da Conti and obtained significant usable information from him, later informed Columbus about this meeting and sent him a map showing the westward route across the Atlantic via Antilia. Toscanelli also passed this information to Behaim of Bohemia (1459–1507) who was working for the Portuguese government.

36 In 1496, the Spanish crown revoked its monopoly and granted licenses to other expeditions organized by Spanish aristocrats and gentry (see Hunt and Murray 1999).

37 Charles was born in the Netherlands, which were a collection of autonomous provinces on the northwest coastline of continental Europe and had formed in the later Middle Ages the core of the state of Burgundy.

38 From 1494 to 1559, Italy became a field of battle for other European rulers. Milan and Naples were the first important cities which fell and passed under foreign control.

39 It was superseded by the 1750 Treaty of Madrid which granted Portugal control of the lands it occupied in South America. However, the latter treaty was immediately repudiated by Spain.

40 The chronicler was the English William Camden (Kamen 2002: 92).

41 When dispatches had to be sent by new routes without an organized postal service, delays were normal. In the margin of a letter informing him of the circumstances, Philip

II wrote, "It is more important that the letters should travel by a safe route than that four or five days be gained, except on occasions when speed is essential" (quoted in Braudel 1972: 358).

3 The emergence of temporalities

1 Next to abstract time and its application in economics, there is also subjective time, i.e., duration (*durée*). While physicists observe events in succession with abstract time being only perceived in its spatial representation, according to the philosopher Henri Bergson (1859–1941), time can also to be perceived as duration, an endless flowing process in which one moment grows from another and yields to the next one. While external observation measures time, duration is an inner experience. This inner presentation shows us the nature of time as a process.

2 The gnomon is a straight pole, which is put vertically in a flat piece of ground and open to the sun's rays. The moment when the gnomon's shadow is shortest is called noon. Watching the shadow and its variety every day taught the priests about the phenomenon of the seasons.

3 The true solar year is some 365.242199 days and one can appreciate the intense frustrations over the centuries of astronomers trying to link up the sun and moon (Duncan 1998).

4 A true lunar month is 29.5306 days long as measured by modern instruments, equal to a precise 12-month lunar year of 354.3672 days.

5 The Tower of Winds at Athens – an octagonal building – erected in the first century BC by a Macedonian astronomer, Andronicus of Cyrrhus, was a model of the cosmos composed and constructed to glorify the beauty and simplicity of heavenly motion. The building's eight sides are decorated with figures in relief representing the winds. The tower served as a kind of water clock. In 1901, a few Greek fishermen found by accident, near the island of Antikythera, a device made by Greeks which could execute complicated astronomical measurements. It was a miniaturized device with more than 30 gears and could measure the position of the sun, moon, and other planets (see Marchant 2008).

6 The partitioning of day and night into 24 hours, a human imposition upon nature, came perhaps from the division of the zodiac into 12 equal parts, each one marked by a constellation through which the sun passed over the course of one moon cycle. Because it was supposed that the sun took the sun approximately 360 days to make a complete annual circuit through the cosmos, an easy system for dividing up seasonal times seems to have suggested itself; that is, intervals divisible by 6 and 12. The sexagesimal notation came to be a part of time reckoning, with 60 minutes to the hour and 60 seconds per minute, 12 hours per day and night and so on. Spatially, the circle representing the round of the sun on its celestial course was also divided into 360 degrees. The Maya applied a reckoning system based on the number 20 (vigesimal).

7 Komziak (2011) mentioned a completely different pattern of seasons: flood time from August to December, vegetation time from January to April, and harvest time from May to August.

8 The Chinese never changed the calendar to start with a fixed date. The national day to celebrate a new year is determined by the position of the moon. 2010 is the year of the dragon.

9 The Incas employed eight-day weeks; at the end of each week the king changed wives (Aveni 1989). Five-day market weeks were kept in parts of Africa and Central America.

10 When the Christians began to reject this kind of sacrifice, they clearly rejected the traditional Roman religion.

11 When preachers say on Christmas Day that Christ was born so many years ago they always give the number of the current year, implying that the nativity took place on December 25, 1 BC. However others insisted upon AD 1. Originally AD stood for Anno Diocletianus, the antichristian Roman emperor who had constructed his own year-reckoning system. The monk Dionysius Exiguus had replaced the existing AD practice with a new one based on the year Jesus was born. "An oddity of the system is that it locates his birth in 1 BC." (Bluedorn 2002: 15). In normal usage AD 1 is preceded by 1 BC. The Christian era is so well established.

12 Sabbath comes from the Jewish concept of a periodic recurrence of a day of abstention from work in the cycle – the seventh day – which the Jews gave over to the worship of the deity. The Sabbath's observance could become totally independent of the lunar or any other natural cycle due to the introduction of an artificial mathematical rhythm. The astrological seven-day week, which evolved in Alexandria during the second century BC, was introduced to the west through Rome some time toward the end of the first century BC. It was Alexander the Great's conquest of Greece, Babylonia and Egypt which brought those three civilizations together and was indirectly responsible for the evolution of the astrological week (Zerubavel 1985). "It was a convergence of both Jewish and astrological weeks around the time Christianity was being introduced into the Roman empire that produced the seven-day cycle that has since spread throughout most of the civilized world" (Ibid.: 23).

13 In 1752, England also introduced the first of January, after which most European countries – including Russia – came into line.

14 365.2425 days or 365 days, 5 hours, 49 minutes, and 29 seconds.

15 During the Counter-Reformation with its inquisition and strict dogmas, not even Galileo Galilei (1564–1642) was able to change prejudices against scientific explorations concerning the issue of time. In 1613, even the astronomer Johannes Kepler argued in support of reforms.

16 They were not practical and did not achieve widespread use until the end of the Middle Ages with the glorious invention of the mechanical clock (Aveni 1989). Around the fourteenth century, when public mechanical clocks began to dominate city activity, the unequal or temporal hours began to give way to what were known as equal hours.

17 According to the astrological week of chronocratories which most probably originated in Alexandria, "each one of the 24 hours of the daily cycle was assigned to one of the seven planet-deities which was regarded as its chronocrator or controller" (Zerubavel 1985).

18 The Jews and Muslims used a lunar Calendar that began its time reckoning from sunset.

19 "Saxons gave us the English word day, which did not come from the Latin dies, but the word in Saxon for 'to burn' during the hot days of summer. Hour is from Latin and Greek words meaning season. Originally, it referred to the fact that the length of the daylight period varies according to the seasons" (Duncan 1998: 63).

20 Florence saw itself confronted with strikes and waves of class struggle culminating in the Ciompi (wool carders) strike. They seized governmental power in 1378 to settle the question of who was to master time. The resistance lasted for some time, but eventually large companies such as the Peruzzi Company, which split its organization in 1335 between an operation in Florence and its foreign branches in Naples, Sicily, Avignon, Bruges, Paris and England, were able to dominate the business. The companies' branches were managed by Italian employees who eased communication and maintained the cardinal importance of trust (Hunt and Murray 1999).

21 Zeitgeber is now a proper English word according to the *Oxford English Dictionary*. The first Zeitgeber is the two-phase cycle of lightness and darkness. Zeitgebers organize human life.

22 It facilitated the imposition of disciplinary power and, in that sense, preceded the establishment of the factory system with its rigid disciplinary practices by several centuries. However, before the new temporal discipline could become part of tacit knowledge, it had to be taught explicitly, a task that took centuries.

23 The Basel clockmaker Heinrich Halder called this mechanism in 1385 a women's disposition (*Frouwen Gemuethe*). The endless movement of the foliot was compared with restlessness (*Unruhe*) and was associated with a general human trait (Rispens 2005, 118).

24 Windmills had been introduced to England in the twelfth century, possibly under the influence of Crusaders returning from the wars in the Middle East but probably more so must have come from Muslim Spain.

25 Although the mechanical clock predated in a technological and cultural sense, modern science, i.e., mechanical philosophy, by several centuries, the invention of the mechanical clock must still be considered as a crucial contribution to the rise of modern science in seventeenth century Europe.

26 Temperance should not be confused with temperedness (*equanimitas*) (cf. Iterson *et al.* 2001).

27 This treatise deals with a Trojan goddess Othea, the virtue representing the wisdom of a woman who advises a young prince about the formation of his character and the creation of good governance.

28 This does not mean that the interest for the virtue of temperance fully waned. Flemish classicist, Justus Lipsius, the man who instructed Prince Maurits, took the comparison between clock and princely behavior one step further and stated that the clock not only symbolized princely, but also divine, authority. He qualified temperance as reflecting a protestant work ethos. The pioneer of modern education, the Calvinist J. A. Comenius, compared the process of a well-planned systematic schooling to clockwork and indicated that as a measure for temperance, eight hours a day for schoolchildren would be enough.

29 The first telescope was designed by the Dutch spectacles maker Hans Lippershey from Middelburg. When he applied for a patent at The Hague, Prince Maurits inspected the device and noticed that standing at the inner square of the Binnenhof at the Hague, he could see the clock tower from the Old Church at Delft (a distance of eight kilometers in a straight line) reading the hours.

30 The Renaissance had already witnessed a growing appreciation for scientific ideals, in particular those related to the "mathematification" of science.

> The scientific revolution is inextricably bound up with the Renaissance and with the Reformation, three peak landmarks that led to the dissolution of feudalism and the rise of capitalism. A set of political, economic and social networks became established which created the material conditions for scientists to do their work.
>
> (J. Pitt in Boersema 2009: 329)

31 Vaucanson constructed elaborate mechanical models of human and animal machines in the firm conviction that it was possible to replicate the mechanical qualities of human beings and animals.

32 He published several different other works about *Volupté* and *Histoire naturelle de l'Ame* anonymously in the Netherlands.

33 The king became nicknamed "Regiments dompteur".

34 It was intended to make La Mettrie a member of the Royal Academy of Science, but his tract *L'Homme plante*, 1748, prohibited it.

35 At one moment he drew an analogy between the human being and the monkey as well as between Huygens' precise planetarium and the Julian calendar to just show how difficult it is to make correct comparisons (see Sutter 1988: 137).

36 Never did he see in Vaucanson's androids more than a superficial association. The last sentence in *Petit Homme à longue queue* (1751) – a fully satirical work and the last text he published during his lifetime – was: "Puisque la vie n'est qu'un jeu de la nature, il faut savoir rire dans la tempete [While life is but a game of nature, one should know how to laugh during a storm]". How nuanced the Prussian king's thoughts were about his personal doctor was demonstrated after La Mettrie's death in 1751. The king wrote a eulogy that was read aloud in 1752 during a gathering of the Berlin Academy of Science in which he drew an analogy between a clockmaker and a statesman. Just like an able mechanic is not satisfied with looking at the outside of a watch but opens it and examines its springs and wheels, so an able politician exerts himself to understand the permanent principles of courts, the engines of the politics of each prince, and the sources of future events.

> He leaves nothing to chance; his transcendent mind foresees the future, and from the chain of causes penetrates even the most distant ages. In a word, it is the part of prudence to know all things in order that all things may be judged and every precaution taken.
>
> (Quoted in Mayr 1986: 108)

4 The age of merchant capitalism

1 The mercantile era drew to a close around 1800, giving way to industrial capitalism. According to Karl Marx, the transformation of merchant capitalism into industrial capitalism shifted the process of primitive accumulation of capital, resulting in the establishment of the factory system and a rapid exploitation of industrial wage labor.

2 Initially, the charter that created the new company was valid for 21 years. An EIC ship from London to Calcutta took about 7 months, a VOC ship leaving the Dutch coast at Texel to Batavia about 8 months. One of the initiatives to reduce transaction costs was to install specific sailing schedules and travel times.

3 They showed a tolerance more motivated by business sense than by an enlightened ideology.

4 William had extolled his decent "van duitsen bloed", i.e., from popular non-Latinised roots (see Sorge 2005: 87).

5 Digges published in 1571 a military work *An Arithmeticall Militaire Treatise Named Stratioticus* which was dedicated to the Earl but spoke in an admiring way about the military activities of William of Orange. The book insisted that the officers had to understand algebra and numbers which would assist them arranging hundreds of men in new battle formations, teach the soldiers to march in step and perform like automatons.

6 A Compagnie van Verre was established in 1594 to find a way across the North to Asia, an expedition which failed.

7 As the British chartered the EIC for a period of 50 years, the Dutch decided that they should respect a similar time frame.

8 A year later he published *Itinerario: Voyage ofte schipvaert van Jan Huyghen van Linschoten naer Oost ofte Portugaels Indien*, 1579–92 (*Travel account of the voyage of the sailor Jan Huyghen van Linschoten to the Portuguese East India*). The volume was translated in English in 1598 and stimulated both the creation of the EIC and the VOC (see Vance 1986: 429–430).

9 In the Italian Po Valley, a complicated network of canals was also constructed to improve low-cost transportation facilities for the entire region, but it was less well established than the Dutch network.

10 Sometimes the *beurtveer* and *trekvaart* networks overlapped (Vries 1981).

11 Two centuries later, the same two cities would establish the first railroad line in the Netherlands.

12 The *trekvaart* fever was not only observable in Holland but in other provinces as well. Frisian requests for the same facility were put forward by stressing the urgent need to travel with greater speed and dependability. Between 1618 and 1622, a canal was dug from the city of Groningen to Zuidbroek as part of a large-scale peat digging industry in the province of Groningen then underway. The first *trekschuiten* in the Low Countries were introduced here.

13 Travel from one end of the Mediterranean to the other took from 40 to 60 days. This time dimension was unaltered from the time of the Roman Empire. The Low Countries was a small world whose size measuring along its longest axis was about eight days of travel.

14 From 1576, The Dutch army was paid every so-called Dutch month, which divided the year into eight blocks of 42 days plus one of 29, thereby reducing the number of pay instalments to nine (Wilson 2009: 145). In 1588, the States-General together with the Council of States had decided to decrease the size of the army, ensuring a decent and regular pay for the soldiers in exchange for obedience.

15 The army was being subjected to the rules of the evolving modern Dutch nation-state.

16 Willem Lodewijk had described this line-up in one of his letters to Maurits (August 12, 1594).

17 This technique was modified around 1595 so that men stood in blocks of five, "peeling off as a group right or left once they had fired so as to reduce the number of gaps required in the line" (Wilson 2009: 87).

18 "Those adopting the Dutch style counter-march needed thinner lines and more shot than pike, deploying a ratio of two to one in a ten-rank line by the 1590s, with the pike men in the center flanked by equal numbers of musketeers. The Spanish and imperial infantry favored the larger, deeper formations that had been the norm earlier in the sixteenth century" (ibid.).

19 Rosa (2005) calls this the superiority of the faster ("Herrschaft des Schnelleren") which was initially developed in military practices.

20 This is why the French historian and philosopher Michel Foucault described Maurits' formula, in which rhythm is determinative, as a "discipline à travers une rythmique du temps."

21 The Dutch painter, Rembrandt, probably used some of the bodily positions of soldiers drawn by De Gheyn for his *Nightwatch* which he painted in 1642 (Westermann 1996).

22 With the introduction of the *achterlader*, soldiers could fire more accurately than with the musket, the standard weapon till the 1770s.

23 With this modern army, Maurits was able to conquer the city of Groningen in 1594 and eventually end the Spanish domination. When he freed the largest city in his military career, he received a new tent: a traditional version of a company home that top managers currently receive.

24 "Maurits can, therefore, be seen as a typical Renaissance figure in his urge to get his house in order" (Kist 1966: 123).

25 Under the favorable conditions of the long European growth cycle of the sixteenth century, the landlords of the Brandenburg nobility had amassed impressive fortunes by producing grain for export. In 1417, the Hohenzollers purchased Brandenburg and worked hard to consolidate and expand their patrimony. The Elector Joachim II converted

to the Lutheran faith and reinforced the training of his army. Ironically, however, military training for the French nobility was abhorred in the seventeenth century; "the noblemen sought to suppress all institutions of the kind that forced them to show themselves worthy of employment" (Vagts 1959). Since the mid-eighteenth century, however, the Ecole Militaire gained much status.

26 He married the Dutch Louise Henriette, daughter of the Dutch *stadhouder* Frederick Hendrik, brother of Maurits who, as *stadhouder* between 1625 and 1647, had employed as art agent and chief diplomat Constantijn Huygens, father of Christiaan, later to become the famous inventor of the pendulum clock.

27 Schooling became a "discrete object of the state action" with standardized certification procedures for teachers and general-issue elementary textbooks for pupils. Orphanage schools put in place close psychological observation of pupils and an emphasis on self-discipline and an acute awareness of time by putting hourglasses in every classroom (Clark 2006: 133).

28 The Low Countries continued to be one of the main schools of warfare for Europe, especially during the period 1672 to 1730 when "the Low Countries were a strategic hub of great struggle between Louis XIV and the European coalition ranged against him" (Israel 1995: 3).

29 The Low Countries continued to be one of the main schools of warfare for Europe, especially during the period 1672 to 1730 when "the Low Countries were a strategic hub of great struggle between Louis XIV and the European coalition ranged against him" (Israel 1995: 3).

30 Directors within the Chamber of Commerce of Amsterdam kept their claim of a managerial position until they died.

31 Thirty-three expeditions took approximately 3,000 slaves to be expedited to the Indies and Americas.

32 The EIC showed much less military involvement in the Asian area than the Dutch VOC. It was much more oriented in diplomatic involvement.

33 In 1750, the number of employees had risen to 12,000 marine men and 17,000 soldiers.

34 In 1672 when the Dutch fought a sea battle with Britain, a craze for calicoes began.

35 After 1639, the Dutch Company was the only European trader with access to Japan.

36 The establishment of the WIC had already been considered in 1606–1607 but had been aborted when truce negotiations with Spain were on its way in 1607 (see Israel 1995: 326).

37 The WIC had known a predecessor established in 1607 but had failed; the Guinea Company as successor had been more successful and built a fortress called Nassau near Elmina (see Thomas 1997: 161).

38 In Greek society, slavery was such a fact of life that some of the greatest philosophers of their age hardly deemed its existence worth mentioning.

39 Although castration was prohibited in the Byzantine Empire, it was practiced extensively while eunuchs were seen as luxury slaves with a higher commercial value (see Rotman 2009).

40 Jewish merchants were, also highly involved during these times in the slave traffic between present-day France and Spain (Rotman 2009: 73).

41 Cyprus, Crete, and Corfu were economically valuable possessions for the Venetians, but Cyprus had to surrender to the Turks in 1573.

42 Expatriate Arabs of Spain had conquered the island in 827.

43 Slave trade resurged in the Mediterranean area after the ravage of the Black Death. Manpower was needed in domestic activities in Italian cites as well as in rural estates.

44 In this respect the "Industrial Revolution was given a colonial trial run" (Abernethy 2000: 11).

45 The first import of black slaves is dated 1510 when King Ferdinand licensed the House of Trade in Seville to send 250 Christian blacks acquired in Lisbon to the New World. They originally had been captured "in aggressive raids on the African coast" (Kamen 2002: 136).
46 Now the United States Virgin Islands. The Danish government sent slaves from the Castle of Christiansburg to their island colonies of Saint Thomas and St Croix
47 Haiti became the principal sugar colony of the French.
48 Although blacks converted to Christianity, the colour of their skin still maintained their lower status. In 1639, for example, the Maryland colony had spelled out that the slave's freedom could not be achieved through Christian baptism.
49 In Spain and Portugal, however, the possibility of freeing the slaves from the burden of supporting the colonial structure was less envisaged.
50 Slavery was outlawed in Britain by a High Court judgment of 1772, but its impact was minimal. In 1833, the British parliament abolished slavery throughout the Empire, and the Royal Navy was required to enforce the law by apprehending slavers and liberating the human cargoes. A serious blow to Christian slave trade came in August 1816 when a large British fleet, assisted by six Dutch vessels, bombarded Algiers to disengage the slave trade in Algeria and Morocco (see Milton 2004). In 1840, an international antislavery convention was held where a heavy presence of America's most fervent abolitionists such as William Lloyd Garrison and Theodore Weld were featured (Schama 2008: 253). The Dutch ceased their activities in 1863, though they had abolished the slave trade on paper already in 1814. In 1876, slavery ended in Puerto Rico and in Cuba in 1884. Slavery was abolished in Brazil in 1888.
51 Over its 200-year history, the average dividend it paid out to its 358 shareholders was 18 percent a year.
52 During the Anglo-Dutch wars, Britain defeated the United Provinces and annihilated what was left of the Dutch seaborne power. The ensuing Batavian Revolution and Orange counter-revolution hastened the displacement of Amsterdam by London as a financial entrepot of the European world economy. This was completed in the course of the Napoleonic wars which wiped the United Provinces off the map of Europe. The trade of sugar and coffee built a global integration during the eighteenth century run by the French and British empires.
53 Until 1707 one talks about England, from 1707 onwards it is Britain due to the Act of Union when Scotland became part of England and gained access to the colonies (see Berg 1994).
54 While Britain possessed 23 colonies in 1792, that amount had risen to 43 by 1816.
55 France extended its grip over Indo-China, while the United States took control over the Philippines after a war with Spain in 1898 (Jones 2000).
56 Between 1660 and 1719, 54 major joint-stock companies received charters: 23 in mining and manufacturing, 11 in overseas trade, and 20 in banking, insurance, fishing and other fields of endeavor.
57 Following a fraudulent action in selling stock, the British parliament passed the Bubble Act in 1820, placing severe restrictions upon the establishment of joint-stock companies. Although they became less attractive, their demise did not retard Britain's economic expansion.
58 In particular, emperor Qianlong (1735–1796) was extremely fond of these expensive musical clocks with exotic designs combining Chinese ornaments with Western elements which were generally produced by producer-exporters like James Cox. The emperor's preference was imitated by local tax collectors and other civil servants and certainly promoted a booming interest for Western clocks.
59 Swiss and French Jesuit clockmakers traveled to China and demonstrated to the Chinese emperor the marvels of their clockwork timepieces. The Chinese, however, looked at the

clocks and watches as luxury goods. They did not perceive them as scientific timekeepers but as artifacts for pleasure. The Jesuits offered them as gifts to the emperor and his "entourage", but they could not notice an interest amongst the Chinese in the mechanistic metaphor nor did they assimilate the language and clock-making technology upon which the mechanical clocks had been built (Jami 1995).

60 He "eliminated the balance-wheel regulator and substituted a pendulum hanging freely from a cord or wire, hence largely isolated from disturbance by the train" (Landes 1983: 116). Huygens wrote an instruction about his invention in 1665: "Kort onderwys aengaende het ghebruyck der horologiën tot het vinden der lenghte van Oost en West."

61 Yet, in Britain there still existed a body of literature that generally expressed negative sentiments about the clock. This had a lot to do with its inaccuracy "which fed negative connotations and pure rejection of whatever the mechanical clock stood for" (Mayr 1986: 124). The English ambivalence to clocks was fed by a comparable ambivalence to other mechanical objects. The English word *engine* and *mechanical* traditionally had a negative connotation amongst the well-to-do. "The frequency with which the disliked clocks were identified as German was significant" (Mayr 1986: 51). Whereas continental authors usually associated the clock with an incentive for purposeful activity, some Englishmen saw it as a reminder of one of life's more somber truths like death.

62 The Royal Society, established in 1660, clearly illustrated the appearance of an experimental, enquiring spirit which motivated scientists and inventors to improve knowledge, a process which preceded the Industrial Revolution. The Society thoroughly discussed Huygens' discoveries of rings around Saturn which earned him a splendid reputation in England and increased society's interest in Huygens' clock-making skills.

63 Yet eventually Britain's watch-making industry fell back technologically "while both entrepreneurs and labor resisted innovation" (Landes 1983: 300–301).

64 "With a sextant and a compass, a mariner can obtain his latitude by ascertaining the angle between the sun, the ship, and the horizon when the sun is directly to the north or south (local noon time). From this he knows how many degrees he is north or south of the equator (his latitude). To find his position east or west is, however, a much more difficult task because the earth, in spinning on its axis, upsets astronomical calculations. However, the earth spins at a relatively even rate through 360 degrees in a period that we have divided into 24 hours." Since the sailor could also ascertain the latitude, he would thereby know his precise position (Macey 1980: 28).

65 Parallel work had been undertaken in France. Pierre le Roy completed a clock in 1766 that could keep accurate time at sea which laid the foundation for today's modern Marine chronometers. Le Roy never received an award for his invention. However, by 1800, the basic design was so sophisticated and production on a large scale so advanced that the design survived largely unaltered until quartz crystal technology revolutionized the entire field of timekeeping.

66 This practice had been devised by British astronomer R. N. Maskelyne in 1767.

5 Time in British industry and railroad transport

1 The sulfur-free form of coal known as coke improved the smelting of iron.

2 The adoption of coal-cutting machinery had to wait till the second half of the nineteenth century.

3 One engine could replace a pumping operation employing 50 horses.

4 With the hot blast in 1829, the expansion of the iron industry moved ahead quickly.

5 As late as 1880, British manufacturers controlled almost 40 percent of the world trade in industrial goods.

6 Although the joint-stock acts of 1856 and 1862 permitted the formation of limited companies, British enterprises largely remained partnerships.
7 The term "punctuality" first appeared in the English language in the eighteenth century (Levine 1997: 57).
8 Many potteries allowed their potters traditional freedoms and although factories were open for 84 hours per week the amount of work done probably did not exceed 60–63 hours (Clark 1994:155). The undisciplined potteries kept large stocks of inventory to accommodate the uncoordinated workers.
9 Around 1800, prescriptions on factory behavior were still scant. In fact, there were few factory instructions available, nor was there a disciplinary attitude. An exception was the Law Book of the Crowley Ironworks – probably the largest works in Europe at the time. It provided a precise guide to the behaviour, on and off duty, of all ranks. "Certain facts suggest that Crowley's laws were normally respected, though others indicate that enforcement left something to be desired" (Flinn 1957: xvi).
10 Max Weber (*The Protestant Ethic and the Spirit of Capitalism* (1904–1905)) saw in Benjamin Franklin the perfect exemplar of the modern capitalist spirit which combined religious fervor with parsimonious behavior. Franklin expressed the middle-class values like diligence, honesty, industry, and temperance and as the prototype of the self-made man expressed the moral maxims underlying the ethic of capitalism.
11 In 1787, Matthew Boulton produced a double-acting steam engine, a fuel-efficient multi-use machine that would stoke all aspects of the Industrial Revolution. Its up-and-down motion would first pump the mines but, within a few years, its rotary motion would run the Lancashire cotton mills.
12 Wedgwood's eldest child Susannah married the scientist Erasmus Darwin's son, Robert.
13 Some critics nicknamed the five friends the "Lunatic Society."
14 Franklin published this dictum in his *Poor Richard's Almanac.* This compendium of weather forecasts and other practical information became one of those peculiar American genres of management books that revealed the secret path to prosperity. Like Wedgwood, Franklin became one of the leading figures in promoting the abolition of slavery.
15 About the same time Thomas Telford and J. L. McAdam improved the surface of the roads to prevent deterioration due to weather circumstances.
16 In August 79 BC, Mount Vesuvius erupted, destroying Pompeii and Herculaneum, some 240 km south of Rome (Beard 2008).
17 In 1734, the excavations and their growing popularity led to the founding of the Society of Dilettanti, to which Hamilton soon became member, which financially supported the publication of books about the excavations.
18 But the period 1760–1830 reflected a paradoxical development: while working hours and prices for consumer goods increased, there was also an increased desire to spend (Voth 2000: 200).
19 During the period of its 223 years of existence, the company had experienced ups and downs. When the last members of the Wedgwood family retired as managing directors in 1966, management control passed into the hands of directors and managers who had no direct links with the family. Wedgwood as a company lost its family atmosphere in Etruria and later on in nearby Barlaston. It was replaced by a more invigorating and more impersonal spirit. The company moved away from the reliance on kinship and personal recommendation and recruited in a more professional and systematic way. In 1987, Wedgwood in Stoke-on-Trent (60 percent of the stocks were in the hands of Sir A. O'Reilly) was taken over by the Irish company Waterford Chrystal, which since 1783 made crystal glass ware. In 2009, the company was taken over by the American investment company KPS Capital Partners.

20 To reduce the incidence of crime, the spread of street lighting increased during the second half of the eighteenth century and diminished the likelihood of thefts.

21 While the length of the working year rose by the introduction of the shift system and an imposition of additional hours per day, per week, or per year, capital services increased without any measured growth of the visible capital. Factories could simply operate their equipment for more hours. The capital input, i.e., the number of days that machinery was available for use, increased in line with any additional labor input. So, longer hours lowered the costs per unit of capital services and boosted the output.

22 British society struggled with two competing versions of liberalism. One concentrated on civil rights and political reform; the other defended the socio-economic version based on market values.

23 The most important development of the 1820s was the penetration of ideas of the political economy among a powerful and influential section of the Tory Party, particularly the allies of Sir Robert Peel. Those who believed in some version of paternalism and protection were contesting the different versions of political economy. Radicals concerned about work conditions and the miseries of the working class allied themselves with Tories like Michael Thomas Sadler who favored legislation regulating working hours and conditions in factories. Overall, the first half of the nineteenth century was a period of profound ideological and religious dispute. The Reform Bill of 1832 did not lead to an expected extension of the voting right for a larger part of the population. Chartists claimed the right of legal political involvement for the male population but Chartism failed to secure reform and was dismissed by the legislature (Hilton 2006).

24 Traditionally the hand-spinner had the ability to draw out the strand of pre-carded cotton to a thickness at which it would be twisted onto a bobbin by a rotating U-shaped flyer powered by a foot-driven wheel.

25 This marvel of technology was an improved version of what the engineer Jacques de Vaucanson had invented around 1750 when he created life-like automata such as the flute player and the duck, which impressed Louis XIV.

26 In German textile factories, time discipline placed more emphasis upon the duration of production itself than on its precise starting and finishing times. German textile directors could shift hours among the days of the week, as long as the total remained the same. German textile mills introduced punch-in clocks on a wide scale at the turn of 1900 (Biernacki 1995).

27 The ideas concerning the influences of the environment on individuals and society came from Rousseau and other thinkers of Enlightenment.

28 Owen became inspired by ideas of William Godwin who had published an *Enquiry Concerning Political Justice* (1793) in which he propagated a new social order stressing justice, freedom, and equality for the individual.

29 By character, Owen did not mean the individual disposition of each person but the make-up of moral and social ideas that might be infused into a community by education and environment.

30 The most marked difference between this signposting and the time-and-motion studies, which became the basis for scientific management in America's factories, is that the silent monitor was not intended to increase output. It only recorded the temperament or behavior of every individual worker and was not meant to reduce waste, improve efficiency, increase output, or enhance profits.

31 A practice Owen's father-in-law had put in place in 1788 when a fire had destroyed one of the mills.

32 In the early 1830s, seven firms in Manchester alone had about the same amount of workers. But these were not the biggest companies. In the same period, some iron works employed over 5,000 workers.

33 Peel, the son of a wealthy cotton manufacturer, was the founder of the Conservative party, later becoming Prime Minister.

34 The increased costs of substituting adult for child labor were partly offset by the fact that Owen was able to pay lower wages than those in the urban counterparts where rents and social provisions were usually more expensive.

35 The first community had been established in 1804 in Pennsylvania but due to disillusionment, Rapp had moved his community north of Pittsburgh (Indiana) and created New Harmony. As followers of Rapp, the Christian Communitarians were active in agriculture, manufacturing and trade. Work was organized communally which included full employment for every one and occasional rotation of jobs to avoid boredom. "The sound of French horns awakened the communitarians each workday morning, but members made no claim to living a rigorous monastic life of self-denial" (Pitzer and Elliott 1979: 237).

36 The first locomotive was built by the French military engineer Nicolas Cugnot in France in 1769.

37 The Stephenson factory sold locomotives to a German railroad company that exploited the line between Nurnberg and Furth.

38 In 1840 Captain Basil Hall, one-time commissioner for longitude, had suggested that all Post Office clocks throughout the different counties should be kept to Greenwich Mean Time, too, which could be realized by means of the time brought from London by the mail coach chronometers. By doing so, one expression of time would prevail over the country.

39 On the continent, the situation was complicated: Belgian trains ran according to Brussels time, Dutch trains to Amsterdam time. In Germany, railway officials kept any one of five times – those of Berlin, Munich, Stuttgart, Karlsruhe, or Ludwigshafen. However, clock time was kept strictly to local time, and there were posts set alongside the rails marking each change of time. It was apparently customary for watches to be altered in 10-minute steps during the journey. When Count von Moltke pointed out to the German parliament in 1891 the military consequences of lack of standardization, which would impede the coordination of military planning, Berlin time was adopted throughout the nation.

40 In 1790–1791, Chappe devised, with the assistance of his brothers, a system based on sound and synchronised clocks to transfer messages. The invention was a failure, and he turned to optical solutions. By 1794, he had built a series of semaphore stations between Paris and important military posts which were equipped with a mast and two cross arms as large movable pointers. The stations were spaced 10–15 kilometers apart and were operated by four to five people each. Messages were being sent by a sequence of arm positions at one tower and read by an observer at the next tower. The system could transmit coded messages at the rate of several hundred kilometers a day via the "telegraph." When Napoleon took charge of the French Republic in 1799, he ordered the completion of the optical telegraph system to link all major cities in France.

Chappe then introduced an elaborate coding system to transmit messages rapidly. Many common phrases and proper names were encoded by only two optical symbols which facilitated a substantial gain in the speed of transmission (Friedel 2007). However, the system was useless in bad weather and at night and still had a substantial error rate. Yet for a long time, military concerns kept the system in operation, and it was the first "real time" network "designed for careful coordination of many elements, both technical and human, over considerable distances" (ibid.: 299).

41 Weber described the idea of the career as a generic feature of a modern, rational society.

42 Even after 1921 when the British government forced the railroads into concentration leading to four quasi monopolistic giants, their management practices were not readily transferred to British industry, which was still predominantly owned by families. When, in the 1930s, British industry contained more large-scale companies, "bankers and accountants

made more promising consultants and executive recruits than did railway managers" (Gourvish 1986: 202).

43 Germany mainly used its steel production to improve its national, industrial infrastructure. Other major innovations occurred there in the chemical, electrical, optical, and precision instrument industries which rapidly developed.

44 The Americans had designed a very fast ship, the clipper, which sailed the trip from New York via Cape Horn to San Francisco – 19,000 kilometers – in just 90 days, where previously it had taken 150–190 days, yet the ship eventually lost out to the steamship. The last wooden British war ship was built in 1860.

45 The *Great Eastern* was used to lay the transatlantic cable in 1865.

46 The committee overseeing its construction included I. Brunel.

47 The company attracted attention of all the royal courts of Europe with its watches which were worn pinned to clothing.

48 Japanese trading companies continued to exist after World War II as general trading companies, or *sogo shosha*.

6 Politics and time

1 This option was inspired by the suggestion of Christiaan Huygens that the length of a simple pendulum that beats the seconds at the equator could be the most convenient magnitude.

2 The distance between the latitude of Perpignan and Dunkirk had been measured around 1787 and its length was known as the 40 millionth part of the circumference of the earth along the meridian. This could serve as a unit with all the accuracy needed for local surveys. Eventually, it was realized that measurement of the meridian was too complicated. Yet, on July 14, 2000, the French honoured the measurement of the meridian with an incredible lengthy *pique-nique*, "*le dejeuner sur le pavé*".

3 De Laplace told Napoleon that the chief benefit of the metric system was not the destruction of the horrific feudal practice of so many different measures but the division by tens. The meter was later defined as the distance between two marks on a metal bar kept at Sèvres near Paris.

4 In this sense, the Republican calendar followed the Egyptian example of 12 months with 30 days each and an addition of five complimentary days.

5 There were still a few exceptions to the new metric system, such as angular and temporal subdivisions continuing to be sexagesimal, with which the revolutionaries were unhappy.

6 Restif de la Bretonne (1990: 1226) also contributed new names to the months. In "Paris le jour, Paris la nuit" of 1793 his reform looked as follows: vindémaire, brumaire, frimaire, nivos, ventos, pluvios, germinal, floréal, préréal, messidore, fervidore, fructidore. The ten days of the week were called: primidi, duodi, tridi, quartidi, quintidi, sextidi, septidi, octidi, nonidi, decadi.

7 During the French Revolution, French military practices inspired American military forces. After the defeat in Jena in 1806 by the French troops under Napoleon, the Prussian army lost its role of inspiration. German military intellectuals asked for military reform and required the professionalization of military planning and leadership; they asked for command functions to be concentrated in one decision-making unit. Within the German Military Society, which was founded in 1802, two figures in particular influenced the debate. G. J. D. von Scharnhorst called for the introduction in Prussia of the Napoleonic divisional system and the establishments of a territorial militia as a reserve force (Clark 2006, 324). He also proposed improvements to training, tactics and weaponry. Karl von Clausewitz stated that the army should not be conceptualized as a machine but as a consciously willed

organism with its own collective genius. Flexibility and self-reliance, especially among junior commands were, according to him, of vital importance.

8 The Russian Orthodox Church still celebrates Christ's resurrection on a different day than Christians in the West. The maintaining of the Julian calendar also led to other remarkable misunderstandings. In 1908, for example, the imperial Russian Olympic team arrived in London 12 days too late for the games as Russia was still using the Julian calendar (see Richards 1998). Russia held out until 1918 when the communists finally complied with the Gregorian calendar. Japan adopted the Gregorian calendar in 1873 during the westernization of the Meiji period. China did so in 1912. The Balkan countries waited until the Great War to chime in. Beyond the general acceptance of the Gregorian calendar, there was the problem of Easter to be harmonized. Initiatives were taken in 1926 by the League of Nations which recommended that Easter should be kept on the Sunday after the second Saturday in April but an agreement was not reached.

9 The American national conference on weights and measures still struggled in 1916 on whether to implement the metric system or not (Shenhav 1995).

10 During the second half of the nineteenth century, some firms like Cadbury Brothers, Rowntree, and Lever Brothers introduced paternalistic employment practices trying to improve the working conditions of their employees.

11 By 1850, raw materials, mainly raw cotton, constituted almost two-thirds of British imports.

12 The Franco-Prussian war of 1871 dealt a serious blow to the movement: workers of different countries suddenly had to fight each other.

7 Speed in transport and industry

1 Another canal that shortened the sea transport time was the Panama Canal, which, with its 77 kilometers, joined the Caribbean Sea to the Pacific Ocean. A ship sailing from New York to San Francisco via the canal could now travel 9,500 kilometers, well under half of the 22,500 km route around Cape Horn. The United States completed the canal in the early 1900s.

2 In some cases, cities were involved in financing infrastructural projects. For instance, the city of Baltimore provided much of the funding for the B&O Railroad.

3 The German steel producer Krupp also presented steel guns, however, they did not attract much attention until the Russian and Prussian government ordered them in large quantities.

4 Initially, Colt exported only his guns to Europe but, in 1852, decided to export his machinery to England as well because no equipment made in that country was exact enough for the work necessary to manufacture the revolvers. "Due to a lack of sufficient control over such a distance, Colt decided to sell his company to a few Englishmen in 1857" (Wilkins 1987: 71).

5 Within a generation, the British watch-making industry was doomed "due to resistance of both entrepreneurs and labor to innovate" (Landes 1983: 300) and was taken over by American watchmakers with their mass-production methods,

6 Franklin brought France into the struggle against Britain and made it possible for the nascent American republic to receive loan after loan from an increasingly impoverished France.

7 Gribeauval's success relied to a large extent on the contributions by Honoré Blanc who, as a skilled armorer and gunsmith, devised the means of uniformity, interchangeability, and instructions for the gun manufacturers (Friedel 2007).

8 Eli Whitney was the inventor of the cotton gin (short for engine) and also had attempted to introduce interchangeability in production of fire arms. In January 1800, he gave a

demonstration of his new musket to Adams and Jefferson and other dignitaries and surprised the observers how quickly he could assemble the bits at random into a fine musket. The observers believed that they had seen the augury of the machine age. What they did not know was that Whitney had secretly marked the parts that were meant to go together. The parts were not interchangeable and were not uniform.

9 The watchmaker Aaron L. Dennison later joined forces with Eli Whitney, who made riffles with specialized machinery at his Springfield plant in Massachusetts.

10 Since Robert Peel, prime minister of Great Britain, had lifted the ban on the export of machinery in 1842, which had been imposed since 1774, the United States was able to import more equipment for railroad transportation.

11 The army used his approach until 1903.

12 They have been among the most important parts of the military management system later borrowed by the business community (O'Connell 1985: 91).

13 Jefferson ardently believed in sovereignty of the populace and each human being having his own right concerning personal principles of life and liberty:

> Jefferson's West Point was founded to deny the United States its Caesars [...] and ensure the permanent victory of liberalism over militarism. Jefferson wanted to prevent the establishment of a "ministerial" army which was the tool of despots and servile to their masters.
>
> (Schama 2008)

He preferred citizen volunteers who would only take to their muskets in defense of heart and home: they were always citizens first and soldiers second.

14 The Ecole Polytechnique began in 1794 and was intended to educate civil and military engineers. It also advocated the founding of mathematical physics and delivered quite some protagonists among the Saint Simonians. Throughout Europe and America, it was the first and the very model of a modern engineering school such as MIT, the English polytechnics, and the German technische Hochschulen.

15 Whistler resigned in 1842 to become the Chief Engineer of the Moscow–St Petersburg railroad project in Russia.

16 Much later, both aspects became key features in American business practice and remained as an accounting method at the heart of managing a business.

17 By 1850, some 14,500 kilometers of track had been laid out. Ten years later, that number had increased to nearly 50,000 kilometers.

18 The Asian immigrants had to cope with serious racism. Although by 1870 a quarter of all miners in the west of United States were Chinese, "the labor movement itself played a leading role in the struggle to exclude Chinese and Japanese workers from American life. Labor Unions like the Knights of Labor demanded for the legal prohibition of Chinese immigrants. This strategy was warmly supported by the Workingmen's Party of California, as well as by San Francisco's governing Union Labor Party" (Montgomery 1987: 85).

19 The paradox of public transport was that the better it was doing its job, the less efficient it might become: to assess its economic performance, non-economic concerns have to be included as well.

20 The new system of standard time zones was not codified for the United States until March 19, 1918, when US Congress passed a standard time act which legalized the system. It authorized the Interstate Commerce Commission to define the boundaries between time zones. The Trans-Siberian Railway, from Moscow to Vladivostok and completed in 1916 operated for long on Moscow time. The Russian railroads ran on a wider gauge which caused time-consuming delays at national borders to accommodate for the switch.

The watchmaker Breitling manufactured the Chronomat GMT which adapts automatically to the switching of time zones.

21 Since that time, America has become, according to Hoopes (2003), the "premier market for management gurus." Businesspersons, engineers, managers, academic management writers, and consultants have constantly launched new views on management practices.

22 Whereas the strike of 1877 had been put down by the military, in 1894, action was ended by the courts but at the same time the American railroad union was formed.

23 It continued the work of the dissolved First International, though excluding the still-powerful anarcho-syndicalist movement and unions and was in existence until 1916.

24 To prevent a strong formation of radical union socialism amongst anarchist groups, several different American states began to institutionalize progressively a so-called Labor Day as a festival and holiday. This Labor Day, the first Monday in September, became a national holiday from 1894 onwards to recognize the "dignity of work." It was meant to reduce May Day to a minority event.

25 Only after 1890 was it decided to commemorate the struggle for the eight-hour day every year on May 1, and this slogan turned into the cry of the turn-of-the-century labor Unionists and socialist parties.

26 For astronomical purposes, Greenwich Mean Time was reckoned by the 24-hour clock until 1925, commencing at noon. Since then, it has been reckoned from midnight and was renamed Universal Time or UT in 1928.

27 Joseph Conrad immortalized the incident in *The Secret Agent.*

28 Not all European countries were included in this convergence circle during this period: the Netherlands, for example, was, in this period, still a rich mercantile economy. Its process of industrialization abounded only at the end of this period.

8 Time and motion

1 "The scandal of capitalism, on the other hand, is that it depends on surplus labor and is thus unable to eliminate the exhaustion it produces" (Rabinbach 1990: 74). According to Rabinbach, Marx's paradigm of work in terms of emancipation became much more a paradigm of production. Labor power, very much in line with Helmholtz, became the engine of historical development.

2 This epistemology is premised on the dualism of a knowing subject gaining knowledge of a known object.

3 Marey's studies of cardiovascular disease brought him fame, and in 1868 he was awarded the chair of Natural History of Organized Bodies at the College de France.

4 Among those invited by Marey were Helmholtz and the portrait photographer F. Nadar.

5 Today it is the area for the annual tennis tournament of Roland Garros.

6 He looked at fatigue as the negative aspect of a body conceived as a thermodynamic machine capable of conserving and deploying energy.

7 Marey, however, was not interested in the commercial possibilities of cinematic illusion.

8 Weber himself reported extensively about the problems of fatigue and monotony in factories in the Archiv für Sozialwissenschaft und Sozialpolitik 1908–09.

9 Their approach intended to improve the profitability of firms by introducing new methods of calculating efficiency and by reducing fatigue at work. Improved technology was a higher priority than motivated workers.

10 This was an essentially different perspective than Taylorism and Fordism which were oriented on compensation of unhappiness through external non work-related material rewards.

11 This practice had been copied from the French General De Gribeauval who had introduced the assembly of guns with replaceable parts in the French army.
12 Taylor turned towards engineering and machine shops as a rejection of his father's career as a lawyer (Kakar 1970).
13 He used the same measurements – foot-pounds per minute – which had been adopted by James Watt when measuring the horsepower of his steam engines. There is a remarkable similarity between the Quakers F.W. Taylor and Benjamin Franklin. Both used watches to inspect workmen's output: "I had the curiosity to look at my watch when two men began to cut at a pine; in six minutes they had it upon the ground, and I found it of fourteen inches diameter" (Franklin 1989: 135).
14 This idea of breaking down work into its component parts became the core of Scientific Management.
15 Taylor also started with experiments to improve the cutting methods of steel, experiments which lasted for 26 years and made him a wealthy man.
16 It always has been dubious whether Taylor was allowed to carry the title "engineer." On April 20, 1878, he was admitted to the engineers' club of Philadelphia. In 1881, he enrolled at Stevens Institute of Technology in Hoboken, but Taylor was a phantom student: he never attended classes and never left his work with its 10 hours a day and six days a week at Midvale.
17 Secretly, the government prompted the army to invest money in the company of S. P. Langley, an astronomer and pioneer of aviation to develop a plane capable of carrying a man.
18 Flat-out meant that one group loaded a car with 16½ tons of iron in 14 minutes. This worked out to 71 tons per man per day without rest periods.
19 Taylor's initial experiments at Bethlehem concerned analyzing labor activities and the speed with which employees of Bethlehem Steel could shovel coal by varying the size of the shovel. Workers who might well have thought that they shoveled in their own natural rhythms found themselves lectured by Taylor on techniques of shoveling (a term of derision when adopted by his opponents). Piece work was introduced and became the spur for the laborers to work more efficiently. Taylor was not interested in the question whether intensifying labor also led to a decrease in working hours to prevent fatigue and stress. He was only interested in fighting slack and soldiering. In every single situation, he found that – contrary to what Saint Simon had stated – intensive supervision, and thus reigning over people, turned out to be the best method. However, his methods to observe and measure what laborers exactly did during the labor process proved very much contestable because they were derived from subjective observations. Despite this subjectivity in his analyses, the shovelers who improved their shoveling earned 60 percent more in wages.
20 Managers had estimated that a daily level of 45 tons per man was deemed achievable, but this was guesswork.
21 Taylor's approach was not very revolutionary: the military would not have found it so far removed from the types of drill that had been in existence since the sixteenth century when Maurits introduced a rigid discipline and close-order maneuvering which gave armies a decisive advantage.
22 Remington Rand was established in 1927 after a merger between Remington Typewriter, Rank Xerox, and Powers Accounting Machine. In 1955, Sperry Corporation acquired Remington Rand and then became known as Sperry Rand. In the 1950s, Sperry Rand rapidly fell behind IBM even after it had bought Univac. Rand Company merged with Burroughs in 1986 to form Unisys.
23 Like the American philosopher William James, he believed that the conflict between capital and labor could be solved through mutual understanding.

24 Brandeis embraced Jefferson's admonition that government is best when it undertakes socially necessary programs that are approved by experimentation to be unworkable when left to the private sector (Schama 2008). There are remarkable links to Pericles and his views on the way Athenian citizens structured their lives. He asked for true citizen participation as much in politics as in economics.

25 However, his adherence to these techniques reflected a lack of realism as he did not notice that there also was "driving" involved as the case of Henry Noll has shown. The pace set by the fastest workers usually became the standard, and those who did not comply with the piecework criteria would find their wages reduced.

26 Europe, on the other hand, was maintaining a major part of its rail infrastructure and services and invested in high-speed trains.

27 Kakar (1970) suggests that, although Taylor's fascination with time studies started when he was still in high school, he was a neurotic personality who remained notoriously unpunctual.

28 The government changed its policy during the Second World War and introduced Taylorism in the war industry.

29 His approach was a remarkable continuation of De Gheyn's *Wapenhandelinghe* introduced by Prince Maurits for the Dutch army around 1600.

30 His father did not appreciate the fact that Henry repaired the watches of his neighbors for free (see Bluedorn 2002: 12).

31 The internal combustion engine was originally developed by French and German inventors. Nikolaus Otto built his first gasoline-powered engine in 1861. A more efficient version was patented by Rudolf Diesel in 1892.

32 This approach is forever contrasted to the Japanese "just-in-time" production system.

33 Also, it is a common misconception that his standardized products like the Model T came in one color. This was actually not true either.

34 With the modern production techniques and organization methods the choreographic approach of Maurits fused with the synchronization of the assembly line.

35 Whereas the largest French company, Peugeot, was making 5,000 cars a year, Ford by 1920 produced approximately 2,000,000 cars per year in the United States.

36 During the interwar period, Britain became a leader in the development of industrial psychology and physiology.

37 In Liverpool the demonstrations and walk-outs went on to epic proportions among workers of warehouses, breweries, oil mills, wool houses, and railway men demanding reduced hours and an increase in pay, finally involving up to 70,000 people.

38 Janouch's book *Conversations with Kafka* is rather hagiographic and Kafka's words are "reported speech," most probably to make it more attractive for publication. The book is full of errors, including the dating of supposed conversations in Prague at times when Kafka was away in the countryside. The author himself has admitted that he never recorded the source of any of the numerous quotations from disparate readings and conversations that he noted in his notebooks.

39 During two decades, Münsterberg had lectured at Harvard University where he had pioneered in advertising and marketing research and the application of the film.

40 The French novelist Jean Giraudoux (1882–1944) who became the first writer ever to be awarded the wartime Legion of Honor in 1915 mocked the industrial development and cynically said: "The only safeguard of order and discipline in the modern world is a standardized worker with interchangeable parts. That would solve the entire problem of management" (Giraudoux quoted in Donkin 2001: 134).

41 "The B is a fraction of a minute of work combined with a fraction of a minute of rest, which were aggregated to the unity B" (Nelson 1992: 161).

42 The war experience did not seem to have had an impact on the Bedaux approach. By 1948, the Bedaux system was well known in British industrial management circles and had been installed in more than 600 firms with varying degrees of success" (Nelson 1992: 165). The Bedaux consultancy continued its practices until the 1960s.
43 In 1936–1941, Stakhanov became a student at the Industrial Academy in Moscow and during the war was appointed director of a mine in Karaganda. Between 1943 and 1957, Stakhanov worked in the Ministry of Coal Industry of the USSR.
44 Sloan had started his career in Hyatt Roller Bearing Company of Newark (New Jersey) where he had become the company's president in 1899. The bearings the company produced became the industry standard, and it grew rapidly. In 1916, the company merged with United Motors Corporation which, in 1918, was acquired by General Motors.
45 During the war, division managers were even more formally represented on committees responsible for strategic planning and resource allocation. The CMP not only served GM but became widely implemented by the American government as well. "CMP was used to organize all war plans in the United States in much the same way" (Beniger 1986: 313).
46 During the mid-1950s, GM began to lose market share to foreign competitors such as Volkswagen in the small market for economy cars. It decided to design a cheap car, Corvair, which was no success. General Motors' long decline began when the company reintroduced a textbook multidivisional form in 1958, which established a firm distinction between strategic planning and daily operations. The new structure created internal dissension and contestation. Management practices became divided in distinctive layers: managing technical core activities such as planning and supervising, managing the social systems which primarily focus on interpersonal relationship within the organization, and managing external relationships focusing on socio-legal and environmental issues (Parsons and Smelser 1956). Between 1964 and 1969, the return on investment for the Chevrolet division – one of their cash cows – dropped from 55.4 percent to 10.3 percent. In the 1960s, centralization was reinforced to reduce production costs but these reductions came at the expense of adaptability and product quality. In the 1970s, GM went back to a decentralized management structure but, after the 1973 oil embargo, GM's overall sales and profits fell dramatically, and its long dominance of the United States auto industry was shattered. Commitments and identification with the firm were required to create an atmosphere of initiative and efficiency. In the complex and uncertain environment, the company did not need "an emphasis on predictability and accountability but rather on flexibility and innovation. In such environments, competitive advantage focuses less on cost reduction and more on value creation" (Freeland 2001: 321). Sloan's model with centralized planning and financial control and with decentralized operations became replaced in the 1970s by the matrix structure which rejected the precept of one man, one boss. In favor of a multiple command structure, corporations like Shell, Philips, and Unilever experimented with the new model but to little avail and replaced it by introducing different forms of networking (Davis 1987).
47 During the Second World War, the Ford Motor Company built Sherman tanks, aircraft engines, and B-24 bombers.

9 Time and emotion

1 Hugo Münsterberg, one of the pioneers of industrial psychology, had suggested in 1914 a potential link between lighting and behavior.
2 The Italian G. Marconi had shown a keen interest in maintaining wireless contact with ships and, in 1896, succeeded in launching the first wireless signaling activities. With

financial support of rich relatives in the whiskey business, the following year he established the Wireless Telegraph and Signal Company, which became Marconi's Wireless Telegraph Company in 1900. British Lloyds became interested in the facilities Marconi offered for maintaining constant contact with ships. The German company Telefunken copied the rapid transmitting of information and, from 1903 onwards, offered similar facilities. The tragic event with the *Titanic* in 1912 could have been lessened if the standard procedure that was later implemented to share rapidly all the incoming signals of ships had been in place at that time. Marconi received, together with Karl Braun, the Nobel Prize for the invention of the radio which transmitted information quickly.

3 It reminded them of the famous demonstrations for the eight-hour day in the vicinity of the McCormick Harvester Works in 1886. This had led to the Haymarket riot which resulted in 13 deaths and the subsequent execution of four German-American anarchists. The Haymarket Square where the events had occurred was near the street of one of the Western Electric plants where many German-Americans worked.

4 In 1889, H. Hollerith invented a method which used a keyboard punch to make holes in predetermined positions in standardized cards; the cards were counted individually by means of hand insertion into an electrical circuit-closing press which had a pin contact for each possible hole location. It was a further improvement of Jacquard's punch card (1801) and Charles Babbage's steam-powered "Analytical Engine," which could perform calculations.

5 In 1908, the Hawthorne Works had already set up a medical department and from 1909 onwards offered live music at lunch once a week. In 1915, AT&T was one of the first companies to offer a stock purchase plan. The Bell company became famous for its "Ma Bell" policy by "mothering" its employees.

6 Another 9,000 employees were working at the company's Kearny plant in New Jersey. By the end of the decade it already employed more than 40,000 workers.

7 Personnel matters were considered so important that, from 1922 onwards, they had to report to a General Personnel Committee at the New York head office, which was placed under the control of a vice president who reported directly to the president.

8 Electricity-based lighting forms would allow for further "colonization of the night" (Melbin 1987).

9 These issues came out of the wartime studies on industrial fatigue which had shown that shorter hours and rest periods sometimes resulted in increased efficiency.

10 When the production of two of the women did not increase quite as much as that of the other women, the researchers decided in January 1928 to remove them from the test room stating that they were not being cooperative enough. The impact of this intervention was that the research outcomes as such had become unreliable.

11 Mayo had tried several times to interview workers about their private lives because he believed their personal lives influenced their work life. Yet, he could never get the workers to cooperate with him fully.

12 The issue of sufficient light for the nature of work had already been mentioned by Jeremy Bentham as one of the "conveniences" which should be considered in the layout of factories to save time in movements, see Hume (2004).

13 His work at AT&T was paralleled by a career in public service. He was, at various times, president of United Services Organization, head of the General Education Board, and later president of the Rockefeller Foundation, chairman of the National Science Foundation, and fellow of the American Association for the Advancement of Science as well as of the American Academy of Arts and Sciences.

14 Herbert Simon disagreed with Barnard and, in his *Administrative Behavior,* argued for the application of positivist science to administrative theory and decision-making (Scott 1996).

15 Several of his assumptions and preconceptions influenced the writers, like the idea that the performance of the workers was a measure of the adaptation of the worker to the industrial environment.
16 The Hawthorne experiments were viewed as a good example for what could be gained in terms of knowledge and production increases. It helped researchers convince management to support and conduct research in their factories.
17 European labor unions did not mention the thirty-hour proposal which had been launched in the United States, but they frequently referred to the positive effects of the National Recovery Act of 1933.
18 Some leaders in the shorter hour's movement in Western Europe believed that shorter hours would increase productivity, while others were convinced that a reduction of working hours implied a sharing of work.
19 As a rule, workers were on the job for seven hours; the first shift was from 4 to 11 a.m.; the second from 11 a.m. to 6 p.m.; the third from 6 p.m. to 1 a.m.; the remaining three hours were used for maintenance, cleaning, and repairs (Chase and Siegelbaum 1988).
20 Periods of intense activity ("storming") alternated with times of full relaxation.
21 The eight-hour workday persisted until the late 1950s when the seven-hour day was gradually reintroduced.
22 This movement was still popular among managers long after its popularity had declined within scientific circles.
23 When, in 1952, the top manager, Charles Wilson, became the new American secretary of Defense, he famously remarked that "what is good for our country is good for General Motors and vice versa" (Chandler and Mazlish 2005: 141).
24 Sloan was not pleased with Drucker's GM analysis and in his own biography, *My Years at GM,* never referred to him.
25 "His celebration of Japanese management anticipated the American Japanophilia of the 1980s" (Nelson 1992: 218). One of his famous epigrams was: "People are time-consumers and most people are time-wasters."
26 European firms, on the other hand, preferred to operate in networks or business groups (see Whittington and Mayer 2000).

10 The accelerated transfer of people and cargo

1 His plane was an adaptation of the Wright Brothers' design.
2 Around 1880, several different Swiss watchmakers provided the first wristwatches for the German Navy. The wristwatch probably first proved itself practical in the South African Boer War of 1899–1902.
3 As dutiful sons of a bishop of the church of the United Brethren, they never worked on Sunday.
4 In September 1908, Wilbur Wright took Hart O. Berg on board his airplane. She became the first American woman to fly as a passenger, soaring for two minutes and seven seconds at Le Mans (France).
5 To facilitate night flights, light beacons were installed at a spacing of 32 kilometers.
6 The company was a merger between three American airlines that all provided air mail services to Central America, Cuba, and Key West. One of the companies was Aviation Corporation of the Americas, which was owned by Juan Trippe.
7 As a student at Yale, Trippe had always been fascinated by the air industry and, with money borrowed from his friends, set up Long Island Airways and arranged cooperation with the United Fruit Company by reducing shipping time. Later, he founded Eastern Air Transport to deliver express mail between Boston and New York. Due to his contacts with the government of Cuba, he obtained landing rights in Cuba and used this for his new

company, Aviation Corporation of the Americas, to ship their products from Honduras to the United States.

8 Before him, some 78 people had crossed the Atlantic by dirigibles.

9 Higher flight speeds were initially limited by the principle that to achieve a certain speed, one would need to increase the power needed as the third power of that speed. By proper streamlining drag could be reduced.

10 It was acquired by American Airlines in 2001.

11 A few years later, four big companies dominated the field: United Transcontinental and Western (a merger between TAT and Western Air Transport), American and Eastern Air Transport.

12 Other German aircraft companies and engine builders were active in jet propulsion such as Junkers Aircraft and the Messerschmitt Company.

13 When the United States entered the war in 1941, there were no American designs available for a jet engine. Copies of the technical drawings of the Rolls-Royce jet engine were secretly flown to General Electric's facilities to produce aero engines, a project which became the genesis of General Electric's aerospace division. The Gloster Meteor became the first British jet fighter in 1943 and started operations in 1944 within the Royal Air Force (RAF). It was aerodynamically not a very advanced aircraft, but it eventually became an effective jet fighter that would serve different air forces for decades. In 1945, it set a new speed record of 954 km/h.

14 Stalin ordered A. N. Tupolev to fully copy B29 bombers, which, on their way to bombing Japan, could not make the return flight and landed in Vladivostok.

15 Boeing as well as European constructors left the production of large jet engines to Rolls-Royce, Pratt & Whitney, and General Electric.

16 As owner of Rolls-Royce, the British government decided to sell off those parts of the business that were not considered strategic to the national interest. Thus, the car division was spun-off as an independent firm and became known as Rolls-Royce Motor Cars Ltd. By the 1990s, Rolls-Royce cars such as the Bentley were still among the world's most exclusive, and certainly most expensive, quality cars. Small batch production and quality control still went hand in hand.

17 Aircraft speed is calculated on the basis of the block time – from engines on to engines off – for a journey.

18 Traditionally, European national railroad companies were involved in transnational shipping. With the integration of the European Union, however, the Community of European Railway and Infrastructure Companies (CER) concluded that there were weaknesses in electronic data interchange to streamline freight traffic between the European Union and countries in the East European region. Transit conditions needed to be improved. The railway system in Russia, Belarus, and Ukraine, for example, still used budget support mechanisms which differed from those in the European Union. Improvements in this domain were delayed by European discussions concerning the privatization of railroads. The vertically integrated national railway companies needed to be dissolved and a separation of operations of the rail infrastructure (tracks, signaling, electrification, and other fixed facilities including stations) from the operation of freight and passenger services needed to occur. A new regulatory framework for a common public regulation of the European rail network system was eventually put in place (OECD 2004).

19 Before the 1960s, logistics was a military term and was an operational activity to which not much attention was paid. Transport just had to be executed as part of company practice.

20 There is currently talk about *slow steaming* as a way to reduce fuel consumption and pollution at sea.

11 Information and communication revolution

1 A US subsidiary of the British-owned and British-managed Marconi's Wireless Telegraph Company.
2 For an adequate description, see Friedel (2007).
3 ITT, the spin-off of Western Electric's international operations, became a dominant player with Harold Geneen imposing a doctrine of management by the numbers and growth through unrelated acquisitions. Cross-Atlantic telephone services made it possible for wealthy J. P. Getty to run his California oil empire by telephone from European hotels where he preferred to stay long-term.
4 Etienne-Jules Marey and Edward Muybridge had experimented with techniques to take sequential photographs; Thomas Edison was so inspired that he developed the kinetoscope, which he presented to the public in 1894.
5 The application of computers in manufacturing was envisaged by John Diebold (1952) in *Automation*. The introduction of micro-computers greatly facilitated the progress of automation in manufacturing.
6 A common standard, the Transmission Control Protocol/Internet Protocol (TCP/IP) was designed by V. Cerf and J. Postel in 1974.
7 RCA, however, could not maintain its position and sold its computer division to Sperry Rand's UNIVAC in 1971.
8 That is, computer and electronic industries with the telecommunications industry.
9 By 1890, Herman Hollerith had extended Jacquard's punch cards to tabulate US census data. This information-processing technology survived to this day owing largely to the corporation to which Hollerith's innovation gave life: IBM.
10 The constituent firms manufactured scales, coffee grinders, meat slicers, and a line of punch card tabulating machines.
11 Subsidiaries of IBM provided services for census data in prewar Central and Eastern Europe which enabled the Nazis to round-up Jews and other minorities and register them for deportation (Black 2001).
12 The president restructured the company at the same time and introduced the M form and implemented a system of internal competition known as "contention management," i.e., as the best way to motivate people by pitting them against one another.
13 Two main competitors, GE and RCA, were the only ones in those years to really challenge IBM. GE had purchased two computer businesses, France's Machines Bull and Italy's Olivetti, and had established cooperation with Japanese Toshiba.
14 Many hobbyists were able to design their own personal computer once availability of components began increasing at low costs. One of the first was H. E. Roberts (1941–2010), who, as an engineer, entrepreneur, and medical doctor, was initially a maker of radio kits for model airplanes. He designed the first commercially successful personal computer in 1975 and is, therefore, known as "the father of the personal computer." He founded Micro Instrumentation and Telemetry Systems (MITS) in 1970, where Bill Gates and Paul Allen started their careers by developing software.
15 IBM could hardly imagine that their AT (advanced technology) desktop model with their own memory devices would be obsolete in five years' time. Intel's co-founder G. E. Moore, however, expected the performance of the integrated circuit to double every 18 months. Moore's vision turned into a "law" which set the standard for the entire industry.
16 EDS and Andersen Consulting were gaining ground in information services, while Intel and Microsoft were more profitable in the PC market than IBM at that time.
17 Compaq developed a new business model, the Optimized Distribution Model (ODM), which started to look at the company' practices from a customer's point of view. Power

shifted to the end-user, and sales of application software began to grow faster than sales of hardware. Gone were the days when IBM could simply take its customer base for granted.

18 The authors claim that "time-tellers" can become "clock-builders" if they spend less time thinking about particular product lines and market strategies, and think more about organization design, shift the image to seeing the firm itself as the ultimate creation, and build something that can never become obsolete: a company, highly adaptable to external change with "spiritual inheritance" of deeply rooted core values passed on from generation to generation.

12 The Japanese miracle and just-in-time

1 In 1987, the railroads were privatized and a new infusion of private managerial talent took place.

2 This decision was followed soon by Egypt in 1875, China in 1912, and Turkey in 1927 under Ataturk.

3 One sector where this transfer took place was electrical and telecommunications, where companies such as Fujitsu, NEC, Toshiba, and Mitsubishi Electric formed joint ventures with Western corporations.

4 The textile industry declined as the Japanese military took control over the firms, and forced the companies to produce armaments. Toyoda Spinning and Weaving merged with other textile firms and from 1942 onwards had to produce aircraft parts (Bernstein 1997a).

5 In 1950, Toyota Motor Sales Company was established and exported trucks to the US military to serve the Korean War.

6 The Soviet Union lost its membership like many South American countries which previously had been members.

7 There he had worked on improving mica transmitters at the Hawthorne plant where he heard for the first time about statistical quality control designed by Walter Shewhart, who was the first to work on Statistical Process Control (SPC) techniques at the Bell Labs to regain better control of the manufacturing operations at the Western Electric plants. Together with Shewhart, Deming formulated a systematic approach to problem solving via the PDCA cycle: Plan, Do, Check, Action, a method which was applied to production during the war.

8 He had visited Japan in early 1947 to help develop a national census for 1951.

9 Kaoru Ishikawa helped transfer quality-control methods to the Japanese industry.

10 During the war, Juran worked for the Lend-Lease project where he reduced paper work and handling time to process a request. The Lend-Lease arrangement allowed the American government to obtain goods which were deemed essential for the defense of America. The government negotiated many lend-lease agreements with Britain and the USSR, the aim of which was to allow trade balances to be ignored among the participating countries during the war. Lend-Lease became an example of a most effective organization in Washington under the guidance of Juran. He published his experiences in *Bureaucracy, a Challenge to Better Management* (1944) to improve the functioning of the federal government and stressed that the principles of business management should be applied to solve the increasing chaos of government. After the publication of *Management of Inspection and Quality Control* (1945), he, like Deming, attended the business school of NYU. Later on, Juran urged Drucker to join the same business school. Concurrent to his teaching job, he also became a consultant.

In 1946, the Gillette company asked Juran to assist in the improvement of its production processes and the quality of its products. There, he encountered a culture highly resistant to change. Juran had success with this change policy and improved the balance between costs involved in improving quality and the attainment of a better value of quality.

11 Whereas in the United States large manufacturing companies were more likely to be vertically integrated, Japanese firms relied on subcontracting. "Toyota and Nissan, for example, turned to subcontracting as a way to meet the growing demand for their cars" (Blackford 1998: 198).
12 Toyota reclined and retook its exports in 1964; by 1967, the USA became Toyota's largest export market and manufacturing operations in the USA started in 1972.
13 The Corolla was launched in 1966 and became Toyota's most successful model. In 1994, it was the best-selling car in the world, with 18 million being sold.
14 T. Ohno had neither a university education nor any experience in car production.
15 At 72 years old, Ohno retired from Toyota but remained chairman of two Toyota group-firms, Toyoda Spinning and Weaving and Toyoda Gosei, until his death.
16 Whereas Japanese companies and governmental bodies embraced the improvement of quality, the majority of American senior managers during the 1950s and 1960s were not listening to Juran, to Deming, or to Armand Feigenbaum, another advocate of total quality control, or to any other quality advocates.
17 American car manufacturers generally did not pay too much attention to the waste-saving methods and quality improvement policies Japanese firms had implemented. Nevertheless, the successful Japanese business practices ignited within the American business community a growing willingness to learn from Japan.
18 Critics of the TPS approach however, have emphasized that the work is much standardized, repetitive and that *kaizen* leads to a constant increase of line speed.
19 Japanese experts made management contributions of their own. Shingo Shigeo, mechanical engineer at Toyota and Mitsubishi shipbuilding, Kaoru Ishikawa, professor at the University of Tokyo, and Genichi Taguchi, a statistician, improved SPC for shop-floor employee use. Shigeo promoted the *zero defects* objective, advancing worker-based total management control systems; Ishikawa simplified SPC theory, creating the "fishbone: cause-effect diagram", which was used in process analyses as part of the continuous improvement (*kaizen*) system. Taguchi formulated the "Quadratic Loss Function", which measured the multiplier effect of minute quality losses along the process line. Due to the efforts of these statistical and industrial specialists, Japanese automobile manufacturers transformed the largely technical SPC methods, imported from America, into highly efficient shop-floor, process control management systems.
20 At the headquarters of Toyota there are three portraits displayed: one of the founder, one of the current president, and one of Deming (Crainer 2000). In order for firms to seriously apply quality policies, Deming designed 14 principles which aim for continuous improvement of the system of production and service by instituting training on the job and self-improvement programs.
21 Although Crosby was an effective self-publicist he used an evangelical style to promote quality management.
22 Ford purchased 25 percent of Mazda in 1979.
23 However, these companies did not all become a success: overwork, injuries, and poor race relations plagued some of the new firms (Kenney and Florida 1993).
24 In 1993, Robert Flood published *Beyond TQM* which offered a critical analysis of TQM based on a thorough analysis of its strengths and weaknesses .While many organizations had experienced unexpected difficulties in implementing TQM, the book intended to advance a new understanding of TQM and to explain its value to practitioners.
25 GM had introduced new forms of work in their American Cadillac plant in 1979 and in its European plant in Aspern near Vienna to experiment with delegating responsibilities to teams. It also explored in a large set of similar "laboratories" how to improve its work

organization and increase flexibility. The functionally specialized, spatially dispersed, scattered plants gradually became integrated production complexes.

26 The NUMMI cooperation lasted until 2009.

27 Deming did not like the phrase.

28 Two journalists of *The Economist*, Micklethwait and Wooldridge (1990: 20), were extremely critical about the behavior of many American managers who rather looked like "Soviet bureaucrats, living in a dual world-the real world and the world of officially sanctioned ideology."

29 Juran had noticed that over the years American engineers and manufacturing people had developed a "language of things," while business owners and managers developed another "language of money" to measure the value of their business (see Butman 1997: 171). None of them paid systematic attention to the employees.

30 Although usually the origins of these management concepts are difficult to locate, once they stand out as recognizable phenomena they share four main characteristics:

- Management concepts usually have a striking label such as Total Quality Management (TQM), Business Process Reengineering (BPR), Strategic Intent and Core Competence (SICC), Balanced Score Card (BSC), Knowledge Management (KM), Total Responsibility Management (TRM), or Customer Relationship Management (CRM). Whenever possible, these concepts are reduced into acronyms, which have become an indispensable trademark of best practice doctrines. (Buchanan and Badham 1999: 155) It was believed that these management concepts help management to overcome ambivalence, incertitude, and anxiety (Townley, 2002).
- Management concepts describe particular management issues in general terms. These issues are usually related to increases in costs or loss of customers. Managers then face an irresolute but pressing problem that calls for new meaning and thus compel a search to develop a more probable course of action to improve the situation. Concepts can frame such a particular organizational problem and make it recognizable for the managers involved. BPR, for example, will be seen as a useful analysis because by using it, managers can recognize that the actual company structure has to be redesigned.
- Management concepts offer general solutions for recognizable problems. Concepts do not furnish principles but instead guidelines, which do not have the degree of "settledness" that principles possess. They are often issued as a provisional measure until more is known about the practical usefulness of a concept. They do not offer rules that prescribe specific actions, but as guidelines bring about mutual orientations between actors and propel action in a certain unspecific direction. These guidelines evolve from the values and practices of a specific community of actors where the concept initially has been developed. For example, BPR will justify its interventionist guidelines by stating that companies with obsolete structures will become more efficient once the organization has been redesigned and modern information and communication technology have been introduced. In order to be persuaded of the quality of certain guidelines, a fourth characteristic comes into play.
- The proposed solution will be promoted by referring to success stories of well-known firms, which already have implemented the concept. General Electric, Boeing, IBM, HP, Shell, Philips, and Toyota are usually portrayed as convincing examples of the success of specific concepts. These examples provide evidence for the plausibility of the solution and establish a convincing new practice. The stories are narratives, which articulate the knowledge employed in situations that have created a new best practice.

31 One such example was the re-engineering fad of the 1990s introduced by James Champy, a consultant from CSC index, and Michael Hammer, a computer science professor at MIT. Their idea on re-engineering was a fundamental rethinking and radical redesign of business processes to achieve dramatic improvements in issues such as costs, quality, and speed. Managers who were inspired by their work began restructuring, downsizing, de-layering, and outsourcing their companies and empowering employees by means of temporary and contractual employment within a global division of labor. Without recognizing that the origins of re-engineering were more based on the view of a corporation as a machine than as a humane system, a host of literature about processes of change, leadership and change masters appeared. This kind of approach was still *managerial* in the sense that it embraced the traditional view that managers get things done through the *actions* of others (Harding 2003).

32 The book was a shift away from McKinsey's previous approach focussing on industrial bureaucracy shaped in a multi-divisional organizational structure with a socially conformist organization man. *In Search of Excellence* propagated new ways of thinking, feeling, and acting based on the market and natural entrepreneurial qualities.

13 Flexitime

1 Portions of this chapter have been excerpted from Bryson and Karsten (2009).

2 The TPS consists of four unwritten rules: all work is highly specified as to content, sequence, timing, and outcome; every customer–supplier connection is a direct one; the pathway for every product and service must be simple; any improvement is made in accordance with the scientific method at the lowest possible level in the organization (Spear and Bowen 2006).

3 *Karoshi*, which is written as 過労死 (かろうし), simply means "work oneself to death."

4 Although, since 2007, the worldwide recession hit Toyota, too, it remained one of the best-selling automotive manufacturers and surpassed GM in 2008 to become the largest car manufacturer in the world at that time (Rowley 2009).

5 However, as Japan moved into the late 1990s, the rising yen made production more expensive, and the prices of Japanese exports rose. The economic crisis and the yen-shocks seriously affected the car companies in Japan. Whereas Toyota could remain profitable, Nissan suffered from huge losses.

6 In the literature, the concept had initially been introduced in the mid-1960s by the American strategy author, Igor Ansoff, who suggested that Fordist-structured firms still needed to develop external and internal flexibility to cope with unforeseeable contingencies. Ansoff (1965: 55–57) had linked external flexibility to "the maxim of not putting all of one's eggs in a single basket." It was mainly a focus on the variety of strategic options corporations had to consider. Concerning internal flexibility, he stated that this issue "is as old as business itself and seeks to provide "a cushion for response to catastrophe." Flexibility in those years had primarily been seen "as an organizational potential, created by flexible configuration strategies and broad strategic schemas" (Volberda 1998: 73). However, since then the range of views on flexibility expanded.

7 This was not a topic much debated at corporate levels during the heydays of Taylorism and Fordism. Market developments could be clearly forecasted, and rigid planning policies secured deliveries against low prices. A number of alternatives to mass standardized production have since been detected: customized production which assumes a certain reduction in volume; diversified quality mass production which purportedly combines the benefits derived from product differentiation with significant quantities of production;

and adaptive or reactive production which builds upon the constant redefinition of market niches and the fastest possible use of new technologies (Boyer and Hollingsworth 1997; Lawrence *et al.* 1999). Some authors, however, pointed out that Taylorism and Fordism did not prevail in all industries. Piore and Sabel (1984) emphasized that in Northern Italy, for example, geographically localized networks of small and medium-sized enterprises manufactured a wide and changing array of customized products in short runs. These industrial districts with their flexible specialization were characterized by flexible production serving a differentiated and segmented market of consumption (Hyman 1991). Uncertainty, fragility, and mutability as constitutive features of economic life strengthened the capacity of these firms to adjust the volume and composition of output. These firms rapidly introduced new products in response to shifting demand and business strategy (Sabel and Zeitlin 1997; Zeitlin 2003).

8 These approaches were similar to "putting-out" arrangements described in Chapter 5, dealing with eighteenth-century industrial relations in Britain. Some American multinational enterprises introduced large networks with subcontracting arrangements. And while concentrating on a single trade, they began to prosper and merge with others that had the same core business. Some corporations embraced Total Quality Management programs which included employee involvement as an explicit component of their operations (Wilkinson *et al.* 1998; Cole and Scott 2000).

9 In such a case their extra effort was likely to be unpaid (particularly for managerial or professional staff).

10 The United Kingdom soon followed extensive overwork and longer working hours (Donkin 2001).

11 Proctor *et al.* (1994), however, struck a counterpoint by suggesting that much of this criticism was ill-founded.

12 In some cases, functional flexibility contributes to the quality of services such as, for example, in healthcare where a reduction of the number of staff a patient has contact with, the more of a holistic service can be provided (Kelliher and Desombre 2005).

13 *Internal* flexibility refers in general to maintaining flexible relations with current permanent personnel, including aspects such as transferral, or making changes to an employee's responsibilities or working hours. It relates to an increased variation in working hours and flexible working-time arrangements of regular workers employed on standard contracts offering alternative work scheduling and multi-skilling.

The concept of *internal* flexibility first appeared in the late 1960s when German employees were allowed some individual choice as to starting and quitting times (Olmsted and Smith 1989: 12). At that time, it was seen as a means of relieving transit and commuting time problems. After some try-out in its German subsidiary, the Hewlett-Packard Company is generally credited with introducing flexitime in the USA in 1972. Workers gaining some form of internal flexibility were assured the continuation of their employment and earnings, whereas the employers had no worries about the quality of their staff. Purcell *et al.* (1999) have used the descriptive terms of "outsourcing" to refer to external flexibility and "insourcing'" to refer to internal flexibility.

14 Destandardization of labor may indicate an emerging risk society and a further commoditization of labor.

15 In fact, no employment contract is permanent – they are either indefinite or until the age of retirement. The term "temporary" is here rather an ambiguous term. Casey (1988) has identified eleven often overlapping categories of temporary workers.

16 The external forms are often associated with a more "unstructured" (Purcell *et al.* 1999) and one-sided approach, i.e., the employer has total control over when the employee works and may vary this at short notice. This may be contrasted with internal flexibility

arrangements such as flexitime when working hours are subject to mutual control within an agreed framework. Studies on organizational learning and technological innovation apply temporal flexibility by drawing a distinction between exploration and exploitation. Exploration includes things captured by terms such as variation, experimentation, innovation and flexibility. Exploitation includes such things as refinement, implementation, execution, and efficiency. These two concepts require different organizational structures and patterns as well as capabilities, and complicate the precise implementation of temporal flexibility (Li *et al.* 2008).

17 Some authors have put forward further categorizations of flexibility. Sparrow and Marchington (1997) proposed no fewer than seven categories. But these additional forms do not provide conceptual clarity as more categorization leads to more overlap and more ambiguity. Some part-time work may not be subject to variation at all (just simply a contract for fewer weekly or annual hours) and should, therefore, not be considered to be flexible employment! Functional and numerical flexibility are prominently used as the description of flexible employment.

18 Despite all the recent evidence that non-standard work forms are increasing – with populist predictions towards a jobless future – we do observe considerable continuity with traditional approaches. Casey *et al.* (1997) found that overtime was the most common form of flexibility, but Dex and McCulloch (1997) claimed that the tide has turned and that even in Britain full-time standard employment was increasing and temporary work declining. Although this might be true for the British private sector with its decreasing use of non-standard forms, in the public sector they are still on the increase. Cost-cutting policies and uncertainty remain very high in the British public sector. In other countries, however, such as the Netherlands, the job machine is based on an expansion of part-time employment. One of the key struggles is how people organize their entry into and exit from various roles over their life course. On the family level, timing involves the synchronization of individual life transitions with collective family ones.

19 The findings of Casey *et al.* (1997) have supported the conclusion by Hunter *et al.* (1993) that employers in general prefer to use full-time permanent employees because they believe that they are easiest to manage, the most committed and the least likely to leave. Geary (1992) examined firms which used a dual strategy with permanent and temporary workers and noted that the managers were averse to this policy because of all the problems that arose. One of these was the tension between temporary and permanent workers, which has been noted in other studies (Filipzcak 1997). Some businessmen (usually supported by consultants and politicians) go so far as to argue that organizations should not offer any sort of career but, instead, the commodity of employability. It is dubious whether such kind of deals is what most employees seek, and there may be serious resultant damage to the employment relationship. This could create problems and costs for employers (Bryson and Barnes 2000).

20 The female American ambassador in Pakistan resigned as she hardly ever saw her two daughters due to security reasons.

21 Bryson and Barnes (2000) found problems that cannot be "managed" away. When dual strategies are used, the problems in the "direct control, low commitment" group seem to spill over into the "indirect control, high commitment" group. It is difficult to assess whether there are managerial solutions because so often the peripheral groups are virtually ignored by HR policy-makers and line managers seem to be just left to cope, often with reluctance and frequently without much success. Another problem is that the types of flexibility that are extensively used seem to be very far from the high-trust, high-commitment models. In other words, the rhetoric about new-style HRM practices and flexible employment includes contradictions and paradoxes (Blyton and Turnbull 1992). Walby (1989) has pointed out – and is echoed by others (Geary 1992;

Legge 1995) – that many of the so-called flexible forms of employment actually create rigidities, the very thing they are supposed to prevent. Such organizational rigidities force American women, for example, to step out of a full-time job when their children are young and return to work part-time when the children are older (Mainiero and Sullivan 2006).

22 Since 1976, Canada has shown a 44-percent increase in total employment growth, but this was mainly due to growth in non-standard jobs (Gonzales-Rendon *et al.* 2005). The largest single employer was no longer Ford, GE, or GM but a temporary employment agency called Manpower Inc. (Sturgeon 2002).

23 The focus on the flexible firm has drawn attention to the possible consequences of too much flexibility. Companies might use people only as they need them, leading to *just-in-time* employment. Rifkin (1995) has warned for the outcome of flexibility *of* the worker: "Temporary workers and outsourcing make up the bulk of today's contingent workforce" (Rifkin 1995: 194). Employers nowadays call in flexible personnel from outside the company for whom only short-term employment relations are initiated and which can be easily broken. This form of external flexibility creates contingent work in the sense that there is a lack of attachment between the workers and the employer. Examples of this type of work are temporary workers employed by temporary work agencies and employees on temporary or fixed-duration contracts without the prospect of a regular employment contract. External flexibility might seek optimal utilization of human resources while adapting the organization to accelerated technological changes, but this requires continuous revision of skills and organizational structures to enable new forms of organizing.

Yet, at an institutional level, a different light is being shed on changes in the nature of work, work organization and institutional embeddings. That light will help to explain why there are still some apparent differences between what is happening in different countries. One problem with the flexible firm is that it is a tight model suggesting an optimal final configuration for organizations. This is somewhat paradoxical in a dynamic, uncertain world; surely adaptability entails being able to manage transitions – to be in a permanent state of transformation. However, the reason organizations are successful is because they are, at least to some extent, organized. They seek to provide some stability and predictability and a measure of control over uncertainty.

24 An OECD report (OECD 2004: 48) underlined that Japan and the USA are countries where longer working hours still prevail. It signaled, however, a growing pressure amongst groups who find it hard to comply with a full-time, nine-to-five routine, and stressed that people have "the greatest difficulty in finding a balance between work and family life, which is very significantly linked to the presence of children in the household, to be younger and working longer hours or in more demanding jobs, or being self-employed."

25 In 1958, the European Economic Community was established with a set of institutional arrangements at a supranational level. During the 1960s and 1970s, it further developed as a "side show" to economic policy making by individual states (Kitschelt *et al.* 1999a, b). With the creation of the European Court of Justice, the European Council as representation of nations and the European Commission as a post-representational body, a concerted attempt was launched to upgrade the competitiveness of the European economies.

26 In the 1970s, the working time regime in Europe had still been bearing the imprint of the "male breadwinner" arrangement. Women were less likely to be employed than men, and those in employment worked fewer average hours.

27 Other Directives of the EU about works councils, parental leave (1995), part-time work (1997), labor contracts for a determinate period (1999), labor arranged by agencies (2001), and telework (2002) are steps which reinforce a process of convergence of labor conditions within the EU.

28 For the seniors, a pro-active labor market policy based on lifelong learning and a "destandardization" of the life course became envisioned.

29 Despite these institutional differences, flexible employment grew throughout the 1990s, especially among women who took up part-time jobs (Delsen 1998).

30 Based on Esping-Andersen's typology, Dulk (2001) compared the UK (liberal), Italy (conservative/corporatist), the Netherlands (conservative-corporatist) and Sweden (social-democratic). Her conclusions indicated that employers in all four countries offer flexible hours to make it possible for employees to balance work with care activities. Many Swedish employers allow working parents with young children to work shorter hours. In Sweden, it is more common to be flexible, have opportunities for telework, a compressed working week or work occasionally at home. Although Italian and Dutch societies are both called conservative-corporatist, Dutch and Italian employers offer different kinds of arrangements for parents to be flexible. Dutch employers more often provide childcare and flexible work arrangements, whereas Italian employers mainly offer leave arrangements (Dulk 2001).

31 Since 2003, the OECD has been echoing the European Union model with its positive supply-side measures more than the previous neo-liberal model with its free market approach.

32 A remarkable socio-cultural result of the interconnectedness between regularity, standardization and co-ordination has been the introduction of the eight-hour day which regulated the division between labor and free time, supported the standardization of the working hours and synchronized shop-floor activities.

14 Time and cultures

1 While the European Union has 43,492, the United States counts 2,418 MNEs.

2 Today, IBM programmers in Beijing work on software and forward it daily by Internet to other IBM programmers, for example, in Seattle. Once they have added some refinements, it is then sent to another IBM operating unit in Europe for some adjustment after which – via India – it ends up being returned to Beijing. Work on a specific programming issue currently proceeds around the clock (Friedman 1999).

3 Worldwide industrialization continued even when the international economic order fell apart after World War I. The global economic regime carried on even after the fall of the Berlin wall in 1989, which symbolized the fall of communism.

4 They function with frequencies of some 2.5 MHz (million cycles per second) checked by cesium-beam resonators vibrating at 9,192,631,830 ± 10 Hz, which was the value made available by the Royal Greenwich Observatory in 1955 (see Essen and Parry 1957).

5 The shift from Greenwich to the atom-time center in Paris once again illustrates how highly politicized the issue of the coordination of social time has become. The debate between Greenwich and Paris which started in 1883 finally has fallen into the hands of Paris.

6 On December 31, 2008, one more second was added in order to match the astronomical time.

7 Members of the ITU decided favorably in 2012 to postpone a time-correcting intervention of a second until 2015.

8 This is certainly the case in the financial markets. Since the dissolution of the Bretton Woods agreements, capital movements have become less and less restricted and have resulted in financial globalization with periodical financial catastrophes such as the East Asian Crisis of 1997–1998, an increasing accumulation of dollar reserves and American Treasury bonds in China, and the collapse in 2008 (see James 2009).

9 The environmental cues that each day reset the rhythms are called Time Givers (from the German *Zeitgeber*).

10 The circadian rhythm is linked to the light–dark cycle. Human beings kept in total darkness for extended periods eventually function with a free-running rhythm of about 25 hours. Each "day" their sleep cycle is pushed backward or forward, depending on whether their endogenous period is shorter or longer than 24 hours. People who suffer with jetlag usually take melatonin as a drug to keep their biological rhythm on track. So, "If our bodies follow the circadian patterns, so to some extent must *we*" (Young 1988: 41).

11 Swatch's progressive marketing approach consisted of inexpensive, fashionable watches that were available in varying colors corresponding to the fashion of the day. A demonstration of the dominance of wristwatches in modern lifestyle was illustrated in an advertisement in 2007 which portrayed a completely nude, female body encased in a set of Swatch wristwatches (Figure 14.1).

12 DST for the United States and its territories is not observed in Hawaii, American Samoa, Guam, Puerto Rico, the Virgin Islands, and most of Arizona.

13 Most countries of the world now observe DST in some form, although beginning and ending dates are different in the southern hemisphere. During summer, the Russian clocks are two hours ahead of Standard Time, but during winter, all eleven Russian time zones are one hour ahead of Standard Time. DST has generally not surfaced near the equator where sunrise times do not vary enough to justify it. Asia and Africa generally do not observe DST. Many computer-based systems can shift clocks automatically to adapt the DST changes based on the time zone settings. Expatriates who are used to working and traveling across time zone boundaries usually have to keep track of multiple rules as not all locations observe DST. In that sense, the length of the day has become variable again.

14 In low-context cultures, people get straight to the point and do not waste any time (cf. Trompenaars and Hampden-Turner 1998). In a low-context culture, time spent while waiting for a meeting is seen as wasted time (cf. Eskerod and Skriver 2007).

15 Clock time should be spent wisely and saved if possible (Arrow *et al.* 2004).

16 McGrath (1988) states that clock time depicts a continuum of linear, indefinitely divisible, objective and quantifiable units, which are homogeneous, uniform, regular, precise, deterministic, and measurable.

17 This concept is fundamentally different from "forecast," which is an external plan to which action has to conform. Amongst the farming communities there was hardly a notion of such an external plan which was based on calculated planning. When farm families eventually became integrated into the wider society where commercialization and mechanization were introduced, the precise organization of time including the future became integrated in the lives of farmers (Wildenbeest 1988).

18 Different time perspectives are "used in encoding, storing, and recalling experienced events, as well as in forming expectations, goals, contingencies, and imaginative scenarios" (Zimbardo and Boyd 1999: 1272–1273).

19 In Chinese agrarian villages, peasants get up in winter when the rooster crows and in summer according to the position of how high the sun has risen. In the Chinese lunar calendar, one year is separated into 24 *Jie Qi* (节气, where *Jie* means season; *Qi* means weather or temperature). Linguist Zheng remarked that the Chinese do not have the habit of indicating precise time schedules in their language when it comes to the future (Zheng and Jia 1993). They use adjectives, such as "in the near future" or "in the not too distant future." There is a saying "*Bu Jian Bu San* (不见不散)," which means "I will keep on waiting until you come no matter how late you will be." It does seem that Chinese people tend to use a vague way to describe time and what matters for the appointment in China is not the time the clock or watch shows but the event that the other party appears. The starting day

of each of *Jie Qi* varies every year. Chinese peasants actually rely more on these 24 *Jie Qi* to carry out their agricultural activities. Chinese people celebrate traditional festivals, such as mid-autumn day and spring festival according to the lunar calendar, which varies every year (information provided by master student Li Yanjie in September 2010).

20 Ethiopia has 13 months of 30 days instead of 12 months like in the West. This feature is an inheritance from colonial times as well as the traditional Italian concept of clock time where the counting of hours started with the rise of the sun (the twilight). The end of the day was heralded by sunset when darkness fell (see Camuffo 2002). In Nigeria, two different calendars exist simultaneously. The one calendar is similar to the Western practice; the other is structured according to the rhythms of the market days which take place every four days. In Zambia, during the season of groundnuts, people spend entire afternoons together sitting cozily under a tree, pealing and eating nuts: "groundnut time."

21 But sometimes certain patterns can be discerned even in event time. Arrow *et al.* (2004) made a distinction between predictable events, for instance paydays, and unpredictable events, such as catastrophes, an earthquake, or a tsunami. From a clock time viewpoint, doing business in a culture where events prevail, all interactions are extremely time-consuming (Trompenaars and Hampden-Turner 1998) and might result in discomfort amongst those who are clock-driven (Macduff 2006).

22 A report in *IPS News* by Godoy (2009) found that there had been 24 suicides at France Telecom in 2009–2010. In the suicide notes of dead workers at French car manufacturers Renault and Peugeot, they found poor working conditions as the main source of their desperation. The General Confederation of Workers (CGT) estimated that 400 French people kill themselves each year because of difficulties due to work pressure. A representative of the European Trade Union Institute found that 27 percent of all European workers have complained about their jobs negatively affecting their health (Godoy 2009). During the last decade, an increasing amount of the phenomenon well known in Japan as *karoshi* was reported in France as well (Overman 2009).

23 He found this pair of time dimensions through observing people's responses toward culture other than their own and discussing those responses with them.

24 Even in high-context cultures, similar behavior can be noticed. Singaporean executives tend to prefer a monochronic processing of activities (Ang and Teo 1997).

25 Yet, on the other hand, technological advances in communication systems (such as answering machines) have also enhanced the opportunity for polychronic behavior.

26 In countries where event time prevails, people tend to be polychronic (Hall 1976; Palmer and Schoorman 1999). When time is treated as a system where the same events return in natural cycles, the character of time is polychronic. Levine (1997) asserted that encountering P-time and event time is more customary in developing economies, while M-time and clock time are more common in industrialized societies where individuals are focused on achievement. But even in the western society where there are many people who prefer a monochronic behavior, there are others who act in a polychronic way.

27 Rubber time, or island time at Bali, is found where inhabitants are not really keen on scheduling. It is telling that the Balinese – whose daily activities are entwined with religious, musical, dramatic, and artistic rituals – often refer to the hour on the clock as rubber time. "When asked when the bus will arrive or leave the typical answer will be 'four o'clock', rubber time" (Levine 1997: 47–48). However, also in low-context cultures, which overemphasize the role of clock time, there are still moments of event time to be discerned.

28 Clark (1978) similarly studied sugar-beet processing where the seasonal work required intensive periods alternated with periods of slow activities.

29 Clark's (1985) research demonstrated that in British organizations both clock-based and event-based times coexisted. These and similar studies illustrate how event times are based on interpersonal relationships which determine the "natural" course of work in sociocultural terms and do not follow any strict clock time pattern (Brislin and Kim 2003). The conclusion was that not all the hours were the same for the engineers. Tinsley (1998) found that American managers are significantly more polychronic than their counterparts in Germany and Japan.

30 Within the field of psychology this distinction has become essential to understand how individuals manage their time. Francis-Smythe and Robertson (1999), for example, introduced a time-personality scale (composed of punctuality, planning, polychronicity, and impatience) and concluded that those who score high on this scale had a higher need for controlling the situation and controlling their work.

31 It might be that this ratio differs across cultures, even though people all around the world have found a personal balance between work and leisure (Brislin and Kim 2003). There is no criterion to decide which orientation is the best, as both M-time and P-time both have weaknesses; M-time does not pay sufficient attention to the human aspects of society, while P-time systems often create inefficient chaos (see Levine 1997). Education, gender, and age might influence someone's way of making use of time. Yet it is important to differentiate between P-culture and M-culture apart from P-individuals and M- individuals. Individuals can be polychronic in a monochronic culture or vice versa. This is also the reason why studies related to time-use are usually carried out on both an individual level and a collective level.

32 McCollum and Sherman (1991) found that companies with a matrix structure which had a large percentage of personnel working in research and development (R&D) assigned to two projects rather than to one project or to three or more projects reached the highest levels of effectiveness. This result suggests that a moderate level of polychronicity may improve the performance of a firm.

33 It is probably true that in non-Western societies this acquaintance with seasonal cycles is much stronger. The more people pick their own rhythms, the better the chance they will pick the right ones (see Young 1988).

34 Polychronic work patterns are not identical to multitasking, while multitasking combines both speed and a multi-activity pattern. Bluedorn (2002) has distinguished four times of behavior patterns:

- quantitative polychronicity, which involves engaging several *similar* tasks simultaneously;
- a quantitative monochronic pattern, which means being engaged in one task and completing it before moving to another *similar* one;
- qualitative polychronicity, which involves being engaged in multiple *dissimilar* tasks simultaneously;
- a qualitative monochronic pattern which is being engaged in a single task and completing it before being engaged in another but *dissimilar* task.

Semco's ideas have been taken up by psychologists who write about the psychological contracts companies arrange with their staff to democratically decide about working hours, content of the job, performance, salaries, ambitions, and training (see Rousseau 2008). However, unlimited flexibility may lead – as Bluedorn (2002: 76) warned – to "unproductive dithering."

35 This topic is, however, of recent date. One might remember that the first joint-stock trading companies of the seventeenth century did not care about cultural distance when they initiated their operations abroad. They were extremely ethnocentric, and once time began

to play an important role, simply spread their Western clock time orientation intensively, and local people internalized this practice. Some Australian overseas Chinese families, who in the nineteenth century had become used to this practice, returned around 1900 as pioneers of the modern department stores in Mainland China. They organized a huge influx of imported goods such as clocks to potential Chinese customers (Hang and Godley 2009). Cultural distance in the use of the clocks did not seem to matter much for the department stores of Shanghai to display these clocks to local city dwellers. The city had become the fifth largest city in the world with over one million inhabitants. Its city transportation infrastructure expanded and facilitated an easy access to these stores.

36 In line with Hofstede's work, Trompenaars and Hampden-Turner (1998) developed a broader framework to understand cultural differences and the problems of cross-cultural communication. They designed seven different value dimensions to clarify the differences between national cultures. One of their dimensions deals with time and draws a distinction between past and future orientations and the extent to which time is viewed as linear versus holistic, integrating past and present together with future possibilities. It is a redefined version of Hall's distinction between monochronic and polychronic and was translated as the distinction between sequential versus synchronic time. It was recognized that organizational cultures are related to national cultures which are based on different societal factors such as political, legal, educational, and labor relations systems, and they indicate a large diversity amongst countries.

37 Information-and-communication technology does not only encourage a worldwide synchronization, but also creates possibilities for de-synchronization. Answering machines, voicemail and e-mail are all dependent on a persistent computer memory but leave it to the user if and when they want to make use of its sealed time. Synchronized time can therefore become less strict and compelling than it used to be. It will, however, depend on the user how they want to deal with these new opportunities.

38 An example of ethnocentric behavior can be found in the ratio between task time versus social time during workdays. In some cultures, social activities, e.g. informal chats, are important even during work hours. In cultures where the division is 50:50, social time is not regarded as wasting time or inefficient, while persons from clock time cultures with an 80–20 ratio often react negatively toward a 50:50 ratio (Brislin and Kim 2003).

39 Even professionals are currently haunted by schedules: medical doctors in the Netherlands only have eight minutes per patient; teachers are required to precisely indicate how much time they spend on class room teaching; accountants have to "write time" to charge their clients their declarable hours.

40 Although Schein (1992) has argued that monochronic values are more suitable for large organizations, it is questionable whether that is the right approach for multicultural organizations. His findings lead to the conclusion that large monochronic and small polychronic organizations should be more effective than the opposites (Bluedorm *et al.* 1999). Although polychromic values are more suitable for small organizations, especially in the early stages of their development, this might also be relevant for multicultural organizations.

41 For example, ethnoscape points out that tourists, immigrant workers, and refugees jointly create a specific space to influence nation-states.

42 This is highly relevant in the context of the transfer of knowledge between headquarters and subsidiaries. When time rules are not formulated explicitly, managers will have to improvise and reconcile clock time and event time, M-time and P-time, high-context and low-context cultures. They will become masters of learning through experience and adequately harmonize differences in language, business practices, and cultures with different concepts of time and living patterns.

43 The timescale within which expats usually have to operate – in terms of 3–4 years of assignment – may cause pressure resulting in a tendency to enforce knowledge transfer within the time of their assignment (Mawdsley *et al.* 2002).

15 Epilog

1 In the area of transport a paradox has been generated: as trains, cars, and planes become progressively faster the time spent in transit has not been compressed at an equal rate. Standstills and traffic jams are key features of today's traffic around urban centers. Endless queues in crowded lobbies are typical of travel by plane.

2 In 1912, the Chinese government implemented the five-time-zone approach but the People's Republic of China established in 1949 only one single time zone. Currently, the western time of Xinjiang and Tibet is two hours later than Beijing time, and middle time – Chongqing time – is one hour later than Beijing time. The regions in the middle and western China resort to the two different times flexibly. Both Yunnan and Inner Mongolia use middle time.

BIBLIOGRAPHY

Abernethy, D. B. (2000) *The Dynamics of Global Dominance: European Overseas Empires, 1415–1980*, New Haven, CT: Yale University Press.

Abo, T. (ed.) (1994) *Hybrid Factory: The Japanese Production System in the United States*, New York: Oxford University Press.

Abrahamson, E. (1991) 'Managerial fads and fashions: the diffusion and rejection of innovations', *Academy of Management Review, 16*(3): 586–612.

Ackroyd, D. (2007) *Estates, Enterprises and Investment at the Dawn of the Industrial Revolution*, Aldershot, HR: Ashgate.

Adam, B. (1994) *Perceptions of Time*, London: Routledge.

Adam, B. (1995) *Timewatch: The Social Analysis of Time*, Cambridge: Polity Press.

Adam, B. (2004) *Time*, Cambridge: Polity Press.

Adams, S. B., and Butler, O. R. (1999) *Manufacturing the Future: A History of Western Electric*, Cambridge: Cambridge University Press.

Adler, P. S., Goldoftas, B., and Levine, D. I. (1997) 'Ergonomics, employee involvement, and the Toyota Production System: a case study of NUMMI's 1993 model introduction', *Industrial and Labor Relations Review, 50*(3): 416–437.

Alis, D., Karsten, L., and Leopold, J. (2006) 'From Gods to Goddesses: Horai management as an approach to coordinating working hours', *Time and Society, 15*(1): 81–104.

Ancona, D. G., Okhuysen, G. A., and Perlow, L. A. (2001a) 'Taking time to integrate temporal research', *Academy of Management Review, 26*(4): 512–529.

Ancona, D. G., Goodman, P. S., Lawrence, B. S., and Tushman, M. L. (2001b) 'Time: a new research lens', *Academy of Management Review, 26*(4): 645–663.

Anderson, B. R. O. (1991) *Imagined Communities: Reflections on the Origin and Spread of Nationalism*, London: Verso.

Anderson, E., and Gatignon, H. (1986) 'Modes of foreign entry: a transaction cost analysis and propositions', *Journal of International Business Studies, 17*(3): 1–26.

Anderson, P. (1974) *Lineages of the Absolutist State*, London: Verso.

Anderson, P. (1987) *Passages from Antiquity to Feudalism*, London: Verso.

Andrew, E. (1981) *Closing the Iron Cage: The Scientific Management of Work and Leisure*, Montreal: Black Rose Books.

Andrews, K. R. (2005) *Introduction to the Thirtieth Anniversary Edition of The Functions of the Executive*, in C. I. Barnard (1968/2005) *The Functions of the Executive: 30th Anniversary Edition* (pp. i–xxxv), Cambridge, MA: Harvard University Press.

Ang, J., and Teo, T. S. H. (1997) 'CSFs and sources of assistance and expertise in strategic IS planning: a Singapore perspective', *European Journal of Information Systems, 6*(3): 164–171.

Ansoff, H. I. (1965) *Corporate Strategy: An Analytic Approach to Business Policy for Growth and Expansion*, New York: McGraw-Hill.

Anxo, D., and O'Reilly, J. (2000) 'Working time regimes and transitions in comparative perspective', in J. O'Reilly, I. Celebrián, and M. Lallement (eds) *Working Time Changes: Social Integration Through Transitional Labour Markets* (pp. 61–92), Cheltenham: Edward Elgar.

Aoki, M. (1990) 'Toward an economic model of the Japanese firm', *Journal of Economic Literature, 28*(1): 1–27.

Appadurai, A. (1996) *Modernity at Large: Cultural Dimensions of Globalization*, Minneapolis, MN: University of Minnesota Press.

Appadurai, A. (1998) 'Globale Ethnische Räume. Bemerkungen und Fragen zur Entwicklung einer Transnationalen Anthropologie', in U. Beck (ed.) *Perspektiven der Weltgesellschaft* (pp. 11–40), Frankfurt am Main: Suhrkamp.

Arendt, H. (1958) *The Human Condition*, Chicago: University of Chicago Press.

Arrighi, G. (2010) *The Long Twentieth Century: Money, Power and the Origins of our Times*, London: Verso.

Arrow, H., Poole, M. S., Henry, K. B., Wheelan, S., and Moreland, R. L. (2004) 'Time, change, and development: the temporal perspective on groups', *Small Group Research, 35*(1): 73–105.

Aslet, C. (1999) *The Story of Greenwich*, London: Fourth Estate.

Atkinson, J. (1984) *Flexibility, Uncertainty and Manpower Management*, Brighton: Institute of Manpower Studies.

Atkinson, J., and Meager, N. (1986) *Changing Working Patterns: How Companies Achieve Flexibility to Meet New Needs*, Brighton: Institute of Manpower Studies.

Aubert, J. J. (1994) *Business Managers in Ancient Rome: a Social and Economic Study of Institores, 200 B.C.–A.D. 250*, Leiden: E.J. Brill.

Aveni, A. F. (1989) *Empires of Time: Calendars, Clocks, and Cultures*, New York: Basic Books.

Bagwell, P. S. (1974) *The Transport Revolution from 1770*, London: Batsford.

Bailyn, L. (2002) 'Time in organizations: constraints on, and possibilities for, gender equity in the workplace', in R. J. Burke and D. L. Nelson (eds) *Advancing Women's Careers: Research and Practice* (pp. 262–272), New York: Blackwell.

Baird, I.S., Lyles, M.A., and Wharton, R. (1990) 'Attitudinal differences between American and Chinese managers regarding joint venture management', *Management International Review, 30*(3): 53–68.

Ballard, D. I., and Seibold, D. R. (2004) 'Communication-related organizational structures and work group temporal experiences: the effects of coordination method, technology type, and feedback cycle on members' construals and enactments of time', *Communication Monographs, 71*(1): 1–27.

Bank, J. (1992) *The Essence of Total Quality Management*, London: Prentice-Hall.

Barkema, H. G., Shenkar, O., Vermeulen, F., and Bell, J. H. J. (1997) 'Working abroad, working with others: how firms learn to operate international joint ventures', *Academy of Management Journal, 40*(2): 426–442.

Barkey, K. (2008) *Empire of Difference: The Ottomans in Comparative Perspective*, Cambridge: Cambridge University Press.

Barnard, C. I. (1968/2005) *The Functions of the Executive: 30th Anniversary Edition*, Cambridge, MA: Harvard University Press.

Bartky, I. R. (2007) *One Time Fits All*, Stanford, CA: Stanford University Press.

Bartlett, C. A., and Ghoshal, S. (1998) *Managing Across Borders: The Transnational Solution*, Boston, MA: Harvard Business School Press.

Bayly, C. A. (2004) *The Birth of the Modern World 1780–1914: Global Connections and Comparisons*, New York: Wiley Blackwell.

Beard, M. (2008) *Pompeii, the Life of a Roman Town*, London: Profile Books.

Beatson, M. (1995) *Labour Market Flexibility*, London: Stationery Office Books.

Beck, U. (2002) 'The cosmopolitan society and its enemies', *Theory, Culture and Society, 19*(1–2): 17–44.

Belkin, L. (2003) *Life's Work: Confessions of an Unbalanced Mom*, London: Simon and Schuster.

Benders, J., and Veen, K. van (2001) 'What's in a fashion? Interpretative viability and management fashion', *Organization, 8*(1): 33–53.

Beniger, J. R. (1986) *The Control Revolution: Technological and Economic Origins of the Information Society*, Cambridge, MA: Harvard University Press.

Bereiter, C. (2009) 'Innovation in the absence of principled knowledge: the case of the Wright brothers', *Creativity and Innovation Management, 18*(3): 234–241.

Berg, M. (1994) *The Age of Manufactures 1700–1820: Industry, Innovation and Work in Britain*, London: Routledge.

Bergreen, L. (2007) *Marco Polo: From Venice to Xanadu*, New York: Alfred A. Knopf.

Bernstein, J. R. (1997a) 'Toyota automatic looms and Toyota automobiles', in T. K. McCraw (ed.) *Creating Modern Capitalism: How Entrepreneurs, Companies and Countries Triumphed in Three Industrial Revolutions* (pp. 398–438), Cambridge, MA: Harvard University Press.

Bernstein, J. R. (1997b) 'Japanese Capitalism', in T. K. McCraw (ed.) *Creating Modern Capitalism: How Entrepreneurs, Companies and Countries Triumphed in Three Industrial Revolutions* (pp. 441–489), Cambridge, MA: Harvard University Press.

Bernstein, M. A., and Adler, D. E. (1994) *Understanding American Economic Decline*, Cambridge: Cambridge University Press.

Bhabha, H. K. (1994) *The Location of Culture*, New York: Routledge.

Bielenski, H. (1999) *Employment Options for the Future: High Demand for New Jobs in Europe. High Interest in Non-Standard Workforms*, Munich: Infratest Burke Sozialforschung.

Biernacki, R. (1995) *The Fabrication of Labor: Germany and Britain, 1640–1914*, Berkeley, CA: University of California Press.

Biesebroeck, J. van (2007) 'The cost of flexibility', *Assembly Automation, 27*(1): 55–64.

Black, E. (2001) *IBM and the Holocaust*, New York: Crown Publishing.

Black, J. (1995) 'A military revolution? A 1660–1792 perspective', in C. J. Rogers (ed.) *The Military Revolution Debate*, Boulder, CO: Westview Press.

Blackburn, R. (1997) *The Making of New World Slavery: From the Baroque to the Modern 1492–1800*, London: Verso.

Blackford, M. G. (1998) *The Rise of Modern Business in Great Britain, the United States and Japan*, Chapel Hill, NC: The University of North Carolina Press.

Blair, A., Karsten, L., and Leopold, J. (2001) 'The European Working Time Directive and its effect on flexibility within organizations: a British–Dutch comparative analysis', in G. Morello, B. Adams, and I. Sabelis (eds) *Time and Management* (pp. 159–218), Palermo: ISIDA.

Blaut, J. M. (ed.) (1992) *1492: The Debate on Colonialism, Euro-Centrism and History*, Trenton, NJ: Africa World Press.

Blount, S., and Janicik, G. A. (2001) 'When plans change: examining how people evaluate timing changes in work organizations', *The Academy of Management Review, 26*(4): 566–585.

Bluedorn, A. C. (1998) 'A. C. Bluedorn, an interview with anthropologist Edward T. Hall', *Journal of Management Inquiry, 7*(2): 109–115.
Bluedorn, A. C. (2002) *The Human Organization of Time: Temporal Realities and Experience*, Stanford, CA: Stanford University Press.
Bluedorn, A. C., and Denhardt, R. B. (1988) 'Time and organizations', *Journal of Management, 14*(2): 299–320.
Bluedorn, A. C., Kaufman, C., and Lane, P. (1992) 'How many things do you like to do at once? An introduction to monochronic and polychronic time', *Academy of Management Executive, 6*(4): 17–26.
Bluedorn, A. C., Kalliath, T. J., Strube, M. J., and Martin, G. D. (1999) 'Polychronicity and the Inventory of Polychromic Values (IPV): the development of an instrument to measure a fundamental dimension of organizational culture', *Journal of Managerial Psychology, 14*(3–4): 215–230.
Blyton, P., and Morris, J. (1991) *A Flexible Future?: Prospects for Employment and Organization*, New York: Walter de Gruyter.
Blyton, P., and Turnbull, P. J. (1992) *Reassessing Human Resource Management*, London: Sage.
Boersema, D. (2009) *Philosophy of Science*, New York: Pearson.
Bos, R. ten (2000) *Fashion and Utopia in Management Thinking*, Tilburg: Tilburg University Press.
Botti, H. F. (1997) 'Going local – the hybridization process of situated learning', in L. Alvarez (ed.) *The Diffusion and Consumption of Business Knowledge* (pp. 48–72), Basingstoke: Macmillan.
Botticelli, P. (1997) 'Rolls-Royce and the rise of high-technology industry', in T. K. McCraw (ed.) *Creating Modern Capitalism: How Entrepreneurs, Companies and Countries Triumphed in Three Industrial Revolutions* (pp. 49–93), Cambridge, MA: Harvard University Press.
Botton, A., de (2009) *The Pleasures and Sorrows of Work*, London: Hamish Hamilton.
Boulin, J. Y. (1992) 'Les politiques du temps de travail en France: la perte du sens', *Futuribles, 165–166*(5): 41–62.
Boulin, J. Y., Lallement, M., Messenger, J. C., and Michon, F. (2006) *Decent Working Time: New Trends, New Issues*, Geneva: ILO.
Bourdieu, P. (1963) 'The attitude of the Algerian peasant toward time', in J. Pitt-Rivers (ed.) *Mediterranean Countrymen* (pp. 55–72), Paris: Mouton.
Boyer, R. (1987) 'Labor flexibilities: Many forms, uncertain effects', *Labor and Society, 72*(1): 107–129.
Boyer, R., and Hollingsworth, H. J. (1997) *Contemporary Capitalism: The Embeddedness of Institutions*, Cambridge: Cambridge University Press.
Boyns, T., and Smith, I. (2005) 'British management theory and practice: the impact of Fayol', *Management Decision, 43*(10): 1317–1334.
Braudel, F. (1972) *The Mediterranean and Mediterranean World in the Age of Philip, Vol. II*, London: William Collins Sons & Co.
Braudel, F. (1975) *Capitalism and Material Life, 1400–1800*, New York: Harper Colophon.
Braudel, F. (1977) *Afterthoughts on Material Civilization and Capitalims*, Baltimore, MD: Johns Hopkins University Press.
Braudel, F. (1984) *Civilization and Capitalism, 15th–18th Century, Vol. III: The Perspective of the World*, New York: Harper & Row.
Braun, M. (1992) *Picturing Time: The Work of Etienne-Jules Marey (1830–1904)*, Chicago: The University of Chicago Press.
Braun, R. (1991) 'The "docile" body as an economic-industrial growth factor', in P. Higgonet, D. S. Landes, and H. Rosovsky (eds) *Favorites of Fortune: Technology, Growth and Economic Development Since the Industrial Revolution* (pp. 120–141), Cambridge, MA: Harvard University Press.

Brendon, P. (2009) *The Decline and Fall of the British Empire 1781–1997*, New York: Knopf.

Brewster, C., Hegewisch, A., Holden, L., and Lockhart, T. (eds) (1992) *The European Human Resource Management Guide*, London: Academic Press.

Brewster, C., Mayne, L., and Tregaskis, O. (1997) 'Flexible working in Europe', *Journal of World Business, 32*(2): 133–151.

Bridges, W. (1994) *Job Shift*, White Plains, NY: Addison-Wesley.

Brinkley, D. (2003) *Wheels for the World: Henry Ford, His Company and a Century of Progress*, New York: Viking.

Brislin, R. W., and Kim, E. S. (2003) 'Cultural diversity in people's understanding and uses of time', *Applied Psychology: An International Review, 52*(3): 262–382.

Brode, H. (2000) *Tippu Tip: The Story of His Carrier in Zanzibar and Central Africa*, Zanzibar: The Gallery Publications.

Bryson, C., and Barnes, N. (2000) 'Working in higher education in the United Kingdom', in M. Tight (ed.) *Academic Work and Life* (pp. 147–186), Amsterdam: Elsevier.

Bryson, C., and Karsten, L. (2009) 'Managing uncertainty or managing uncertainly?', in L. Harris and J. Leopold (eds) *The Strategic Managing of Human Resources* (pp. 34–69), London: Pearson.

Buchanan, D., and Badham, R. (1999) *Power, Politics and Organizational Change*, London: Sage.

Buckley, P. J. (2009) 'Business history and international business', *Business History, 51*(3): 307–333.

Buckley, P. J., and Casson, M. C. (1998a) 'Analyzing foreign market entry strategies: extending the internationalization approach', *Journal of International Business Studies, 29*(3): 539–561.

Buckley, P. J., and Casson, M. C. (1998b) 'Models of the multinational enterprise', *Journal of International Business Studies, 29*(1): 21–44.

Buecher, C. (2006) *Christoph Kolumbus: Korsar und Kreuzfahrer*, Darmstadt: Primus Verlag.

Burleigh, M. (2008) *Blood and Rage: A Cultural History of Terrorism*, London: Harper Press.

Business Europe (2007) *Key Challenges Facing European Labor Markets: A Joint Analysis of European Social Partners*, Brussels: Business Europe, ETUC, UEAPME, and CEEP.

Butman, J. (1997) *Juran: a Lifetime of Influence*, New York: John Wiley.

Cahill, M. C. (1932) *Shorter Hours: A Study of the Movement Since the Civil War*, New York: AMS Press.

Cairncross, F. (1999) *The Death of Distance*, Boston, MA: Harvard Business School Press.

Camuffo, D. (2002) 'Errors in early temperature series arising from changes in style of measuring time, sampling schedule and number of observations', *Climatic Change*, (1–3): 331–352.

Canclini, N. G. (2005) *Hybrid Cultures: Strategies for Entering and Leaving Modernity*, Minneapolis, MN: University of Minnesota Press.

Casey, B. (1988) *Temporary Employment: Practice and Policy in Britain* (PSI Report No. 678), London: Policy Studies Institute.

Casey, B., Metcalf, H., and Millward, N. (1997) *Employers' Use of Flexible Labour* (PSI Report No. 837), London: Policy Studies Institute.

Cass, B. (1995) 'Overturning the male breadwinner model in the Australian social protection system', in P. Saunders and S. Shaver (eds) *Social Policy and the Challenge of Social Change* (pp. 47–66), Sydney: University of New South Wales.

Castells, M. (1996) *The Information Age, Vol. I: The Rise of the Network Society*, London: Blackwell.

Castells, M. (1997) *The Information Age, Vol. II: The Power of Identity*, London: Blackwell.

Castells, M. (1998) *The Information Age, Vol. III: End of Millennium*, London: Blackwell.

Castree, N. (2009) 'The spatio-temporality of capitalism', *Time and Society, 18*(1): 26–61.

Cavendish, R. (1982) *Women on the Line*, London: Routledge & Kegan Paul.

Chandler, A. D. (1967) *The Campaigns of Napoleon*, London: Weidenfeld & Nicolson.

Chandler, A. D. (1977) *The Visible Hand: The Managerial Revolution in American Business*, Cambridge, MA: Belknap Press.

Chandler, A. D., and Mazlish, B. (2005) *Leviathans: Multinational Corporations and the New Global History*, Cambridge: Cambridge University Press.

Chase, W., and Siegelbaum, L. (1988) 'Work time and industrialization in the USSR, 1917–1941', in G. Cross (ed.) *Work Time and Industrialization: An International History* (pp. 183–216), Philadelphia, PA: Temple University Press.

Chaudhuri, K. N. (1981) 'The English East India Company in the seventeenth and eighteenth centuries: a pre-modern multinational organization', in L. Blussé, and F. Gaastra (eds) *Companies and Trade* (pp. 29–47), Amsterdam: Leiden University Press.

Chaunu, P. (1979) *European Expansion in the Later Middle Ages*, Amsterdam: North-Holland Publishing Company.

Child, J., and Rodrigues, S. (1996) 'The role of social identity in the international transfer of knowledge through joint ventures', in S. R. Clegg, and G. Palmer (eds) *The Politics of Management Knowledge* (pp. 46–68), London: Sage.

Christopher, E. (2006) *Slave Ship Sailors and Their Captive Cargoes, 1730–1807*, Cambridge: Cambridge University Press.

Chua, A. (2007) *Day of Empire: How Hyper Powers Rise to Global Dominance and Why They Fall*, New York: Doubleday.

Clair, W. S. (2007) *The Door of No Return: The History of Cape Coast Castle and the Atlantic Slave Trade*, New York: BueBridge.

Clapham, J. H. (1950) *An Economic History of Modern Britain: The Early Railway Age 1820–1850*, Cambridge: Cambridge University Press.

Clark, G. (1994) 'Factory discipline', *The Journal of Economic History, 54*(1): 128–161.

Clark, G., and Werf, Y. van der (1998) 'Work in progress? The industrious revolution', *The Journal of Economic History, 58*(3): 830–843.

Clark, P. (1978) 'Temporal innovations and time structuring in large organizations', in J. T. Fraser, N. Lawrence, and D. Park (eds) *The Study of Time III* (pp. 391–416), New York: Springer Verlag.

Clark, P. (1985) 'A review of the theories of time and structure for organizational sociology', *Research in the Sociology of Organizations, 4*: 35–80; reprinted in S. B. Bacharach and S. M. Mitchell (eds) *Research in the Sociology of Organizations*, Greenwich, CT: JAI Press.

Clark, T. (1995) *Managing Consultants: Consultancy as the Management of Impressions*, Buckingham: Open University Press.

Clark, T., and Mallory, G. (1996) *The Cultural Relativity of Human Resource Management: Is There a Universal Model?*, Oxford: Oxford, Blackwell.

Clarke, D. (2006) *Descartes: A Biography*. Cambridge: Cambridge University Press.

Clawson, D. (1980) *Bureaucracy and the Labor Process: The Transformation of US industry 1860–1920*, New York: Monthly Review Press.

Cohen, D. (2003) *Our Modern Times: The New Nature of Capitalism in the Information Age*, Cambridge, MA: MIT Press.

Cohen, F. (1994) *The Scientific Revolution*, Chicago: The University of Chicago Press.

Cohen, M. L., and Quinter, J. L. (1996) 'The derailment of railway spine: a timely lesson for post-traumatic fibromyalgia syndrome', *Pain Review, 3*: 181–202.

Cole, R. E., and Scott, W. R. (2000) *The Quality Movement and Organization Theory*, London: Sage.

Collins, J. C., and Porras, J. I. (1997) *Built to Last: Successful Habits of Visionary Companies*, New York: Harper Business.

Crafts, N. F. R., and Venables, A. J. (2003) 'Globalization in history: a geographical perspective', in M. D. Bordo, A. M. Taylor, and J. G. Williamson (eds) *Globalization in Historical Perspective* (pp. 323–364), Chicago: University of Chicago Press.

Crainer, S. (2000) *The Management Century: A Critical Review of 20th Century Thought and Practice*, San Francisco, CA: Jossey Bass.

Crosby, A. W. (1997) *The Measure of Reality: Quantification and Western Society, 1250–1600*, Cambridge: Cambridge University Press.

Crosby, P. B. (1980) *Quality Is Free*, New York: McGraw-Hill.

Cross, G. (1988) 'Work time in international discontinuity', in G. Cross (ed.) *Work Time and Industrialization: An International History* (pp. 155–182), Philadelphia, PA: Temple University Press.

Cross, G. (1989) *A Quest for Time: The Reduction of Work in Britain and France 1840–1940*, Berkeley, CA: University of California Press.

Crossan, M., Pina e Cunha, M., Vera, D., and Cunha, J. (2005) 'Time and organizational improvisation', *Academy of Management Review, 30*(1): 129–145.

Czarniawska, B., and Sevon, G. (1996) *Translating Organizational Change*, Berlin: Walter de Gruyter.

Dan, T. N. (1992) 'The spatialization of metric time: the conquest of land and labour in Europe and the United States', *Time and Society, 1*(1): 29–50.

Das, T. K. (1986) *The Subjective Side of Strategy Making: Future Orientations and Perceptions of Executives*, Santa Barbara, CA: Praeger.

Davis, D. R. (1997) 'Critical evidence on comparative advantage? North–North trade in a multilateral world', *Journal of Political Economy, 105*(5): 1051–1060.

Davis, S. M. (1987) *Future Perfect*, Boston, MA: Addison-Wesley.

Dawson, M. (2003) *The Consumer Trap: Big Business Marketing in America*, Urbana, IL: University of Illinois Press.

Dawson, P. (1991) 'The historical emergence and changing role of supervisors', *Asia Pacific HRM, 29*(2): 36–50.

Deardorff, A.V. (2003) 'What might globalization's critics believe?', *The World Economy, 26*(5): 639–658.

Deery, S., and Mahoney, A. (1994) 'Temporal flexibility: management strategies and employee preferences in the retail industry', *Journal of Industrial Relations, 36*(3): 332–352.

Dehé, J. (2005) *Een Slaafsch en Ongezond Bedrijf: De Geschiedenis van het Openbaar Vervoer in Waterland, 1630–1880*, Hilversum: Verloren.

Delsen, L. (1998) 'Why do men work part-time?', in C. Fagan and J. O'Reilly (eds) *Part-time Prospects: An International Comparison of Part-Time Work in Europe, North America, and the Pacific Rim* (pp. 57–76), London: Routledge.

Delsen, L. (2000) *Exit Polder Model: Sociaal-Economische Ontwikkelingen in Nederland*, Assen: Van Gorcum.

Deming, W. E. (1986) *Out of the Crisis*, Cambridge, MA: Cambridge University Press.

Demorgon, J., and Molz, M. (1996) 'Bedingungen und Auswirkungen der Anayse von Kultur(en) und interkulturellen Interaktionen', in A. Thomas (ed.) *Psychologie Interkulturellen Handelns* (pp. 43–80), Göttingen: Hogrefe.

Deng, G. (1999) *Maritime Sector, Institutions, and Sea Power of Pre-Modern China*, Westport, CT: Greenwood Press.

Dew, T. R. (1832/1981) 'Abolition of Negro slavery', in D. G. Faust (ed.) *The Ideology of Slavery: Proslavery Thought in the Antebellum South, 1830–1860* (pp. 23-27), Baton Rouge, LA: Louisiana State University Press.

Dex, S., and McCulloch, A. (1997) *Flexible Employment: The Future of Britain's Jobs*, London: Macmillan.

Dia, M. (1996) *Africa's Management in the 1990s and Beyond*, Washington, DC: World Bank.

Diamond, J. (1997) *Guns, Germs and Steel: The Fates of Human Society*, New York: W.W. Norton.

Dias, J. (2007) 'Empire-building in the Old World and the New', in R. Aldrich (ed.) *The Age of Empires* (pp. 68–111), London: Thames & Hudson.

Dicken, P. (2007) *Global Shift: Mapping the Changing Contours of the World Economy*, London: Sage.

Diebold, J. (1952) *Automation: The Advent of the Automatic Factory*, New York: Van Nostrand.

Digges, L. (1571/1968) *Stratioticos*, Amsterdam: Theatrum Orbis Terrarum.

Ditton, J. (1979) 'Baking Time', *Sociological Review, 27*(1): 157–167.

Dittrich, K., Duysters, G., and De Man, A. P. (2004) 'Using Networks for Changing Innovation Strategy: The Case of IBM, Rotterdam', *ERIM Report Series*, Reference No. ERS-2004-111-ORG.

Doganis, R. (2002) *Flying off Course*, London: Routledge.

Dohrn-van Rossum, G. (1992) *Die Geschichte der Stunde*, München: Carl Hanser Verlag.

Dolan, B. (2004) *Wedgwood, the First Tycoon*, New York: Viking.

Donkin, R. (2001) *Blood, Sweat and Tears: The Evolution of Work*, London: Texere.

Donnachie, I. (2000) *Robert Owen: Owen of New Lanark and New Harmony*, East Linton: Tuckwell Press.

Doorne-Huiskes, J. van, Dulk, L. den, and Peper, A. (2005) 'Organizational change, gender and integration of work and private life', in A. Peper, J. van Doorne-Huiskes, and L. den Dulk (eds) *Flexible Working and Organizational Change: The Integration of Work and Personal Life* (pp. 39-61), Cheltenham: Edward Elgar.

Dowrick, S., and DeLong, J. B., (2003) 'Globalisation and convergence', in M. D. Bordo, A. M. Taylor, and J. G. Williamson (eds) *Globalization in Historical Perspective* (pp. 191–226), Chicago: University of Chicago Press.

Drescher, S. (2009) *Abolition: A History of Slavery and Antislavery*, Cambridge: Cambridge University Press.

Driel, H., and Dolfsma, W. A. (2009) 'Path dependence, initial conditions and routines in organizations: the Toyota Production System reexamined', *Journal of Organizational Change Management, 22*(1): 49–72.

Drucker, P. (1946) *Concept of the Corporation*, Princeton, NJ: John Day.

Drucker, P. (1955) *The Practice of Management*, Melbourne: Heinemann.

Drucker, P. (1979) *Adventures of a Bystander*, Melbourne: Heinemann.

Du Plessis, A., and Venter, F. (2007) 'You say tick, I say tock', *Monash Business Review, 3*(3): 23–24.

Dulk, L. den (2001) *Work–Family Arrangements in Organizations: A Cross-National study in the Netherlands, Italy, the United Kingdom and Sweden*, Amsterdam: Rozenberg.

Duncan, D. E. (1998) *Calendar: Humanity's Epic Struggle to Determine a True and Accurate Year*, London: Fourth Estate.

Dyson, R. (1992) 'Changing labour utilization in NHS trusts: the reprofiling paper', University of Keele.

Earley, P. C., and Mosakowski, E. (2004a) 'Cultural intelligence', *Harvard Business Review, 82*(10): 139–146.

Earley, P. C., and Mosakowski, E. (2004b) 'Toward culture intelligence: turning cultural differences into a workplace advantage', *The Academy of Management Executive, 18*(3): 151–157.

Echteld, P. (2007) 'Time-greedy employment relationships: four studies on the time claims of post-Fordist work', PhD thesis, University of Groningen.

Edwards, R. (1979) *Contested Terrain: The Transformation of the Workplace in the Twentieth Century*, New York: Basic Books.

Elias, N. (1985) *Über die Zeit*, Frankfurt: Suhrkamp.

Elias, N. (1992) *Time: An Essay*, Oxford: Blackwell.

Emmer, P. C. (1981) 'The West India Company 1621–1791: Dutch or Atlantic?, in L. Blussé and F. Gaastra (eds) *Companies and Trade: Essays on Overseas Trading Companies During the Ancient Régime* (pp. 71–95), Leiden: Leiden University Press.

Ericson, S. J. (1989) 'Private railroads in the Meiji Era: forerunners of modern Japanese management', in T. Yui and K. Nakagawa (eds) *Japanese Management in Historical Perspective* (pp. 51–77), Tokyo: University of Tokyo Press.

Eskerod, P., and Skriver, H. J. (2007) 'Organizational culture restraining in-house knowledge transfer between project managers – a case study', *Project Management Journal 38*(1): 110–112.

Esping-Andersen, G. (2000) 'Two societies, one sociology, and no theory', *British Journal of Sociology, 51*(1): 59–77.

Essen, L., and Parry, J. V. L. (1957) 'The Caesium resonator as a standard of frequency and time', *Philosophical Transactions of the Royal Society London, Ser. A, 250*(973): 45-69.

Ettinghoffer, F. F., and Blanc, G. (1998) *Le Syndrome de Chronos: Du Mal Travailler au Mal Vivre*, Paris: Dunod.

European Commission (1999) *Equal Opportunities for Women and Men in the European Union* (COM, 106 final), Luxembourg: Publications Office of the European Union.

European Commission (2000) *Employment in Europe 2000*, Luxembourg: Publications Office of the European Union.

European Commission (2004) *Proposition de Directive du Parlement Europeen et du Conseil* (SEC 1154, 22-09-2004), Luxembourg: Publications Office of the European Union.

European Commission (2007) *Flexicurity Pathways: Turning Hurdles into Stepping Stones*, report by the European Expert Group on Flexicurity, Luxembourg: Publications Office of the European Union.

European Commission (2008) *Report on Equality Between Women and Men – 2008*, Luxembourg: Publications Office of the European Union.

Evans, H. (2004) *They Made America*, New York: Little, Brown.

Ezzamel, M., Hoskin, K., and Macve. R. (1990) 'Managing it all by numbers: a review of Johnson and Kaplan's "Relevance Lost"', *Accounting and Business Research 20*(78): 153–166.

Fagan, C. (2001) 'Time, money and the gender order: work orientations and working time preferences in Britain', *Gender, Work and Organization 8*(3): 239–266.

Fagan, C. (2003) *Working Time Preferences and Work Life Balance in the EU: Some Policy Considerations for Enhancing the Quality of Life*, Dublin: European Foundation for the Improvement of Living and Working Conditions.

Fagan, C., and O'Reilly, J. (eds) (1998) *Part-time Prospects: An International Comparison of Part-Time Work in Europe, North America, and the Pacific Rim*, London: Routledge.

Felstead, A., and Jewson, N. (eds) (1999) *Global Trends in Flexible Labor*, London: Macmillan.

Felstead, A., Jewson, N., and Walters, S. (2005) *Changing Places of Work*, New York: Palgrave Macmillan.

Ferguson, N. (2011) *Civilization: The West and the Rest*, London: Allen Lane.

Fernandez-Armato, F. (2006) *Pathfinders: A Global History of Exploration*, Oxford: Oxford University Press.

Filarski, R. (1997) 'Opkomst en verval van vervoersystemen: de ontwikkeling vanuit een historisch perspectief', *Tijdschrift Vervoerswetenschap, 2*: 116–117.

Filipzcak, B. (1997) 'Managing a mixed workforce', *Training, 34*(10): 96–101.

Fincham, R., and Roslender, R. (2004) 'Rethinking the dissemination of management fashion', *Management Learning 35*(3): 321–336.

Findlay, R., and O'Rourke, K. H. (2003) 'Commodity market integration, 1500–2000', in M. D. Bordo, A. M. Taylor, and J. G. Williamson (eds) *Globalization in Historical Perspective* (pp. 13–64), Chicago: University of Chicago Press.

Finley, M. I. (1983) *Ancient Slavery and Modern Ideology*, London: Pelican.

Flinn, M. W. (ed.) (1957) *The Lava Book of the Crowley Iron Works*, London: Surtees Society.

Flood, R. (1993) *Beyond TQM*, Chichester: John Wiley.

Fogel, R. W., and Engerman, S. L. (1974) *Time on the Cross: The Economics of American Negro Slavery*, London: Little, Brown.

Foucault, M. (1975) *Surveiller et Punir*, Paris: Gallimard.

Foucault, M. (1977) *Discipline and Punish*, London: Penguin Books.

Francis-Smythe, J., and Robertson, I. (1999) 'Time-related individual differences', *Time and Society*, *8*(2): 273–292.

Franklin, B. (1989) *The Autobiography and Other Writings*, New York: Bantam Books.

Freeland, R. F. (2001) *The Struggle for Control of the Modern Corporation: Organizational Change at General Motors, 1924–1970*, Cambridge: Cambridge University Press.

Freeman, C., and Louca, F. (2001) *As Time Goes By: From the Industrial Revolution to the Information Revolution*, New York: Oxford University Press.

Frenkel, M. (2008) 'The multinational corporation as a third space: rethinking international management discourse on knowledge transfer through Homi Bhabha', *Academy of Management Review*, *33*(4): 924–942.

Fridenson, P. (1987) 'Un tournant taylorien de la société francaise (1904–1918)', *Annales Economies Societes et Civilisation*, *43*(5): 1031–1060.

Fridenson, P., and Reynaud, B. (2004) *La France et le Temps du Travail, 1814–2004*, Paris: Odile Jacob.

Friedel, R. (2007) *A Culture of Improvement*, Cambridge, MA: MIT Press.

Friedman, T. L. (1999) *The Lexus and the Olive Tree*, New York: Farrar, Strauss and Giroux.

Fritze, R. H. (2009) *Invented Knowledge: False History, Fake Science and Pseudo-Religions*, London: Reaction Books.

Gaastra, F. S. (1991) *De Geschiedenis van de VOC*, Zutphen, Walburg Press.

Gaastra, F. S. (2003) *The Dutch East India Company: Expansion and Decline*, Zutphen: Walburg Press.

Gabel, M., and Bruner, H. (2003) *Global Inc.: An Atlas of the Multinational Corporation*, New York: The New Press.

Galambos, L., and Abrahamson, E. J. (2002) *Any Time, Anywhere: Entrepreneurship and the Creation of a Wireless World*, Cambridge: Cambridge University Press.

Garfield, S. (2002) *The Last Journey of William Huskisson*, London: Faber & Faber.

Gater, S., and Vincent, D. (1988) *The Factory in a Garden: Wedgewood from Etruria to Barlaston – The Transitional Years*, Stoke-on-Trent: Keele Live Histories Centre.

Geary, D. (1981) *European Labour Protest, 1848–1939*, London: Croom Helm.

Geary, D. (1992) *European Labour Politics from 1900 to the Depression*, London.

Geissler, K. A. (1999) *Vom Tempo der Welt: Am Ende der Urzeit*, Freiburg: Herder.

Geneen, H., and Moscow, A. (1984) *Managing*, Garden City, NY: Doubleday.

Genovese, E.D. (1976) *Roll Jordan Roll*, New York: Vintage Books.

Gerstner, L.V. (2002) *Who Says Elephants Can't Dance?*, New York: HarperCollins.

Geyer, M., and Bright, C. (1995) 'World history in a global age', *American Historical Review*, *100*(4): 1034–1061.

Ghoshal, S., and Bartlett, C.A. (1990) 'The multinational corporation as an interorganizational network', *The Academy of Management Review*, *15*(4): 603–625.

Gibson, C. B., Waller, M. J., Carpenter, M. A., and Conte, J. M. (2007) 'Antecedents, consequences, and moderators of time perspective heterogeneity for knowledge management in MNO teams', *Journal of Organizational Behaviour*, *28*(8): 1005–1034.

Giddens, A. (1979) *Central Problems in Social Theory*, London: Macmillan Press.

Giddens, A. (1987) *Social Theory and Modern Sociology*, London: Polity Press.

Giddens, A. (2006) *Sociology* (5th edn), London: Polity Press.

Giddens, A. (2007) *Europe in the Global Age*, London: Polity Press.

Gilbreth, F. P. (1953) 'Motion study and time study: instruments of precision', in W. R. Spriegel and C. E. Mijers (eds) *The Writings of the Gilbreths*, Homewood, IL: R. D. Irwin.

Gillespie, R. (1991) *Manufacturing Knowledge: A History of the Hawthorne Experiments*, Cambridge: Cambridge University Press.

Gillispie, C. C. (2004) *Science and Polity in France: The Revolutionary and Napoleonic Years*, Princeton, NJ: Princeton University Press.

Gipouloux, F. (2011) *The Asian Mediterranean*, Cheltenham: Edward Elgar.

Glennie, P., and Thrift, N. (1996) 'Reworking E.P. Thompson's "Time, Work-Discipline and Industrial Capitalism"', *Time and Society, 5*(3): 275–299.

GMAC (2008) expatica.com/hr/story/reporting-on-global-relocation-trendsin2008.html, retrieved 20 December 2011.

Godoy, J. (2009) 'This job is killing me: authoritarian corporate model spurring suicides in Europe', *IPS News Report*.

Golden, L. (2005) 'The flexibility gap: employee access to flexibility in work schedules', in I. U. Zeytinoglu (ed.) *Flexibility in Workplaces* (pp. 38–56), Geneva: ILO.

Gonzales-Rendon, M., Acaide-Castro, M., and Florez-Saborido, I. (2005) 'Is it difficult for temporary workers to get a permanent job?' Empirical evidence from Spain', in I. U. Zeytinoglu (ed.) *Flexibility in Workplaces* (pp. 22–37), Geneva: ILO.

Gordon, J. S. (2003) *A Thread Across the Ocean: The Heroic Story of the Transatlantic Cable*, New York: Perennial.

Gorski, P. S. (1993) 'The protestant ethic revisited: disciplinary revolution and state formation in Holland and Prussia', *American Journal of Sociology, 99*(2): 265–316.

Gourvish, T. R. (1986) 'The railways and the development of managerial enterprise in Britain 1850–1939'. *Development of Managerial Enterprise: Proceedings of the Fuji Conference*, The International Conference on Business History, Tokyo.

Gowler, D., and Legge, K. (1983) 'The meaning of management and the management of meaning: a view from social anthropology', in M. J. Earl (ed.) *Perspectives on Management: A Multidisciplinary Analysis* (pp. 197–233), Oxford: Oxford University Press.

Grandin, G. (2009) *Fordlandia: The Rise and Fall of Henry Ford's Forgotten Jungle City*, New York: Metropolitan Books.

Grint, K. (1997) *Fuzzy Management: Contemporary Ideas and Practices at Work*, Oxford: Oxford University Press.

Grint, K. (2000) *The Arts of Leadership*, Oxford: Oxford University Press.

Grint, K., and Case, P. (1998) 'The violent rhetoric of reengineering: management consultancy on the offensive', *Journal of Management Studies, 35*(5): 557–577.

Guillen, M. F. (1994) *Models of Management: Work, Authority and Organization in a Comparative Perspective*, Chicago: University of Chicago Press.

Gupta, A. K., and Govindarajan, V. (2000) 'Knowledge flows within multinational corporations', *Strategic Management Journal, 21*(4): 473–496.

Gurevich, A. J. (1985) *Categories of Medieval Culture*, London: Routledge & Kegan Paul.

Hagstrom, R. G. (2002) *The Essential Buffett: Timeless Principles for the New Economy*, New York: John Wiley & Sons.

Hakim, C. (2004) *Key Issues in Women's Work: Female Diversity and the Polarization of Women's Employment*, London: Routledge.

Hall, E. T. (1959) *The Silent Language*, Garden City, NY: Doubleday.

Hall, E. T. (1976) *Beyond Culture*, Garden City, NY: Doubleday.

Hall, E.T. (1983) *The Dance of Life: The Other Dimension of Time*, Garden City, NY: Doubleday.
Hall, E. T., and Hall, M. R. (1990) *Understanding Cultural Differences: Germans, French, and Americans*, Yarmouth, ME: Intercultural Press.
Hall, M., and Sission, K. (1997) *Time for Change? Coming to terms with the EU Working Time Directive*, London: Eclipse Group.
Hall, P. A. (1999) 'The political economy of Europe in an era of interdependence', in H. Klitschelt, P. Lange, G. Marks, and J. Stephens (eds) *Continuity and Change in Contemporary Capitalism* (pp. 135–163), New York: Cambridge University Press.
Hallion, R. (2009) 'Streamline revolution', *Aviation History*, The Weider History Group.
Hammer, M., and Champy, J. (1993) *Reengineering the Corporation: A Manifesto for Business Revolution*, New York: HarperCollins.
Hang, H., and Godley, A. (2009) 'Revisiting the psychic distance paradox: international retailing in China in the long run (1840–2005)', *Business History, 51*(3): 383–400.
Harding, N. (2003) *The Social Construction of Management*, London: Routledge.
Harvey, D. (1989) *The Condition of Postmodernity: An Enquiry into the Origins of Cultural Change*, Cambridge, MA: Blackwell.
Hassard, J. (1989) 'Time and organization', in P. Blyton, J. Hassard, S. Hill, and K. Starkey (eds) *Time, Work and Organization* (pp. 79–104), London: Routledge.
Hatchuck, A., and Weil, B. (1995) *Experts in Organization*, Berlin: Walter de Gruyter.
Hayes, R. H., and Abernathy W. J. (1980) 'Managing our way to economic decline,' *Harvard Business Review*, July/August: 67–77.
Heijer, H. (2005) *De Geoctrooieerde Compagnie: De VOC en de WIC als Voorlopers van de Naamloze Vennootschap*, Deventer: Kluwer.
Heller, A. (1974) *The Theory of Need in Marx*, London: Allison & Busby.
Heppenheimer, T. A. (1995) *Turbulent Skies*, New York: John Wiley and Sons.
Heusinkveld, S. (2004) 'Surges and sediments: organizational concepts between transience and continuity', PhD thesis, University of Nijmegen.
Hibou, B. (2011) 'Work discipline, discipline in Tunisia: complex and ambiguous relations', in G. Sabar, L. Bethlehem, and L. Schier (eds) *Rethinking Labor in Africa: Past and Present*, London: Routledge.
Hilton, B. (2006) *A Mad, Bad, and Dangerous People?*, Oxford: Clarendon Press.
Hobsbawm, E. (1974) *Laboring Men: Studies in the History of Labor*, London: Weidenfeld & Nicolson.
Hobsbawm, E. (1995) *Age of Extremes: The Short Twentieth Century, 1914–1991*, London: Abacus.
Hofstede, G. (2001) *Culture's Consequences: Comparing Values, Behaviors, Institutions and Organizations Across Nations*, Thousand Oaks, CA: Sage.
Holden, N. J. (2002) *Cross-Cultural Management: A Knowledge-Management Perspective*, London: Prentice Hall.
Holford-Strevens, L. (2005) *The History of Time: A Very Short Introduction*, Oxford: Oxford University Press.
Hong, P., Park, Y., Roh, J. J., and Tomido, T. (2009) 'Market flexible customizing system (MFCS) of Japanese vehicle manufacturers: an analysis of Toyota, Nissan and Mitsubishi', *International Journal of Production Economics, 118*(2): 375–386.
Hongladarom, S. (2002) 'The web of time and the dilemma of globalization', *The Information Society, 18*(24): 241–249.
Honour, H (1975) *The New Golden Land: European Images of America from the Discoveries to the Present Time*, New York: Pantheon Books.
Hoopes, J. (2003) *False Prophets: The Gurus Who Created Modern Management and Why Their Ideas Are Bad for Business Today*, New York: Basic Books.

Hope, W. (2006) 'Global capitalism and the critique of real time', *Time and Society, 15*(2–3): 275–302.

Hopkins, T. K. (1990) 'Note on the concept of hegemony', *Review, 13*(3): 409–411.

Hopper, T., and Makintosh, N. (1998) 'Management accounting numbers: freedom or prison: Geneen versus Foucault', in A. McKinlay and K. Starkey (eds) *Foucault, Management and Organization Theory: From Panopticon to Technologies of Self* (pp. 126–150), London: Sage.

Hoskin, K. (1998) 'Examining accounts and accounting for management: inverting understandings of the "economic"', in A. McKinlay and K. Starkey (eds) *Foucault, Management and Organization Theory: From Panopticon to Technologies of Self* (pp. 93–110), London: Sage.

Hoskin, K., and Macve, R. H. (1986) 'Accounting and the examination: a genealogy of disciplinary power', *Accounting, Organizations and Society, 1*(2): 105–136.

Hoskin. K., and Macve, R.H. (1988) 'The genesis of accountability: the West Point connections', *Accounting, Organizations and Society, 13*(1): 37–73.

Hotho, J. (2009) 'Internationalization decisions: the effects of country differences and familiarity perceptions', PhD thesis, University of Groningen.

Hounshell, D. A. (1984) *From the American System to Mass Production, 1800–1932: The Development of Manufacturing Technology in the United States*, Baltimore, MD: Johns Hopkins University Press.

Howse, D. (1980) *Greenwich Time and the Discovery of the Longitude*, Oxford: Oxford University Press.

Huczynski, A. A. (1993) *Management Gurus*, London: Routledge.

Hume, L. J. (2004) *Bentham and Bureaucracy*, Cambridge: Cambridge University Press.

Hunnicutt, B. K. (1988) 'The new deal: the salvation of work and the end of the shorter-hour movement', in G. Cross (ed.) *Worktime and Industrialization: An International History* (pp. 217–243), Philadelphia: Temple University Press.

Hunt, E. S., and Murray, J. M. (1999) *A History of Business in Medieval Europe, 1200–1550*, Cambridge: Cambridge University Press.

Hunter, L., McGregor, A., MacInnes, J., and Sproull, A. (1993) 'The flexible firm: strategy and segmentation', *British Journal of Industrial Relations, 31*(3): 383–407.

Huygens, C. (2005) *In De Onvoltooid Verleden Toekomende Tijd*, Leiden: Historische Uitgeverij.

Hyman, R. (1991) 'Plus can change. The theory of production and the production of theory', in A. Pollert (ed.) *Farewell to Flexibility* (pp. 259–283), Oxford: Blackwell.

Ikeo, A. (2000) 'Economists and economic policies', in A. Ikeo (ed.) *Japanese Economics and Economists Since 1945* (pp. 143–185), London: Routledge.

Ikuko, N. (1997) 'The 'civilization' of time: Japan and the adoption of the Western time system', *Time and Society Journal, 6*(2/3): 237–259.

Israel, J. I. (1989) *Dutch Primacy in World Trade, 1585–1740*, Oxford: Clarendon Press.

Israel, J. I. (1995) *The Dutch Republic: Its Rise, Greatness, and Fall, 1477–1806*, Oxford: Oxford University Press.

Iterson, A. van, Mastenbroek, W., and Soeters, J. (2001) 'Civilizing and informalizing: organizations in an Elisian context', *Organization, 8*(3): 497–514.

Iyer, A. V., Seshadri, S., and Vasher, R. (2009) *Toyota Supply Chain Management*, New York: McGraw-Hill.

Jackson, B. (2004) *Management Gurus and Management Fashions*, London: Routledge.

Jackson, N., and Carter, P. (1998) 'Labour as dressage', in A. McKinlay and K. Starkey (eds) *Foucault, Management and Organization Theory: From Panopticon to Technologies of Self* (pp. 49–64), London: Sage.

Jackson, T. (1997) *Inside Intel: Andy Grove and the Rise of the World's Most Powerful Chip Company*, New York: Dutton.

Jacoby, S.M. (2004) *Employing Bureaucracy, Managers, Unions, and the Transformation of Work in the 20th Century*, Mahwah, NJ: Lawrence Erlbaum Associate Publishers.

Jacques, M. (2009) *When China Rules the World*, London: Allen Lane.

James, H. (2009) *The Creation and Destruction of Value: The Globalization Cycle*, Cambridge, MA: Harvard University Press.

Jami, C. (1995) 'Western Devices for measuring time and space: clocks and Euclidian geometry in late Ming and Ch'ing China', in C. Huang and E. Zuercher (eds) *Time and Space in Chinese Culture* (pp. 169 –200), Leiden: Brill.

Janouch, G. (1951/1971) *Gesprekken met Kafka*, Westbroek: Harlekijn.

Jauch, U. P. (1998) *Jenseits der Maschine: Philosophie, Ironie and Aesthetik bei Julien Offray de La Mettrie*, München: Carl Hanser Verlag.

Johanson, J., and Vahlne, J. (1990) 'The mechanism of internationalization', *International Marketing Review*, 7(4): 11–24.

Johnson, H.T. (1992) *Relevance Regained: From Top Down Control to Bottom Up Empowerment*, New York: Free Press.

Jones, E. (1981/2003) *The European Miracle: Environments, Economies and Geopolitics in the History of Europe and Asia*, Cambridge: Cambridge University Press.

Jones, G. (2000) *Merchants to Multinationals: British Trading Companies in the 19th and 20th Centuries*, New York: Oxford University Press.

Jones, G. (2007) 'Globalization', in G. Jones and J. Zeitlin (eds) *The Oxford Handbook of Business History* (pp. 141–170), Oxford: Oxford University Press.

Jones, J. M. (1988) 'Cultural differences in temporal perspectives: instrumental and expressive behaviors in time', in J. E. McGrath (ed.) *The Social Psychology of Time: New Perspectives* (pp. 21–38), Newbury Park, CA: Sage.

Jonnes, J. (2009) *Eiffel's Tower*, New York: Penguin.

Jospin, L. (1999) *New York Times*, 22 November.

Judt, T. (2011) 'Bring back the rails', *New York Review of Books*, 13 January.

Juran, J. M. (1944) *Bureaucracy: A Challenge to Better Management*, New York: Harper & Bros.

Juran, J. M. (1945) *Management of Inspection and Quality Control*, New York: Harper & Bros.

Jürgens, U., Malsch, T., and Dohse, K. (1993) *Breaking from Taylorism: Changing Forms of Work in the Automobile Industry*, Cambridge: Cambridge University Press.

Kahlert, H., Muehe, R., and Brunner, G.L. (1999) *Wristwatches: History of a Century's Development*, Kew Gardens, Surrey: Bushwood Books.

Kahn, K. B. (2004) 'Approaches to new product forecasting', in K. B. Kahn, G. Castellion, and A. Griffin (eds) *The PDMA Handbook of New Product Development* (pp. 173–187), New York: John Wiley and Sons.

Kakar, S. (1970) *Frederick Taylor: A Study in Personality and Innovation*, Cambridge, MA: MIT Press.

Kamen, H. (2002) *Empire: How Spain Became a World Power 1492–1763*, New York: HarperCollins.

Kamoche, K. N. (1999) *Sociological Paradigms and Human Resources*, Aldershot, HR: Ashgate.

Kandolin, I., and Kauppinen, K. (1996) 'New working time models in comparison: 6+6 hour shifts and compressed workweek', *Finnish Work Research Bulletin, 4*: 31–35.

Kanigel, R. (1997) *The One Best Way: Frederick Winslow Taylor and the Enigma of Efficiency*, New York: Viking.

Kapuscinski, R. (2001) *The Shadow of the Sun: My African Life*, London: The Penguin Press.

Karabell, Z. (2004) *Parting the Dessert*, New York: Vintage Books.

Karagozoghu, N., and Brown, W. (1993) 'Time-based management and the new production development process,' *Journal of Product Innovation Management, 10*: 204–215.

Karsten, L. (1990) *De Achturendag*, Amsterdam: International Institute of Social History Publications.

Kasarda, J. D., Appold, S. J., and Mori, M. (2006) 'The impact of the air cargo industry on the global economy', paper prepared for the International Air Cargo Association Air Cargo Forum, Calgary.

Kaufman, C. F., Lane, P. M., and Lindquist, J. D. (1991) 'Exploring more than 24 hours a day: a preliminary investigation of polychronic time use', *Journal of Consumer Research, 18*(3): 392–401.

Kelliher, C., and Desombre, T. (2005) 'Breaking down boundaries: functional flexibility and occupational identity in health care', in I. U. Zeytinoglu (ed.) *Flexibility in Workplaces* (pp. 99–106), Geneva: ILO.

Kenney, M. (1999) 'Transplantation?: Comparing Japanese television assembly plants in Japan and the US', in J. K. Liker, W. M. Fruin, and P. S. Adler (eds) *Remade in America: Transplanting and Transforming Japanese Management Systems* (pp. 256–293), Oxford: Oxford University Press.

Kenney, M., and Florida, R. (1993) *Beyond Mass Production: The Japanese System and Its Transfer to the US*, New York: Oxford University Press.

Kern, S. (1983) *The Culture of Time and Space 1880–1918*, Cambridge, MA: Harvard University Press.

Kern, S. (2001) *The Culture of Time and Space, 1880–1918* (2nd edn, p. 13). Cambridge, MA: Harvard University Press.

Kieser, A. (1997) 'Rhetoric and myth in management fashion', *Organization,* 4(1): 49–74.

Kim, D., Pan, Y., and Park, H.S. (1998) 'High- and low-context culture: a comparison of Chines, Korean, and American cultures', *Psychology and Marketing, 15*(6): 507-521.

Kist, B. (1966) 'De erfenis van prins Maurits', *Armamentaria, 35*: 23–31.

Kitschelt, H., Lange, P., Marks, G., and Stephens, J.D. (1999a) 'Convergence and divergence in advanced capitalist democracies', in H. Kitschelt, P. Lange, G. Marks, and J. D. Stephens (eds) *Continuity and Change in Contemporary Capitalism* (pp. 427–260), Cambridge: Cambridge University Press.

Kitschelt, H., Lange, P., Marks, G., and Stephens, J. D. (eds) (1999b) *Continuity and Change in Contemporary Capitalism*, Cambridge: Cambridge University Press.

Klein, P. W. (1981) 'The origins of trading companies', in L. Blussé and F. Gaastra (eds) *Companies and Trade: Essays on Overseas Trading Companies During the Ancient Régime* (pp. 17–28), Leiden: Leiden University Press.

Kleinknecht, A., and ter Wengel, J. (1998) 'The myth of economic globalization', *Cambridge Journal of Economics, 22*: 637–647.

Kluckhohn, C. (1961) 'Universal categories of culture', in F. W. Moore (ed) *Readings in Cross-Cultural Methodology* (pp. 89–105), New Haven, CT: Human Relations Area Files Press.

Kobrin, S. J. (2005) 'Multinational corporations, the protest movement, and the future of global governance', in A. D. Chandler and B. Mazlish (eds) *Leviathans: Multinational Corporations and the New Global History* (pp. 219–136), Cambridge: Cambridge University Press.

Koehn, N. F. (1997) 'Josiah Wedgwood and the first industrial revolution', in T. K. McCraw (ed.) *Creating Modern Capitalism: How Entrepreneurs, Companies, and Countries Triumphed in Three Industrial Revolutions* (pp. 17–48), Cambridge, MA: Harvard University Press.

Kogut, B., and Zander, U. (1992) 'Knowledge of the firm, combinative capabilities, and the replication of technology', *Organization Science, 3*(3): 383–397.

Kogut, B., and Zander, U. (1993) 'Knowledge of the firm and the evolutionary theory of the multinational corporation', *Journal of International Business Studies, 24*(4): 625–645.

Komzsik, L. (2011) *Cycles of Time: From Infinity to Eternity*, Bloomington, IN: Trafford Publishing.

Koning, H. (1991) *Columbus: His Enterprise*, London: Latin America Bureau.

Korten, D. C. (1995) *When Corporations Rule the World*, West Hartford, CT: Kumarian Press and Berrett-Koehler Publishers.

Kostova, T., and Roth, K. (2003) 'Organizational coping with institutional upheaval in transition economies', *Journal of World Business, 38*(4): 314–330.

Kremer, M. (2005) 'How welfare states care', PhD thesis, University of Utrecht.

Ladner, S. (2009) 'Agency time', *Time & Society, 18*(2/3): 284–305.

Lancaster, J. (2004) *Making Time*, Boston, MA: North-Eastern University Press.

Lander, E., and Liker, J. K. (2007) 'The Toyota production system and art: making highly customized and creative products the Toyota way', *International Journal of Production Research, 45*(16): 3681–3698.

Landes, D. (1983) *Revolution in Time*, Cambridge, MA: Belknap.

Lang, M. (2006) 'Globalization and its history', *The Journal of Modern History, 78*(4): 899–931.

Langlois, R. N. (2007). *The Dynamics of Industrial Capitalism: Schumpeter, Chandler, and the New Economy*, London: Routledge.

Langlois, R. N., and Steinmüller, W. E. (1999) 'The evolution of competitive advantage in the worldwide semiconductor industry, 1947–1996', in D. C. Mowery and R. R. Nelson (eds) *The Sources of Industrial Leadership* (pp. 19–78), New York: Cambridge University Press.

Lash, S., and Urry, J. (1994) *Economies of Signs and Space*, London: Sage.

Laurent, A. (1983) 'The cultural diversity of Western conceptions of management', *International Studies of Management and Organization, 13*(1): 75–96.

Law, J. (1986) 'On the methods of long-distance control: vessels, navigation and the Portuguese route to India', in J. Law (ed.) *Power, Action and Belief: A New Sociology of Knowledge?* (pp. 234–263), London: Routledge.

Law, J. (1989) 'Technology and heterogeneous engineering: the case of Portuguese expansion', in W. E. Bijker, T. P. Hughes, and T. Pinch (eds) *The Social Construction of Technological Systems* (pp. 111–124), Cambridge, MA: MIT Press.

Lawrence, S., Giles, C. L., and Bollacker, K. D. (1999) 'Digital libraries and autonomous citation indexing', *IEEE Computer, 32*: 67–71.

Le Goff, J. (1977) *Pour un Autre Moyen Age*, Paris: Gallimard.

Leamer, E. E. (2007) 'A flat world, a level playing field, a small world after all, or none of the above? A review of Thomas L. Friedman's "The World is Flat"', *Journal of Economic Literature, 45*(1): 83–126.

Lee, H., and Liebenau, J. (1999) 'Time in organizational studies: towards a new research direction', *Organization Studies, 20*(6): 1035–1058.

Legge, K. (1995) *Human Resource Management: Rhetorics and Realities*, London: Macmillan.

Lesger, C. (2006) *The Rise of the Amsterdam Market and Information Exchange: Merchants, Commercial Expansion and Change in the Spatial Economy of the Low Countries*, c. *1550–1630*, Farnham: Ashgate.

Levine, R. (1997) *A Geography of Time: The Temporal Misadventures of a Social Psychologist, or How Every Culture Keeps Time Just a Little Bit Differently*, New York: Basic Books.

Levine, R., and Norenzayan, A. (1999) 'The pace of life in 31 countries', *Journal of Cross-Cultural Psychology, 30*(2): 178–205.

Levine, R., Fast, N., and Zimbardo, P. (2004) 'The power of persuasion: a field exercise', *Teaching of Psychology, 31*: 136–138.

Levinson, M. (2006) *The Box: How the Shipping Container Made the World Smaller and the World Economy Bigger*, Princeton, NJ: Princeton University Press.

Levitt, T. (1983) 'The globalization of markets', *Harvard Business Review, 83*(3): 92–102.

Lewis, L. D. (2008) *God's Crucible: Islam and the Making of Europe, 570 to 1215*, London: W. W. Norton.

Lewontin, R. C. (1997) 'The Cold War and the transformation of the academy', in N. Chomsky, L. Nader, I. Wallerstein, R. C. Lewontin, and R. Ohmann (eds) *The Cold War and the University: Toward the Intellectual History of the Postwar Period* (pp. 1–34), New York: The New Press.

Li, Y., Vanhaverbeke, W., and Schoenmakers, W. (2008) 'Exploration and exploitation in innovation: reframing the interpretation', *Creativity and Innovation Management, 17*(2): 107–126.

Lillrank, P. (1995) 'The transfer of management innovations from Japan', *Organisation Studies, 16*(6): 971–989.

Lindert, P. H., and Williamson, J. G. (2003) 'Does globalization make the world more unequal?', in M. D. Bordo, A. M. Taylor, and J. G. Williamson (eds) *Globalization in Historical Perspective*, Chicago: University of Chicago Press.

Lindquist, J. D., and Kaufman-Scarborough, C. (2007) 'The polychronic-monochronic tendency model: PMTS scale development and validation', *Time and Society, 16*(2/3): 253–285.

Lippe, T. van der, den Dulk, L., Doorne-Huiskes, A. van, Schippers, J., Lane, L., and Bäck-Wiklund, M. (2009) *Quality of Life in a Changing Europe*, Deliverable of EU-project Quality, Utrecht: Utrecht University.

Lister, R. (2003) *Citizenship: A Feminist Perspective*, London: Palgrave.

Locke, R. M., and Thelen, K. (1995) 'Apples and oranges revisited: contextualized comparisons and the study of comparative labor politics', *Politics and Society, 23*(3): 337–367.

Locke, R. R. (1996) *The Collapse of the American Management Mystique*, Oxford: Oxford University Press.

Love, P. (2006) 'Melbourne celebrates the 150th anniversary of the eight-hour day', *Labour History, 91*: 193–196.

McCollum, J. K., and Sherman, J. D. (1991) 'The effects of matrix organization size and number of project assignments on performance', *IEEE Transactions on Engineering Management, 38*(1): 3875–3878.

McCraw, T. K. (1984) *Prophets of Regulation*, Cambridge, MA: Belknap.

McCraw, T. K. (ed.) (1997) *Creating Modern Capitalism: How Entrepreneurs, Companies, and Countries Triumphed in Three Industrial Revolutions*, Cambridge, MA: Harvard University Press.

McCrum, R. (2010) *How the English Language Became the World's Language*, New York: Viking.

Macduff, I. (2006) 'Your pace or mine? Culture, time, and negotiation', *Negotiation Journal, 22*(1): 31–45.

Macey, S. L. (1980) *Clocks and the Cosmos: Time in Western Life and Thought*, Hamden, CT: Archon Books.

McGowan, C. (2004) *Rail, Steam, and Speed: The Rocket and the Birth of Steam Locomotion*, New York: Columbia University Press.

McGrath, J. E., and Rotchford, N. (1983) 'Time and behavior in organizations', in L. Cummings and B. Staw (eds) *Research in Organizational Behavior, Vol. 5* (pp. 57–101), Greenwich, CN: JAI Press.

McGregor, A., and Sproull, A. C. (1991) 'Employer Labor Utilization Strategies', Employment Department Research Paper no. 5.

McKinlay, A., and Taylor, P. (1998) 'Through the looking glass: Foucault and the politics of production', in A. McKinlay and K. Starkey (eds) *Foucault, Management and Organization Theory: From Panopticon to Technologies of Self* (pp. 173–190), London: Sage.

McKitterick, R. (2008) *Charlemagne: The Formation of a European Industry*, Cambridge: University Press.

McNeill, J. R., and McNeill, W. H. (2003) *The Human Web: A Bird's-Eye View of World History*, New York: W.W. Norton.

McNeill, W. H. (1974) *Venice, the Hinge of Europe, 1081–1797*, Chicago: University of Chicago Press.

McNeill, W. H. (1984) *The Pursuit of Power: Technology, Armed Force, and Society Since AD 1000*, Chicago: University of Chicago Press.

Madrick, J. (2009) *The Case for Big Government*, Princeton, NJ: Princeton University Press.

Maier, C. S. (1975) *Recasting Bourgeois Europe*, Princeton, NJ: Princeton University Press.

Mainiero, L. A., and Sullivan, S. E. (2006) *The Opt-Out Revolt: Why People are Leaving Companies to Create Kaleidoscope Careers*, Mountain View, CA: Davies-Black Publishing.

Makeba, M. (1988) *Makeba: My Story*, London: Bloomsbury.

Manning, P. (2003) 'Why Africans? The rise of the slave trade to 1700', in G. Heuman and J. Walvin (eds) *The Slavery Reader* (pp. 30–41), London: Routledge.

Marchant, J. (2008) *Decoding the Heavens: Solving the Mystery of the World's First Computer*, London: William Heinemann.

Marginson, P. (1989) 'Employment flexibility in large companies: change and continuity', *Industrial Relations Journal, 29*(2): 101–118.

Martin, A. (1971) *Enterprise Denied: Origins of the Decline of American Railroads, 1897–1917*, New York: Columbia University Press.

Masood, E. (2009) *Science and Islam*, London: Icon Books.

Mawdsley, E., Townsend, J. G., Porter, G., and Oakley, P. (2002) *Knowledge, Power and Development Agendas: NGOs North and South*, Oxford: INTRAC.

Mayo, A. J., Nohria, N., and Rennella M. (2009) *Entrepreneurs, Managers, and Leaders: What the Airline Industry Can Teach Us About Leadership*, New York: Palgrave Macmillan.

Mayo, E. (1933) *The Human Problems of an Indus Trial Civilisation*, New York: Macmillan.

Mayr, O. (1986) *Authority, Liberty and Automatic Machinery in Early Modern Europe*, Baltimore, MD: The Johns Hopkins University Press.

Mayrhofer, W. (1997) 'Warning: flexibility can damage your organizational health!', *Employee Relations, 19* (6): 519–534.

Mbiti, J. S. (1969) *African Religions and Philosophy*, Oxford: Heinemann.

Meager, R. (1985) 'Connoisseurship', *British Journal of Aesthetics, 25*(2): 137–152.

Meier, C. (2005) *From Athens to Auschwitz*, Cambridge, MA: Harvard University Press.

Meijer, F. (2010) *De Middellandse Zee*, Amsterdam: Atheneum-Polak & Van Gennep.

Melbin, M. (1987) *Night as Frontier: Colonizing the World After Dark*, New York: The Free Press.

Mense-Petermann, U. (2006) 'Micro-political or inter-cultural conflicts? An integrating approach', *Journal of International Management, 12*(3): 302–317.

Menzies, G. (2003) *Fourteen Twenty-One*, New York: William Morrow.

Micklethwaut, J., and Wooldridge, A. (1990) *The Witch Doctors: Making Sense of Management Gurus*, New York: Three Rivers Press.

Milton, G. (2004) *White Gold: The Extraordinary Story of Thomas Pellow and North Africa's One Million European Slaves*, London: Sceptre.

Minoguchi, T. (2000) 'From reconstruction to rapid growth', in A. Ikeo (ed.) *Japanese Economy and Economists Since 1945* (pp. 238–251), London: Routledge.

Mintz, S.W. (1985) *Sweetness and Power: The Place of Sugar in Modern History*, New York: Viking.

Mokyr, J. (2002) *The Gifts of Athena*, Princeton, NJ: Princeton University Press.

Molleman, H. B. M., and Slomp, J. (2003) 'The impact of team and work characteristics on team functioning', *Proceedings of the Group Technology/Cellular Manufacturing World Symposium*, Columbus, OH.

Mommsen, W. J. (1974) *The Age of Bureaucracy*, Oxford: Basil Blackwell.

Montaigne, M. de (1972) *Essais, Tome 1*, Paris: Livre de Poche.

Montgomery, D. (1987) *The Fall of the House of Labor*, Cambridge: Cambridge University Press.

Moore, K., and Lewis, D. (1999) *Birth of the Multinational: 2000 Years of Ancient Business History*, Copenhagen: Copenhagen Business School Press.

More, T. (1516/1978) *Utopia*, London: Penguin Classics.

Morgan, G. (1986) *Images of Organizations*, London: Sage.

Mosakowski, E., and Earley, P. C. (2000) 'A selective review of time assumptions in strategy research', *Academy of Management Review, 25*(4): 796–812.

Mumford, L. (1934) *Technics and Civilization*, New York: Harcourt, Brace and World.

Mumford, L. (1964) *Myth of the Machine, Vol. 1: Technics and Human Development*, New York: Harcourt Brace Jovanovich.

Munro, J. (2000) 'English "backwardness" and financial innovations in commerce with the low countries, 14th to 16th centuries', in P. Stabel, E. B. Blond, and A. Greve (eds) *International Trade in the Low Countries (14–16th Centuries): Merchants, Organisation, Infrastructure* (pp. 105–167), Studies in Urban, Social, Economic, and Political History of the Medieval and Early Modern Low Countries (M. Boone, general editor), no. 10, Leuven-Apeldoorn: Garant.

Munro, J. (2010) 'Tawney's century, 1540–1640: the roots of modern capitalist entrepreneurship', in D. Landes, J. Mokyr, and W. J. Baumol (eds) *The Invention of Enterprise: Entrepreneurship from Ancient Mesopotamia to Modern Times* (pp. 107–155), Princeton, NJ: Princeton University Press.

Nadwordny, M. J. (1957) 'Frederick Taylor and Frank Gilbreth: competition in scientific management', *The Business History Review, 31*(1): 23–34.

Nagel, J. G. (2007) *Abenteuer Fernhandel: die Ostindienkompanien*, Darmstadt: Wisseschaftliche Buchgesellschaft.

NBC (1980) *If Japan Can, Why Can't We?* New York: NBC.

Needham, J. (1969) 'Science and China's influence on the world', in J. Needham, *The Grand Titration*, London: George Allen & Unwin.

Needham, J., Ling, W., and De Solla Price, D. J. (1960) *Heavenly Clockwork*, Cambridge: Cambridge University Press.

Nelson, D. (1992) *A Mental Revolution: Scientific Management Since Taylor*, Columbus, OH: Ohio State University Press.

Nishizawa, T. (2000) 'Ichiro Nagayama and stabilization of labor–management relations', in A. Ikeo (ed.) *Japanese Economics and Economists Since 1945* (pp. 227–238), London: Routledge.

Nohria, N., and Eccles, R. G. (1997) 'Where does management knowledge come from?', in J. L. Alvarez (ed.) *The Diffusion and Consumption of Business Knowledge* (pp. 278–304), London: Macmillan.

Nohria, N., and Ghoshal, S. (1997) *The Differentiated Network: Organizations Knowledge Flows in Multinational Corporations*, San Francisco, CA: Jossey-Bass.

Nohria, N., and Eccles, R. G. (1998) *Beyond the Hype*, Cambridge, MA: Harvard Business School Press.

Nohria, N., and Mayo, A. (2007) *Juan Trippe and Pan American World Airways*, Harvard Business School Case 406–086, Boston, MA: Harvard Business School Press.

North, D. C. (1990) *Institutions, Institutional Change and Economic Performance*, Cambridge: Cambridge University Press.

North, D. C., and Thomas, R. P. (1973) *The Rise of the Western World: A New Economic History*, Cambridge: Cambridge University Press.

North, J. D. (2005) *God's Clockmaker: Richard of Wallingford and the Invention of Time*, London: Hambledon.

Nunn, N. (2008). 'The long-term effects of Africa's slave trades', *The Quarterly Journal of Economics, 123*(1): 139–176.

O'Connell, F. (1985) The corps of engineers and the rise of modern management 1827–1856, in M. R. Smith (ed.) *Military Enterprise and Technological Change: Perspectives on the American Experience* (pp. 87–116), Cambridge, MA: MIT Press.

OECD (1994) *The OECD Jobs Study: Facts, Analysis, Strategies*, Paris: OECD Publication Service.

OECD (2004) 'Regulatory Reform of Railways in Russia', paper presented at European Conference of Ministers of Transport (ECMT), Paris: OECD Publications Service.

Oeij, P., and Wiezer, N. (2006) *New Work Organisation, Working Conditions and Quality of Work: Towards the Flexible Firm?*, Report ef0274, Dublin: European Foundation for the Improvement of Living and Working Conditions.

Ohmae, K. (1975/1982) *The Mind of the Strategist*, New York: McGraw-Hill.

Ohmae, K. (1990) *The Borderless World: Power and Strategy in the Interlinked Economy*, London: HarperCollins.

Oliver, N., Deldridge, R., and Barton, H. (2002) 'Lean production and manufacturing performance improvement in Japan, the UK and US, 1994–2001', Working paper no. 232, ESRC Center for Business Research, Cambridge: University of Cambridge.

Olmsted, B., and Smith, S. (1989) *Creating a Flexible Workplace: How to Select and Manage Alternative Work Options*, New York: Amacom.

Olssen, E. (1978) 'The making of a political machine: the railroad unions enter politics', *Labour History, 19*: 373–396.

O'Malley, M. (1990) *Keeping Watch: A History of American Time*, New York: Viking.

O'Reilly, J. (1992), 'Where do you draw the line? Functional flexibility, training and skill in Britain and in France', *Work, Employment and Society, 6*(3): 369–396.

O'Rourke, K. H., and Williamson, J. G. (2002) 'When did globalization begin', *European Review of Economic History, 6*: 23–50.

Ortmann, G. (1995) *Formen der Produktion. Organisation und Rekursivität*, Opladen: Verlag für Sozialwissenschaften.

Osterhammel, J., and Petersson, N. P. (2004) *Geschichte der Globalisierung: Dimensionen, Prozess, Epochen*, München: Beck Verlag.

O'Toole, J. (1995) *Leading Change: Overcoming the Ideology of Comfort and the Tyranny of Custom*, San Francisco, CA: Jossey-Bass.

Ouchi, W. G. (1981) *Theory Z: How American Business Can Meet the Japanese Challenge*, Boston, MA: Addison-Wesley.

Overman, D. (2009) 'Karoshi and the Economic Crisis', *The Daily Censored*, 30 November.

Palmer, D. K., and Schoorman, F. D. (1999) 'Unpackaging the multiple aspects of time in polychronicity', *Journal of Managerial Psychology, 14*(3/4): 323–344.

Panaccione, A. (1989) *The Memory of May Day*, Venice: Marsilio Editori.

Parker, G. (1998) *The Grand Strategy of Philip II*, New Haven, CT: Yale University Press.

Parsons, T., and Smelser, N. J. (1956) *Economy and Society*, London: Routledge & Kegan Paul.

Pascale, R. T. (1990) *Managing on the Edge*, London: Penguin.

Pascale, R. T., and Athos A. G. (1981) *The Art of Japanese Management: Applications for American Executives*, London: Simon & Schuster.

Pedersen, T., and Petersen, B. (2004) 'Learning about foreign markets: are entrant firms exposed to a "shock effect"?', *Journal of International Marketing, 12*(1): 103–123.

Peper, B., Doorne-Huiskes, J. van, and Dulk, L. den (2005) *Flexible Working and Organisational Change*, Cheltenham: Edward Elgar.

Perlow, L. (1998) 'Boundary control: the social ordering of work and family in a high tech corporation', *Administrative Science Quarterly, 43*(2): 328–357.

Peters, T. J., and Waterman, R. H. (1982) *In Search of Excellence: Lessons from America's Best Run Companies*, New York: Harper & Row.

Piore, M. J., and Sabel, C. (1984) *The Second Industrial Divide*, New York: Basic Books.

Pirenne, H. (1967) 'Stages in the social history of capitalism', in R. Bendix and S. M. Lipset (eds) *Class, Status, and Power: Social Stratification in Comparative Perspective* (pp. 97–131), London: Routledge & Kegan Paul.

Pisan, C. de (1970) *The Epistle of Othea*, edited by C. F. Buehler, Oxford: Oxford University Press.

Pitzer, D. E., and Elliott, J. M. (1979) 'New Harmony's first Utopians, 1814–1824', *Indiana Magazine of History, 75*: 224–300.

Polanyi, K. (1977) *The Great Transformation*, Boston, MA: Beacon Press.

Pollard, S. (1968) *The Genesis of Modern Management*, London: Penguin Books.

Pollert, A. (1991) *Farewell to Flexibility*, Oxford: Blackwell.

Pollock, F. (1955) 'Automation in the USA', in T. W. Adorno and W. Dirks (eds) *Frankfurter Beiträge zur Soziologie, Band 1* (pp. 77–156), Frankurt am Main: Europäische Verlaganstalt.

Pollock, F. (1962) *Automation: Materialien zur Beurteilung der oekonomischen und sozialen Folgen*, Frankfurt/Main: Europäische Verlaganstalt.

Postma, J. (1990) *The Dutch and the Atlantic Slave Trade, 1600–1815*, Cambridge: Cambridge University Press.

Postone, M. (1993) *Time, Labor, and Social Domination*, Cambridge: Cambridge University Press.

Prescott, E. C. (2004) 'Why do Americans work so much more than Europeans?', *Federal Reserve Bank of Minneapolis Quarterly Review, 28*(1): 2–13.

Presser, H. B. (2003) *Working in a 24/7 Economy: Challenges for American Families*, New York: Russell Sage.

Presser, H. B., Gornick, C., and Parashar, S. (2008) 'Gender and nonstandard work hours in 12 European countries', *Monthly Labor Review, 131*(2): 83–103.

Priester, L. R. (1987) *De Nederlandse houding ten aanzien van de slavenhandel en slavernij, 1596–1863: het gedrag van de slavenhandelaren van de Commercie Compagnie van Middelburg in de 18de eeuw* (pp. 43 and 74 n. 28), Middelburg.

Proctor, S., Rowlinson, M., McCardle, L., Hassard, J., and Forrester, P. (1994) 'Flexibility, political and strategy: in defense of the world of the flexible firm', *Work ,Employment and Society,* 8(2): 221-242.

Pugh, E.W. (1995) *Building IBM: Shaping an Industry and its Technology*, Cambridge, MA: MIT Press.

Purcell, J., Purcell, K., and Tailby, S. (1999) 'Temporary work agencies: here today, gone tomorrow?', *British Journal of Industrial Relations, 42*(4): 705–725.

Putzger, I. (2009) 'Air freight carriers', *Canadian Transportation and Logistics, 112*(11): 22–24.

Rabinbach, A. (1990) *The Human Motor: Energy, Fatigue and the Origins of Modernity*, Berkeley, CA: University of California Press.

Rediker, M. (2007) *The Slave Ship*, New York: Viking.

Reinmoeller, P. R., and Chong, L. C. (2002) 'Managing the knowledge-creating context: a strategic time approach', *Creativity and Innovation Management, 11*(3): 165–174.

Restif de la Bretonne, N. E. (1990) 'Les nuits de Paris', in M. Deon and D. Baruch (eds) *Paris le jour, Paris la nuit*, Paris: Robert La Font.

Richards, E. G. (1998) *Mapping Time: The Calendar and its History*, Oxford: Oxford University Press.

Rifkin, J. (1987) *Time Wars*, New York: H. Holt and Company.

Rifkin, J. (1995) *The End of Work: The Decline of the Global Work Force and the Dawn of the Post-Market Era*, New York: G. P. Putnam's Sons.

Rispens, S. I. (2005) 'Machine reason: a history of clocks, computers and consciousness', PhD thesis, University of Groningen.
Robbins, S. P., and Barnwell, N. (2002) *Organization Theory: Concepts and Cases*, Frenchs Forest, NSW: Prentice Hall.
Roberts, M . (1967) *Essays in Swedish History*, Minneapolis: University of Minnesota Press.
Rodrigues, M. J. (2010) 'Beyond Lisbon', in A. Hemerijck, B. Knapen, and E. van Doorne (eds) *Aftershocks: Economic Crisis and Institutional Choice* (pp. 234–244), Amsterdam: Amsterdam University Press.
Roediger, D. R., and Rosemont, F. (1986) *Haymarket Scrapbook*, Chicago: Charles H. Kerr.
Roediger, D. R., and Foner, P. S. (1989) *Our Own Time: A History of American Labor and the Working Day*, London: Verso.
Roethlisberger, E. J., and Dickson, W. J. (1939) *Management and the Worker, An Account of a Research Program Conducted by the Western Electric Company, Hawthorne Works, Chicago*, Cambridge, MA: Harvard University Press.
Rollins, T., and Roberts, D. (1998) *Work Culture, Organizational Performance, and Business Success: Measurement and Management*, Westport, CT: Quorum Books.
Rosa, H. (2003) 'Social acceleration: ethical and political consequences of a desynchronized high-speed society', *Constellations, 10*(1): 4–33.
Rosa, H. (2005) *Beschleunigung: die Veränderung der Zeitstrukturen in der Moderne*, Frankfurt am Main: Suhrkamp.
Rosa, H. (2010) *Alienation and Acceleration: Towards a Critical Theory of Late-Modern Temporality*, Malmö/Arhus: NSU Press.
Rossabi, M. (1992) *Voyage from Xanadu: Rabban Sauma and the First Journey from China to the West*, Tokyo: Kodansha International.
Rossum, M. van (2011) 'Intra-Aziatische Vaart', *Tijdschrift voor Sociale en Economische Geschiedenis, 8*(3):32-69.
Rotman, Y. (2009) *Byzantine Slavery and the Mediterranean World*, Cambridge, MA: Harvard University Press.
Rousseau, D. M. (2005) *I-deals: Idiosyncratic Deals Employees Bargain for Themselves*, Armonk, NY: M. E. Sharpe.
Rowley, I. (2009) 'Toyota's cost-cutting drive: making the tough choices', *Business Week Online*, 29 January.
Roy, D. (1960) 'Banana time: job satisfaction and informal interaction', *Human Organization, 18*(4): 156–168.
Ruggie, J. G. (1993) 'Territoriality and beyond: problematising modernity in international relations', *International Organisation, 47*(1): 139–174.
Sabel, C. F., and Zeitlin, J. (eds) (1997) *Worlds of Possibility*, Cambridge: Cambridge University Press.
Sabelis, I. H. J. (2004) 'Global speed: a timescape view on transnationality', *Culture and Organization, 10*(4): 291–301.
Saee, J. (2005) *Managing Organizations in a Global Economy*, Mason, OH: South-Western.
Sandler, M. W. (2008) *Atlantic Ocean, the Illustrated History of the Ocean that Changed the World*. New York: Sterling.
Saunders, C., Slyke, C. van, and Vogel, D. R. (2004) 'My time or yours? Managing time visions in global virtual teams', *Academy of Management Executive, 18*(1): 19–31.
Savage, M. (1998) 'Discipline, surveillance and the "Career": employment on the Great western Railway 1833–1914', in McKinlay and K. Starkey (eds) *Foucault, Management and Organization Theory: From Panopticon to Technologies of Self* (pp. 65–92), London: Sage.
Schama, S. (2008) *The American Future*, London: The Bodley Head.
Schein, E. H. (1992) *Organizational Culture and Leadership*, San Francisco, CA: Jossey-Bass.

Scheuerman, W. E. (2003) 'Speed, states and social theory: a response to Hartmut Rosa', *Constellations, 10*(1): 42–48.

Scheuerman, W. E. (2004) *Liberal Democracy and the Acceleration of Time*, Baltimore, MD: Johns Hopkins University Press.

Schor, J. B. (1992) *The Overworked American*, New York: Basic Books.

Schriber, J. B., and Gutek, B. A. (1987) 'Some time dimensions of work: measurement of an underlying aspect of organization culture', *Journal of Applied Psychology, 72*(4): 642–650.

Scott, W. G. (1996) 'Do historical facts speak for themselves? Imagination and intuition, in management history', *Journal of Management History, 2*(3): 85–89.

SCP and CBS (2006) *Emancipatiemonitor 2006*. The Hague: Sociaal Cultureel Planbureau/ Centraal Bureau voor de Statistiek.

Selmer, J., Chiu, R. K., and Shenkar, O. (2007) 'Cultural distance asymmetry in expatriate adjustment', *Cross Cultural Management: An International Journal, 14*: 150–160.

Sels, L., and Hootegem, G. van (2001) 'Seeking the balance between flexibility and security: a rising issue in the low countries', *Work, Employment and Society, 15*(2): 327–352.

Semler, R. (2003) *The Seven-Day Weekend: Finding the Work/Life Balance*, London: Century.

Semler, R. (2007) 'Out of this world: doing things the Semco way', *Global Business and Organizational Excellence, 26*(5): 13–21.

Sennett, R. (1998) *The Corrosion of Character: The Personal Consequences of Work in the New Capitalism*, London: W.W. Norton.

Sennett, R. (2000) 'Fighting the flexible firm', *Workers Online*, Issue 73. Available at: http://workers.labor.net.au/73/c_historicalfeature_sennett.html (accessed 20 July 2012).

Sennett, R. (2006) *The Culture of the New Capitalism*, New Haven, CT: Yale University Press.

Sennett, R. (2009) 'The moral bankruptcy of new capitalism', in A. Hemerijck, B. Knapen, and E. van Doorne (eds) *Aftershocks: Economic Crisis and Institutional Choice* (pp. 163–168), Amsterdam: Amsterdam University Press.

Shenhav, Y. (1995) *From Chaos to Systems: The Engineering Foundations of Organization Theory, 1879–1932*, Ithaca, NY: Cornell University Press.

Silver, B. (2003) *Forces of Labor: Workers' Movements and Globalization Since 1870*, Cambridge: Cambridge University Press.

Simmel, G. (1900/2002) 'The metropolis and mental life', reprinted in G. Bridge and S. Watson (eds) *The Blackwell City Reader* (pp. 11–19), Malden, MA: Wiley-Blackwell.

Simmel, G. (1990) *The Philosophy of Money*, edited by D. Frisby, London: Routledge.

Skinner, Q. (1978) *The Foundations of Modern Political Thought*, Cambridge: Cambridge University Press.

Sklair, L. (2002) *Globalization: Capitalism and its Alternatives*, New York: Oxford University Press.

Slaven, G., and Totterdell, P. (1993) 'Time management training: does it transfer to the workplace?', *Journal of Managerial Psychology, 8*(1): 20–28.

Sloterdijk, P. (1999) *Sphären Bd II: Globen Makrosphärologie*, Frankfurt am Main: Suhrkamp.

Sloterdijk, P. (2005) *Im Weltinnenraum des Kapitals*, Frankfurt am Main: Suhrkamp.

Smith, A. (1776/1991) *The Wealth of Nations*, Amherst, NY: Prometheus.

Smith, M. R. (1985) 'Army ordnance and the American system of manufacturing, 1815–1861', in M. R. Smith (ed.) *Military Enterprise and Technological Change*, Cambridge MA: MIT Press.

Smith, P. G. (2005) 'Accelerated product development: techniques and traps', in K. B. Kahn (ed.) *The PDMA Handbook of New Product Development* (2nd edn, pp. 173–187), Hoboken, NJ: John Wiley and Sons.

Sobel, D., and Andrewes, W. J. H. (1998a) *Longitude*, London: Fourth Estate.

Sobel, D., and Andrewes, W. J. H. (1998b) *The Illustrated Longitude*, New York: Walker.

Sobel, R. (2000) *The Entrepreneurs: Explorations Within the American Business Tradition*, Washington, DC: Beard Books.

Sombart, W. (1916) *Der Moderne Kapitalismus*, München: Duncker and Humbolt.

Sorge, A. (2005) *The Global and the Local: Understanding the Dialectics of Business Systems*, Oxford: Oxford University Press.

Sorge, A., and Iterson, A. van (1995) 'Human resource management in the Netherlands', in I. Brunsteing (ed.) *Human Resource Management in Western Europe* (pp. 191–209), Berlin: Walter de Gruyter.

Sorge, A., and Witteloostuijn, A. van (2004) 'The (non)sense of organizational change: an essay about universal management hypes, sick consultancy metaphors and healthy organization theories', *Organization Studies, 25*(7): 1205–1231.

Soskice, D. (1999) 'Divergent production regimes: coordinated and uncoordinated market economies in the 1980s and 1990s', in H. Kitschelt, P. Lange, G. Marks, and J. D. Stephens (eds) *Continuity and Change In Contemporary Capitalism* (pp. 101–134), Cambridge: Cambridge University Press.

Soskice, D. (2009) 'Varieties of capitalism: varieties of reform', in A. Hemerijck, B. Knapen, and E. van Doorne (eds) *Aftershocks: Economic Crisis and Institutional Choice* (pp. 133–142), Amsterdam: Amsterdam University Press.

Sparrow, P. R. (1998) 'The pursuit of multiple and parallel organizational flexibilities: reconstituting jobs', *European Journal of Work and Organizational Psychology, 7*(1): 79–95.

Sparrow, P. R., and Marchington, M. (eds) (1997), *Human Resource Management: The New Agenda*, London: Financial Times Pitman Publication.

Spear, S. J., and Bowen, H. K. (1999) 'Decoding the DNA of the Toyota production system', *Harvard Business Review, 77*(5): 97–106.

Spear, S. J., and Bowen, H. K. (2006) 'Decoding the DNA of the Toyota Production System', *Harvard Business Review* (HBR OnPoint Enhanced Edition), 2904-PDF-ENG.

Spicer, A. (2006) 'Beyond the convergence–divergence debate: The role of spatial scales in transforming organizational logic', *Organization Studies, 27*(10): 1467–1483.

Spriegel, W. R., and Myers, C. E. (1953) *The Writings of the Gilbreths*, Chicago: Richard D. Irwin.

Stakhanov, A. (1939) *The Stakhanov Movement Explained*, Moscow: Foreign Languages Publishing House.

Stalk, G. (1988) 'Time – the next source of competitive advantage', *Harvard Business Review, 66*: 41–51.

Stalk, G. (1998) 'The time paradigm', *Forbes*, 30 November.

Stalk, G., and Hout, T. M. (1990) *Competing Against Time: How Time-Based Competition is Reshaping Global Markets*, New York: Free Press.

Standage, T. (1998) *The Victorian Internet*, London: Weidenfeld & Nicolson.

Stannard, D. E. (1992) *American Holocaust: Columbus and the Conquest of the New World*, Oxford: Oxford University Press.

Starkey, K. (1988) 'Time and work organization: a theoretical and empirical analysis', in M. Young and T. Schuller (eds) *The Rhythms of Society* (pp. 95–117), London: Routledge.

Starkey, K., and McKinley, A. (1993) *Strategy and Human Resource*, Oxford: Blackwell.

Steengaard, N. (1981) 'Violence and the rise of capitalism: Frederic C. Lane's theory of protection and tribute', *Review (Fernand Braudel Center), 5*(2): 247–273.

Stock, J. R., and Lambert, D. M. (2001) *Strategic Logistics Management*, New York: McGraw-Hill.

Stopford, J. M., and Dunning, J. H. (1983) *Multinationals: Company Performance and Global Trends*, New York: Macmillan.

Streeck, W. (1988) 'Comment on Ronald Dore "Rigidities in the labour market"', *Government and Opposition, 23*(4): 413–423.

Strum, P. (1993) *Brandeis: Beyond Progressivism*, Westbrooke Circle, KS: University Press of Kansas.

Sturgeon, T. J. (2002) 'Modular production networks: a new American model in industrial organization', *Industrial and Corporate Change, 11*(3): 451–496.

Supiot, A. (1999) *Au-de la de l'Emploi: Transformations du Travail et Devenir du Droit du travail en Europe*, Paris: Flammarion.

Supiot, A. (2001) *Beyond Employment: Changes in Work and the Future of Labour Law in Europe*, Oxford: Oxford University Press.

Sutter, A. (1988) *Göttliche Maschinen: Die Automaten für Lebendiges bei Descartes, Leibniz, La Mettrie und Kant*, Frankfurt am Main: Athenaeum.

Swift, J. (1726/1965), *Gulliver's Travels*, London: The Folio Society.

Taylor, B. (1999) 'Japanese management style in China? Production practices in Japanese manufacturing plants', *New Technology, Work and Empowerment, 142*(2): 129–142.

Taylor, F. W. (1904) *Shop Management*, New York: Harper.

Taylor, F. W. (1911) *The Principles of Scientific Management*, New York: Harper.

Teagarden, M. B., Meyer, J., and Jones, D. (2008) 'Knowledge sharing among high-tech MNCs, in China and India: invisible barriers, best practices and next steps', *Organizational Dynamics, 37*(2): 190–202.

Thevenet, M. (2001) 'Vie professionelle, vie privee, et developpement personnel', *Revue Francaise de Gestion, 134*: 106–119.

Thomas, H. (1997) *The Slave Trade: The History of the Atlantic Slave Trade 1440–1870*, London: Picador.

Thompson, C. A., Beauvais, L. L., and Lyness, K. S. (1999) 'When work–family benefits are not enough: the influence of work family culture on benefit utilisation, organisational attachment and work family conflict', *Journal of Vocational Behaviour, 54*(3): 392–415.

Thompson, E. P. (1963) *The Making of the English Working Class*, London: Victor Gollancz.

Thompson, E. P. (1967) 'Time, work-discipline and industrial capitalism', *Past and Present, 38*: 56–97.

Thrift, N. (1988) 'Vivos voco: Ringing the changes in the historical geography of time consciousness', in M. Young and T. Schuller (eds) *The Rhythms of Society* (pp. 53–94), London: Routledge.

Thrift, N. (1996) 'Inhuman geographies: landscapes of speed, light and power', in N. Thrift (ed.) *Spatial Formations*, London: Sage.

Tietze, S., Musson, G., and Scurry, T. (2006) 'Improvising services, balancing lives?', in P. Blyton (ed.) *Work–Life Integration* (pp. 180–195), London: Palgrave.

Tinsley, C. (1998) 'Models of conflict resolution in Japanese, German, and American cultures', *Journal of Applied Psychology, 83*(2): 316–323.

Tisdall, P. (1982). *Agents of Change: The Development and Practice of Management Consultancy*, London: Heinemann.

Tosi, H. L., and Carroll, S. J. (1973) *Management by Objectives: Applications and Research*, New York: Macmillan.

Townley, B. (2002) 'Managing with modernity', *Organization, 9*(4): 549–573.

Trompenaars, F., and Hampden-Turner, C. (1998) *Riding the Waves of Cultures: Understanding Diversity in Global Business*, Saddle River, NJ: Prentice-Hall.

Tsai, W. (2002) 'Social structure of "coopetition" within a multiunit organization: coordination, competition and intraorganizational knowledge sharing', *Organization Science, 13*(2): 179–190.

Tsuji, Y. (2006) 'Railway time and rubber time: the paradox in the Japanese conception of time', *Time and Society, 15*(2–3): 177–195.

Tsutsui, W. M. (1998) *Manufacturing Ideology: Scientific Management in Twentieth Century*, Princeton, NJ: Princeton University Press.

Tyson, T. N., Fleischman, R. K., and Oldroyd, D. (2004) 'Theoretical perspectives of accounting for labor on slave plantations of the USA and British West Indies', *Accounting, Auditing and Accountability Journal, 17*(5): 758–778.

Uglow, J. (2007) *The Lunar Men: Five Friends Whose Curiosity Changed the World*, New York: Farrar, Straus and Giroux.

UNCTAD (2009) *World Investment Report 2009: Transnational Corporations, Agricultural Production and Development*, Geneva: UNCTAD.

UNDP (2000) *Human Development Report 2000: Human Rights and Human Development*, New York: UNDP.

UNDP (2005) *Human Development Report 2005: International Cooperation at a Crossroads: Aid, Trade and Security in an Unequal World*, New York: UNDP.

US Bureau of the Census (1975) *Historical Statistics of the United States: Colonial Times to 1970*, 2 Volumes, Washington, DC: US Government Printing Office.

US Bureau of the Census (2011) www.census.gov/, retrieved 20 December 2011.

Usunier, J. C. (2000) *Marketing Across Cultures*, Englewood Cliffs, NJ: Prentice Hall.

Vagts, A. (1959) *A History of Militarism*, New York: Free Press.

Vance, J. E. (1986) *Capturing the Horizon: The Historical Geography of Transportation*, New York: Harper & Row.

Vickery, S. K., Droge, C. L. M., Yeomans, J. M., and Markland, R. E. (1995) 'Time-based competition in the furniture industry: an empirical study', *Production and Inventory Management Journal, 36*(4): 14–21.

Vink, M. (2003) 'The world's oldest trade: Dutch slavery and slave trade in the Indian Ocean in the 17th century', *Journal of World History, 14*(2): 131–177.

Visser, J. (2000) 'The first part-time economy in the world: does it work?', AIAS Working Paper, University of Amsterdam.

Visser, J., and Hemerijck, A. (1997) *The Dutch Miracle*, Amsterdam: Amsterdam University Press.

Volberda, H. W. (1998) *Building the Flexible Firm: How to Remain Competitive*, Oxford: Oxford University Press.

Voth, H. J. (2000) *Time and Work in England, 1750–1830*, Oxford: Clarendon Press.

Vries, J. de (1978) 'Barges and capitalism: passenger transportation in the Dutch economy 1632–1839', *A.A.G. Bijdragen, 21*: 33–398.

Vries, J. de (1982) *Barges and capitalism: Passenger Transportation in the Dutch Economy, 1632–1839*, Utrecht: Hes Publishers.

Vries, J. de (1994) 'The industrial revolution and the industrious revolution', *The Journal of Economic History, 54*(2): 249–270.

Vries, J. de (2010) 'The limits of globalization in the early modern world', *Economic History Review, 63*(3): 710–733.

Vries, J. de, and Woude, A. van der (1997) *The First Modern Economy*, Cambridge: Cambridge University Press.

Walby, S. (1989) 'Flexibility and the changing sexual division of labour', in S. Wood (ed.) *The Transformation of Work?* (pp. 127–140), London: Unwin Hyman.

Warouw, J. N. (2004) 'Assuming modernity: migrant industrial workers in Tangerang, Indonesia', PhD thesis, The Australian National University.

Waters, M. (1995) *Globalization*, London: Routledge.

Watson, G. H. (2009) 'It's the people, stupid, not the stupid people!', *The Journal for Quality and Participation, 31*(4): 42–44.

Weber, M. (1904–05/1989) *The Protestant Ethic and the Spirit of Capitalism*, New York: Scribners.

Weber, M. (1908) 'Die Grenznutzlehre und das psychophysische Grundgesetz', *Archiv fuer Sozialwissenschaft und Sozialpolitik, 27*(2): 546–558.

Weber, M. (1909–1910) 'Zur Psychophysik der industriellen Arbeit', *Archiv Fuer Sozialwissenschaft und Sozialpolitik, 28*(1): 219–277; 719–761.

Weber, M. (1921/1968) *Economy and Society*, New York, Bedminster Press.

Weber, M. (1931) *Verkürzung der Arbeitszeit*, Gewerkschaftliche Schriften 4, Zürich: Kommissionsverlag der Genossenschaftsbuchhandlung.

Weber, M. (1947) 'The meaning of "ethical neutrality" in sociology and economics', in E. A. Shils and H. A. Finch (eds/trans.) *The Methodology of the Social Sciences* (pp.1–47), New York: The Free Press.

Weber, M. (1956) *Wirtschaft und Gesellschaft*, 2 volumes, Köln: Kiepenheuer and Witsch.

Webster, J. (2001) 'Reconciling adaptability and equal opportunities in European workplaces', Report for DG-Employment of the European Commission.

Weishaupt, T. J. (2011) *From the Manpower Revolution to the Activation Paradigm: Explaining Institutional Continuity and Change in an Integrating Europe*, Amsterdam: Amsterdam University Press.

Weiss, S. E. (1994a) 'Negotiating with "Romans" – Part 1', *Sloan Management Review, 35*(2): 51–61.

Weiss, S. E. (1994b) 'Negotiating with "Romans"– Part 2', *Sloan Management Review, 35*(3): 85–99.

Wershler-Henry, D. (2007) *The Iron Whim: A Fragmented History of Typewriting*, Ithaca, NY: Cornell University Press.

Westermann, M. (1996) *A Worldly Art: The Dutch Republic 1585–1700*, London: Weidenfeld & Nicolson.

Whipp, R., Adam, B., and Sabelis, I. (2002) *Making Time: Time and Management in Modern Organizations*, New York: Oxford University Press.

White, D. W., Absher, R. K., and Huggins, K. A. (2011) 'The effect of handiness and cultural distance on socio-cultural adaptation in an expatriate sales population', *Journal of Personal Selling and Sales Management, 3*: 325–337.

White, L. (1962) *Medieval Technology and Social Change*, Oxford: Clarendon Press.

White, L. (1969) 'The iconography of temperantia and the virtuousness of technology', in T. K. Rabb and J. E. Seigel (eds) *Action and Conviction in Early Modern Europe: Essays in Memory of E. H. Harbison* (pp. 197–219), Princeton, NJ: Princeton University Press.

Whitrow, G. J. (1975) *The Nature of Time*, London: Harmondsworth.

Whittington, R., and Mayer, M. (2000) *The European Corporation*, Oxford: Oxford University Press.

Wickens, P. (1987) *The Road to Nissan: Flexibility, Quality, Teamwork*, New York: Macmillan.

Wijnolst, N., and Wergeland, T. (2009) *Shipping Innovation*, Amsterdam: IOS Press.

Wildenbeest, G. (1988) 'Keeping up with the times: time and its rhythms in the countryside', *Netherlands Journal of Sociology, 24*: 134–145.

Wilkins, M. (1987) 'Defining a firm: history and theory', in P. Hertner and G. Jones (eds) *Multinationals: Theory and History* (pp. 80–96), Aldershot: Gower.

Wilkinson, A., Redman, T., Snape, E., and Marchington, M. (1998) *Managing With TQM: Theory and Practice*, London: Palgrave Macmillan.

Willan, T. S. (1953) *The Muscovy Merchants of 1555*, Manchester: Manchester University Press.

Wilson, J. F. (2001) *Business Cultures and Business Performance: A British Perspective* (Jaarboek No. 64), Amsterdam: NEHA.

Wilson, P. H. (2009) *Europe's Tragedy: A History of the Thirty Years War*, London: Allen Lane.

Winchester, S. (2008) *The Man Who Loved China*, New York: HarperCollins.

Winchester, S. (2010) *Atlantic*, New York: HarperCollins.

Wittmann, M., Dinich, J., Merrow, M., and Roenneberg, T. (2006) 'Social jetlag: misalignment of biological and social time', *Chronobiology International, 23*(1–2): 497–509.

Wolf, M. (2005) *Why Globalization Works*, New Haven, CT: Yale University Press.

Wolveren, K. van (1989) *The Enigma of Japanese Power*, New York: Random House.

Womack, J. T. (1991) *The Machine That Changed the World: The Story of Lean Production*, New York: Free Press.

Womack, J. T., and Jones, D. T. (1996) *Lean Thinking: Banish Waste and Create Wealth in Your Corporation*, New York: Simon & Schuster.

Wrege, C. D., and Perroni, A. G. (1974) 'Taylor's pig-tale: a historical analysis of Frederick W. Taylor's pig-iron experiments', *Academy of Management Journal, 17*(1): 6–27.

Wren, D. (1987) *The Evolution of Management Thought*, New York: Wiley.

Yakura, E. K. (2002) 'Charting time: timelines as temporal boundary objects', *Academy of Management Journal, 45*(5): 956–970.

Yeh, R. T., Pearlson, K., and Kozmetsky, G. (2000) *Zero Time: Providing Instant Customer Value – Every Time, All the Time!*, Hoboken, NJ: John Wiley and Sons.

Yost, E. (1949) *Frank and Lillian Gilbreth*, New Brunswick, NJ: Rutgers University Press.

Young, M. (1988) *The Metronomic Society*, London: Thames & Hudson.

Zakour, A. B. (2004) 'Cultural differences and information technology acceptance', *SAIS Conference Proceedings*, Paper 26.

Zan, L. (2004) 'Accounting and management discourse in protoindustrial settings: the Venice Arsenal in the turn of the XVI Century', *Accounting and Business Research, 32*(2): 145–175.

Zanden, J. L. van (1996) 'Over de rationalitiet van het ondernemersgedrag van de VOC: enkele empirische bevindingen', in C. A. David, W. Fritschy, and L. A. Van der Valk (eds) *Kapitaal, Ondernemerschap en Beleid* (pp 409–422), Amsterdam: NEHA.

Zanden, J. L. van (2009) *The Long Road to the Industrial Revolution*, Leiden: Brill.

Zeitlin, J. (2003) 'Productive alternatives: flexibility, governance and strategic choice in industrial history', in F. Amatori and G. Jones (eds) *Business History Around the World* (pp. 62–82), Cambridge: Cambridge University Press.

Zeitlin, J. (2007) 'The historical alternatives approach', in G. Jones and J. Zeitlin (eds) *The Oxford Handbook of Business History* (pp. 120–140), Oxford: Oxford University Press.

Zeng, X. Q., Lu, L., and Idris, S. U. (2005) *Working Time in Transition: The Dual Task of Standardization and Flexibilisation in China*, Geneva: ILO.

Zerubavel, E. (1981) *Hidden Rhythms: Schedules and Calendars in Social Life*, Chicago: University of Chicago Press.

Zerubavel, E. (1982) 'The standardization of time: a socio-historical perspective', *American Journal of Sociology, 88*(1): 1–23.

Zerubavel, E. (1985) *The Seven Day Circle*, New York: The Free Press.

Zheng, L., and Jia, G. (1993) *American-English and American Culture* (In Chinese), Changsha: Hunan Education Publisher.

Zimbardo, P. G., and Boyd, J. N. (1999) 'Putting time in perspective: a valid, reliable individual differences metric', *Journal of Personality and Social Psychology, 77*(6): 1271–1288.

Zuboff, S. (1984) *In the Age of the Smart Machine*, New York: Basic Books.

INDEX